Time,
The Harmonizer of All Things

MARIUS ALBU

Cover image from

Kanstantsin Shcharbinski, Dreamstime.com

Foreword

In recent times, the quantity and diversity of information extended faster than the comprehension of an individual mind, despite the basically human wish to understand the world as a whole, though not necessarily in detail. Such a wish can be accomplished by a multidisciplinary review of prehistoric traces and remains; ancient and medieval works and findings; Renaissance and post-Renaissance innovations, inventions and creations; modern styles, researches, discoveries, theories, technologies, and later computers, and mass communication; as well as by an attentive examination of advance in various fields of interest or activity with their timelines and successive stages; and moreover by a thorough analysis of their running times.

Each field of interest or activity has its own historical roots and stages of advance, which make it distinct from others. However, these fields can be approached in a similar manner because they are based on corresponding laws, principles, or concepts, such as the law of conservation of total energy and the principle of increasing entropy, governing their development or evolution. Meanwhile, the stages of advance in each field are succeeding one another faster and faster, i.e. with an increasing frequency, in course of its own time that seems to run unevenly, thus contradicting the common sense of an evenly running time.

All fields of interest or activity belong to the great field of the universe, with a difference that the existences of first ones are significantly shorter than the existence of the last one, which may have expanding stages following each other faster and faster with an increasing frequency, so that its own time would also unevenly running but on a scale so long lasting as it seems to be even, whereby the unjustified idea of infinite time.

The reason for unevenly running time is presented in an introductory background to time on a thermodynamic basis, whence the natural process underlying any field of interest or activity takes place by conserving its total energy that means the sum of its effective (orderly) energy and its dissipative (disorderly) energy is maintained constant; and meanwhile by increasing its power efficiency that implies an increasing frequency or, equivalently, a decreasing period in the course of a process. This period is expressed as either a function of time, or a function of instantaneous angular speed defined by time rate of an ever increasing angle. The equality of these two complementary functions leads to a differential equation that, integrated and then inversed, results in a general *time function*, according to which the ratio of running time to final time equals the hyperbolic tangent of the ever increasing angle divided by a double-cycle. This time function governs all processes underlying the fields of interest or activity, which are lasting between billions and tens of years, and therefore it is convenient that from the multitude of values taken by the time function to select only those corresponding to values of argument equalling integers, or semi-integers, quarter-integers, eighth-integers, and so on, as numbers of completed double-cycles, or cycles, semi-cycles, quarter-cycles, etc., and only one kind of such subdivisions to be selected for calculating the discrete times of transition between the successive sequences of each process. In order to identify what kind of sequences are fitting the evolution or development of a studied process, the more reliably established times given in years ago must be first converted in times from the origin of the process, and only the last ones compared to the discrete values of time function for values of hyperbolic tangent of numbers counting completed double-cycles, or cycles, semi-cycles, quarter-cycles, and so on, which are displayed together with their ratios and differences in six tables useful to find out the kind of sequences suitable for a studied process.

In each chapter, the studied field of interest or activity is defined, unfolded in chronological order, divided into sequences with times of transition from one to another, either estimated by previous researches or indicated by examination of their chronological display, processed in the manner described above, and finally presented with its timeline, sequences, and intrinsic characteristics from the initial time to the final time of the process underlying that studied field.
Excepting the introductory background to time, in all chapters the use of mathematical symbols, equations, expressions, formulae, and demonstrations is reduced to minimum and often avoided, so that the work can be easily read and understood by people unfamiliar with them, simply crossing over such passages, without losing the general course of presentation.

The entire work is an attempt to open or enlarge the interest for processing the development, evolution or advance of creativity, cosmology, geology, climatology, biology, anthropology, humanity, linguistics, art, theology, musicology, exploration, inhabitation, knowledge, navigation, meteorology, medicine, sociology, construction, science, industry, civilization, legislation, philosophy, aero-astronautics, and high technology. Such an enlarged view was inspired and promoted by the Romanian Academician <u>Gheorghe Murgeanu</u>, whose teaching guided many of his disciples in their activities.

Contents

Background to time

In general, time (Latin *tempus* 'time, period', Old English *tīma* 'limited space of time') is defined as a dimension in which events can be ordered from the past through the present into the future, and also the measure of durations of events and the intervals between them. However, 'what exactly is time?' remains one of the biggest unanswered questions in physics.

As one of the most profound concepts, time was always subject of reflection, measure and division, using *sundials*, *clepsydras* (from Greek κλέπτειν 'to steal' and χιδορ water'), *horologes* (from Greek ὥρα 'hour' and λέγειν 'to tell'), and *clocks* (a derivation from Celtic *clagan* or *clocca* 'bell'), which were improved from water clocks through mechanical and electric clocks to digital clocks. By these devices, the units of time, such as hours, minutes, seconds and their submultiples, as well as the succession of days, months, seasons, years, centuries and millennia, were more and more precisely measured.

In marking and dividing time, the *calendar* (Latin *calendārium* 'an account-book' from *Kalendae* 'the first day of the month, calends') played an important role in social and religious life all over the world. In course of the mentioned achievements, there were numerous calendars counting time from their beginnings such as: *Byzantine* from 5509-5508BC, *Assyrian* from 4750BC, *Hebrew* from 3761-3760BC, *Old Mayan* from 3372BC, *Hindu* from 3102BC-AD78, *Chinese* from 2637-2636BC,
Berber from 950BC, *Roman* (*Ab urbe condita* 'AUC') from 753BC, *Buddhist* from 544BC, *Ethiopian* from AD7-8, *Coptic* from AD283-284, *Armenian* from AD551, *Early Islamic* from AD578-579, *Bengali* from AD593, *Iranian* from AD621-622, *Current Islamic* from AD622, and *Bahá'i* from AD1843-1844.

Thoughts, inventions, mechanisms and statements concerning time, achievements in measuring and dividing time, and thermodynamic basis for natural processes developing or evolving in time were due to remarkable peoples, rulers, spiritual leaders, calenderers, writers, philosophers, scientists, physicists, inventors, clockmakers, producers, users and organizations, some of them mentioned below in chronological order with their contributions or participations.

3500-2300BC: *Sumerians* have divided a year into 12 lunar months of

1

29 or 30 days, keeping the lunar year of 354 days in step with the solar year of 325.25 days by an extra month added periodically; and also every six years their calendar included another extra month of 62 days;

3372BC: *Mesoamericans* began the oldest *Mayan calendar*, which later developed on the standard base of a day-night cycle known as *kin* 'day, person or sun', and grouped in *uinal* (20 days), *tun* (360 days), *koltun* (1800 days), *katun* (7200 days), *baktun* (144,000 days), and so on up to a precessional cycle account of 9,360,000 days;

3000-2500BC: *Prehistoric British* celebrated the *winter solstices*, quite accurately marked by a horseshoe arrangement of five trilithons and an upright pillar 'altar stone' on the axis of the horseshoe at the open end, within a megalithic monument, now known as *Stonehenge*;

2685-2560BC: *Old Kingdom Egyptians* marked the beginning of *pyramid age* by constructing pyramid-shaped structures (burial monuments) with usually a west-east orientation, and with markers for the Sun's positions at the summer and winter *solstices*, and sometimes at equinoxes;

c.2000BC: *Ancient Hindu Indians* believed, according to the later collection of sacred texts entitled *Sanatana Dharma* 'The Eternal Teaching', that the duration of time changed from world to world, and that our time was not absolute, distinguishing a cosmic time and a earthly time; time, known as *kala* 'time, or death' was perceived as cyclical, and also introduced eight divisions of cosmic time with durations ranging from 8.64 trillion to 432 thousand earthly years;

c.1550BC: *Mesopotamians* and *Egyptians* used a *bowl-shaped outflow* as the *simplest water clock*;

c.1500BC: *Hittites* built in their capital Hattusha (now Boğazköy, in central Turkey) a central temple so oriented to observe astronomical moments related to the Sun, especially the *solstices*;

1400-1300BC: Moses (Môsheh), leader of Israelites, addressing God, prayed such as *Teach us to number our days aright* (Psalm 90:12);

1300-1200BC: *Babylonian star catalogues*, with a sexagesimal place-value number system, were used for studying astronomy and *measuring time*;

c.1200BC: *Jewish prophets* wrote in the first book of the Old Testament, called *Genesis*, that in the Fourth Day of Creation 'God made two great lights; the greater light to rule the day, and the lesser light to rule the night: he made the stars also. And God set them in the firmament of the heaven to give light upon the earth, and to rule over the day and over the night, and *to divide the light from the darkness*: and God saw that it was good.'

After 1000BC: *Greeks* and then *Romans* measured time by *clepsydras*, which were in use for centuries;

776BC-AD394: *Greeks* were reckoning time from the first Olympic Games (Greek *τά Ὀλύμπια* 'The Olympics') in *Olympiads*, as periods of four years between two successive Olympic Games; their calendar comprised 354-day years, each with twelve months of 29 or 30 days, and has been kept in line with the solar year of 365.25 days by adding an extra, intercalary, month in every other year;

510BC-27BC: *Romans* divided time into yearly *consulates* named after the elected consuls, and marked the beginning of year by gathering the citizens on the Field of *Mars* (the god of war), from where the name *March* was given to the first month of springtime;

c.330BC: Aristotle, Greek philosopher and scientist, wrote that *Tempus est mensura motus rerum mobilium* 'Time is the measure of movement';

c.140BC: *Antikythera mechanism*, a brilliant engineering and a conception of genius, was constructed as an *analogue computer* designed to calculate the positions of the Sun and Moon with associated phases and eclipse cycles, as well as the orbital positions of Mercury, Venus, Mars, Jupiter and Saturn; it comprised 30 bronze gears, mounting on offset axes, including a Sun gear, a Moon train, and Metonic, Olympiad, Callippic, Saros and Exeligmos trains, altogether using ratios from Babylonian astronomy (now in National Archaeological Museum of Athens);

140-129BC: Hipparchus, Greek astronomer and mathematician, discovered the *precession of the equinoxes*, developed a theory of *Sun's eccentric motion*, and improved calculations for the *prediction of eclipses*;

59-58BC: Julius Caesar, Roman general and statesman, during his first consulate effectuated drastic changes in *Roman calendar* counting from Rome's foundation as *ab urbe condita* (AUC, assumed to be 753BC), so that the year 709AUC began on 1 January and ran over 365 days until 31 December;

30BC: Publius Virgil, Roman poet, wrote that *Sed fugit interea, fugit inreparabile tempus* 'But meanwhile it is flying, irretrievable time is flying';

29BC-AD4: <u>Octavianus Augustus</u>, the first Roman emperor, adjusted J Caesar's calendar, introducing the *leap year* in 737AUC, and the resultant *Julian calendar* was almost everywhere used in Europe until 1582;

AD8: <u>Publius Ovid</u>, Roman poet, finished his Latin poem *Metamorphoseon libri* 'Books of Transformations', where in Book XV, 234, wrote *Tempus edax rerum* 'Time that devours all things';

27-30: <u>Jesus Christ</u>, central figure of Christian faith, set the question *Are there not twelve hours of daylight?* (John 11:9);

49: <u>Lucius Annaeus Seneca</u>, Roman Stoic philosopher, statesman and tragedian, wrote the famous essay *On the Shortness of Life*, showing that *life is long if know how to use it*;

70-80: <u>Marcus Aurelius</u>, Roman emperor and philosopher, thought that *There is a sort of river of things passing into being, and Time is a violent torrent. Every instant of time is a pinprick of eternity, All things are petty, easily changed, vanishing away*, his *Meditationes*, book 6;

525: <u>Dionysius Exiguus</u>, Scythian Christian scholar, fixed the dating of the Christian era in his *Cyclus Paschalis*, and devised the *Anno Domini* 'AD' system, dating from the Incarnation of Jesus Christ; this system was adopted by all European countries until 1422 when Portugal finally rallied to it;

622: <u>Muhammad</u>, Arab prophet and founder of Islam, left Mecca for Medina, and this event marked the *beginning of Muslim era*;

630-present: <u>*Arabs*</u> have practiced *star gazing* to determine the *times of water distribution* to the communities for irrigation;

724-727: <u>Yi Xing</u>, Chinese inventor, constructed a *water-driven celestial sphere*, using *concentric gears* and shafts as part of mechanism, and the first *Chinese clock* to strike hours and half-hours;

731: <u>The Venerable Bede</u>, Anglo-Saxon scholar, theologian and historian, completed the great work *Historia Ecclesiastica Gentis Anglorum* 'Ecclesiastical History of the English People', using the Latin term *ante uero incarnationis dominicae tempus* 'the time before the Lord's true incarnation', equivalent to *before Christ* 'BC' to identify years before the first year of this era;

750-1050: <u>*Vikings*</u> divided the years into seasons, namely *summer* and *winter*, with twelve lunar months in the *Old Icelandic lunar calendar*, the summer being from the modern month of April until the modern month of October, so that winter lasted from October to April;

797: <u>Harun al-Rashid</u>, Abbasid caliph of Bagdad, presented to King Charlemagne of Franks a *particularly elaborate example of water clock*;

1000-10: <u>Abu al-Rayhan al-Biruni</u>, Persian-Muslim scholar and polymath, collected information regarding Persian, Sogdian, Chorasmian, Greek, Jewish, Syrian, Harranian, Arabic, Christian, and Muslim traditions and calendars, which were assembled and explained in his work *Alāthār Albākiya 'an-il-Kūrūn Alkhāliya*, as 'vestiges preserved from ancient generations'; this work was translated into English by the German chronologist Edward Sachau (1879) under the title 'The Chronology of Ancient Nations', W.H. Allen & Co., London;

c.1050: <u>Ibn Khalaf al-Muradi</u>, Arab engineer, invented *first geared mechanical clock*, a mercury powered automata clock;

1088-94: <u>Su Song</u>, Chinese horologist and engineer during the Song Dynasty, created a *clock with escapement mechanism*, probably accurate to within 100 seconds a day, which was incorporated in an *astronomical clock-tower* of 33 feet in height and driven by a water wheel of 10 feet in diameter, specially built at Kaifeng, Henan, on the Huang He River, in East central China;

1176: *French Gothic architects* installed an early European *horologe* at Sens Cathedral, in Burgundy;

1267: <u>Roger Bacon</u>, English philosopher and scientist, completed the *Opus Majus* 'Greater Work', stating that times of full moons comprise numbers of *horae* 'hours', *minuta* 'minutes', *secunda* 'seconds', *tertia* 'thirds', and *quarta* 'fourths';

1277: *Spanish authors* compiled the book *Libros del Saber* of Alfonso X of Castile, containing translations and paraphrases of Arabic works, in which a *mercury clock* is described according to Muslim knowledge of mechanical clocks;

1283: *English Norman architects* installed a *clock with verge (crown wheel) escapement* above the rood screen of Dunstable Priory, Bedfordshire;

1290-92: *English Early Gothic architects* installed a *large horologe* at Westminster, London, and the *great horologe* in Canterbury Cathedral, Kent;

1336: <u>Richard of Wallingford</u>, English engineer, installed in St Albans, Hertfordshire, a *clock with a large astrolabe-type dial*, showing the Sun, moon's age, phase and node, and a star map;

1348-64: <u>Giovanni de'Dondi</u>, Italian horologist and physician, installed a *sophisticated astronomical clock*, which was later enlarged;

1386: *<u>English Gothic architects</u>* installed a large *iron-framed clock* in the aisle of Salisbury Cathedral, Wiltshire, which is considered to be the *world's oldest surviving mechanical device that strikes the hours*, without a dial;

1430: <u>Phillip the Good</u>, Duke of Burgundy, was endowed with a *chamber clock*, the earliest existing *spring driven clock* (now in the Germanisches Nationalmuseum, Nuremberg, Bavaria);

1475-85: <u>Paulus Almanus</u>, German churchman and clockmaker, was author of the *Almanus Manuscript*, in which is described a *clock with a dial indicating minutes* (now preserved in Augsburger Staatsbibliotek, Augsburg, Bavaria);

1556-59: <u>Taqi al-Din</u>, Ottoman engineer, published his book *The Brightest Stars for the Construction of Mechanical Clocks*, where it was described a *mechanical weight-driven astronomical clock* with a verge-and-foliot escapement, a striking train of gears, an alarm, and a representation of the Moon's phases;

1560: <u>*German clockmakers*</u> invented a *clock that recorded seconds* (now in the Fremersdorf Collection, Germany);

1582: <u>Gregory XIII</u>, Italian pope, reformed the Julian calendar, and in his Bull *Inter gravissimas* stated 'Every year that is exactly divisible by four is a leap year, except for years that are exactly divisible by 1000; the centurial years that are exactly divisible by 400 are still leap years'; from then, his calendar was adopted by European countries; Greece rallied to this reform last, in 1923;

1584-94: <u>Jost Bürgi</u>, Swiss clockmaker, invented the *cross-beat escapement*, and then the *remontoire* (from French *remonter* 'to wind up'), a small secondary source of power, a weight or spring, which runs the timekeeping mechanism, and used in *precision clocks and watches* to place the source of power closer to the escapement (now preserved in the Naturwissenschaftlich-Technische Sammlung, Kassel);

1595-1608: <u>William Shakespeare</u>, English playwright, dramatist and poet, versified *When, spite of cormorant devouring Time, / The endeavour of this present breath may buy / That honour which shall bate his scythe's keen edge, / And make us heirs of all eternity*; and also *What custom wills, in all things should we don't, / The dust on antique time would lie unwept, / And mountainous error be too highly heaped / For truth to o'erpeer*;

1604-20: <u>Cornelius Jacobszoon Drebbel</u>, Dutch-born British inventor, devised a *clock driven by changes in atmospheric pressure*;

1620-27: <u>Johannes Kepler</u>, German astronomer, first stated the *equation of time*, expressing the difference between apparent solar time and mean solar time;

1625: <u>Joseph Hall</u>, English prelate and thinker, mentioned in his 'Works' that *Perfection is the child of Time*;

1650-54: <u>James Ussher</u>, Irish prelate, produced the *Annales Veteris et Novi Testamenti*, which fixed the Creation 'precisely' at 4004BC;

1652-73: <u>Christiaan Huygens</u>, Dutch physicist, invented the *pendulum clock*, and then published *Horologium Oscillatorium*, showing that the wide pendulum swings of verge clocks caused them to be inaccurate, because it made the period of oscillation vary with unavoidable changes in drive force, and therefore only small pendulum swings were *isochronous*; also he published the first tables to give the *equation of time* in an essentially correct way;

1670-80: <u>William Clement</u>, British clockmaker, created the first commercial *anchor escapement clocks*, and then tall freestanding clocks with one metre second pendulum, that came to be called *long-cases* or 'grandfather' clocks;

1675-80: <u>Olaus Roemer</u>, Danish astronomer, gave the first estimate of the speed of light, and invented a *telescope movable only in the meridian*, increasing the accuracy in determination of both *time and right ascension in planetary motion*;

1719: <u>Isaac Watts</u>, English hymn writer, was author of *The Psalms of David Imitated* which includes the stanza '*Time, like an ever-rolling stream, / Bears all its sons away; / They fly forgotten, as a dream / Dies at the opening day*';

1739-61: <u>John Harrison</u>, English inventor and horologist, perfected his *H-2* as the *H-4 marine chronometer*, that was in error by less than 5 seconds over ten weeks, being tested in a voyage to Jamaica, in which *longitude* was determined within 18 geographical miles (the drawing of this device is now preserved in the Library of the Worshipful Company of Clockmakers, London); he also invented the *gridiron pendulum*, a *going fuse*, and a *remontoir escapement*;

1748: <u>Benjamin Franklin</u>, US politician, publisher, inventor and scientist, was quoted with *Remember that time is money*;

1750-74: <u>Philipp Matthäus Hahn</u>, German priest, inventor and clockmaker, devised a *wooden-gear clock* for St Peter's Monastery, Black Forest, and a precision sundial called *heliochronometer* that incorporated a *correction for the equation of time* (expressing the discrepancy between apparent and mean solar time); then designed the first functional *mechanical calculator* and constructed the *Copernicus planetarium*, with gears for Mercury, Venus, Earth, Mars and Jupiter, which was installed at Kornwestheim (now exposed in Deutsches Uhrenmuseum 'German Clock Museum' in Furtwangen im Schwarzwald, Germany);

1797: <u>Eli Terry</u>, US clockmaker and businessman, received his first patent for a practicable clock, and developed *mass-produced box clocks*, becoming the founder of the US clock-making industry;

1824: <u>Sadi Carnot</u>, French physicist, published his remarkable *Réflexions sur la puissance motrice du feu et sur les machines propres à developer cette puissance* 'Reflections on the Motive Power of Fire, and on the Machines Appropriate to develop this Power', concerning on scientific principles to an analysis of *working cycle* and *efficiency of steam engine* and establishing *Carnot's theorem* as a principle that specifies limits on the maximum efficiency any heat engine can obtain, as solely depending on the *difference between the hot and cold temperature reservoirs*; thus arriving at an early expression of the *second law of thermodynamics*;

1840-41: <u>Alexander Bain</u>, Scottish clockmaker, obtained the patent for an *electric clock*, and an *electromagnetic pendulum*, but they were not widely manufactured until electric power became available in 1890s;

1843-78: <u>James Prescott Joule</u>, English natural philosopher, showed experimentally that heat is a form of energy, determined quantitatively the amount of mechanical, and later electrical, energy to be expended in the propagation of heat; then he established the *mechanical equivalent of heat*, and expressed the *principle of conservation of energy* as *first law of thermodynamics*;

1848-50: <u>Rudolf Julius Emmanuel Clausius</u>, German physicist, gave another formulation of the *second law of thermodynamics*, postulating that *heat cannot of itself pass from a colder body to a hotter one*;

1848-52: <u>William Thomson Kelvin</u>, Scottish physicist and mathematician, concerning the dissipation of energy, gave a later formulation of the *second law of thermodynamics* in terms of 'no process is possible whose only result is the abstraction of heat from a single heat reservoir and the performance of an equivalent amount of work';

1849: <u>Edgar Allan Poe</u>, US poet and short-story writer, versified *Keeping time, time, time, / In a sort of Runic rhyme, / To the tintinnabulation that so musically wells / From the bells, bells, bells, bells*;

1868-77: <u>Ludwig Boltzmann</u>, Austrian physicist, established the famous *Boltzmann equation*, showing how *increasing entropy* corresponds to increasing molecular randomness;

1878-84: <u>Sandford Fleming</u>, Canadian railway engineer, divided the world into 24 time zones, and devised the *internationally adopted system of standard time*;

1881: <u>Gerard Manley Hopkins</u>, English poet and priest, in his manuscript 'Creation and Redemption: the Great Sacrifice' there is written that '*Time has three dimensions and one positive pitch or direction.* It is therefore not so much like any river or any sea as like the Sea of Galilee, which has the Jordan running through it and giving a current to the whole';

1889-1900: <u>Max Karl Ernst Planck</u>, German theoretical physicist, worked on the *laws of thermodynamics* and black body radiation, and formulated the *quantum theory*, relying on Boltzmann's statistical interpretation of the second law of thermodynamics and showing that energy changes take place in small discrete instalments, called *quanta*;

1903: <u>Eugene Fitch</u>, US clock designer, patented the *digital clock*, a type of clock that displays the time digitally, i.e. in numerals or other symbols;

1904-05: <u>Hendrik Antoon Lorentz</u>, Dutch physicist, derived a mathematical transformation explaining the apparent absence of relative motion between the Earth and the 'ether', and preparing the way for <u>Albert Einstein</u>, German-Swiss-US mathematical physicist, to elaborate the special theory of relativity, which was associated with time *dilation*, expressing the increase in time interval between two events when they occur in a reference frame which is moving at very high speed relative to the observer's reference frame rather than in the observer's rest time, according to the formula $t = t_0 \cdot [1 - (v/c)^2]^{1/2}$, where t is the time of a clock moving with speed v, t_0 is the elapsed time of a stationary clock and c is the speed of light;

1906: <u>Walter Hermann Nernst</u>, German physical chemist, enunciated his *heat theorem*, which was regarded as a statement of the *third law of thermodynamics*, enabling equilibrium constants for chemical reactions to be calculated from heat data;

1913-27: <u>Marcel Proust</u>, French novelist, produced the 13-volume *À la recherché du temps perdu* 'In Search of Lost Time';

1927: <u>Martin Heidegger</u>, German philosopher, published his major work *Sein und Zeit* 'Being and Time', giving an ontological classification of 'Being' and an examination of the distinctively human mode of existence 'Dasein' in the world of objects; and <u>*Bell Telephone Laboratories*</u>, New Jersey, USA, built the first *quartz clock*, a clock using an electronic oscillator that is regulated by a quartz crystal to keep time;

1928: <u>Arthur Stanley Eddington</u>, English astrophysicist, in his work *The Nature of the Physical World*, chapter 4, wrote that 'I shall use the phrase *time's arrow* to express this one-way property of time which has no analogue in space';

1930s: <u>*Peoples around the world*</u> replaced the mechanical clocks, as the most used type of clock, with *synchronous electric clocks*;

1936: <u>Thomas Stearns Eliot</u>, US-born British poet, critic and dramatist, in his writing *Four Quartets* 'Burnt Norton' versified *Time present and time past / Are both perhaps present in time future, / And time future contained in time past*;

1956: <u>Josef Pallweber</u>, Swiss timepiece maker, created a *mechanic-digital clock model*, which was soon perfected by US clockmakers into *digital clocks*;

1967: <u>*International System of Units*</u> defined the *second* as duration of 9,192,631,770 periods of the radiation corresponding to the transition between two hyperfine levels of the ground state of *caesium-133*;

1970: <u>*Hamilton Watch Company*</u> produced the first truly *digital wristwatch* with a light emitting display, called the *Pulsar*;

1988: <u>Stephen William Hawking</u>, English theoretical physicist, published *A Brief History of Time*, exploring various concepts behind modern cosmology, and raised tantalizing prospect that humankind might one day discover 'the mind of God'; and <u>Gerald James Whitrow</u>, British mathematician, cosmologist and science historian, wrote *Time in History. The evolution of our general awareness of time and temporal perspective*, Oxford University Press;

1995-2000: *US and European scientists and engineers* invented clocks based on *laser cooling and trapping atoms, Fabry-Pérot cavities, precision laser spectroscopy*, and *optical combs*;

2004-07: Robin Le Poidevin, British metaphysicist, wrote 'The Experience and Perception of Time', included in Edward N. Zalta, *The Stanford Encyclopedia of Philosophy*; and *The Images of Time: An Essay on Temporal Representation*, Oxford University Press;

2005-11: Harriet Nash, English hydrogeologist, studied the Pre-Islamic and Islamic practice of *star gazing* for marking the *times* of water distribution to community-managed falaj irrigation systems, showing that this practice is rapidly disappearing; and completed the book *Water Management: The Use of Stars in Oman*, BAR International Series 2237;

2006: Carlo Rovelli, Italian theoretical physicist and writer, based of his expertise in physics, wrote *Che cos'è il tempo? Che cos'è lo spazio?* 'What is time? What is space?', Rome, Di Renzo Editore.

Most of the above works were carried out in order to mark and predict periodical events such as solstices, Moon's phases, sea's tides, river-floods, annual sowing and harvesting times, religious celebrations, sacrifices, and other purposes. These works also included attentive observations on the movements of planets, stars, quasars and galaxies, which led to discovery that the universe is expanding, to the Big Bang theory, to the identification of microwave background radiation anisotropy, and to the models of inflation of the universe, aiming to explain its origin, evolution, and eventual fate.

Meanwhile, the cosmic, geological, biological, social, cultural, and many other timelines were more accurately established and divided in intervals of time indicating notable decreases in their succession, which could be explained by a more thorough approach of time itself, based on the viewpoints and statements presented below.

I ¶ In the course of scientific development, two contrasting viewpoints on time prevailed: one, that time is part of the fundamental structure of the universe, as a dimension independent of events, in which *events occur in sequence* (Newton, 1684-87); and another, that time is part of a fundamental intellectual structure, together with space and number, within which humans *sequence and compare events* (Leibniz, 1680-1716; Kant, 1755-95). Later viewpoints on time derived from its representation in a reference frame moving at significant high speed relative to a stationary reference frame, known as time *dilation* (Einstein, 1904-05), or from its one-way direction, or asymmetry, described as time*'s arrow* (Eddington, 1928).

An operational approach of time is related to the observation of a certain number of repetitions of one or another standard *cyclical event*, e.g. the passage of a free-swinging pendulum, which constitutes one standard unit, such as the second or year, useful in the conduct of advanced experiments or everyday affairs of life, and in the analysis of natural processes.

II ¶ Albert Einstein stated that 'A theory is the more impressive the greater the simplicity of its premises, the more different kinds of things it relates, and the more extended its area of applicability. Therefore, the deep impression that classical thermodynamics made upon me. It is the only physical theory of universal content which I am convinced will never be overthrown, within the framework of applicability of its basic concepts' (1905-09). Indeed, any natural process takes place according to the *laws of thermodynamics*:

Zeroth law - if two systems are each in thermal equilibrium with a third system then they are in thermal equilibrium with each other;

First law - the total energy of a thermodynamic system remains constant although it may be transformed from one form to another, that means the conservation of energy (Joule, 1843-78);

Second law - heat can never pass spontaneously from a body at a lower temperature to one at a higher temperature (Clausius, 1848-50, validating the theorem of Carnot, 1824), or equivalently no process is possible whose only result is the abstraction of heat from a single heat reservoir and the performance of an equivalent amount of work (Kelvin, 1850; Planck, 1900, relying on the statistical interpretation of Boltzmann, 1868); and

Third law - the entropy of a substance approaches zero as its temperature approaches absolute zero (Kelvin, 1848; Boltzmann, 1877; Nernst, 1906).

III ¶ Any field of interest or activity evolves as a natural process always limited between its beginning, marked by the initial time $t_0 = 0$, and its end, marked by the final time $t_\bullet$, which represent boundaries deflecting the flow of energy and inducing in it undulations of period $\tau(t)$ decreasing due to the dissipation of energy. Otherwise the process should be perpetual, that means unlimited and then lasting forever with no external source of energy, thereby violating the first law of thermodynamics; or with equal output and input of energy, meaning no dissipation of energy, and then violating the second law of thermodynamics.

Therefore, the *total energy* [joules] of any natural process is a function of both time t and period τ, i.e. $E = E(t, \tau)$, with the total differential $dE = (\partial E/\partial t)\cdot dt + (\partial E/\partial \tau)\cdot d\tau = P_t\cdot dt + P_\tau\cdot d\tau$, where the terms $(\partial E/\partial t)\cdot dt$ and $(\partial E/\partial \tau)\cdot d\tau$ express the infinitesimal changes in *effective (orderly) energy* and in *dissipative (disorderly) energy* respectively; and the corresponding partial derivatives $\partial E/\partial t = P_t$ and $\partial E/\partial \tau = P_\tau$ express the instantaneous effective power and the dissipative power [watts = joules per second]. In agreement with the first law of thermodynamics, the conservation of energy implies the annulment of total differential of energy, that is $dE = P_t\cdot dt + P_\tau\cdot d\tau = 0$, and then $P_t\cdot dt = -P_\tau\cdot d\tau$, leading to power efficiency given by the equality

$$-d\tau/dt = P_t/P_\tau.$$

Meanwhile, according to the second law of thermodynamics, the power efficiency of a process must always be less than 100% which would imply perpetual process, i.e. the ratio $R = P_t/P_\tau$ must be less than or equal to unity, or $R = P_t/P_\tau \leq 1$; and this ratio would increase with the running time t, that is consistent with the principle of increasing entropy, so as $R\cdot t = $ constant and then $d(R\cdot t) = t\cdot dR + R\cdot dt = 0$, whence the equation $dR/R = -dt/t$ defined for $R \leq 1$ and $t \leq t_\bullet$. This equation can be integrated as $\int^1_R dR/R = -\int^{t_\bullet}_t dt/t$, whence $ln(1) - ln(R) = ln(t_\bullet) - ln(/t)$ or $ln(R) = ln(t) - ln(t_\bullet) = ln(t/t_\bullet)$, where ln is the natural logarithm, i.e. logarithm to base $e \approx 2.7182818$. As a result, $R = t/t_\bullet$ or

$$P_t/P_\tau = t/t_\bullet.$$

Substituting P_t/P_τ for $t/t_\bullet$, the above equality $-d\tau/dt = P_t/P_\tau$ becomes $-d\tau/dt = t/t_\bullet$, or

$$-d\tau = (1/t_\bullet)\cdot t\cdot dt,$$

which, by integration $-\int^0_\tau d\tau = (1/t_\bullet)\cdot\int^{t_\bullet}_t t\cdot dt$, gives an expression of the period

$$\tau = (t_\bullet^2 - t^2)/(2t_\bullet),$$

where $0 \leq t \leq t_\bullet$ and $t_\bullet/2 \geq \tau \geq 0$.

IV ¶ On the other hand, the undulations in flow of energy are cyclically repeated at a rate of increasing frequency f that corresponds to decreasing period τ, each of them being related to an instantaneous angular speed ω defined by the time rate of an ever increasing angle θ, as

$$\omega = d\theta/dt = 2\pi\cdot f = 2\pi/\tau,$$

whereby the equality

$$d\theta = (2\pi/\tau)\cdot dt,$$

and then another expression of the period

$$\tau = 2\pi \cdot dt/d\theta,$$

Therefore, the period, frequency, and angular speed are *intrinsic characteristics* of any natural process.

V ¶ The equality of the above complementary expressions of the period

$$(t_\bullet^2 - t^2)/(2t_\bullet) = 2\pi \cdot dt/d\theta$$

leads to the differential equation

$$d\theta = 4\pi \cdot t_\bullet \cdot dt/(t_\bullet^2 - t^2) = 4\pi \cdot d(t/t_\bullet)/[1 - (t/t_\bullet)^2]$$

which, integrated as $\int_0^\theta d\theta = 4\pi \cdot \int_0^{t/t_\bullet} d(t/t_\bullet)/[1 - (t/t_\bullet)^2]$, results in the function

$$\theta(t) = 4\pi \cdot tanh^{-1}(t/t_\bullet),$$

where the angle θ [radians] is given by the product of a double-cycle $4\pi = 2 \cdot (2\pi)$ and the inverse hyperbolic tangent $tanh^{-1}$ of the ratio of running time t to final time $t_\bullet$, remembering that $tanh^{-1}(t/t_\bullet) = (1/2) \cdot ln\{[1 + (t/t_\bullet)]/[1 - (t/t_\bullet)]\} = (1/2) \cdot ln[(t_\bullet + t)/(t_\bullet - t)]$.

The equality $\theta/4\pi = tanh^{-1}(t/t_\bullet)$ is the inverse of $t/t_\bullet = tanh(\theta/4\pi)$ whence the *time function*

$$t = t_\bullet \cdot tanh(\theta/4\pi),$$

in which $tanh(\theta/4\pi) = (e^{\theta/4\pi} - e^{-\theta/4\pi})/(e^{\theta/4\pi} + e^{-\theta/4\pi}) = (e^{\theta/2\pi} - 1)/(e^{\theta/2\pi} + 1)$ is the hyperbolic tangent of the increasing angle θ divided by a double-cycle 4π, and t is the time running from 0 at the beginning of a process to $t_\bullet$ at the end of the process.

VI ¶ This function can be used to re-express the intrinsic characteristics of any natural process as the period

$$\tau = (t_\bullet^2 - t^2)/(2t_\bullet) = [t_\bullet^2 - t_\bullet^2 \cdot tanh^2(\theta/4\pi)]/(2t_\bullet)$$
$$= (t_\bullet/2) \cdot [1 - tanh^2(\theta/4\pi)],$$

frequency

$$f = 1/\tau = (2t_\bullet)/(t_\bullet^2 - t^2) = 2/\{t_\bullet \cdot [1 - tanh^2(\theta/4\pi)]\},$$

and angular speed

$$\omega = 2\pi \cdot f = 2\pi/\tau = 4\pi/\{t_\bullet \cdot [1 - tanh^2(\theta/4\pi)]\},$$

where $0 \le t \le t_\bullet$; $t_\bullet/2 \ge \tau \ge 0$; $2/t_\bullet \le f \le \infty$; and $4\pi/t_\bullet \le \omega \le \infty$.

VII ¶7 The time function governs a wide range of processes taking place at scales of the universe, Earth, climate, life, humanity, human abilities, etc. As these processes and also their subdivisions are lasting between billions and tens of years, it is convenient that from the multitude of values taken by the time function are selected only those corresponding to values of argument equalling integers, or semi-integers, quarter-integers, eighth-integers, and so on. Main subdivisions of time with durations decreasing from the earlier to the later ones have been identified in many chronological scales or timelines, such as 'eras' in geology, 'eras/periods' in musicology, 'periods/ages' in climatology, 'epochs/ages' in anthropology and civilization, 'ages' in sociology and art, etc., which are comprehensively called *time sequences*.

Noticing that the argument $\theta/4\pi$ takes the values $z = 0, 1, 2, 3...$, as numbers of completed double-cycles, it follows that their subdivisions
$$z/2 = i = 0, 1/2, 1, 3/2...; \ z/4 = j = 0, 1/4, 1/2, 3/4...;$$
$$z/8 = k = 0, 1/8, 1/4, 3/8...; \ z/16 = l = 0, 1/16, 1/8, 3/16...;$$
$$z/32 = m = 0, 1/32, 1/16, 3/32...; \ z/64 = n = 0, 1/64, 1/32, 3/64...$$
are numbers of completed cycles, semi-cycles, quarter-cycles, eighth-cycles, etc., corresponding to values of the time function
$$t_i = t_{\bullet} \cdot tanh(i); \ t_j = t_{\bullet} \cdot tanh(j); \ t_k = t_{\bullet} \cdot tanh(k);$$
$$t_l = t_{\bullet} \cdot tanh(l); \ t_m = t_{\bullet} \cdot tanh(m); \ t_n = t_{\bullet} \cdot tanh(n).$$

VIII ¶ Time sequences can be of various kinds in accordance with different subdivisions of the integers taken by the argument $\theta/4\pi$ and their corresponding values of the time function.

As keys to identify the kind of sequences of a studied process can be mentioned the ratios of discrete times delimiting two successive sequences, and the differences of these discrete times divided by the final time $t_{\bullet}$ are both independent of $t_{\bullet}$; so that some of more reliable time limits, previously estimated by dating them back from the present time t_σ (Greek σήμερα or σημερον 'today') as $t_\sigma - t$, and then converted in times from the origin of studied process as $t = t_\sigma - (t_\sigma - t)$, can be easier compared to those calculated with the time function, whereby the kind of sequences $i, j, k, l, m,$ or n better fitting the evolution or development of the studied process from its beginning to its end.

The numbers i, j, k, l, m, n with their corresponding discrete values of the hyperbolic tangent, together with ratios and differences of these values in their succession are displayed in six tables below.

Table *(i)*

$z/2 = i$	$t_i/t_{\bullet} =$ $tanh(i)$	$t_{i-1/2}/t_i =$ $tanh(i - 1/2)/tanh(i)$	$(t_i - t_{i-1/2})/t_{\bullet} =$ $tanh(i) - tanh(i - 1/2)$
∞	1	1	0
...	...	...	...
4	0.9993293	0.9988478	0.0011514
3.5	0.9981779	0.9968712	0.0031231
3	0.9950548	0.9915176	0.0084405
2.5	0.9866143	0.9771068	0.0225867
2	0.9640276	0.9389236	0.0588793
1.5	0.9051483	0.8414027	0.1435541
1	0.7615942	0.6067761	0.2994770
0.5	0.4621172	0	0.4621172
0	0	-	-

Table (j)

z/4 = j	$t_j/t_\bullet =$ tanh(j)	$t_{j-1/4}/t_j =$ tanh(j - 1/4)/tanh(j)	$(t_j - t_{j-1/4})/t_\bullet =$ tanh(j) - tanh(j - 1/4)
∞	1	1	0
...	...	...	...
3.25	0.9969976	0.9980513	0.0019429
3	0.9950548	0.9967891	0.0031950
2.75	0.9918597	0.9947115	0.0052454
2.5	0.9866143	0.9912953	0.0085882
2.25	0.9780261	0.9856870	0.0139985
2	0.9640276	0.9765027	0.0226520
1.75	0.9413755	0.9615166	0.0362273
1.5	0.9051483	0.9371765	0.0568646
1.25	0.8482836	0.8978060	0.0866895
1	0.7615942	0.8339730	0.1264452
0.75	0.6351490	0.7275729	0.1730318
0.5	0.4621172	0.5299926	0.2171985
0.25	0.2449187	0	0.2449187
0	0	-	-

Table (k)

z/8 = k	$t_k/t_\bullet =$ tanh(k)	$t_{k-1/8}/t_k =$ tanh(k - 1/8)/tanh(k)	$(t_k - t_{k-1/8})/t_\bullet =$ tanh(k) - tanh(k - 1/8)
∞	1	1	0
...	...	...	...
2.5	0.9866143	0.9961796	0.0037693
2.375	0.9828450	0.9950970	0.0048189
2.25	0.9780261	0.9937084	0.0061534
2.125	0.9718727	0.9919278	0.0078452
2	0.9640276	0.9896452	0.0099823
1.875	0.9540453	0.9867199	0.0126698
1.75	0.9413755	0.9829725	0.0160293
1.625	0.9253462	0.9781726	0.0201979
1.5	0.9051483	0.9720249	0.0253216
1.375	0.8798267	0.9641485	0.0315431
1.25	0.8482836	0.9540454	0.0389825
1.125	0.8093011	0.9410517	0.0477069
1	0.7615942	0.9242528	0.0576886
0.875	0.7039056	0.9023213	0.0687566
0.75	0.6351490	0.8731805	0.0805493
0.625	0.5545997	0.8332446	0.0924825
0.5	0.4621172	0.7754686	0.1037598
0.375	0.3583574	0.6834481	0.1134387
0.25	0.2449187	0.5077317	0.1205657
0.125	0.1243530	0	0.1243530
0	0	-	-

Table (l)

$z/16 = l$	$t_l/t_\bullet =$ $tanh(l)$	$t_{l-1/16}/t_l =$ $tanh(l - 1/16)/tanh(l)$	$(t_l - t_{l-1/16})/t_\bullet =$ $tanh(l) - tanh(l - 1/16)$
∞	1	1	0
...	...	...	...
1.75	0.9413755	0.9919825	0.0075475
1.6875	0.9338280	0.9909172	0.0084818
1.625	0.9253462	0.9897101	0.0095217
1.5625	0.9158245	0.9883424	0.0106763
1.5	0.9051483	0.9867923	0.0119549
1.4375	0.8931933	0.9850350	0.0133666
1.375	0.8798267	0.9830420	0.0149201
1.3125	0.8649066	0.9807806	0.0166230
1.25	0.8482836	0.9782128	0.0184817
1.1875	0.8298019	0.9752943	0.0205008
1.125	0.8093011	0.9719730	0.0226823
1.0625	0.7866188	0.9681871	0.0250247
1	0.7615942	0.9638618	0.0275226
0.9375	0.7340715	0.9589060	0.0301659
0.875	0.7039056	0.9532060	0.0329385
0.8125	0.6709671	0.9466172	0.0358181
0.75	0.6351490	0.9389507	0.0387754
0.6875	0.5963736	0.9299536	0.0417738
0.625	0.5545997	0.9192756	0.0447697
0.5625	0.5098300	0.9064143	0.0477128
0.5	0.4621172	0.8906184	0.0505471
0.4375	0.4115701	0.8707081	0.0532127
0.375	0.3583574	0.8447145	0.0556477
0.3125	0.3027097	0.8090877	0.0577911
0.25	0.2449187	0.7567131	0.0595855
0.1875	0.1853332	0.6709699	0.0609802
0.125	0.1243530	0.5019477	0.0619343
0.0625	0.0624187	0	0.0624187
0	0	-	-

Table (m)

$z/32 = m$	$t_m/t_\bullet =$ $tanh(m)$	$t_{m-1/32}/t_m =$ $tanh(m - 1/32)/tanh(m)$	$(t_m - t_{m-1/32})/t_\bullet =$ $tanh(m) - tanh(m - 1/32)$
∞	1	1	0
...	...	...	...
1	0.7615942	0.9823532	0.0134397
0.96875	0.7481545	0.9811764	0.0140830
0.9375	0.7340715	0.9799147	0.0147440
0.90625	0.7193275	0.9785607	0.0154219
0.875	0.7039056	0.9771057	0.0161154
0.84375	0.6877902	0.9755403	0.0168231
0.8125	0.6709671	0.9738534	0.0175435
0.78125	0.6534236	0.9720325	0.0182746
0.75	0.6351490	0.9700629	0.0190145

0.71875	0.6161344	0.9679277	0.0197609
0.6875	0.5963736	0.9656069	0.0205112
0.65625	0.5758624	0.9630768	0.0212627
0.625	0.5545997	0.9603093	0.0220124
0.59375	0.5325873	0.9572703	0.0227573
0.5625	0.5098300	0.9539181	0.0234940
0.53125	0.4863360	0.9502014	0.0242189
0.5	0.4621172	0.9460561	0.0249284
0.46875	0.4371888	0.9414013	0.0256187
0.4375	0.4115701	0.9361321	0.0262861
0.40625	0.3852840	0.9301123	0.0269266
0.375	0.3583574	0.9231597	0.0275363
0.34375	0.3308211	0.9150254	0.0281114
0.3125	0.3027097	0.9053611	0.0286481
0.28125	0.2740616	0.8936630	0.0291429
0.25	0.2449187	0.8791746	0.0295924
0.21875	0.2153263	0.8607086	0.0299931
0.1875	0.1853332	0.8362814	0.0303425
0.15625	0.1549907	0.8023256	0.0306377
0.125	0.1243530	0.7517012	0.0308767
0.09375	0.0934763	0.6677489	0.0310576
0.0625	0.0624187	0.5004878	0.0311789
0.03125	0.0312398	0	0.0312398
0	0	-	-

Table (n)

$z/64 = n$	$t_n/t_\bullet =$ $\tanh(n)$	$t_{n-1/64}/t_n =$ $\tanh(n-1/64)/\tanh(n)$	$(t_n - t_{n-1/64})/t_\bullet =$ $\tanh(n) - \tanh(n-1/64)$
∞	1	1	0
...	...	...	...
0.25	0.2449187	0.9398045	0.0147430
0.234375	0.2301757	0.9354867	0.0148494
0.21875	0.2153263	0.9305723	0.0149496
0.203125	0.2003767	0.9249239	0.0150435
0.1875	0.1853332	0.9183584	0.0151309
0.171875	0.1702023	0.9106264	0.0152116
0.15625	0.1549907	0.9013786	0.0152854
0.140625	0.1397053	0.8901094	0.0153523
0.125	0.1243530	0.8760617	0.0154121
0.109375	0.1089409	0.8580460	0.0154646
0.09375	0.0934763	0.8340767	0.0155099
0.078125	0.0779664	0.8005846	0.0155477
0.0625	0.0624187	0.7504274	0.0155780
0.046875	0.0468407	0.6669371	0.0156009
0.03125	0.0312398	0.5001216	0.0156161
0.015625	0.0156237	0	0.0156237
0	0	-	-

18

IX ¶ The time function can be applied by its discrete values corresponding to integers or their subdivisions of its argument, thus covering fields of study such as creativity, cosmology, geology, climatology, biology, anthropology, humanity, linguistics, art, theology, musicology, exploration, inhabitation, knowledge, navigation, meteorology, medicine, sociology, construction, science, industry, civilization, legislation, philosophy, aero-astronautics, and high technology', each of them being analyzed in the following chapters.

For each field of study, the times delimiting its successive sequences are usually given in years ago (before the present time t_σ), and therefore they must be subtracted from the initial time of the process also expressed in years ago in order to find the values suitable for applying the time function. For example, if the given time is 6000 years ago, and the origin of the processed field of study was 20000 years ago, then the time from its origin is 20000 - 6000 = 14000 years as the value accordant to the time function.

X ¶ Taking into account that the available data referring to time limits and durations are more certain for the later sequences, less certain for the earlier sequences, and uncertain or non-existent for future ones; they must be attentively selected, interpreted and used in the most suitable way on one of the above variants, in order to evaluate the times of transition and durations of the sequences both backwards to the earlier and forwards to the later ones, as well as the values of period, frequency and angular speed, thus modelling the processed field of study from its beginning to its end. Such models allow us, for example, to calculate how long the present sequence and the following ones will last, and eventually to predict the end of the last sequence when the processed field of study will reach its final time $t_\bullet$.

XI ¶ In order to avoid a fatalistic misinterpretation, the final time can be interpreted as the fate of one state in development of a studied process, not necessarily meaning that the process simply vanishes, but that it reaches the end of an extending quarter-super-cycle and the beginning of a shrinking quarter-super-cycle, together completing a semi-super-cycle at a time approximately equalling $2t_\bullet$ as the initial time of another semi-super-cycle.

XII ¶ The time function $t = t_\bullet \cdot tanh(\theta/4\pi)$ can be extensively used to simulate the running time anterior and posterior to the known set of sequences, in accordance with its discrete values $\theta/4\pi$ that range unrestrictedly between $-\infty$ and $+\infty$ both backwards and forwards by intervals

$$...t_{\bullet-2} < t_{\bullet-1} < t_{\bullet 0} < t_{\bullet+1} < t_{\bullet+2}...$$

The durations of these intervals

$$...t_{\bullet-1} - t_{\bullet-2};\ t_{\bullet 0} - t_{\bullet-1};\ t_{\bullet+1} - t_{\bullet 0};\ t_{\bullet+2} - t_{\bullet+1}...$$

depend on their decreasing, unchanging, or increasing rates

$$...t_{\bullet-2}/t_{\bullet 0};\ t_{\bullet-1}/t_{\bullet 0};\ 1;\ t_{\bullet+1}/t_{\bullet 0};\ t_{\bullet+2}/t_{\bullet 0}...,$$

and mark the stages of an ultimately closing, flattening, or opening development of the process respectively. Therefore, the successive intervals can be less than, equal to, or more than the reference interval of time with duration $t_{\bullet}$. The intervals between two successive final times represent semi-super-cycles of extending and shrinking process, so that their couple completes a super-cycle virtually repeating on an indefinite duration.

1. Creativity

The process of producing something that is both original and worthwhile, or is characterized by originality and expressiveness is called *creativity* (from Latin *creo* 'to create, make, produce', *creatio* 'choice, election'). An accepted definition of creativity is: the ability to produce something new through imaginative skill, in order to find out a new solution to a problem, a new method or device, or a new artistic object or form. Generally, the term 'creativity' refers to a richness of ideas and originality of thinking.

Psychological studies of highly creative people have indicated that many of them have a strong interest in apparent disorder, contradiction and imbalance, which are perceived as challenges. Such individuals may possess an exceptionally deep, broad and flexible awareness of themselves. Researches also show that intelligence has little correlation with creativity; thus, a highly intelligent person may not be very creative. *Intelligence* is the property of mind that encompasses the capacities to reason, plan, problem solve, think, comprehend ideas, use languages, communicate, and learn. The evolution of human intelligence is closely tied to the evolution of the human brain, and to the emergence of human language. Among the theories concerning the evolution of intelligence can be mentioned: the social brain hypothesis; the sexual selection hypothesis; ecological dominance-social competition; intelligence as a signal of good health and resistance to disease; the group selection theory contending that organism characteristics provide benefit to a group; and intelligence connected with nutrition. Research based on data from functional imaging and structural imaging shows that intelligence arises from a distributed and integrated neural network in some regions of the frontal and parietal lobes (Haier and Jung, 2007). In neurobiology, highly creative people excel at creative innovation, and tend to differ from others in three ways: high level of specialized knowledge; capability of divergent thinking mediated by the *frontal lobe*; and aptitude to modulate neurotransmitters, such as norepinephrine, in their frontal lobe, and limbic stimulative substances, such as serotonin, dopamine, and noradrenaline. Therefore, the frontal lobe seems to be the most important part of the cortex for creativity. The process of creativity involves new combinations of associative elements, which are useful or required, and aided by sleep, especially REM (rapid eye movement). It is proposed that REM sleep adds creativity by allowing 'neocortical structures to reorganize associative hierarchies, in which information from the hippocampus would be reinterpreted in relation to previous semantic representations or nodes.'

Creativity is often associated with *genius* seen as person of extraordinary intellectual power, who displays originality, creativity and the ability to think and work in areas not previously explored. Though geniuses have usually left their unique mark in a particular field, studies have shown that the general intelligence of geniuses is also exceptionally high. Genius appears to be a function of both hereditary and environmental factors.

Scholarly interest in creativity includes a multitude of definitions and approaches in disciplines such as psychology, cognitive science, education, philosophy, technology, theology, sociology, linguistics, art, musicology, business studies, and economics, taking in the relationship between creativity and general intelligence, mental and neurological processes associated with creativity, the relationships between personality type and creative aptitudes or skills, and between creativity and mental health, the potential for fostering creativity through education and training, and the application of creative resources to improve the effectiveness of learning and teaching processes.

Investigation of why some people are more creative than others have focused on a variety of aspects, which are included in a series of theories of creativity. The dominant factors in creativity seem to be the *four Ps*: *process, product, person* and *place*. Creative lifestyles are characterized by nonconforming attitudes and behaviour as well as flexibility.

Historically, in Bible's the first book of Moses, *Genesis* was the greatest *God's work of creation*, consisting of six sequences, called 'days', when 'the heavens and the earth were finished, and all the host of them. And on the seventh day God ended his work which he had made; and he rested on the seventh day from all his work which he had made. And God blessed the seventh day, and sanctified it: because that in it he had rested from all his work which *God created* and made.' Later, in ancient Greek and Roman world, creativity was the concept of an external creative δαιμων 'divinity' or *genius* 'guardian spirit of a man or place, a genius' respectively, in relation to *something sacred or divine*. In European culture, the notion of 'creativity' originated through Christianity, as a *matter of divine inspiration*. By the Renaissance, creativity was first seen, not as a conduit for the divine, but as the *aptitudes or skills of great men*. The meaning of creativity developed by contributions of a series of writers and researchers with their works presented as follows:

1386-88: <u>Geoffrey Chaucer</u>, English poet and founder of the literary English language, gave a modern meaning of creativity as an *act of human creation*, according to his *The Parson's Tale*, the final tale of the poetic cycle *The Canterbury Tales*;

1896-1908: <u>Hermann von Helmholtz</u>, German physiologist and physicist, after his work on conservation of energy, and <u>Jules Henri Poicaré</u>, French mathematician, after his studies of celestial mechanics; based on their scientific experience they began to reflect on and publicly discuss their *creative processes*;

1926: <u>Graham Wallas</u>, English political psychologist, wrote *Art of Thought*, where he exposed one of the first models of *creative process* explained by five successive stages, namely *preparation, incubation, intimation, illumination,* and *verification*;

1927-29: <u>Alfred North Whitehead</u>, English mathematician and Idealist philosopher, in his Edinburgh Gifford Lectures *Process and Reality*, attempted a metaphysics comprising psychological as well as physical experience, with events as the ultimate components of reality, preferring currency of exchange among literature, science, and the arts, etc.;

1950: <u>Joy Paul Guilford</u>, US psychologist, is remembered for his psychometric study of human intelligence, including the distinction between convergent and divergent production, published in *Creativity*, American Psychologist 5(9), and based on statistical analysis whereby the recognition of (measured) creativity as a separate aspect of human cognition to *IQ*-type intelligence, into which it had previously been subsumed; accordingly, the *creative process* consists of *incubation, convergent and divergent thinking, creative cognition approach, explicit-implicit interaction (EII) theory, conceptual blending,* and *honing theory*;

1962-66: <u>Ellis Paul Torrance</u>, US psychologist, first grouped the different subtests of the *Minnesota Tests of Creative Thinking* into three categories, namely *verbal tasks using verbal stimuli, verbal tasks using non-verbal stimuli,* and *non-verbal tasks*; then the *Torrance Tests of Creative Thinking*, that involved tests of divergent thinking and other problem-solving skills, which were scored on four scales, namely *fluency, flexibility, originality,* and *elaboration*;

1967: <u>Joy Paul Guilford's group</u>, US psychologists, approached the *modern psychometric study of creativity*, by several tests to measure creativity - *plot titles, quick responses, figure concepts, unusual uses, remote associations*, and *remote consequences*;

2003: Michael Mumford, US doctor and director of Center for Applied Social Research, wrote a summary of scientific research into creativity, entitled *Where have we been, where are we going? Taking stock in creativity research*, Creativity Research Journal, 15, where he suggested 'Over the course of the last decade, however, we seem to have reached a general agreement that creativity involves the production of novel, useful products';

2005: Daniel Pink, US researcher, published his book *A Whole New Mind: Why Right-Brainers Will Rule the Future*, showing that in present conceptual age we will need to foster and encourage *right-directed thinking* (representing creativity and emotion) over *left-directed thinking* (representing logical, analytical thought); and Alice Flaherty, US neurologist, wrote *Frontotemporal and Dopaminergic Control of Idea Generation and Creative Drive*, presenting a three-factor model of the creative drive, described as resulting from an interaction of the *frontal lobes*, the *temporal lobes*, and *dopamine from the limbic system*;

2007: Richard Haier and Rex Jung, US psychologists, argued that human intelligence arises from a distributed and integrated neural network comprising brain regions in the frontal and parietal lobes, and published their work *The Parieto-Frontal Integration Theory (P-FIT) of Intelligence: Converging neuroimaging evidence*, Cambridge University Press;

2008: Malcom Gladwell, English-Canadian journalist, published a book entitled *Outliers: The Story of Success*, Little, Brown and Company, mentioning a 'divergence test', as opposed to 'convergence tests';

2008-09: Peter Meusburger, Austrian professor at Heidelberg University, elaborated 'Milieus of Creativity: The Role of Places, Environments and Spatial Contexts'; which was published together with Joachim Funke and Edgar Wunder in a comprehensive work entitled *Milieus of Creativity: An Interdisciplinary Approach to Spatiality of Creativity*, Springer, Knowledge & Space, vol.2;

2009: Robert Sternberg, US psychologist, within the work *Creativity Research Journal*, Jaime A. Perkins, Dan Moneypenny, Wilson Co., Cognitive Psychology, CENGAGE Learning, he suggested that there are eight types of creative contribution, namely *replication, redefinition, forward incrementation, advance forward movement*, *redirection, redirection from a point in the past, starting over/re-initiation*, and *integration*.

A convenient approach of creativity first implies a selection of remarkably creative personalities or recognizable geniuses in various fields of interest or activity, counting to a number not too large but just enough for a simple statistical operation needed for a proper analysis of the *creative process*. Despite the subjectivity in selection of such personalities or geniuses, 56 of them with data related to the time-intervals of their activity and life are briefly presented below in chronological order.

335-323BC: <u>Alexander the Great</u> (356-323BC), King of Macedonia and military genius, invaded *Persia*, won a major victory at *Granicus*, defeated King Darius III at a pass near Issus, in Cilicia; occupied *Damascus* and *Tyre*, and marched on to *Palestine*; was welcomed in *Egypt* where he founded the city of *Alexandria*; met again the Persian king at *Gaugamela* near Arbela where he won a decisive victory; entered the cities of *Babylon, Susa* and the capital *Persepolis*; overthrew *Scythians*, and subdued *Sogdiana*; proceeded to the conquest of *India* winning a costly battle against King Porus at *Hydaspes* (Jhelum); marched through *Gedrosia* (Balochistan) suffering heavy losses on the way; and died at Babylon;

260-212BC: <u>Archimedes</u> (287-212BC), Greek mathematician and scientist, formulated the *lever and pulley functions*; founded the science of *hydrostatics*, studying equilibrium positions of floating bodies and discovering that a body weighed when immersed in a fluid shows a loss of weight equal to weight of fluid it displaces, called *principle of Archimedes*; discovered formulae for areas and volumes of *spheres, cylinders, parabolas*, and other plane and solid figures, by methods *anticipating theories of integration*; and used mechanical arguments *involving infinitesimals* as a heuristic tool to obtain results prior a rigorous proof;

AD1480-1509: <u>Leonardo da Vinci</u> (1452-1519), Italian painter, sculptor, architect and engineer, produced paintings including *Baptism of Christ, Adoration of the Magi* (Uffizi Gallery, Florence); famous mural *Il Cenacolo* or *L'Ultima Cena* 'The Last Supper' (Refectory of Santa Maria delle Grazie, Milan); portrait *La Belle Ferronnière* (Louvre, Paris), cartoons *Madonna and Child with St Anne* (Royal Academy, London), and *The Battle of Anghiari* (Palazzo delle Signoria, Florence); as well as celebrated easel picture *Mona Lisa* (Louvre, Paris), and geometric figures *De divina Proportione*; in addition, he made remarkable studies on human and comparative anatomy, metabolism (compared by him to 'a burning candle'), mathematics, botany, weaponry (cannons, tank, parachute), musical instruments, flying and drilling machines, etc., which were gathered together in *Codex Atlanticus*, by the sculptor Pompeo Leoni in the late 16th century, within a twelve-volume atlas-like breadth (Biblioteca Ambrosiana, Milan);

1490-1555: Michelangelo di Lodovico Buonarroti (1475-1564), Italian sculptor, painter and poet, created the famous sculptures *Battle of the Centaurs*, and *Madonna of the Steps* in Florence, marbles *Cupid* in Bologna, *Bacchus* (Museum of Florence), and *Pieta* (St Peter's, Rome), colossal marble block *David* in Florence, *Holy Family of the Tribune*, *Madonna* (National Gallery, London), and the sublime statue of *Moses* in Rome; the almost superhuman *Decorations of the Sistine Chapel ceiling* (Vatican, Rome); then the solar symbolistic fresco *Il Giudizio Universale* 'The Last Judgement' (on the altar wall of the Sistine Chapel in the Vatican), and the completion of *St Peter's* in Rome;

1492-1504: Christopher Columbus (1451-1506), Genoese explorer and discoverer of the New World, based on the strong intuitive reason of a western maritime route to India, he became famous for his four voyages: first, on the flagship *Santa Maria* attended by caravels *Pinta* and *Niña*, to *San Salvador*, *Cuba*, and *Haiti*; second to *Guadeloupe*, *Montserrat*, *Antigua*, *Puerto Rico*, and *Jamaica*; third, to *Trinidad*, and the *main land of South America*; and fourth, to *Honduras*, and *Nicaragua*;

1582-1638: Galilei Galileo (1564-1642), Italian astronomer, mathematician and natural philosopher, discovered that *all bodies fall at the same rate* when air resistance is not present, i.e. acceleration of gravity is constant (at ground level), and a body moving along an inclined plane has a constant acceleration; demonstrated the parabolic trajectories of projectiles; defended the Copernican system in his *Dialogue on the Two Principal Systems of the World*, and completed *Discourses on the Two New Sciences*, discussing at length the principles of mechanics;

1590-1609: William Shakespeare (1564-1616), English playwright, poet and the greatest English dramatist, created famous Renaissance works, including early plays *The Two Gentlemen of Verona*, *Henry VI*, *Titus Andronicus*, *The Taming of the Shrew*, *The Comedy of Errors*, *Love's Labour's Lost*, *Romeo and Juliet*; histories *Richard III*, *Richard II*, *King John*, *Henry IV*, *Henry V*; later comedies *A*

Midsummer Night's Dream, *The Merchant of Venice*, *The Merry Wives of Windsor*, *Much Ado About Nothing*, *As You Like It*, *Twelfth Night*, *Troilus and Cressida*, *Measure for Measure*, *All's Well That Ends Well*; Roman plays *Julius Caesar*, *Antony and Cleopatra*, *Coriolanus*; later tragedies *Hamlet*, *Othello*, *Timon of Athens*, *King Lear*, *Macbeth*; late plays *Pericles*, *The Winter's Tale*, *Cymbeline*, *The Tempest*, *Henry VIII*; and non-dramatic works *Venus and Adonis*, *The Rape of Lucrece*, *'The Phoenix and the Turtle'*, *Sonnets*, and *'A Lover's Complaint'*;

1605-25: <u>Francis Bacon</u> (1561-1626), English philosopher and statesman, wrote the philosophical works *The Advancement of Learning*, *De Augmentis Scientiarum*, and *Novum Organum*, by which he created the method of scientific induction giving an impetus to future scientific investigation; the essays such as *History of Henry VII*, as well as legal and constitutional works including *Maxims of the Law*, *Reading on the Statute of Uses*, and *Elements of the Common Laws of England*; he is also remembered by thoughts such as 'The *knowledge* of man is as the waters, some descending from above, and some springing from beneath; the one informed by the light of nature, the other inspired by divine revelation', 'A prudent question is one-half of *wisdom*', and '*Wise men* make more opportunities than they find';

1609-19: <u>Johannes Kepler</u> (1571-1630), German astronomer and mathematical physicist, recorded that the distances from the Sun of the six planets including the Earth could be related to the five regular solids of geometry, of which the cube is the simplest, which was published in *Misterium Cosmographium*; and then he stated the *three laws of planetary motion*: 1st, planets move in ellipses with Sun at one focus; 2nd, radius vector of each planet describes equal areas of ellipse in equal times; and 3rd, ratio of square period of revolution and cubic mean distance from Sun is same for any planet; these three laws were published in his influential work *Epitome astronomie Copernicanae* 'Epitome of Copernican Astronomy' whereby the *Keplerian universe* (heliocentric with elliptical planetary orbits) originated;

1637-49: <u>René Descartes</u> (1596-1650), French philosopher and mathematician, produced *Discours de la méthode* 'Discourse on Method' with the famous quotation *Dubito, ergo cogito; Cogito, ergo sum* 'I doubt, therefore I think; I think, therefore I am'; *Meditationes de prima Philosophia* 'Meditations on First Philosophy'; *Principia Philosophiae* 'Principles of Philosophy', setting out the fundamental Cartesian doctrines; wrote other philosophical works including

Regulae ad directionem ingenii 'Rules for the Direction of the Mind', and *Les Passions de l'âme* 'Passions of the Soul', thus founding *modern philosophy*; he also introduced *Cartesian co-ordinates* and *Cartesian ovals* in mathematics, as well as *Cartesian diver* and *Cartesian hydrometer* in physics;

1684-87: <u>Isaac Newton</u> (1642-1727), English scientist and mathematician, demonstrated the whole gravitation theory as expounded in *De Motu Corporum* showing that the force of gravity between two bodies is directly proportional to the product of their masses and inversely proportional to the square of distance between them, by which he founded the *Static Newtonian universe* – static but evolving, steady state, infinite; then produced the great *Philosophiae Naturalis Principia Mathematica* 'The Mathematical Principles of Natural Philosophy' where *three laws of motion* are unveiled: 1st, a body in a state of rest or uniform motion will remain in that state until a force acts on it; 2nd, an applied force is directly proportional to the acceleration it induces; and 3rd, for every 'action' force which one body exerts on another, there is an equal and opposite 'reaction' force exerted by second body on first;

1726-62: <u>John Harrison</u> (1693-1776), English inventor and horologist, after the British government offered prizes for discovery of a method to determine the longitude accurately, he invented a series of devices including the *gridiron pendulum*, the *going fusee*, and a *remontoir escapement*; and then developed the *marine chronometer H-4* that, in a voyage to Jamaica, determined the *longitude* within 18 geographical miles (or 29 km), thus becoming the first to solve the longitude problem;

1767-91: <u>Wolfgang Amadeus Mozart</u> (1756-91), Austrian composition genius, created the opera buffa *La finta semplice* 'The Feigned Simpleton'; singspiel *Bastien and Bastienne*; symphonies and quartets; *La finta giardiniera* 'The Feigned Gardener Girl'; violin concertos; *Haffner* Serenade; *Paris Symphony*; *Coronation Mass*; opera series *Idomeneo, rè di Creta* 'Idomeneo, King of Crete'; opera *Die Entführung aus dem Serail* 'The Abduction from the Seraglio'; *Linz*, and *Prague* Symphonies; *Le Nozze di Figaro* 'The Marriage of Figaro', *Don Giovanni*, and *Così fan tutte* 'Women are all Like That'; serenade *Eine kleine Nachtmusik*; operas *Die Zauberflöte* 'The Magic Flute', and *La Clemenza di Tito*; and at last his unfinished *Requiem* 'which seems to concentrate the grief of entire world';

1767-1832: <u>Johann Wolfgang von Goethe</u> (1749-1832), German poet, dramatist, scientist and one of the greatest figures in European literature, was author of plays *Die Laune des Verliebten* 'The Beloved's Whim', and *Die Mitschuldigen* 'The Accomplices'; masterpiece drama *Götz von Berlichingen*; novels *Die Lieden des jungen Werthers* 'The Sorrows of Young Werther', and *Wilhelm Meisters Theatralische Sendung* 'Wilhelm Meister's Theatrical Mission'; verse dramas *Iphigenie auf Tauris*, *Egmont*, and *Torquato Tasso*; epic idyll *Hermann und Dorothea*; then literary works *Wilhelm Meisters Lehrjahre* 'Wilhelm Meister's Apprenticeship', and *Wilhelm Meisters Wanderjahre* 'Wilhelm Meister's Travels'; epic poem *Reineke Fuchs*; drama *Die natürliche Tochter* 'The Natural Daughter'; novel *Die Wahlverwandtschaften* 'The Elective Affinities'; and finally the masterpiece two-part drama *Faust*;

1795-1827: <u>Ludwig van Beethoven</u> (1770-1827), German composer and pianist, was recognized as a genial composer, his reputation being due to *B flat piano concerto*, *Opus 1 piano trios*, *Opus 2 piano sonatas*, three *piano concertos*, *String Quartet Op. 29*, and *Op. 31*; symphonies including the *Eroica Symphony* (No.3), *Destiny Symphony* (No.5), and *Choral Symphony* (No.9) with the *Ode to Joy* 'expressing the gladness of life'; opera *Fidelio*; piano sonatas including the *Waldstein*, *Appassionata*, and *Lebewohl*; *Rasumovsky Quartets*, music to Goethe's *Egmond*, the *Archduke Trio* (dedicated to the Archduke Rudolf of Austria); great achievements the *Diabelli Variations*, six string quartets, 32 piano sonatas, 17 string quartets, and the Mass in D *Missa solemnis*; as well as the *Moonlight Sonata* 'with its drops of light';

1796-1825: <u>Pierre Simon Laplace</u> (1749-1827), French mathematician and astronomer, completed the five monumental volumes of *Mécanique céleste*, a great treatise on celestial mechanics; the influential work on the gravitational attraction of spheroids; published *Système du monde* 'The System of the World', exposing his astronomical theories and famous *nebular hypothesis of planetary origin*; and, in his study of the gravitational attraction of spheroids, formulated the fundamental differential *Laplace equation*; he also brought important contributions to the *theory of probability*;

1801-17: <u>Georges Léopold Cuvier</u> (1769-1832), French anatomist, originated the natural system of animal classification which anticipated the modern division of the animal kingdom into phyla, studied *animal* and *fish fossils*, reconstructing the *extinct giant vertebrates* of the Paris Basin, and linking palaeontology to comparative anatomy; posited 'catastrophism' as a series of extinctions due to periodic global floods after which new forms of life appeared;

and published works including *Leçons d'anatomie comparée, Les Ossements fossiles des quadrupèds* 'The Fossilized Bones of Quadrupeds', *L'Anatomie des mollusques, Le Règne animal distribué d'après son organisation*, and *Histoire naturelle des poissons* 'The Natural History of Fish'; becoming known as the father of *comparative anatomy* and *palaeontology*;

1801-25: <u>Carl Friedrich Gauss</u> (1777-1855), German mathematician, astronomer and physicist, wrote *Disquisitiones arithmeticae*, containing wholly new advances in number theory; first used the *method of least squares* in statistics; studied the *theory of errors* of observation; as well as worked on pure mathematics, including *differential equations, hypergeometric function, curvature of surfaces*, four different *proofs of fundamental theorem of algebra*, and six *quadratic reciprocity*; and also developed celestial mechanics, resulting in the treatise *Theoria motus corporum coelestium* 'The Theory of the Motion of Celestial Bodies';

1810-22: <u>Jean Baptiste Joseph Fourier</u> (1768-1830), French great mathematician and physicist, introduced the expansion of functions in trigonometric series, now known as *Fourier series*, by which almost any function of real variable can be expressed as a sum of sines and cosines of integral multiples of variable; stated the *Fourier transform* used in operational calculus; wrote *Théorie analytique de la chaleur* 'Analytical Theory of Heat', applying the technique to the solution of partial differential equations to describe heat conduction in solid bodies; discovered the greenhouse effect presented in *Mémoire sur la température de globe terrestre et des espaces planétaires*; deduced an important *theorem on the roots of algebraic equations*; wrote *Remarques générales sur l'application du principe de l'analyse algébrique aux équations transcendantes*, and finally *Mémoire d'analyse sur le mouvement de la chaleur dans les fluides*; before all these creative works, he is also remembered for bringing from Egypt an *ink pressed copy of the Rosetta Stone*, availabe for detailed study;

1819-29: <u>Jacob Ludwig Carl Grimm</u> (1785-1863), German philologist, published *Deutsche Grammatik* 'German Grammar', considered 'first great scientific linguistic work of the world'; formulated *Grimm's Law*, which states that in Proto-Indo-European voiceless stops become voiceless fricatives, voiced stops become voiceless stops, and voiced aspirates become voiced stops or fricatives (depending on context); then his chief work *Die deutsche Heldensage* 'The German Heroic Myth'; and together with his brother Wilhelm Carl Grimm wrote the monumental *Deutsches Wörterbuch* 'German

Dictionary';

1822-32: <u>Jean François Champollion</u> (1790-1832), French linguist and founder of Egyptology, grown up under Joseph Fourier's support at Grenoble, and highly interested in history and linguistics, he became the genius who first translated the *Rosetta Stone* hieroglyphs, showing that the Egyptian writing system was a *combination of phonetic and ideographic signs*; wrote a series of works including *Panthéon égyptien, collection des personnages mythologiques de l'ancienne Égypte, Précis du système hiéroglyphique des anciens Égyptiens*, two volume *Monuments de l'Égypte et de la Nubie, Grammaire égyptienne*, and *Dictionnaire égyptien en écriture hiéroglyphique*; he also promoted studies of early Egyptian history and culture;

1823-27: <u>Niels Henrik Abel</u> (1802-29), Norwegian mathematician, was a mathematical genius by the age of 15, proved that there is no algebraic formula for the solution of a general polynomial equation of the 5th degree; developed the concept of elliptic functions; and pioneered its extension to the theory of *Abel's integrals* and *functions*, which became a central theme of later analysis, emphasizing the analogy of *elliptic functions* with familiar *trigonometric functions*; thus influencing the development of complex analysis in mathematics;

1824: <u>Sadi Carnot</u> (1796-1832), French physicist, published the remarkable *Réflexions sur la puissance motrice du feu et sur les machines propres à developer cette puissance* 'Reflections on the Motive Power of Fire, and on the Machines Appropriate to develop this Power', concerning on scientific principles to an analysis of *working cycle* and *efficiency of steam engine*; this work leading to *Carnot's theorem*, and laying the foundation of the *second law of thermodynamics*, which was later established in its final form by R Clausius and WT Kelvin;

1827-34: <u>Michael Faraday</u> (1791-1867), English chemist, physicist and creator of classical field theory, wrote *Chemical Manipulation*, isolated *benzene*, and synthesized *chlorocarbons*; published the great series of *Experimental Researches on Electricity*, describing his many discoveries, and including *electromagnetic induction*, whereby the electromotive force produced around a closed path is proportional to the rate of charge of magnetic flux through any surface bounded by that path; defined the *laws of electrolysis*: 1st, mass of substance altered at an electrode is directly proportional to quantity of electricity transferred at that electrode; and 2nd, for a given quantity of electricity, mass of an elemental material altered at an electrode is directly proportional to element's equivalent weight; discovered the *rotation of polarized light by magnetism*; and founded *electric motor technology*; being generally considered the greatest of all experimental physicists;

1829-59: <u>Isambard Kingdom Brunel</u> (1806-59), English engineer and inventor, designed and constructed the *Thames Tunnel, Clifton Suspension Bridge, Hungerford Suspension Bridge* over Thames at Charing Cross; designed the *Great Western*, the first steamship built to cross the Atlantic Ocean, and the *Great Britain*, the first ocean screw-steamer; the *Great Eastern*, the largest vessel ever built; constructed all the tunnels, bridges and viaducts for the *Great Western Railway*; and also *docks* such as those of Bristol, Monkwearmouth, Cardiff and Milford Haven; his achievements in the *Transport Revolution* and *modern engineering* were highly celebrated both in the UK and overseas;

1831-32: <u>Évariste Galois</u> (1811-32), French mathematician, while still in his teens, was able to determine a necessary and sufficient condition for a *polynomial to be solvable by radicals*, then he sent papers to the Academy of Science where mathematician AL Cauchy refused to accept them for publication, and later to mathematical physicist SD Poisson who declared Galois's work as 'incomprehensible'; at the age of 20 he protested on the July 14 Bastille Day and was arrested for six months; shortly after he was provoked to a duel, and so convinced of his impending death that he stayed up all night writing letters, and composing some 60 pages of what would become his *mathematical testament*; the next day he was severely wounded and a day later died, but his 'testament' laid the foundations for *Galois groups* and *Galois theory*, two major branches of abstract algebra, and the subfield of *Galois connections*; which included the essentials of his discoveries on the theory of algebraic equations and Abelian integrals, thus founding the *Theory of groups*, that was published so late as 1846 due to the French mathematician J Liouville;

1844-81: <u>Charles Robert Darwin</u> (1809-82), English naturalist and one of the originators of the theory of evolution by natural selection, published *The Origin of Species by Means of Natural Selection*; as well as the great series of supplemental treatises including *The Fertilisation of Orchids, The Variation of Plants and Animals under Domestication, The Expression of the Emotions in Man and Animals, Insectivorous Plants, Climbing Plants, The Effects of Cross and Self Fertilisation in the Vegetable Kingdom, Different Forms of Flowers in Plants of the same Species, The Power of Movement in Plants*, and

The Formation of Vegetable Mould through the action of Worms; he also left invaluable knowledge regarding variation and interbreeding in organisms, and formulated the theory of *pangenesis*;

1846-73: James Clerk Maxwell (1831-79), Scottish physicist of genius, at the age of 15 devised a method for drawing oval curves, then published papers on *kinetic theory of gases*, and theoretically established the *nature of Saturn's rings*; investigated colour perception and demonstrated *colour photography* with a picture of tartan ribbon; and overall founded the *theory of electromagnetism*, that was extended in his *Treatise on Electricity and Magnetism*, where he mathematically stated M Faraday's theory of electrical and magnetic forces; this treatise paving the way for the theory of relativity and quantum mechanics, and thus became one of the greatest theoretical physicists the world has known;

1868-84: Ludwig Boltzmann (1844-1906), Austrian physicist, is most celebrated for the application of statistical methods to physics and the relation of kinetic theory to thermodynamics; formulated *Boltzmann's equation*, a fundamental diffusion equation based on particle conservation, showing that the rate of losses, including leakage out of region of interest and rate of disappearance by reactions of all kind, is equal to the rate of production from sources within region and rate of scattering into region, i.e. the *increasing entropy* corresponding to molecular randomness; also he worked on electromagnetism, viscosity, diffusion, and derived the *law for black-body radiation*;

1869-90: Heinrich Schliemann (1822-90), German archaeologist, at the age of 46 he set out in *Ithaka, der Peloponnes und Troja* 'Ithaca, the Peloponnese, and Troy', excavated at the mound of Hisarlic in West Asia Minor, where he discovered nine superimposed city sites at the Homeric place of *Troy*, one of which with a considerable treasure, eventually housed in the Ethnological Museum in Berlin; and also excavated the site of *Mycenae*, then worked in Ithaca, at *Orchomenos* and at *Tiryns*, unveiling the essential characteristics of *Troyan and Mycenaean civilizations*, and giving an impetus to the following archaeological discoveries all-over the world;

1871-91: Thomas Alva Edison (1847-1931), US inventor and physicist, after devising an electric vote-recording machine, he invented the *paper ticker-tape automatic repeater for stock exchange prices*, and became the astonishing inventive genius that won him the name 'the Wizard of Menlo Park'; took out more than 1000 patents in all, including the *gramophone*, the *incandescent light bulb*, and the *carbon granule microphone*; then invented a *system for generating and distributing electricity*, designed the *first power plant*; and also invented *megaphone, storage battery, electric valve* and *kinetoscope*; as well as discovered *thermionic emission*, formally called the 'Edison Effect'; thus he was the most prolific inventor the world has ever seen;

1874-76: <u>Karl Adolph Verner</u> (1846-96), Danish linguist, upon getting up one morning, he was puzzled by the question of why the Gothic words *fadar* and *broþar* have different *consonants* after the root *vowel*, and afterwards carried out linguistic works resulting in the formulation of *Verner's Law*, according to which the Proto-Germanic fricatives become voiced if the next conditions are met: they are not initial, but preceded and followed by voiced, and accent is not on immediately preceding syllable; his important contribution to comparative-historical linguistics appeared in the article *Eine Ausnahme der ersten Lautverschiebung* 'An Exception to the First Sound Shift';

1890-99: <u>David Hilbert</u> (1862-1943), German mathematician, was author of the definitive work on *invariant theory*, removed the need for further work on a subject that had occupied so many mathematicians, at the same time laying the foundations for *modern algebraic geometry*; published a report on *algebraic number theory*; gave the first abstract *axiomatic foundations of geometry*, which made no attempt to define the 'meaning' of basic terms but only to prescribe how they could be used; then studied integral equations, the *calculus of variations*, theoretical physics, and *mathematical logic*;

1899-1901: <u>Max Karl Ernst Planck</u> (1858-1947), German theoretical physicist, became known for his study of the laws of thermodynamics, and *Planck's law of black body radiation*; formulated *quantum theory*, according to which the emission and absorption of energy take place in small discrete instalments or *quanta*, and thus predicted phenomena inexplicable in classical Newtonian theory; also wrote important papers including *Entropy and Temperature of Radiant Heat*, and *On the Law of Distribution of Energy in the Normal Spectrum*; remarkably, his name was given to a series of units (e.g. Planck length, mass, time, temperature, charge), derived units (e.g. Planck current, power, density), and physical notions (e.g. Planck epoch, particle, postulate, principle);

1903-29: <u>Konstantin Eduardovich Tsiolkovsky</u> (1857-1935), Russian and Soviet rocket scientist and pioneer of astronautic theory, after the derivation of so-called 'formula of aviation', he published his chief work *The Exploration of Cosmic Space by Means of Reaction Devices*, using the 'Tsiolkovsky equation' for a *multi-stage rocket* fuelled by liquid oxygen and liquid hydrogen, which established his reputation as 'the father of space flight theory';

1905-19: <u>Albert Einstein</u> (1879-1955), German-Swiss-US mathematical physicist, discovered the *photoelectric effect*, a phenomenon resulting from absorption of photon energy by electrons, leading to their release from a surface, when photon energy exceeds the work function, or otherwise allowing conduction when incident energy exceeds an atomic binding energy; studied *Brownian motion*, explaining empirical evidence for atomic theory and supporting the application of statistical physic; established the *special theory of relativity*, evidencing that the speed of light is independent of the motion of an observer, and postulating the relative motion and constant velocity of light or zero acceleration; discovered the *mass-energy equivalence* (energy is equal to mass multiplied by square speed of light), which has been dubbed 'the world's most famous formula'; developed the *general theory of relativity*, extending relativity from constant to varying velocities or non-zero acceleration; and initiated *relativistic cosmology*, applying general relativity to the universe as a whole;

1913-50: <u>Niels Henrik David Bohr</u> (1885-1962), Danish physicist, greatly extended the *theory of atomic structure* when he explained the spectrum of hydrogen by means of E Rutherford's atomic model and the quantum theories of A Einstein and MKE Planck; initiated *Bohr's model* of atomic structure, later shown to be a solution of E Schrödinger's equation; developed the *liquid drop model of the nucleus*, and introduced the *quantized energy level theory of atom*;

1914-26: <u>Arthur Stanley Eddington</u> (1882-1944), English astronomer, published *Stellar Movements and the Structure of the Universe*, deduced a theoretical relationship between the *mass of a star and its total density* may exist in stars such as white dwarfs; *Space, Time and Gravitation*, which was extended to his *Mathematical Theory of Relativity*, starting from fundamental postulates and deriving in well-known equations and predictions at the scale of universe; and also published *Internal Constitution of the Stars*;

1915: <u>Alfred Lothar Wegener</u> (1880-1930), German meteorologist and geophysicist, published *Die Entstehung der Kontinente und Ozeane* 'The Origin of Continents and Oceans', based on his observations that continents may once have been joined into one supercontinent called *Pangaea* which later broke up, as a hypothesis of *continental drift*; his hypothesis remained controversial until the 1960s, when the structure of oceans became understood and plate tectonics was founded, by which his outstanding scientific prediction was finally proved;

1919-29: Edwin Powell Hubble (1889-1953), US astronomer, was author of fundamental investigations of the realm of the nebulae, finding that spiral nebulae are independent stellar systems and that Andromeda nebula in particular is very similar to our own Milky Way galaxy; then he discovered that galaxies recede from us with speeds increasing with their distance, so that the linear relation between speed of recession and distance was called *Hubble's Law*; thus the *expansion of universe* became the observational basis of *modern cosmology*;

1923-28: Louis-Victor Pierre Raymond de Broglie (1892-1987), 7th duke de Broglie and French physicist, based on the fact that waves can behave as particles, he put forward the converse idea that particles can behave as waves, and became famous for his *wave-particle duality*, thus opening the way to wave mechanics and evidencing that a nobleman like him can be a high ranking scientist; his main works include *Recherches sur la théorie des quanta* 'Researches on the quantum theory', *Pilot Wave theory*, a hidden variable theory, and *La méchanique ondulatoire* 'The Wave Mechanics'; later he wrote *Matière and lumière* 'Matter and Light', and *Une tentative d'interprétation causale et non linéaire de la méchanique ondulatoire: la théorie de la double solution* 'Non-linear Wave Mechanics: A Causal Interpretation';

1925-27: Werner Karl Heisenberg (1901-76), German theoretical physicist, developing NHD Bohr's theory of atomic structure, he re-interpreted classical mechanics and introduced a *matrix-based quantum mechanics* where phenomena must be describable both in terms of wave theory and quanta; as well as formulated the revolutionary *principle of indeterminacy* or *uncertainty principle*, showing that there is a fundamental limit to the accuracy to which certain pairs of variables (such as position and momentum) can be determined;

1926-58: Erwin Schrödinger (1887-1961), Austrian physicist, inspired by L-V Broglie's wave-particle duality, he formulated the celebrated *Schrödinger's wave equation* (stationary and time-dependent *Schrödinger equation*), originating the science of *wave mechanics* as part of quantum theory, which was published in his *Quantisierung als Eigenwertproblem* 'Quantization as an Eigenvalue Problem'; later he wrote *What is Life? & Mind and Matter*, and also *Science and Man*;

1928-30: Paul Adrien Maurice Dirac (1902-84), English mathematical physicist, formulated the *Dirac equation*, which described the behaviour of fermions, and predicted the *existence of antimatter*; derived the *Dirac delta function* (that is zero everywhere except at the single point of origin), which have a great utility in the fields of classical and quantum mechanics; his main published works include the classic *The Principles of Quantum Mechanics*, deducing the *Dirac equation* and predicting the existence of the magnetic monopole, as well as *Spinors in Hilbert Space*, which deals with the basic aspects of spinors from a real Hilbert space formalism; he is regarded as one of the most significant physicists of the 20th century;

1932-48: Norbert Wiener (1894-1964), US mathematician, worked on stochastic processes and harmonic analysis, invented the concepts later called *Wiener integral* and *Wiener measure*; then studied the feedback in the handling of information by electronic devices, which led him to compare this with analogous mental processes in animals, as synthesized in *Cybernetics, or control and communication in the animal and the machine*; thus founding *cybernetics*;

1932-50: Jan Hendrik Oort (1900-92), Dutch astronomer, first evidenced *dark matter*, when he found the mass of the galactic plane must be more than the mass of material that can be seen; suggested that the long-period comets are coming from a common region, now called the *Oort cloud*, of the Solar System; he also discovered the *galactic halo*, a group of stars orbiting the Milky Way but outside the main disk; calculated that the *centre of the Milky Way* is 19200 light-years from the Earth; and found that the light from the Crab Nebula is polarized and produced by *synchrotron emission*;

1936-70: Wernher von Braun (1912-77), German and US rocket pioneer, perfected rockets using *liquid oxygen* and *alcohol-water mixture*; led the operation of launching rockets at Peenemunde, where *V-1* and *V-2 rockets* were perfected in order to be sent against Great Britain during the Second World War; and, after surrendering to the USA, developed *nuclear ballistic missiles*, constructed the first *US artificial earth satellite* 'Explorer I', and the *Saturn rocket* used in 'Apollo 11' moon landing;

1938: Lise Meitner (1878-1968), Austrian physicist, under O Hahn's influence, became one of the forerunners in *nuclear fission*; and Otto Robert Frisch (1904-79), British physicist, studying fission that occurred in experiments of neutron bombardment, became one of the initiators for detonation of an *atomic bomb*; together they hypothesized that the uranium nucleus had split in two, explaining the process, estimating the energy released, and postulating that *mass has been converted into energy*; this postulation was an essential stimulus in physical studies, for either peace or war purposes;

1953: <u>Francis Harry Compton Crick</u> (1916-2004), English molecular biologist, and <u>James Dewey Watson</u> (1928-), US biologist, worked on the *structure of DNA* (deoxyribonucleic acid), finding that biological molecule contained in cells carries genetic information; elaborated the famous model of a *double-helical molecule*, consisting of two strands of nucleotide bases wound around a common axis in opposite directions, and suggesting a simple method for duplication, i.e. if strands are separated, new partner strands are reconstructed for each based on sequence of the old strand; this discovery has an enormous importance in understanding the mechanisms of inheritance in organisms; later, Francis Crick examined the essence of soul, and published *The Astonishing Hypothesis: The Scientific Search for the Soul* (1990);

1963-67: <u>John Tuzo Wilson</u> (1908-93), Canadian geophysicist, had the ideas of *hotspots* in Earth's mantle, oceanic *transform faults*, and *mountain building*, as major steps towards *plate tectonic theory*, he being one of its founders; formulated the 'Wilson cycle' unfolding by magma – crystallization (freezing of rock) – igneous rocks – erosion – sedimentation – sediments and sedimentary rocks – tectonic burial and metamorphism – metamorphic rocks – melting; his name was given to *Wilson Range in Antarctica*, and also to the *Wilson ocean cycle*;

1965: <u>Arno Allan Penzias</u> (1933-), German-born US physicist and radio astronomer, and <u>Robert Woodrow Wilson</u> (1936-), US astronomer, using a large radio telescope, discovered *cosmic microwave background radiation*, a remnant of the Big Bang, coming from all directions with an energy distribution corresponding to that of a black body at a temperature of 2.725 kelvin; their discovery helped establish the *Big Bang theory* of cosmology, and was later published as *Isotropy of Cosmic Background Radiation at 4080 Megahertz*, Science, 156;

1974-2005: <u>Stephen William Hawking</u> (1942-), English theoretical physicist, predicted that a black hole could evaporate through loss of thermal radiation, and that mass can escape from its gravitational pull, now known as *Hawking process*; then published *A Brief History of Time - Black Holes and Baby Universes*; *God Created the Integers: The Mathematical Breakthroughs That Changed History*; and then *The Universe in a Nutshell*;

1982-2000: <u>Bill Gates</u> (1955-), US computer scientist and businessman, had a licence for computer operating system to *International Business Machines* (IBM), fledgling *personal computer* (PC) industry; this system (MS-DOS) was phenomenally successful, and its updated versions, such as *Windows 2000*, allowed the maintenance of Microsoft's PC hegemony, and showed how somebody could became a billionaire and world's most wealthy private individual.

According to above data, the selected creative personalities or geniuses are displayed with their ages, in years, at the first achievement, the second achievement, and at death, as follows:

Name	Age at the first achievement	Age at the second achievement	Age at death
Alexander	21	-	33
Archimedes	27	45	75
da Vinci	28	46	67
Buonarroti	15	45	89
Columbus	41	-	55
Galileo	18	30	78
Shakespeare	26	45	52
Bacon	44	-	65
Kepler	38	-	59
Descartes	41	-	54
Newton	42	60	85
Harrison	33	55	83
Mozart	11	20	35
Goethe	18	40	83
Beethoven	25	45	57
Laplace	47	70	78
Cuvier	32	55	63
Gauss	24	55	78
Fourier	42	-	62
Grimm	34	55	78
Champollion	32	-	42
Abel	21	-	27
Carnot	28	-	36
Faraday	36	55	76
Brunel	23	45	53
Galois	20	-	21
Darwin	35	60	73
Maxwell	15	35	48
Boltzmann	24	45	62
Schliemann	47	-	68
Edison	24	60	84
Verner	28	45	50

Hilbert	28	55	81
Planck	41	60	89
Tsiolkovsky	46	75	78
Einstein	26	45	76
Bohr	28	50	77
Eddington	32	50	62
Wegener	35	-	50
Hubble	30	50	64
de Broglie	31	55	95
Heisenberg	24	40	75
Schrödinger	39	60	74
Dirac	26	55	82
Wiener	38	63	70
Oort	32	60	92
von Braun	24	55	65
Meitner	60	-	90
Frisch	34	60	75
Crick	37	65	88
Watson	25	40	-
JT Wilson	55	-	85
Penzias	32	55	-
RW Wilson	29	50	-
Hawking	32	55	-
Gates	27	50	-
Average	**31.3**	**51.4**	**67.5**

Therefore, the first sequences of creativity are approximately delimited by average times, in years from birth as the origin of creation, as

$$0 \setminus Early\ creations \setminus 31.3 \setminus Intermediate\ creations$$
$$\setminus 51.4 \setminus Intermediate\text{-}mature\ creations... \setminus 67.5.$$

Consequently, the process of creativity is limited by the average age of death, as the final time $t_\bullet = 67.5$ years, and the ratio of transitional times is $(31.3)/(51.4) \approx 0.6089$. A similar value of this ratio is found in Background to time, Table *(i)* for $z/2 = i$ sequences, as $t_{0.5}/t_1 = tanh(0.5)/tanh(1) = 0.6067761$, and then the approximate tranitional times, in years from birth, are

$$t_{0.5} = t_\bullet \cdot tanh(0.5) = (67.5) \cdot (0.4621172) \approx 31.2;$$
$$t_1 = t_\bullet \cdot tanh(1) = (67.5) \cdot (0.7615942) \approx 51.4.$$

Applying the time function $t_i = t_\bullet \cdot tanh(i)$ for the following arguments $i = 1.5; 2; 2.5;$ the timeline of creativity with its sequences and intrinsic characteristics are calculated and presented in the tables below.

$z/2 = i$	Average time (years)		Sequences of creativity
	from origin $t_i = t_\bullet \cdot tanh(i)$	from end $t_i - 67.5$	
	67.5	0	
...	...	...	
2.5	66.6	0.9	_
2	65.1	2.4	_ Late creations
1.5	61.1	6.4	_ Mature-late creations
1	51.4	16.1	_ Intermediate-mature creations
0.5	31.2	36.3	_ Intermediate creations
0	0	67.5	_ Early creations

$z/2 = i$	Time from origin $t_i = t_\bullet \cdot tanh(i)$ (years)	Period $\tau_i = (t_\bullet^2 - t_i^2)/(2t_\bullet)$ (years)	Frequency $f_i = (2t_\bullet)/(t_\bullet^2 - t_i^2)$ (years)$^{-1}$	Angular speed $\omega_i = 2\pi/\tau_i = 2\pi \cdot f_i$ (years)$^{-1}$
	67.5	0	∞	∞
...	...	...	...	...
2.5	66.6	0.89	1.1185682	7.0281715
2	65.1	2.36	0.4242081	2.6653784
1.5	61.1	6.10	0.1640260	1.0306061
1	51.4	14.18	0.0705222	0.4431042
0.5	31.2	26.54	0.0376799	0.2367499
0	0	33.75	0.0296296	0.1861685

2. Cosmology

The study of the origin, evolution, and eventual fate of the universe is called *cosmology* (from Greek *κοσμος* 'order, world'). It emerged and developed in course of human history in three major stages: 1st, direct observation of the Sun and Moon, stars and the Milky Way, meteorites and comets, eclipses and other celestial events, constellations and groups of stars, changes in positions of celestial bodies referring to the Earth, interpretations and simple representations of the universe; 2nd, telescopic observations of the planets and their satellites, discoveries of other stars, movements of planets around the Sun, laws of planetary motion and of gravity, hypotheses of planetary and stellar origins; and 3rd, discovery of the expanding universe, deduction of the thermodynamic processes in the early universe, evidence of cosmic microwave background radiation, inflation of the universe and its dark matter and dark energy, identification of black holes, quasars, pulsars and accelerating expansion of the universe.

Some of the historical views, cosmological variants and conceptions, achievements in disclosing and understanding the universe, as well as studies and researches carried out by communities and personalities, are displayed in chronological order below.

3500-2110BC: *Sumerian knowledge of the universe* - Solar System made of the Sun circled by 11 spherical celestial bodies including the Earth, and maps of many stars and constellations;

c.3000BC: *Babylonian cosmology* – flat earth floating in infinite 'waters' of chaos;

c.2000BC: *Hindu cosmology* – cyclical or oscillatory, infinite in time universe;

c.550BC: Leucippus, Milesian-born Greek philosopher, initiated *atomistic cosmology*, later developed by Democritus and expounded in T Lucretius's poem on the nature of things;

510-400BC: *Pythagorean universe* (named after the Greek philosopher and mathematician Pythagoras) – 'central fire' at the centre of universe;

c.500BC: *Biblical cosmology* – flat earth floating in infinite 'waters' of chaos;

435-280BC: *Atomist universe* (due to Greek philosophers Leucippus and Democritus) – infinite in extent universe;

420-380BC: Democritus, Thracian-born Greek philosopher, initiated the *atomic theory of the universe*, asserting that *all things originate from a vortex of atoms*, and differ by shape and arrangement of their atoms;

335-323BC: *Aristotelian universe* (named after the Greek philosopher and scientist Aristotle) – geocentric, static, steady state, finite extent, infinite time universe;

300BC-AD200: *Stoic universe* (named after the Greek Stoic school of philosophy) – island model of universe;

280-275BC: Aristarchos of Samos, Greek astronomer, produced the first theory of the Earth's motion, according to which *the Earth revolves on its axis and travels in a circle around the Sun*, thus anticipating the theory of N Copernicus; his name was given to the *Aristarchean universe* – heliocentric universe;

230-200BC: Eratosthenes, Greek mathematician, astronomer and geographer, measured the *obliquity of the ecliptic* and calculated the *circumference of Earth* with considerable accuracy;

c.140BC: Hipparchos, Greek astronomer and mathematician, estimated the *relative distances to Sun and Moon*, the *Sun's eccentric movement*; and predicted their *eclipses*;

65-55BC: Titus Lucretius, Roman poet and philosopher, wrote the didactic poem *De rerum natura* 'On the Nature of Things', in six volumes of hexameters, concerning theories on the origin of the universe;

c.AD150: Claudius Ptolemy, Egyptian astronomer and geographer, elaborated the *Earth-centred description* for positioning celestial bodies, and wrote *Megiste* in Greek, translated as *Almagest* in Arabic; whereby the *Ptolemaic model* – geocentric universe;

499: *Aryabhatan model* (named after the Indian mathematician and astronomer Aryabhata) – geocentric or heliocentric universe;

500-1200: *Medieval universe* – finite in time universe;

c.524: *Jain cosmology* (named after *Jain Agamas* teachings of the Indian Jainist Mahavira) – cyclical or oscillatory, eternal and finite universe;

c.1200: *Multiversal cosmology* (due to the Islamic astronomer Fakhr al-Din al-Razi) – multiple worlds in the universe;

1259-1528: *Maragha models* (named after the Islamic *Maragha school* of astronomy) – geocentric universe;

c.1500: *Nilakanthan model* (named after the Indian mathematician Nilakantha Somayaji) – geocentric or heliocentric universe;

1512-43: Nicolaus Copernicus, Polish astronomer, considering Ptolemy's description of world as unsatisfactory, he elaborated the work *De Revolutionibus Orbium Coelestium* 'The Revolutions of the Celestial Spheres', which became the *theory of Sun-centred universe*; whence the *Copernican universe* – heliocentric with circular planetary orbits in the Solar System;

1572-96: Tycho Brahe, Danish astronomer, observed a new star in Cassiopeia, namely a supernova now known as *Tycho's star*, measured the positions of *777 stars*, and created a catalogue of them with such an accuracy that it provided a vital source of information for centuries; whereby the *Tychonic system* – geocentric or heliocentric universe;

1584: Giordano Bruno, Italian hermetic thinker, was author of a pantheistic philosophy, whereby God animated the whole of creation as 'word-soul', and produced his remarkable work *De l'infinito universo et mondi* 'On the Infinite Universe and Worlds', so much hated by the Inquisition;

1604-10: Galileo Galilei, Italian astronomer, mathematician and natural philosopher, demonstrated that a bright new star which had appeared in the constellation *Ophiuchus* was more distant than the planets, confirming Tycho Brahe's conclusion that changes take place in celestial regions beyond the planets; and improved the refracting telescope, which was used for his astronomical revelations published in *Sidereus Nuncius* 'Sidereal Messenger', including the mountains of the Moon, multitude of stars in the Milky Way, and existence of Jupiter's four satellites, namely Io, Europa, Callisto, and Ganymede;

1609-19: Johannes Kepler, German astronomer, stated the *three laws of planetary motion*: 1st, planets move in ellipses with Sun at one focus; 2nd, radius vector of each planet describes equal areas of ellipse in equal times; and 3rd, ratio of square period of revolution and cubic mean distance from Sun is same for any planet; whence the *Keplerian universe* – heliocentric with elliptical planetary orbits in the Solar System;

1637-44: *Cartesian Vortex universe* (named after the French philosopher and mathematician René Descartes) – static but evolving, steady state, infinite universe;

1675-80: Olaus Roemer, Danish astronomer, observing the intervals and knowing the rates of motion of Jupiter and Earth, he obtained the first *estimation of the light speed*, and invented a *telescope movable only in the meridian*, which greatly increased the accuracy attainable in determination of both time and right ascension of celestial bodies;

1676-1715: Edmond Halley, English astronomer and mathematician, made the first catalogue of stars in the southern hemisphere, called *Catalogus Stellarum Australium*, observed a comet, later called *Halley's comet*, and predicted its date of return;

1684-87: Isaac Newton, English scientist and mathematician, demonstrated the whole gravitation theory as expounded in *De Motu Corporum* showing that the force of gravity between two bodies is directly proportional to the product of their masses and inversely proportional to the square of distance between them; whereby the *Static Newtonian universe* – static but evolving, steady state, infinite universe;

1755-64: *Hierarchical universe* (due to the German philosopher Immanuel Kant, and the Swiss mathematician Johann Heinrich Lambert) – static but evolving, steady state, infinite universe;

1764: Johann Heinrich Lambert, Swiss mathematician and philosopher, wrote the philosophical work *Neues Organon* 'New Organ', greatly valued by I Kant, made studies on *astronomy* and the *Milky Way*, and produced a successful popular *book on cosmology*;

1781: William Herschel, British astronomer, discovered the planet *Uranus*, first to be found telescopically, which was initially named 'Georgium Sidus' in honour of King George III;

1796-1825: Pierre Simon Laplace, French mathematician and astronomer, wrote and published *Système du monde* 'The System of the World', exposing his astronomical theories and the famous *nebular hypothesis of planetary origin*;

1846: Urbain Jean Joseph Leverrier, French astronomer, studied the disturbances in motions of planets, inferred the existence of an *undiscovered planet*, and calculated *the point where it can be found*, soon after being identified and called Neptune;

1846-72: Johann Gottfried Galle, German astronomer, discovered three new comets, computed the ephemerides of comets and minor planets for *Astronomisches Jahrbuch*, and moreover, following the predictions of existence of a planet beyond Uranus, he discovered the planet *Neptune*;

1848: <u>Armand Hippolyte Louis Fizeau</u>, French physicist, demonstrated the use of the shift in light frequency, called *red shift*, in determining a star's velocity; the red shift happening when light or other electromagnetic radiation from an object moving away from the observer is increased in wavelength, or shifted to the red end of the spectrum;

1889-1909: <u>Jules Henri Poincaré</u>, French mathematician, was author of works including *Analysis Situs*, the first systematic study of topology; *New Methods of Celestial Mechanics*, *The Principles of Mathematical Physics*, and *Mathematical Creation*; becoming known for *Poincaré conjecture*, *Three-body problem*, *Poincaré duality*, *Poincaré metric*, *Bifurcation theory*, and *Chaos theory*;

1910-33: <u>James Hopwood Jeans</u>, English physicist and astronomer, made significant advances in the theory of *stellar dynamics*, studied the formation of *binary stars*, *stellar evolution*, nature of *spiral nebulae*, and origin of *stellar energy*; as well as published his works *The Universe around Us*, and *The New Background of Science*;

1914-23: <u>Arthur Stanley Eddington</u>, English astronomer, published *Stellar Movements and the Structure of the Universe*, and *Space, Time and Gravitation*; as well as formulated the *Mathematical Theory of Relativity*, starting from fundamental postulates and deriving in his well-known equations and predictions at the scale of universe;

1917: *<u>Einstein universe with cosmological constant</u>* – static nominally, bounded but finite universe; and *<u>De Sitter universe</u>* – expanding flat space, steady state, with positive curvature universe;

1919-20: <u>Albert Einstein</u>, German-Swiss-US mathematical physicist, made the first description of a *static universe* with curved space, later known as 'matter with no motion';

1920-22: <u>Willem de Sitter</u>, Dutch astronomer and cosmologist, demonstrated that an *expanding universe* of constantly decreasing curvature is another possible solution as 'motion with no matter';

1920-25: *<u>MacMillan universe</u>* (named after the US mathematician and astronomer <u>William Duncan MacMillan</u>) – static and steady state universe;

1921-27: <u>Émile Félix Édouard Justin Borel</u>, French mathematician, worked on *measure theory and probability*, and formulated the *theorem* and *product of convolution*, enabling to operate integral transforms for the functions of time, including those related to the universe;

1922: *Friedmann spherical universe* (named after the Russian and Soviet physicist <u>Alexander Friedmann</u>) – spherical expanding space, closed and no curvature universe;

1924: *Friedmann hyperbolic universe* – hyperbolic expanding space, open and no curvature universe;

1927-29: *Original Big Bang model* (due to the US physicists <u>George Gamow</u>, <u>Ralph Alpher</u> and <u>Hans Bethe</u>) – expanding universe with positive curvature greater than gravity;

1928-30: <u>Clyde William Tombaugh</u>, US astronomer, devised a blink comparator enabling to detect if anything had moved in sky between taking of two celestial photographs at a few days apart, and discovered the planet *Pluto*;

1929: <u>Edwin Powell Hubble</u>, US astronomer, discovered that galaxies recede from us with speeds increasing with their distance, called *Hubble's Law*, revealing the *expansion of universe*, as the observational basis of *modern cosmology*;

1929-30: <u>Robert Julius Trumpler</u>, US astronomer, studied the dimensions and brightnesses of *open star clusters in the Milky Way*, and explained the disproportionate faintness of the more distant ones as the effect of *absorption of light in interstellar space*;

1930: *Eddington universe* (named after the English astronomer <u>Arthur Stanley Eddington</u>) – first static, then expanding universe;

1930-35: *Dirac large numbers hypothesis* (named after the English mathematical physicist <u>Paul Adrien Maurice Dirac</u>) – expanding universe;

1931-38: <u>Albrecht Otto Johannes Unsöld</u>, German astrophysicist, discovered the *hydrogen convection zone*, which explains heat transport in the Sun's outer layers, and published *Physik der Sternatmosphären* 'Physics of Stellar Atmospheres';

1932: *Friedmann zero-curvature model* – expanding flat universe with critical density;

1932-50: <u>Jan Hendrik Oort</u>, Dutch astronomer, calculated the mass of galactic material interior to the Sun's orbit, identified the *Oort cloud* as source of long-period comets, and revealed that there is *dark matter* in the centre of our Galaxy;

1933-35: *Milne universe* (named after the British astrophysicist Edward Milne) – kinematic with no expansion universe;

1933-57: Fritz Zwicky, US-Swiss astronomer and physicist, deduced the existence of *missing mass* in the total velocities of galaxies and clusters, made a catalogue of such clusters of galaxies, and published his original book *Morphological Cosmology*;

1939-56: Richard van der Riet Woolley, English astronomer, studied the solar and stellar atmospheres, and published works including *Eclipses of the Sun and Moon*, and *The Outer Layers of a Star*;

1944: Walter Baade, German-born US astronomer, discovered two discrete stellar types, namely the younger *Population I* of *blue stars* in spiral galaxies, and the older *Population II* of fainter *red stars* in elliptical galaxies, with ages ranging from one to ten million years after the Big Bang;

1945-65: Martin Ryle, English physicist radio astronomer, investigated the emission of *radio waves from the Sun*, followed by studies of *radio waves from the universe*, pointing to an evolving universe starting with a 'Big Bang', and mapped the radio sources by an ingenious method of *aperture synthesis*;

1948: George Gamow, Ralph Alpher and Hans Bethe, US physicists, interpreted the thermonuclear processes in the early universe, resulting in the *Big Bang theory* that was elaborated for creation of the universe;

1960: Allan Sandage and Thomas Matthews, US astronomers, detected a faint optical object at the same location as the compact *radio source 3C 48* with an unusual spectrum, which later became known as 'quasar', and identified many other such objects, showing that most of them are not radio emitters;

1964: Hong-Yee Chiu, Chinese-born US astrophysicist, coined the term *quasar*, given to a quasi-stellar radio source, as high red shift origin of electromagnetic energy;

1965: Arno Allan Penzias and Robert Woodrow Wilson, US physicists, discovered *cosmic microwave background radiation* as a residual relic of intense heat associated with the birth of the universe following the hot Big Bang;

1965-76: Vera Cooper Rubin, US astronomer, studied galaxy rotation rates, detected the discrepancy between the predicted angular motion of galaxies and the observed motion by studying galactic rotation curves, called *galaxy rotation problem*, and deduced the existence of *dark matter*;

1970-81: <u>Eric Becklin</u> and <u>Gerald Neugebauer</u>, US astronomers, discovered a strange infrared object radiating intensely in the *Orion* nebula, known as *Becklin-Neugebauer object*, and thought to be a young massive star blowing gases outwards at high speed;

1970-82: <u>Subrahmanyan Chandrasekhar</u>, Indian-born US astrophysicist, made theoretical studies of physical processes of importance to structure and evolution of stars, concluding that stars with masses greater than about 1.4 solar mass will be unable to evolve into white dwarfs, and thus limiting stellar mass, known as *Chandrasekhar limit*;

1974-76: <u>Stephen William Hawking</u>, English theoretical physicist, predicted that a black hole could evaporate through loss of thermal radiation, and that mass can escape from its gravitational pull, now known as *Hawking process*;

1974-82: <u>William Alfred Fowler</u>, US physicist, worked on details of *stellar nucleosynthesis*, including solar neutrino flux calculations, and on *nuclear reactions of importance in formation of chemical elements in universe*;

1977-98: <u>Gary William Gibbons</u>, British theoretical physicist, used the thermal Green's functions to prove the universality of *thermodynamic properties of horizons*, including cosmological event horizons; and then contributed to the research of *supergravity, p-branes*, and *M-theory*, mainly motivated by *string theory*; publishing his work *Born-Infeld particles and Dirichlet p-branes*, Nuclear Physics, B 514;

1978: <u>Yakov Borisovich Zeldovich</u>, Soviet astrophysicist, studied the initial hydrogen-to-helium ratio and isotropy in the early universe, warning on serious *flatness* and *horizon problems* of Big Bang cosmology;

1979-97: <u>Alan Guth</u>, US theoretical physicist and cosmologist, formally proposed the idea of cosmic inflation, showing that the nascent universe passed through a phase of exponential expansion driven by a positive *vacuum energy* density (negative vacuum pressure), known as the inflationary hypothesis, which was named *inflation*, later called 'old inflation'; and published some results of his work in *The Inflationary Universe: The Quest for a New Theory of Cosmic Origins*;

1980-81: <u>Andrei Linde</u>, Russian-born US theoretical physicist, was author of the theory known as *chaotic inflation* in which the conditions for inflation are actually satisfied quite generically, and inflation will occur in virtually any universe that begins in a chaotic, high energy state, and has a scalar field with unbounded potential energy; by his work promoting *Cosmic inflation* – Big Bang modified to solve horizon and flatness problems in the universe;

1980-92: <u>Joseph Hooton Taylor</u>, US astronomer and physicist, and <u>Hulse Russel</u>, US physicist, made a systematic research for *pulsars* as rapidly rotating dense stars which appear to emit regular pulses of radio waves, and first discovered a *binary pulsar* as a pulsar in orbit of a dense neutron star;

1981: <u>Katsuhiko Sato</u>, Japanese theoretical physicist, made another proposal of *inflation*, showing that, as a direct consequence, any observable universe originated in a *small causally connected region*;

1981-82: <u>Andrei Linde</u>, <u>Andreas Albrecht</u> and <u>Paul Steinhardt</u>, US theoretical physicists, introduced a model called 'new inflation' or *slow-roll inflation*, as an alternative of 'old inflation';

1982-83: <u>Don Page</u>, Canadian physicist, wrote *Thermodynamics of Black Holes in Anti- de Sitter Space*, opening new ways for understanding the evolution of the universe;

1983-86: *Eternal inflation* (initially proposed by the US cosmologist <u>Alan Guth</u>) – Big Bang with cosmic inflation and multiple model of the universe;

1983-88: <u>Stephen William Hawking</u>, English, and <u>Don Page</u>, Canadian, theoretical physicists, attempted to compute probability of inflation in *Hartle-Hawking initial state*, with ambiguous results;

1984-94: <u>Aleksei Aleksandrovich Starobinsky</u>, USSR-Russian astrophysicist, published on *stochastic inflation* and *multiple-inflation*; and also observed that quantum corrections to general relativity should be important in the early universe, resulting in curvature-squared corrections to the Einstein-Hilbert setting;

1985-87: <u>Michael Boris Green</u>, <u>John Schwarz</u> and <u>Edward Witten</u>, English physicists, had the idea that ultimate constituents of nature, when inspected at very small scales, do not exist as point-like particles but as *strings* in more than three dimensions, leading to the foundation of *string theory*, later developing as superstring theory, and opening a new way for cosmic studies;

1988-2005: <u>Stephen William Hawking</u>, English theoretical physicist, published *A Brief History of Time - Black Holes and Baby Universes*; *God Created the Integers: The Mathematical Breakthroughs That Changed History*; and *The Universe in a Nutshell*;

1989-93: George Smoot and John Mather, US researchers of Cosmic Background Explorer (COBE), investigated the cosmic microwave background radiation of the universe, and discovered its *blackbody feature and anisotropy*, indicating 'ripples' which show that the early universe was not smooth and uniform, so that matter could concentrate to form galaxies and stars;

1989-2000: Riccardo Giacconi, Italian-born US astrophysicist, and Masatoshi Koshiba, Japanese physicist, discovered *cosmic X-ray sources*, and laid the foundations of *X-ray astronomy*;

1990-2001: Raymond Davis, Jr, US chemist and astrophysicist, Masatoshi Koshiba, Japanese physicist, and Ricardo Giacconi, Italian-born US astrophysicist, did pioneering research in astrophysics, especially in *detection of cosmic neutrinos*, trying to solve the *solar neutrino problem* in the Homestake Experiment;

1995-2010: Saul Perlmutter, US, Brian Paul Schmidt, Australian-US, and Adam Guy Riess, US, astrophysicists, discovered the *accelerating expansion of the universe through observation of distant supernovae*;

1999-2011: Brian Greene, US theoretical physicist and string theorist, published a series of works on string theory, including *The Elegant Universe: Superstrings, Hidden Dimensions, and the Quest for the Ultimate Theory*; *The Fabric of the Cosmos: Space, Time, and the Texture of Reality*; *Icarus at the Edge of Time*; and *The Hidden Reality: Parallel Universes and the Deep Laws of the Cosmos*;

2001-07: *Cyclic models* – universe expanding and contracting in *cycles with 11-dimensional string theory*; and then *readjustment according to the solution of the entropy problem* in the universe;

2007: Lefteris Papantonopoulos, Greek editor, published *The Invisible Universe: Dark Matter and Dark Energy*, Springer, Berlin Heidelberg;

2008: Scott Trager, Dutch, Sandra Faber, US, and Alan Dressler, US, astronomers, published a paper entitled *The stellar population histories of early-type galaxies – III. The Coma Cluster*, showing that low-mass galaxies formed the bulk of their stars later than high-mass galaxies, that is 8 rather than 12 billion years from the Big Bang, and afterwards most of the galaxies had a smaller star formation episode; also estimating single-stellar populations with ages of 1.26; 1.58;

2.00; 2.51; 3.16; 3.98; 5.01; 6.31; 7.94; 10.00; and 12.59 billion years;

2010-11: Richard Panek, US astronomer, wrote and published works such as *Dark Energy: The Biggest Mystery in the Universe*, Smithsonian Magazine; and *The 4% Universe: Dark Matter, Dark Energy, and the Race to Discover the Rest of Reality*, Mariner Books: Houghton Mifflin Harcourt;

2011: Sabino Matarrese, Vittorio Gorini and Ugo Moschella, Italian editors, published their book *Dark Matter and Dark Energy: A Challenge for Modern Cosmology*, Springer, Berlin Heidelberg;

2012: NASA's *Wilkinson Microwave Anisotropy Probe (WMAP) project* resulted in establishing that the universe would be 13.772 ± 0.059 billion years old;

2012-13: Clara Moskowitz, Scientific US editor, prepared and published her work *Dark Matter Still Hidding: Latest Experimental Sweep Comes Up Empty*, Scientific American;

2013: *Lambda-CDM concordance model* and *European Space Agency's Planck satellite team* showed that the universe's age would be 13.798 ± 0.037 billion years and about 13.82 billion years respectively, so that combining the Planck data with previous missions, the best combined estimate of the universe's age was established to be 13.798 ± 0.037 billion years;

NASA current missions: 1) *NuSTAR* (Mission to search for *black holes*, to map *supernova explosions*, and to study the most *extreme active galaxies*); 2) *Juno* (Mission to discover Jupiter's secrets about the *Solar System's early history*, Juno being endeavoured to unlock these secrets); 3) *IBEX* (Mission to achieve the first global observations of the region beyond the termination shock at the very *edge of the Solar System*); 4) *Gravity Probe B* (Mission to use the relativity gyroscope for testing two *unverified predictions of Albert Einstein's general theory of relativity*), 5) *Galex* (Mission to map the *history of star formation* in the universe), and 6) *Cassini-Huygens Mission* (Research to unlock the *secrets of Saturn*).

From the Big Bang, the very early universe evolved through the Planck epoch up to 10^{-43} seconds, grand unification epoch up to 10^{-36} seconds, and electroweak epoch up to 10^{-12} seconds. This was followed by the electroweak symmetry breaking and quark epoch between 10^{-12} and 10^{-6} seconds, and the Hadron epoch between 10^{-6} seconds and 1 second when particles formed, and also the differentiation between ordinary and dark matters started. At this stage, space was filled with a mishmash of electrons and atomic nuclei, and continued to be opaque to radiation from about 10 seconds, when the Photon epoch started, to 378-380 thousand years after the Big Bang, when most of the electrons had been captured by protons to make up electrically neutral hydrogen atoms in a process called 'recombination' by which radiation was no longer blocked and could travel freely across the growing universe. Once photons decoupled from matter, they spread through the universe without interacting with matter, and constituted what is observed now as cosmic microwave background radiation.

The further evolution of the universe depended on its early 'ripples' revealed by the Cosmic Background Explorer (COBE, 1989-1993), showing that the early universe was not smooth and uniform, so that ordinary matter could concentrate to form structures later developed as the first stars about 400-500 million years and first galaxies around 1000 million years from the origin of the universe.

Recent researches and calculations indicate that the present observable universe has a volume of $(3.5...3.6) \cdot 10^{80}$ cubic metres, and mass of $(1.25...1.70) \cdot 10^{53}$ kilograms evaluated for ordinary matter. Besides this *ordinary matter* (together with neutrinos) accounting for only 4.9%, the universe is made up of cold *dark matter* representing 26.8%, and *dark energy* representing 68.3% of the total mass-energy of the universe. Within the observable universe, the density of ordinary matter (with critical density $4.08 \cdot 10^{-28}$ kg/m^3), dark matter, and dark energy all lumped together is estimated to be $9.9 \cdot 10^{-27}$ kg/m^3. Dark matter is a type of matter of which existence and properties are inferred from its gravitational effects on visible matter and radiation, as well as the large-scale structure of the universe. It was first postulated by Jan Hendrik Oort (1932) to account for the orbital velocities of stars in the Milky Way, and by Fritz Zwicky (1933) to account for evidence of 'missing mass' in the total velocities of galaxies in clusters; subsequently its presence was attested by Vera Cooper Rubin (1965-76) who studied rotational speeds of the galaxies; later it was unveiled by gravitational lensing of background objects in galaxy clusters, temperature distribution of hot gas in galaxies and clusters of galaxies; and recently by the pattern of anisotropies in the cosmic microwave background.

Dark energy permeates all space and tends to accelerate the expansion of the universe, and does not interact with any of the fundamental forces other than gravity. Its evidence is indirectly deduced from: distance measurements and their relation to red shift, which suggest a faster expansion of the universe in its later stage; the theoretical reasons for an additional energy, other than ordinary or dark matter, to explain the absence of any detectable global curvature of the universe; and the distribution of mass in the universe to account for the large scale wave-patterns of its mass density.

In terms of introductory Background to time, the *total energy E* of the universe consists of an *effective (orderly) energy* $E_e(t)$ as equivalent to ordinary and dark matters lumped together, and a *dissipative (disorderly) energy* $E_d(\tau)$ as dark energy, so that $E = E_e(t) + E_d(\tau) =$ constant; meanwhile the rate $P_e(t)$ of increasing effective energy $E_e(t)$ divided by the rate $P_d(\tau)$ of decreasing dissipative energy $E_d(\tau)$ is equal to the ratio of running time t to final time $t_\bullet$, i.e. $P_e(t)/P_d(\tau) = t/t_\bullet$. Assuming that the ratio $P_e(t)/P_d(\tau)$ is approximately equal to the ratio $E_e(t)/E_d(\tau)$, the resulting relationship $E_e(t)/E_d(\tau) \approx t/t_\bullet$ written in the form $t_\bullet \approx t/[E_e(t)/E_d(\tau)]$ can be used to estimate the final time of the expanding universe (the Big Crunch). For example, at the present time $t_\sigma \approx 13.798$ billion years from the Big Bang, when $E_e(t_\sigma)/E \approx 31.7\%$ and $E_d(\tau_\sigma)/E \approx 68.3\%$, the final time of the expanding universe can be evaluated as $t_\bullet \approx t_\sigma/[E_e(t_\sigma)/E_d(\tau_\sigma)] \approx (13.798$ billion years$)/[(31.7\%)/(68.3\%)] \approx 29.729$ billion years from is origin (the Big Bang).

From the initial matter of the very early universe stars were born, developed, differentiated, transformed, and died leaving traces after them. The currently observable universe is home to about $6 \cdot 10^{22}$ stars, which are of various types, including:

Single stars, such as dwarf, giant, and super-giant stars, with red, orange, yellow, bluish, white, or black colours;

Multiple and *double stars*, as components moving together round their common centre of gravity;

Variable stars, as pulsating, eruptive, or cataclysmic stars;

Novae, appearing from outburst, as white dwarf components accompanied by other normal stellar components, forming binary systems;

Supernovae, resulting either from destruction of former stars, as binary systems of bigger and smaller components alternatively transferring mass to one another until one of them becomes carbon-dominated white dwarf in a process ending by carbon detonation; or from sudden collapse of very massive super-giants exhausting their nuclear fuel, developing nickel-iron cores surrounded by layers of silicon and sulphur, neon and magnesium, carbon-neon and oxygen, helium and hydrogen, and ending when all energy production stops, the outer layers crushing down on to the cores that collapse, while the protons and neutrons are forced together to release neutrinos, thus generating *neutron stars* (*pulsars*) with diameters no more than a few kilometres, rates of rotation very fast, and radio radiation from their magnetic poles;

Black holes, emerging from stars too massive to explode when their energy is running out and gravity is taking over, starting to collapse with no outburst, and becoming so dense that even light cannot escape from their attraction;

Stellar clusters, as either open or loose bunches containing tens or hundreds of stars not arranged in definite structures, or symmetrical systems of huge globular clusters containing up to a million stars;

Nebulae, identified as either remnants of supernovae or old and highly evolved planetary nebulae, or true nebulae consisting mainly of hydrogen and dust.

A single generation of stars characterized by a common age and chemical composition is called *stellar population*. A galaxy can be composed of a large number of individual populations, such as: an 'old' generation of more than 3000 million years, containing no massive, hot stars, only cool low-mass stars on the main sequence and red giants, with yellow-red colour; a 'young' generation of less than 100 million years, containing massive stars, with blue colour; and a 'very young' generation of less than 10 million years, producing much ionizing radiation that creates H II ('Ionized Hydrogen' in spectral notation, as large amounts of ionized atomic hydrogen) regions in the surrounding gas, with pinkish tint colour.

There were indirectly deduced or directly evidenced mainly three stellar populations, which are briefly described below.

Stellar population III, also called metal-free stars, was hypothetically formed from 400-500 to 1500-1700 million years after the Big Bang as an oldest population, which was indirectly deduced by a gravitationally lensed galaxy in a very distant part of the universe, and also by some components of faint blues galaxies. The existence of this population was also suggested by the recent discoveries of stars forming galaxies about 13200 million light-year away, small galaxies merging to form larger ones at about 13000 million light-years away, and a quasar at 12700 million light-years away, indicating that they were formed at 600; 800; and 1100 million years after the Big Bang respectively. This hypothetical population of stars in the early universe exhausted their fuel long time ago, and exploded in extremely energetic pair-instability supernovae.

Stellar population II, or metal-poor stars, includes a halo population II with age of 13000-12000 million years, halo-population II of stars older than 10000 million years, and an intermediate population II younger than 10000 million years (for example the oldest stars in the Milky Way are about 10000 million years old) but not less than 3000 million years. Population II is common in the bulge near the centre of our galaxy, and in globular clusters.

Stellar population I, also called metal-rich stars, includes an older disc population I of ages from 1000 to 100 million years, and a younger spiral arm population I of ages from 100 to 10 million years. It is common in the spiral arms of the Milky Way and also in many other galaxies.

In another classification, Stellar Populations are grouped such as: *Old* - more than 3000 million years (cool low-mass stars); *Young* - less than 100 million years old (massive stars); and *Very Young* - less than 10 million years (with 'H II' regions in the surrounding gas).

Despite the uncertainties in delimiting the cosmic sequences between the universe's origin 13798 million years ago and the present, there are two main margins in the formation of Stellar Populations: one between 10000 and 3000 million years ago when most of the Population II stars were born (including intermediate and later H II stars); and another between 100 and 10 million years ago when most of the Population I stars were born (including earlier and intermediate stars). Therefore, the universe's expansion can be sequenced, in million years ago, as:

Early universe (including pre-stellar, stellar population III, and earlier H II stellar population) \ 10000...3000 ≈ 6500 \ *Late universe* (including stellar population II, together with intermediate and later H II stellar populations, and older disc population I of galaxies) \ 100...10 ≈ 55 \ *Current universe* (including younger spiral arm population I of galaxies, and recently formed stellar population I) through present onward.

Referring to the universe's origin 13798 million years ago, these time limits become 13798 - 6500 ≈ 7298; and 13798 - 55 ≈ 13743 million years respectively, whereby the ratio 7298/13743 ≈ 0.531. A similar value of this ratio can be found in Background to time, Table *(j)*, such as $t_{0.25}/t_{0.5} = tanh(0.25)/tanh(0.5) = 0.5299926$, indicating that cosmic timeline is divided into $z/4 = j$ sequences. Because $t_{0.25} \approx 7298$ is less accurately estimated than $t_{0.5} \approx 13743$ million years, the final time $t_{\bullet}$ can be better calculated as

$t_\bullet = t_{0.5}/tanh(0.5) \approx (13743)/(0.4621172) \approx 29739$ million years,

so that the transitional times, also in million years, are:

$$t_{0.25} = t_\bullet \cdot tanh(0.25) \approx (29739)\cdot(0.2449187) \approx 7284;$$
$$t_{0.5} = t_\bullet \cdot tanh(0.5) \approx (29739)\cdot(0.4621172) \approx 13743;$$
$$t_{0.75} = t_\bullet \cdot tanh(0.75) \approx (29739)\cdot(0.6351490) \approx 18889.$$

Continuing to apply the function $t_j = t_\bullet \cdot tanh(j)$, the timeline, sequences, and intrinsic characteristics of the expanding universe can be reconstituted and predicted in a manner shown in the tables below.

$z/4 = j$	Time (million years)		Cosmic sequences
	from origin $t_j = t_\bullet \cdot tanh(j)$	from present $t_j - 13798$	
	29739	+15941	
...	...	...	
0.75	18889	+5091	–
0.5	13743	-55	– Current universe
0.25	7284	-6514	– Late universe
0	0	-13798	– Early universe

$z/4 = j$	Time from origin $t_j = t_\bullet \cdot tanh(j)$ (million years)	Period $\tau_j = (t_\bullet^2 - t_j^2)/(2t_\bullet)$ (million years)	Frequency $f_j = (2t_\bullet)/(t_\bullet^2 - t_j^2)$ (million years)$^{-1}$	Angular speed $\omega_j = 2\pi/\tau_j = 2\pi \cdot f_j$ (million years)$^{-1}$
	29739	0	∞	∞
...	...	...	...	...
0.75	18889	8870.7	0.0001127	0.0007083
0.5	13743	11694.0	0.0000855	0.0005373
0.25	7284	13977.5	0.0000715	0.0004495
0	0	14869.5	0.0000673	0.0004226

According to the above model, the universe is not eternal, but limited between the initial time *0* at the Big Bang and the final time $t_\bullet \approx$ 29739 million years at the Big Crunch, as a close universe in its expanding quarter-super-cycle (with *red shift* in spectral lines), which will be followed by a shrinking one (with *blue shift* in spectral lines), together constituting a semi-super-cycle of about 59.5 billion years. If this succession of semi-super-cycles will continue, then they would be gradually shorting.

Although the expansion of universe is a natural (non-perpetual) process, its sequences are lasting so long as time seems to flow linearly during the relatively short existence of humanity; however the universe timeline lasts from its beginning 13798 million years ago to its end 15941 million years in the future.

Time *t* is an indissoluble characteristic as the multiplicand of the light speed c within the intrinsic coordinate $c \cdot t$ representing the base of any imaginable multidimensional space. Moreover, $c \cdot t$ is related to other coordinates $\{x_i\}$ according to the inequality

$$c^2 \cdot t^2 \geq \Sigma_i\, x_i^2, \text{ or } t \geq (1/c) \cdot (\Sigma_i\, x_i^2)^{1/2},$$

where t is defined by the function $t = t_\bullet \cdot tanh(\theta/4\pi)$, with $t \rightarrow 0$ when $\theta \rightarrow 0$, and $t \rightarrow t_\bullet$ when $\theta \rightarrow \theta_{t\bullet}$. Therefore, the above inequality can be rewritten as

$$t_\bullet \cdot tanh(\theta/4\pi) \geq (1/c) \cdot (\Sigma_i\, x_i^2)^{1/2}, \text{ or}$$
$$\theta \geq 4\pi \cdot tanh^{-1}[(c \cdot t_\bullet)^{-1} \cdot (\Sigma_i\, x_i^2)^{1/2}].$$

Taking into account that the angle θ ranges extensively in the interval $-\infty < \theta < +\infty$, while the function $t(\theta)$ ranges alternatively in intervals $-t_\bullet \leq t \leq +t_\bullet$, the multidimensional spaces could be either positive or negative. Thus the anti-time sequences justify the corresponding multidimensional anti-spaces and meanwhile the extensive and integrated understanding of the universe in its evolution.

3. Geology

The science concerning the history and development of the Earth with its associated processes is called *geology* (from Greek γη 'land, earth' and λογος 'word, speech, story'), the scientific study of the Moon is called *selenology* (from Greek σεληνη 'moon' and λογος), and of the other planets is called *planetology* (from Greek πλανήτης 'planet' and λογος), all of them intended to give insights into the origins, compositions, structures and history of these studied bodies.

Geology is a major academic discipline that includes many fields and related disciplines, such as: biogeochemistry, Earth science, economic (mining, petroleum, natural gas) geology, engineering geology, environmental geology, geoarchaeology, geochemistry, geochronology, geodynamics, geological modelling, geomagnetism, geometallurgy, geomicrobiology, geomorphology, geomythology, geophysics, geostatistics, glaciology, historical geology, hydrogeology, marine geology, mineralogy, natural hazards, palaeoclimatology, palaeontology, palynology, pedology, petrology, petrophysics, planetary geology, plate tectonics, sedimentology, seismology, speleology, stratigraphy, structural geology, thermochronology, and volcanology.

The foundation and development of geology and its branches was based on observations, studies, experiments, researches, theories, methods, equipments, missions, operations, and so on, accomplished by a series of personalities, from whom some are mentioned below.

c.300BC: <u>Theophrastus</u>, Greek philosopher, completed his work *Peri Lithon* 'On Stones', as probably the first study of Earth's physical material;

AD77: <u>Pliny the Elder</u>, Roman scholar, pointed out details on many *minerals* and *metals* in practical use, and made a correct note on the *origin of amber*, which were described in his universal encyclopaedia *Historia Naturalis*;

1020-30: <u>Avicenna</u> ('Abd Allah ibn Sina), Persian philosopher and physician, gave explanations for the *formation of mountains*, *origin of earthquakes*, and other topics central to modern geology;

1020-40: <u>Abu al-Rayhan al-Biruni</u>, Muslim scholar and geologist, was author of the earliest writings on *geology of India*, including the hypothesis that *Indian subcontinent was once a sea*;

59

1082-95: <u>Shen Gua</u>, Chinese administrator, engineer and scientist, observed fossil animal shells in a geological stratum of a mountain located hundreds of miles from the ocean, and inferred that the *land was formed by erosion of mountains and by deposition of silt*, such as presented in his remarkable compilation *Brush Talks from Dream Brook*;

1546-55: <u>Georgius Agricola</u>, German mineralogist and metallurgist, was author of *De Natura Fossilum*, in which a systematic classification of minerals was attempted, and *De Re Metallica*, as a detailed record of 16th century mining, ore-smelting and metal working;

1600: <u>William Gilbert</u>, English physician, emitted the hypothesis that the Earth is a giant magnet, which was presented in his work *De Magnete*, and included the conjecture of terrestrial magnetism with electricity as two allied emanations of a single force;

1675-85: <u>Nicolaus Steno</u>, Danish physician, naturalist and theologian, made studies in crystallography, formulating *Steno's law* of crystal structure, and also in geology and palaeontology, establishing the *law of superposition*, *principle of original horizontality*, and *principle of lateral continuity*;

1784: <u>Richard Kirwan</u>, Irish chemist, did valuable research on *chemical affinity* and *composition of salts*, and published the first systematic work on *mineralogy*;

1785-99: <u>James Hutton</u>, Scottish geologist, after many journeys into Scotland, England and Wales, he published the three-volume *A Theory of the Earth*, demonstrating that the Earth's internal heat caused intrusions of molten rock into crust, and that granite was produced by cooling of molten rock, and not by precipitating in the primeval ocean;

1790-1801: <u>Déodat Guy Gratet de Dolomieu</u>, French geologist, soldier and traveller, journeyed in Italy, Sicily, Portugal, Alps and Pyrenees, and wrote about *earth tremors in Calabria* and *Italian volcanoes*; studied the mineral *dolomite*, called after him, that later was used by extension for the *Dolomite mountain range* of Italy;

1794-1801: <u>Andrés Manuel Del Rio</u>, Spanish geologist and mineralogist, worked on the *origin of mineral veins*, *paragenesis of sulphide minerals*, and effects of *trace elements*; published *Elementos de Orictognosia*, as the first textbook of mineralogy in the Americas; and discovered a new metallic element called *panchromium*, subsequently known as vanadium;

1794-1824: <u>William Smith</u>, English civil engineer and geologist, did surveying work during construction of the Somerset Coal Canal, studied a variety of rock sequences of different ages; produced a *coloured map of Bath area* and collected fossils to identify strata and to fix their position in succession; followed by drawing the *first geological map of English counties* that became known as 'The Map That Changed the World';

1795-1801: <u>René Just Haüy</u>, French crystallographer and mineralogist, published *Traité de minéralogie* 'Treatise of Mineralogy', by which *crystallography* was properly founded;

1801-17: <u>Georges Léopold Cuvier</u>, French anatomist, studied *animal and fish fossils*, reconstructing *extinct giant vertebrates* of Paris Basin, and linking palaeontology to comparative anatomy; he published *Les Ossements fossiles des quadrupèds* 'The Fossilized Bones of Quadrupeds', and *Histoire naturelle des poissons* 'The Natural History of Fish', thus founding *palaeontology*;

1809-17: <u>William Maclure</u>, US geologist, drew the first *geological map of the USA*, and published his work *Observations on the Geology of the United States*;

1822-38: <u>Gideon Algernon Mantell</u>, English palaeontologist, wrote about *The Fossils of the South Downs*, discovered several dinosaur types, including *Iguanodon*, introduced the notion of *age of reptiles*; and also published his works *Geology of the South-east of England*, and *The Wonders of Geology*;

1830-63: <u>Charles Lyell</u>, Scottish geologist, was author of the treatises *Principles of Geology*, *The Elements of Geology*, and *The Geological Evidence of the Antiquity of Man*;

1835-42: <u>James Dwight Dana</u>, US mineralogist, crystallographer and geologist, wrote his works *On the condition of Vesuvius*, and *System of Mineralogy*, also providing scientific observation on a US *exploring expedition in Antarctic and Pacific*;

1835-45: <u>Roderick Impey Murchison</u>, Scottish geologist, established the *Silurian System*, and made an extensive geological survey of Russian Empire, which was followed by publication of *The Geology of Russia in Europe and the Urals*;

1835-47: <u>Henry Thomas De la Beche</u>, English stratigrapher and geologist, began the first national *Geological Survey* of his country; and published *Manual of Geology*, and *Researches in Theoretical Geology*;

1835-54: <u>Adam Sedgwick</u>, English geologist, established the *Cambrian System* in Wales, and published *British Palaeozoic Fossils*;

1851-97: <u>John William Dawson</u>, Canadian geologist, discovered some of the earliest known *terrestrial vertebrate fossils* inside carboniferous fossil tree stumps at Joggins, Nova Scotia, and published *Acadian Geology, The Story of Earth and Man, Origin of the World, Fossil Men*, and *Relics of Primeval Life*;

1854-82: <u>Josiah Dwight Whitney</u>, US geologist, produced works such as *Mineral Wealth of the United States, Auriferous Gravels of the Sierra Nevada*, and *Climate Changes of Later Geological Time*;

1856-88: <u>Frederick Augustus Genth</u>, US mineralogist and chemist, did important research on *meteorites*, chemistry, and *rare minerals*; wrote a monograph on *ammonium-cobalt compounds*, and discovered *23 new mineral species* including *genthite*, named in his honour;

1875-90: <u>Grove Karl Gilbert</u>, US geomorphologist, stratigrapher, structural geologist and cartographer, recognized the nature of intrusions named *laccoliths* in Henry Mountains; and published *Monograph on Lake Bonneville*, describing the history of Pleistocene climate and hydrography of the Great Basin, and discussing subsequent deformation of old shore levels, thus throwing light on *isostatic readjustments of the Earth's crust*;

1879-1918: <u>Charles Lapworth</u>, English geologist, introduced the *Ordovician System*, so resolving the controversy about the junction between Cambrian and Silurian sedimentary rocks, and published *Monograph of British Graptolites*;

1883-1906: <u>Richard Dixon Oldham</u>, Irish geologist and seismologist, wrote *Catalogue of Indian Earthquakes, Bibliography of Indian Geology, On the Propagation of Earthquake Motion to Great Distances*, and deduced the *Earth's core* from seismological records, as well as made for the first time the distinction between primary and secondary seismic waves;

1885-1909: <u>Eduard Suess</u>, Austrian geologist, studied the evolution of various features of the Earth's surface, particularly the problem of mountain building, also gave a special attention to volcanic islands and their associated deep-sea trenches in Pacific Ocean, leading to the theory that there had once been a great supercontinent made up of present southern continents, and forerunning the modern theories of continental drift, as well as published the four-volume *Das Antlitz der Erde* 'The Face of the Earth';

1896-1949: <u>Josiah Edward Spurr</u>, US geologist, established the *age of Tertiary period* between 45 and 60 million years; explored Alaska where his name was later given to *Mt Spurr*; did major research on *lunar topography and geology*; also wrote *Geology Applied to Mining*, and *Geology Applied to Selenology*;

1897-1925: <u>Gheorghe Munteanu Murgoci</u>, Romanian geologist and pedologist, identified the *Getic Thrust Nappe* of Southern Carpathians; published *The Geological Synthesis of the South Carpathians*; studied the geology of Dobruja and its role in the *Cimmerian orogeny*; and produced the first *soil map* of his country;

1899-1924: <u>John Joly</u>, Irish geologist and physicist, published *An Estimate of the Age of the Earth, Radium and the Geological Age of the Earth, Radioactivity and Geology*, and *The Surface History of the Earth*;

1902-23: <u>Alfred Lacroix</u>, French mineralogist, petrologist and structural geologist, investigated eruptive rocks, studying eruptions of Mont Pelée volcano, and recognizing *nuée ardente* as a glowing cloud type of eruption; published the three-volume *Minéralogie de Madagascar* 'Mineralogy of Madagascar', and also worked on igneous and volcanic rocks of the Massif Central, Etna, Vesuvius, and on meteorites;

1904-05: <u>Ludovic Mrazec</u>, Romanian geologist, was founder of the *Theory of diapirism*, resulting from his study of *salt plastic behaviour* 'diapirism' in several oil fields of his country;

1905-24: <u>Gerhard Jacob de Geer</u>, Swedish Quaternary geologist, devised a novel and valuable *method for dating deposits* by comparing sequences of *varves* (annual deposits of sediment under glacial melt-water), and deciphered an *annual chronology* reaching back 15,000 years from the present day; then demonstrated global climatic events and advanced the knowledge of the last *Ice Age*;

1906-27: <u>John Walter Gregory</u>, English geologist and explorer, published *The Dead Heart of Australia, The Rift Valleys and Geology of East Africa, To the Alps of Chinese Tibet*, and *Elements of Economic Geology*;

1909: <u>Andrija Mohorovičić</u>, Yugoslav seismologist, identified the *Mohorovičić discontinuity* 'Moho' between the Earth's crust and mantle, and calculated its depth;

1910-35: <u>Henry Fairfield Osborn</u>, US palaeontologist and zoologist, published *The Age of Mammals, Man of the Old Stone Age, The*

Origin and Evolution of Life, and a vast monograph on *Proboscidea*;

1913-50: <u>Gheorghe Iosif Macovei</u>, Romanian geologist, studied the Cretaceous formations, oil fields in his country, the *geology of South Dobruja* and *Eastern Carpathians*, also wrote the comprehensive treatise *Geology of Romania*;

1915: <u>Alfred Lothar Wegener</u>, German meteorologist and geophysicist, published *Die Entstehung der Kontinente und Ozeane* 'The Origin of Continents and Oceans', based on observations that continents may once have been joined into one supercontinent called *Pangaea* which later broke up, as a hypothesis of *continental drift*, later developed by plate tectonics;

1926-30: <u>Leopold von Buch</u>, German geologist and traveller, elaborated the first coloured *geological map of Germany* in 42 sheets, studied *Alpine geology*, introduced the term *gabbro*, described other igneous rocks, and produced a *classification of cephalopods*;

1927-59: <u>Gheorghe Murgeanu</u>, Romanian geologist, researched the *Dacic Nappe System*, and identified remnant elements of green schist from a former mountainous chain, called by him *Cuman Cordillera*, which were ingested into the Carpathian Flysch; he published works such as *Recherches géologiques dans la Vallée de Doamnei et la Vallée de Valsan (Mounténie occidentale)*, *Sur l'âge des schistes ménilitiques et des gypses inferieurs de la Mounténie occidentale*, and *Cretaceous Flysch of Predeal Pass*;

1928: <u>Norman Levi Bowen</u>, US geologist, did pioneering work of experimental petrology, particularly the study of silicates and igneous rocks, which was published as *The Evolution of Igneous Rocks*;

1930-68: <u>Alexandru Codarcea</u>, Romanian geologist, crystallographer and mineralogist, identified the *Nappe of Severin* in south-east Romania; published *Vues nouvelles sur la tectonique du Banat méridional et du Plateau de Mehedinti*, and *Structure géologique du massif des roches alkalines de Ditrau*; also discovered a new variety of mica, called after him *codarcit* (green mica);

1933-50: <u>Robert Broom</u>, South African palaeontologist, studied human ancestry, leading to his published works *The Coming of Man*, and *Finding the Missing Link*;

1935-47: <u>Victor Moritz Goldschmidt</u>, Norwegian geologist and crystallographer, worked on *petrology of southern Norway*, made extensive *X-ray study of binary compounds* of elements, analyzed the *distribution of elements in the Earth*, and published a massive book entitled *Geochemistry*;

1940: <u>Reginald Aldworth Daly</u>, US geologist, accomplished a seminal work describing lithosphere and asthenosphere, which was entitled *Strength and Structure of the Earth*;

1941-60: <u>Harry Hammond Hess</u>, US marine geophysicist and geologist, discovered the flat-topped seamounts, called *guyots*, and described the *oceans as young, ephemeral and with constant renewal by magma flowing into mid-ocean ridges*;

1950-54: <u>Victor Hugo Benioff</u>, US geophysicist, and <u>Kiyoo Wadati</u>, Japanese seismologist, studied earthquake sources on sub-ducting plates, and evidenced the *Wadati-Benioff zones*;

1954-89: <u>Georges Matheron</u>, French mathematician and geologist, founded *geostatistics* and *mathematical morphology*; coining of eponym *krigeage* 'Kriging' in his *Krigeage d'un Panneau Rectangulaire par sa Périphérie*; and published *Traité de géostatistique appliquée, The theory of regionalised variables and its applications, Random sets and integral geometry*, and *Estimating and Choosing: An Essay on Probability in Practice*;

1963-67: <u>John Tuzo Wilson</u>, Canadian geophysicist, had the ideas of *hotspots* in Earth's mantle, oceanic *transform faults*, and *mountain building*, as major steps towards *plate tectonic theory*, and formulation the 'Wilson cycle'; the author's name being also given to *Wilson Range in Antarctica*, and to the *Wilson ocean cycle*;

1966-69: Five <u>*US Orbiters*</u> to *map the Moon's surface*; and then *Apollo 8*, *9* and *10* to test the lunar module, followed by <u>*US Apollo 11 mission*</u> for Moon-landing, achieved by astronauts <u>Neil Alden Armstrong</u>, <u>Buzz Aldrin</u> and <u>Michael Collins</u> who successfully landed there and collected *rock-samples* to be later analyzed;

1968: <u>Thomas Henry Clark</u> and <u>Colin Stearn</u>, Canadian geologists, produced the first work concerning the entire North American continent, entitled *The Geological Evolution of North America*;

1969-80: <u>Sigurdur Thorarinsson</u>, Icelandic geologist and glaciologist, was the first to use *dating of ash layers* for studying the eruption history of volcanoes, and the first to analyse catastrophic *glacier outburst floods*;

1970-80: <u>Luis Walter Alvarez</u>, US experimental physicist, studied the *catastrophe which killed the dinosaurs*, deducing that its cause was the impact on Earth of an asteroid or comet, at the Cretaceous-Palaeogene limit;

1977-93: <u>Stephen Jay Gould</u>, US palaeontologist, wrote essays, including *Ever Since Darwin, The Panda's Thumb, The Mismeasure of Man, Hens' Teeth and Horses' Toes, The Flamingo's Smile, Wonderful Life, Bully for Brontosaurs*, and *Eight Little Piggies*.

1989: <u>Walter Brian Harland</u>, British geologist, <u>Richard Armstrong</u>, US-Canadian scientist, <u>Allen Cox</u>, US geophysicist, <u>Lorraine Craig</u>, British scientist, <u>Alan Smith</u>, British geologist, and <u>David Smith</u>, Canadian editor, published *A Geologic Time Scale 1989*, a work of reference for many years;

1990-93: <u>*US Magellan mission*</u> launched from *Shuttle Atlantis* for radar mapping of Venus;

1996-97: <u>US *Pathfinder mission*</u> with *Sojourner* 'Rover' for landing on Mars and *collecting rocks* from it;

2005: <u>Felix Gradstein</u>, Norwegian scientist, <u>James Ogg</u>, US geologist and climatologist, and <u>Alan Smith</u>, British geologist, edited *A Geologic Time Scale 2004*, Cambridge University Press;

2013: <u>Kim Cohen</u>, <u>Stan Finney</u> and <u>Phil Gibbard</u>, US-English geochronologists, produced *International Chronostratigraphic Chart*, International Commission on Stratigraphy;

<u>*NASA current missions*</u>: *Artemis* (Study of the Moon's interaction with the Sun), and *Curiosity, the Mars Science Laboratory* (Mission to assess whether Mars ever was, or is still, an environment able to support microbial life, which from August 2012 has used a rover landed for *collecting and analyzing samples*).

The galactic thin disc of the Milky Way was formed 8.8±1.7 billion years ago corresponding to about 5 billion years after the Big Bang. Within the Milky Way, the formation of the Solar System took place about 4.6 billion years ago corresponding to approximately 9.2 billion years from the origin of the universe, and its fate is predicted to be between 1 and 5 billion years from the present time.

In our Solar System, the planets came into existence at about the same time by accretion from a cloud material (solar nebula) crossed at first by the youthful Sun on its trajectory intersecting that cloud. Later on, the Solar System was cyclically crossing, every cosmic year of 225 million years, through the remains of same source cloud laid in its slightly different plane of rotation within the Galaxy.

The Earth was formed 4568 million years ago and then carried together with the other planets through the source cloud every cosmic year, at those times being possibly submitted at impacts contributing to geological and climatic changes.

The early Earth was submitted to a heavy bombardment with smaller meteorites and comets, as well as with large rocky bodies, the biggest being that of a body the size of Mars, striking the proto-Earth a glancing blow. The collision between this impacting body, called *Theia*, and the Earth released about 100 million times more energy than the impact that later caused the extinction of dinosaurs. According to the giant impact hypothesis, a portion of the mantle material was ejected into orbit around the Earth and condensed within a couple of weeks into the single spherical body of Moon. Based on the radiometric dating of rocks brought during the *Apollo program*, the Moon seems to be about 4500 million years old, i.e. less than 100 million years after the birth of our Solar System.

Initially, the Earth had no water and atmospheric gases. Water was later supplied by icy meteorites from the outer asteroid belt and some large planetary embryos, as well as by comets, altogether intensively bombarding the young Earth and carrying meteoritic water, possibly containing some non-cellular living forms, to our planet. The presence of the detrital zircon crystals dated to 4400 million years shows evidence of having undergone contact with liquid water, suggesting that the planet already had accumulations of water, as seas or even ocean at that time.

Liquid water resulted from the Earth's cooling, evaporation as clouds, falling down as rain and accumulation to form the earliest ocean beginning as early as 4400 million years ago, and partially covering the Earth.

Some models suggest that the Earth would have been covered in ice which melted when the carbon dioxide and methane were enough to induce a greenhouse effect, the carbon dioxide being ejected by volcanoes, and the methane being produced by early bacteria. Another greenhouse gas represented by ammonium would have been ejected by volcanoes, but quickly destroyed by ultraviolet radiation.

Inside the early Earth, gravitational compaction resulted in increasing pressure and temperature, associated with melting of oxides of iron and other heavy metals that separated from the still solid silicates and moved to the centre of gravity together with their sulphur compounds, where the core was forming with release of thermal energy to the covering mantle where the silicates were then also melted, generating basic and ultrabasic magmata able to erupt and to form rocks at the Earth's surface. Within the mantle and crust, this process includes the *Wilson rock cycle* (JT Wilson, 1963-67), which unfolds as: magma –

crystallization (freezing of rock) – igneous rocks – erosion – sedimentation – sediments and sedimentary rocks – tectonic burial and metamorphism – metamorphic rocks – melting. Correspondingly, the tectonic plates, floating on a ductile mantle, rifted into pieces diverging and forming new ocean basins, followed by motion reversal, convergence back together, plate collision, and mountain building.

Today the Earth has a mean radius of 6371 kilometres, surface area of 510,100,000 square kilometres, volume of 1083 billion cubic kilometres, mass of $5.974 \cdot 10^{24}$ kilograms, and average mass density of 5515 kilograms per cubic metre. Its temperature was originally 2.725 kelvin, and is now between 5778 kelvin in the Earth's centre and 287.2 kelvin at the Earth's surface. From the centre to the Earth's surface, there are the following layers: the *inner core* at 0-1180 kilometres with temperatures of 5778-5492 kelvin; the *outer core* at 1180-3480 kilometres with temperatures of 5492-4250 kelvin; the *lower mantle* at 3480-5720 kilometres with temperatures of 4250-1973 kelvin; the *upper mantle* at 5720-6331 kilometres with temperatures of 1973-873 kelvin; and the *crust* at 6331-6371 kilometres with temperatures of 873-287.2 kelvin.

Excepting the thermal energy intercepted by the Earth from the Sun's radiation at a rate of $4.38 \cdot 10^{17}$ watts taken through the atmosphere, hydrosphere and upper crust; the internal thermal energy at the present time, estimated at about $1 \cdot 10^{31}$ joules, corresponds to the upward outflow of $4.20 \cdot 10^{13}$ watts, from which $2.88 \cdot 10^{13}$ watts due to planetary accretion; $1.05 \cdot 10^{13}$ watts due to radioactive decay of minerals in the continental crust and mantle; and $0.27 \cdot 10^{13}$ watts due to bombardment with extra-terrestrial bodies (asteroids, comets, meteors and meteorites), the currents of convection in the outer core and mantle, terrestrial tides, exothermal chemical reactions, changes of physical state, and so on.

The Earth's geological time is usually divided in eons, eras, periods, epochs, and ages succeeding each other in an apparently unregulated manner. However, during Earth's evolution, they had a general decreasing tendency, while the certainty in delimiting them roughly increases from the earlier to the later ones, according to the precision of data resulted from analyses of rocks and meteorites, fossil records, dating techniques, environmental chemistry, climate change, evolution of life, etc.

Although the results dating the transitions from an era to another are quite widely varying (Harland *et al.*, 1989; Gradstein and Ogg, 2005;

Cohen *et al.*, 2013), recent Geological Time Scales indicate that the Earth's evolution was marked by four eons, roughly delimited in million years, as: Hadean from 4568 to 3800-3000, Archaean from 3800-3000 to 2520-2200, Proterozoic from 2520-2200 to 545-480, and Phanerozoic from 545-480 to present time.

Hadean and Archaean eons have been somehow doubtingly divided in 2-4, and 3-4 so-called 'eras' respectively, while Proterozoic and Phanerozoic eons were divided in 3 proper eras each, separated by transitional times, in million years ago, estimated as:

2520...2200 ≈ 2360 \ *Palaeoproterozoic* \ 1660...1400 ≈ 1530 \
Mesoproterozoic \ 1000...850 ≈ 925 \ *Neoproterozoic*
\ 545...480 ≈ 512 \ *Palaeozoic* \ 250...230 ≈ 240
\ *Mesozoic* \ 66 \ *Cenozoic* to present.

Despite disputes in estimating the limits of eras comprised in Proterozoic and Phanerozoic eons, it seems that these eras are succeeding each other in a quite regular manner of shortening from the earlier to the later ones. From the Earth's origin about 4568 million years ago, as the initial time *0*, the limits of completed eras are displayed, in million years, as follows: 4568 - 2360 ≈ 2208; 4568 - 1530 ≈ 3038; 4568 - 925 ≈ 3643; 4568 - 512 ≈ 4056; 4568 - 240 ≈ 4328; 4568 - 66 ≈ 4502. The manner in which eras are shortening can be disclosed by their successive ratios 2208/3038 ≈ 0.727; 3038/3643 ≈ 0.834; 3643/4056 ≈ 0.898; 4056/4328 ≈ 0.937; 4328/4502 ≈ 0.961. Such consecutive ratio values can be found in Background to time, Table *(j)* for $z/4 = j$ sequences, column $t_{j-¼}/t_j = tanh(j-¼)/tanh(j)$, where the more accurate values are displayed in the following order $t_{1.5}/t_{1.75} = 0.9615166$; $t_{1.25}/t_{1.5} = 0.9371765$; $t_1/t_{1.25} = 0.8978060$; $t_{0.75}/t_1 = 0.8339730$; $t_{0.5}/t_{0.75} = 0.7275729$.

Thus, identifying that the geological timeline has $z/4 = j$ sequences, and knowing that the most recent and then more accurately determined change of eras is the transition from Mesozoic to Cenozoic about 66 million years ago, i.e. $t_{1.75}$ ≈ 4568 - 66 ≈ 4502 million years from the Earth's origin, the final time can be calculated using the formula $t_{1.75}/t_• = tanh(1.75)$ which leads to $t_• = t_{1.75}/tanh(1.75)$ ≈ (4502)/(0.9413755) ≈ 4782 million years from the Earth's origin, or 4782 - 4568 ≈ 214 million years from the present into the future.

Consequently, the earlier eras are delimited by times calculated, in years from the Earth's origin, as

$t_{0.5} = t_• \cdot tanh(0.5)$ ≈ (4782)·(0.4621172) ≈ 2210;
$t_{0.75} = t_• \cdot tanh(0.75)$ ≈ (4782)·(0.6351490) ≈ 3037;

$$t_1 = t_\bullet \cdot tanh(1) \approx (4782) \cdot (0.7615942) \approx 3642;$$
$$t_{1.25} = t_\bullet \cdot tanh(1.25) \approx (4782) \cdot (0.8482836) \approx 4056;$$
$$t_{1.5} = t_\bullet \cdot tanh(1.5) \approx (4782) \cdot (0.9051483) \approx 4328.$$

So calibrated, the geological timeline, sequences, and intrinsic characteristics are displayed as follows:

$z/4 =$ j	Time (million years)		Geological sequences
	from the origin $t_j = t_\bullet \cdot tanh(j)$	from present t_j - 4568	
	4782	+214	
...	...	...	
2	4610	+42	− Cenozoic era
1.75	4502	-66	_ Mesozoic era
1.5	4328	-240	_ Palaeozoic era
1.25	4056	-512	_ Neoproterozoic era
1	3642	-926	_ Mesoproterozoic era
0.75	3037	-1531	_ Palaeoproterozoic era
0.5	2210	-2358	_ Archaean eon
0.25	1171	-3397	_ Hadean eon
0	0	-4568	

$z/4 =$ j	Time from origin $t_j = t_\bullet \cdot tanh(j)$ (million years)	Period $\tau_j =$ $(t_\bullet^2 - t_j^2)/(2t_\bullet)$ (million years)	Frequency $f_j =$ $(2t_\bullet)/(t_\bullet^2 - t_j^2)$ (million years)$^{-1}$	Angular speed $\omega_j =$ $2\pi/\tau_j = 2\pi \cdot f_j$ (million years)$^{-1}$
	4782	0	∞	∞
...	...	...	...	...
2	4610	168.9	0.0059204	0.0371991
1.75	4502	271.8	0.0036791	0.0231167
1.5	4328	432.4	0.0023124	0.0145293
1.25	4056	670.9	0.0014906	0.0093655
1	3642	1004.1	0.0009959	0.0062574
0.75	3037	1426.6	0.0007010	0.0044043
0.5	2210	1880.3	0.0005318	0.0033415
0.25	1171	2247.6	0.0004449	0.0027955
0	0	2391.0	0.0004182	0.0026278

4. Climatology

The study of *climate* (from Greek κλίμα 'place, zone'), scientifically defined as weather conditions averaged over a period of time, is called *climatology* (from Greek κλίμα and λογος 'word, speech, story'), representing a branch of atmospheric sciences, and including aspects of oceanography and biogeochemistry.

Observations and studies on climate and weather were carried out by a series of personalities, expeditions and missions, some of them being mentioned bellow, in chronological order, with their contributions.

335-325BC: Aristotle, Greek philosopher and scientist, accomplished the textbook *Meteorology*, describing what is now known as *hydrologic cycle*, and thus founding the scientific study of meteorology;

330-310BC: Pytheas of Massilia, Greek navigator and geographer, travelling to the British Isles, had the idea to *relate the spring tides to the phases of Moon*;

315-290BC: Theophrastus, Greek philosopher, compiled a work on climate and weather forecasting, called the *Book of Signs*, with dominant influence in studies of atmospheric phenomena for about two millennia;

c.150BC: Seleucus of Seleucia, Babylonian astronomer, described the *phenomenon of tides*, in order to support his heliocentric theory;

AD15-20: Strabo, Greek geographer and Stoic, in his *Geographica* 'Geography', described the *tides in Persian Gulf* having their greatest range when the Moon was furthest from the plane of equator;

25: Pomponius Mela, Latin geographer, was author of the three-volume compendium *De Situ Orbis* 'A Description of the World', and formalized the *climatic zone system*;

77: Pliny the Elder, Roman scholar, collated many *tidal observations*, which were presented in his work *Historia Naturalis*;

870-890: Abu Hanifah Dinawari, Muslim Kurdish astronomer and historian, wrote *Kitâb al-anvâ'* 'Book of Climate', pointing the seasons and rain, *anvâ* meaning 'heavenly bodies of rain', and atmospheric phenomena such as winds, snow, and floods;

1075-90: Shen Kua, Chinese administrator, engineer and scientist, knowing that a dry-climate area was unsuitable for the growth of bamboo, and observing petrified bamboos found underground near Yanzhou, Shaanxi province, he inferred that *climates naturally shifted* over an enormous span of time;

1262-80: St Albertus Magnus (Doctor Universalis), German philosopher and cleric, was the first to propose that each drop of falling rain had the form of a small sphere, and this form meant that the *rainbow* was produced by light interacting with each raindrop;

1266-76: Roger Bacon, English philosopher and scientist, was the first to calculate the *angular size of the rainbow*, stating that the rainbow summit cannot appear higher than 42° above the horizon;

1686-1716: Edmond Halley, English astronomer and mathematician, presented a systematic study of the trade *winds* and *monsoons*, identifying solar heating as the cause of atmospheric motions; and suggested that *auroras* are caused by 'magnetic effluvia' moving along the Earth's magnetic field lines;

1750-60: Benjamin Franklin, North-American inventor and scientist, drew the first *mapped course of the Gulf Stream* for use in sending and receiving mail to and from Europe;

1801-17: Georges Léopold Cuvier, French anatomist, posited the *catastrophism* as a series of extinctions due to *periodic global floods*, after which new forms of life appeared;

1803: Luke Howard, British chemist and meteorologist, published *Essay on the Modifications of Clouds*, assigning cloud types Latin names, as *cumulus*, *stratus*, and *cirrus*, as well as intermediate and compound modifications, such as *cirrostratus*, and *stratocumulus*;

1822-25: Joseph Fourier, French mathematician, was author of the well known treatise *Théorie analytique de la chaleur* 'Analytical Theory of Heat'; and described the *greenhouse effect* in his *Mémoire sur la température de globe terrestre et des espaces planétaires* 'Memoir on the temperature of terrestrial globe and of the planetary spaces';

1840-1847: Jean Louis Agassiz, Swiss-born US naturalist and glaciologist, published *Études sur les glaciers* 'Study on Glaciers', and *Système Glacière* 'Glacier System';

1863: Francis Galton, English scientist, wrote and published his work *Meteorographica*, as the basis for modern weather maps, and coined the term *anticyclone*;

1867: <u>William Thomson Kelvin</u>, Scottish physicist and mathematician, made the first systematic harmonic analysis of tidal records, resulting in a *tide-predicting machine* with a system of pulleys to add together six harmonic time functions;

1872-76: *British Challenger Expedition*, achieved the first oceanographic investigation for studying *global distribution of sediments* in the world ocean;

1875-85: <u>James Croll</u>, Scottish physicist and geologist, wrote and published *Climate and Time, in their Geological Relations*, providing a credible explanation for the *causes of ice ages*; and also *Climate and Cosmology*;

1898-1910: <u>Vilhelm Friman Koren Bjerknes</u>, Norwegian mathematician and meteorologist, formulated the *circulation theorem*, applying it to a study of atmospheric and ocean processes; then deduced equations enabling *calculation for a developing cyclone*, based on both thermal energy and baroclinicity; and developed *numerical models* for predicting the changes of weather;

1920s: <u>Milutin Milankovich</u>, Yugoslav geophysicist, gave an astronomical explanation for the long-term, cyclical global climate changes that caused the Pleistocene *ice ages*, according to so-called *Milankovich cycles*, consisting of precession, obliquity, and eccentricity, which are repeating approximately every 21, 41, and 100 thousand years respectively;

1925-27: *German Meteor Expedition*, collected sediment samples, showing that the absence of foraminifer skeletons in the lower layers was caused by the *low temperatures of sea water during the Ice Age*;

1951-54: <u>Helmut Erich Landsberg</u>, German-born US climatologist, organized the first *numerical weather prediction* efforts, fostering the use of *statistical analysis in climatology*, which led to its promotion into a physical science;

1952-60: <u>Guy Stewart Callendar</u>, English engineer and inventor, proposed that eventually became known as the *Callendar effect*, the theory that linked rising carbon dioxide concentrations in the atmosphere to global temperature;

1958-88: <u>James Ephraim Lovelock</u>, English chemist, invented the 'electron capture detector', a high-sensitivity device, which was used in the first measurements of the *accumulations of chlorofluorocarbons in atmosphere*; put forward 'Gaia' hypothesis, indicating that the

Earth's climate is constantly regulated by plants and animals; and proposed the *biotic feedback*;

1960: <u>*Weather Satellite TIROS-1*</u>, was successfully launched, marking the beginning of age when the *weather information became globally available*;

1969-80: <u>Syukuro Manabe</u>, Japanese meteorologist and climatologist, pioneered the use of computers to simulate *global climate change*, and *natural climate variations*;

1990-2012: <u>Peter Gleick</u>, US hydro-climatologist, studied the *hydrologic impacts of climate change, snowfall/snowmelt responses, water adaptation strategies*, and *consequences of sea-level rise*;

1997-2007: <u>*Analogue techniques for weather forecasting*</u>, represented by: 1) *El Niño – Southern Oscillation* (ENSO), a global coupled ocean-atmosphere phenomenon; 2) *Madden – Julian Oscillation* (MJO), an equatorial travelling pattern of anomalous rainfall at planetary scale; 3) *North Atlantic Oscillation* (NAO), based on the difference of normalized sea level pressure between Azores and Iceland; 4) *Northern Annual Mode* (NAM), defined for the northern hemisphere winter sea level pressure from the tropics and subtropics; 5) *Northern Pacific* (NP) *Index*, for area-weighted sea level pressure over the region 30°N - 65°N and 160°E - 140°W; 6) *Pacific Decadal Oscillation* (PDO), a pattern of Pacific climate variability that shifts phases on at least inter-decadal time scale, usually about 20 to 30 years; and 7) *Inter-decadal Pacific Oscillation* (IPO), displaying sea surface temperature and sea level pressure patterns with a cycle of 15-30 years, and affecting both the north and south Pacific;

2008-11: <u>*Pangaea Expedition*</u>, led by <u>Mike Horn</u>, South African-born Swiss explorer, was a worldwide voyage starting from Punta Arenas towards the South Pole, continuing with Australia, Asia, and Russia, then with Iceland, Greenland and towards the North Pole, and ending back to Punta Arenas; this voyage was associated with an economic study aimed at a better understanding of the *climate crisis*;

<u>*NASA current missions*</u>: *Arctas* (Research of the composition of the *troposphere* from aircraft and satellites), and *Aeronomy of Ice in the Mesosphere* (Two-year mission to study Polar Mesospheric Clouds).

Looking back so far as the earliest stage of our planet, soon after its origin 4568 years ago, the *first atmosphere* emerged about 4560...4540 ≈ 4550 million years ago and consisted of gases in the solar nebula, primarily hydrogen, and secondarily water vapour, methane and ammonia, which were mostly driven off by the solar wind.

Subsequently, it was replaced by the *second atmosphere* consisting largely of nitrogen, and additionally of carbon dioxide and inert gases produced by out gassing from volcanoes, as well as gases released during the heavy bombardment of Earth by big asteroids including water vapour that condensed and formed the first accumulations of water as proved by the existence of detrital zircon crystals dated to 4400 million years indicating contact with liquid water; its composition changed around 4000 million years ago when early basic carbon isotropy came into existence, suggesting that the fundamental features of the carbon cycle were established, and again about 3400 million years ago when nitrogen became the major component of the second atmosphere.

The *third atmosphere* developed by transferring carbon dioxide to and from continental carbon stores during the re-arrangement of continents by plate tectonics; by apparition of free oxygen from about 1800 million years ago indicated by the end of banded iron formations consuming oxygen to reduce elements such as iron, and the beginning of the oxygenated atmosphere by exceeding the capacity of consumption by reducible materials; then the amount of oxygen in the atmosphere increased, reaching a peak of c.30% about 280 million years ago, significantly higher than today's 21%; the changes in composition of atmosphere being governed by two main processes, namely the use of carbon dioxide from and release of oxygen to atmosphere by plants' photosynthesis, and oxidation of pyrite and volcanic release of sulphur into atmosphere which tend to reduce the content of atmospheric oxygen.

In order to date the climate changes on our planet, there were carried up a series of geodynamical, sedimentological, paleontological, palynological, palaeobiological, palaeoclimatological, dendrochronological, and hydrochemical studies, as well as radiocarbon and other radioisotopes dating, and climate change modelling; by which the climate fluctuations or oscillations were identified as so-called minor, minor-intermediate, intermediate, intermediate-major, and major, lasting ten-hundred, hundred-thousand, thousand-million, ten-hundred million, and hundred-thousand million years respectively.

At the global scale, the major fluctuations were alternatively associated with average temperature varying from 12-16°C to 22-23°C and ocean level ranging from -150 to +150 metres respectively.

The major climate fluctuations consist in alternating *ice ages*, or *glacial ages*, and *interglacials*; the ice age being a period of long-term reduction in temperature of the Earth's surface and atmosphere, which resulted in the presence or expansion of continental ice sheets, polar ice sheets, and alpine glaciers, while interglacial age being an intermittent warm period. Among the causes of ice ages can be mentioned: changes in Earth's atmosphere, position of the continents, fluctuations in ocean currents, uplift of the Tibetan plateau and surrounding mountain areas above the snowline, variations in Earth's orbit (Milankovich cycles), variations in the Sun's energy output, and volcanism.

Referring only to later major ice ages in the Earth's history, their limits have been dated in various variants, the recent ones issued in 1980-1990, and in 2005-2013.

The outdated 1980-1990 variants display the following climate periods delimited, in million of years ago, as follows:

585-460: *Interglacial period* during the transition from Neoproterozoic to Palaeozoic, with the *interglacial maximum* at 510 when the average temperature was 22°C;

460-440: *Glacial period* in the early Palaeozoic, with the *glacial minimum* at 450 when the average temperature was 12°C;

440-315: *Interglacial period* in the mid-Palaeozoic, with the *interglacial maximum* at 420 when the average temperature was 22°C;

315-265: *Glacial period* by the end of Palaeozoic, with the *glacial minimum* at 295 when the average temperature was 12°C;

265-151: *Interglacial period* at the beginning-mid-Mesozoic, with the *interglacial peak* at 250 when the average temperature was 23°C;

151-133: *Glacial period* in the late Mesozoic, with the *glacial minimum* at 142 when the average temperature was 16°C;

133-22: *Interglacial period* during the transition from Mesozoic to Cenozoic, with the *interglacial maximum* at 65 when the average temperature was 22°C;

22-present: *Last glacial period* by the mid-Cenozoic, with the *glacial minimum* at 5.5 when the average temperature was 12°C; within this last glacial period being identified the most recent glacial 20,000-10,000 years ago when environmental conditions were worsening, ice sheets covered large areas at higher latitudes and altitudes both in North and South hemispheres, and oceanic level lowered to more than 100 metres below its present level; as well as the current interglacial from 10,000 years ago to present-day.

Although outdated and doubtful, the above periods are presented in a continuous succession, i.e. without undetermined intervals between them.

The 2005-2013 variants of dating the last ice ages or glaciations indicate their existence at times such as 40 (mid Eocene), 26 (late Oligocene), 14 (Antarctica-Himalayan rise-Greenland-Arctic), and 2.58 (late Pliocene-Quaternary) million years ago. However, these ice ages are not major ones. The proper major ice ages are named after the places where they have been identified, such as *Huronian, Cryogenian* or *Sturtian-Varangian, Andean-Saharan*, and *Karro*, which were recently dated as presented below in chronological order with their intervals of time in million years ago: 2400-2100: *Huronian* Ice Age; 800-635: *Cryogenian* or *Sturtian-Varangian* Ice Age; 450-420: *Andean-Saharan* Ice Age; and 360-260: *Karro* Ice Age.

Between these major ice ages should been not only major interglacial intervals, but also other ice ages not yet identified. Following the same course and being closely associated with the geological eras, the major *climatic periods* originated 4560...4540 ≈ 4550 million years ago, as the climate initial time *0*, and delimited by times, in million years ago, as:

4560...4540 ≈ 4550 \ *Hadean* (*Earlier atmospheric*) \ 3800...3000 ≈ 3400 \ *Archaean* (*Later atmospheric*) \ 2500...2200 ≈ 2350 \ *Palaeoproterozoic* (including *Huronian Ice Age*) \ 1600...1450 ≈ 1525 \ *Mesoproterozoic* (*Interglacial*) \ 950...900 ≈ 925 \ *Neoproterozoic* (including *Cryogenian Ice Age*) \ 525...495 ≈ 510 \ *Palaeozoic* (including *Andean-Saharan* and *Karoo Ice Ages*) \ 250...230 ≈ 240 \ *Mesozoic* (*Interglacial* and *Glacial*) \ 75...63 ≈ 69 \ *Cenozoic* (including last *Interglacials* and *Glacials*) up to the present.

Referring to the climate initial time, these limits become 4550 - 4550 = 0; 4550 - 3400 = 1150; 4550 - 2350 = 2200; 4550 - 1525 = 3025; 4550 - 925 = 3625; 4550 - 510 = 4040; 4550 - 240 = 4310; and 4550 - 69 = 4481 million years from the origin of climate respectively; so that their ratios are succeeding as follows: 0/1150 = 0; 1150/2200 ≈ 0.523; 2200/3025 ≈ 0.727; 3025/3625 ≈ 0.834; 3625/4040 ≈ 0.897; 4040/4310 ≈ 0.937; and 4310/4481 ≈ 0.962. The corresponding figures of ratios are given in Background to time, Table *(j)* for $z/4 = j$ sequences, as: $t_0/t_{0.25} = 0$; $t_{0.25}/t_{0.5} = 0.5299926$; $t_{0.5}/t_{0.75} = 0.7275729$; $t_{0.75}/t_1 = 0.8339730$; $t_1/t_{1.25} = 0.8978060$; $t_{1.25}/t_{1.5} = 0.9371765$; $t_{1.5}/t_{1.75} = 0.9615166$.

Taking into account that the most recent time is $t_{1.75} = t_\bullet \cdot tanh(1.75) \approx$ 4481 million years from the origin of climate, as the most accurately determined, it follows that the final time can be calculated as $t_\bullet = t_{1.75}/tanh(1.75) \approx (4481)/(0.9413755) \approx 4760$ years from climate origin, or 4760 - 4550 $\approx$ 210 million years from present into the future. Therefore, the earlier transitional times, in million years from the origin of climate, are accurately calculated as

$$t_{0.25} = t_\bullet \cdot tanh(0.25) \approx (4760) \cdot (0.2449187) \approx 1166;$$
$$t_{0.5} = t_\bullet \cdot tanh(0.5) \approx (4760) \cdot (0.4621172) \approx 2200;$$
$$t_{0.75} = t_\bullet \cdot tanh(0.75) \approx (4760) \cdot (0.6351490) \approx 3023;$$
$$t_1 = t_\bullet \cdot tanh(1) \approx (4760) \cdot (0.7615942) \approx 3625;$$
$$t_{1.25} = t_\bullet \cdot tanh(1.25) \approx (4760) \cdot (0.8482836) \approx 4038;$$
$$t_{1.5} = t_\bullet \cdot tanh(1.5) \approx (4760) \cdot (0.9051483) \approx 4309.$$

Thus calibrated, the climatic timeline, sequences, and intrinsic characteristics are reconstituted in the tables below.

$z/4 = j$	Time (million years)		Climatic sequences
	from origin $t_j = t_\bullet \cdot tanh(j)$	from present $t_j - 4550$	
	4760	+210	
...	...	...	
2	4589	+39	_Cenozoic (with Interglacials+Glacials)
1.75	4481	-69	_Mesozoic (Interglacial and Glacial)
1.5	4309	-241	_Palaeozoic (incl. Andean and Karoo)
1.25	4038	-512	_Neoproterozoic (incl. Cryogenian)
1	3625	-925	_Mesoproterozoic (Interglacial)
0.75	3023	-1527	_Palaeoproterozoic (incl. Huronian)
0.5	2200	-2350	_Archaean (Later atmospheric)
0.25	1166	-3384	_Hadean (Earlier atmospheric)
0	0	-4550	

$z/4 = j$	Time from origin $t_j = t_\bullet \cdot tanh(j)$ (million years)	Period $\tau_j = (t_\bullet^2 - t_j^2)/(2t_\bullet)$ (million years)	Frequency $f_j = (2t_\bullet)/(t_\bullet^2 - t_j^2)$ (million years)$^{-1}$	Angular speed $\omega_j = 2\pi/\tau_j = 2\pi \cdot f_j$ (million years)$^{-1}$
	4760	0	∞	∞
...	...	...	...	...
2	4589	167.9	0.0059549	0.0374158
1.75	4481	270.8	0.0036924	0.0232003
1.5	4309	429.6	0.0023276	0.0146245
1.25	4038	667.2	0.0014987	0.0094166
1	3625	999.7	0.0010003	0.0062852
0.75	3023	1420.1	0.0007042	0.0044246
0.5	2200	1871.6	0.0005343	0.0033571
0.25	1166	2237.2	0.0004470	0.0028085
0	0	2380.0	0.0004202	0.0002675

5. Biology

Terrestrial time, studied by geology, was closely associated with life evolution, studied by *biology* (from Greek *βιος* 'life, livelihood' and *λογος* 'word, speech, story'). The evolution of life on Earth has been described, studied, reconstituted and assembled by prominent physicians, philosophers, biologists, biophysicists, biochemists, botanists, zoologists, anatomists, surgeons, physiologists, geneticists, gynaecologists, embryologists, haematologists, pathologists, virologists, paediatricians, pharmacologists, immunologists, histologists, cytologists, bacteriologists, palaeontologists, speleologists, physicists, chemists, crystallographers, astronomers, and mathematicians, such as those briefly presented below, in chronological order, with their contributions.

c.450BC: <u>Empedocles</u>, Greek philosopher, believed that some creatures less well-adapted to life on Earth had perished in past; while the *surviving ones resulted from joining in different combinations by intermixing*, and wherever 'everything turned out as it would have if it were on purpose, there creatures survived, being accidentally compounded in a suitable way';

330-322BC: <u>Aristotle,</u> Greek philosopher and scientist, thought and wrote on biology and zoology, believed that sense perception is only means of human knowledge, and made the first scientific *classification of life forms*;

AD1377: <u>Ibn Khaldun</u>, Arab philosopher, asserted that humans developed from 'the world of the monkeys', as pointed out in his *Muqaddima*, an introduction to the monumental history of Arabs;

1542: <u>Leonhard Fuchs</u>, German botanist, pioneering German botany, published *Historia stirpium* 'History of Rootstocks', describing how plants function;

1735-37: <u>Carolus Linnaeus</u>, Swedish naturalist and physician, published his major works *Systema Naturae*, *Fundamenta Botanica*, *Genera Plantorum*, and *Critica Botanica*; as well as introduced the *binomial nomenclature* of generic and specific names for animals and plants, which permitted the hierarchical organization later known as systematics;

1749-67: <u>George-Louis Leclerc Buffon</u>, French naturalist, was author of the monumental 44-volume *Histoire naturelle*, as an influential work in arousing interest in natural history, and foreshadowing the theory of evolution;

1778-1826: <u>Antoine Laurent de Jussieu</u>, French botanist, reorganized the National Museum of Natural History, and elaborated *Genera Plantarum*, adopting family names from C Linnaeus and from himself, and finishing with over 100 *family names*, of which 76 are still in current use;

1779-96: <u>Jan Ingenhousz</u>, Dutch chemist and biologist, worked on *photosynthesis*, and discovered that *all parts of plants give off carbon dioxide both in darkness and light*;

1790-1817: <u>Georges Léopold Cuvier</u>, French anatomist, originated the *natural system of animal classification*, anticipating modern division of animal kingdom into phyla; and the *theory of catastrophism*, as series of extinctions due to periodic global floods after which new forms of life appeared; as well as founded the *comparative anatomy* and *palaeontology*;

1798-1830: <u>Bernard de Laville de Lacépède</u>, French naturalist, produced important works, including *Histoire naturelle des poissons* 'The Natural History of Fishes', and *Les Âges de la nature* 'The Ages of Nature';

1805-43: <u>Alexander Humboldt</u>, German naturalist, published his the monumental 23-volume *Voyage de Humboldt et Bonpland aux Regions Equinoxiales* 'Personal Narrative of Travels to the Equinoctial Regions', as well as *Géographie du nouveau continent*, and *Asie Centrale*, showing great interest in biology;

1809-22: <u>Jean-Baptiste Lamarck</u>, French naturalist and evolutionist, published his *Philosophie zoologique* 'Zoological Philosophy', postulating that acquired characters can be inherited by later generations, and breaking with old notion of immutable species; and then *Histoire naturelle des animaux sans vertèbres* 'Natural History of Invertebrates', recognizing that species needed to adapt in order to survive environmental changes, so preparing the way for the theory of evolution;

1810: <u>Robert Brown</u>, Scottish botanist, published *Prodromus Florae Novae Hollandiae et insulae Van-diemen*, and first noticed that, in general, *living cells contain a nucleus*;

1818-42: <u>Étienne Geoffroy Saint-Hilaire</u>, French zoologist, published his *Philosophie anatomique*, *L'Histoire naturelle des mammifères* 'Natural history of Mammals', *Philosophie zoologique* 'Zoological

Philosophy', and *Études progressives d'un naturaliste* 'Progressive Studies of a Naturalist';

1835-70: Hugo von Mohl, German botanist, researched plant cell structure and physiology, as first attempts at cytochemistry, and differentiated the cell membrane, nucleus, cellular fluid, and utricle, coining the term *protoplasm* in plant cell biology;

1859-81: Charles Robert Darwin, English naturalist, published *The Origin of Species by Means of Natural Selection*, and great series of supplemental treatises including *The Fertilisation of Orchids, The Variation of Plants and Animals under Domestication, The Expression of the Emotions in Man and Animals, Insectivorous Plants, Climbing Plants, The Effects of Cross and Self Fertilisation in the Vegetable Kingdom, Different Forms of Flowers in Plants of the same Species, The Power of Movement in Plants*, and *The Formation of Vegetable Mould through the action of Worms*; he also left invaluable knowledge regarding variation and interbreeding in organisms, and formulated the theory of *pangenesis*, concerning acquired and inherited aspects;

1862-1906: Ernst Heinrich Philipp August Haeckel, German biologist, naturalist and physician, discovered, described and named thousands of *new species*, mapped a *genealogical tree* relating to all life forms; coined many terms in biology, including *anthropogeny, ecology, phylum, phylogeny, stem cell, Protista* (kingdom); and developed the *recapitulation theory* 'ontogeny recapitulates phylogeny';

1867-1902: August Friedrich Leopold Weismann, German biologist, investigated the development of two-winged flies *Diptera*, describing neurohumoral organ known as *Weismann ring*; stated the *germ-plasm theory*, deducing that information required for development and final form of an organism must be contained within germ cells, eggs and sperm, and transmitted unchanged from generation to generation; as well as published *Essay upon Heredity and Kindred Biological Problems*, and *Vorträge über Descendenztheorie* 'Lectures on Evolutionary Theory';

1869-89: Francis Galton, English scientist, made studies on *heredity*, founded *eugenics* as science of creating superior offspring; and reformulated Darwin's pangenesis as being *both particulate and inherited*;

1874-1936: Ivan Petrovich Pavlov, Russian physiologist, worked into three main areas of physiology, namely *circulatory system, digestive system*, and *higher nervous activity* including brain; made famous research showing that if a bell is sounded whenever food is presented to a dog, it will eventually begin to salivate when bell is sounded without food being presented, this process being termed as *conditioned* or *acquired reflex*, and leading to theories of animal and human behaviour;

1877-1908: <u>Victor Babes</u>, Romanian physician, biologist and bacteriologist, discovered a parasitic sporozoan of ticks, named *Babesia* (*Babesiidae* family), which causes a rare and severe disease called *babesiosis*; published the world's first treatise of bacteriology *Les bactéries et leur rôle dans l'anatomie et l'histologie pathologiques des maladies infectieuses* 'Bacteria and their role in the histopathology of infectious diseases'; discovered cellular inclusions in rabies-infected nerve cells, which were named *Babes-Negri bodies*, and introduced the *rabies vaccination*;

1886-1905: <u>Ilya Ilyich Mechnikov</u>, Russian embryologist and immunologist, and <u>Paul Ehrlich</u>, German bacteriologist, studied the action of *phagocytes* (mobile cells attacking foreign bodies) in animal infections; and initiated the *side-chain theory* in immunology, respectively;

1890-1905: <u>Albrecht Kossel</u>, German physiological chemist, investigated the chemistry of cells and proteins, explaining that, in blood leukaemia, the *guanide* derived from decomposed young nucleated erythrocytes; and also discovered the *histidine* in spermatozoa;

1892-1910: <u>Grigore Antipa</u>, Romanian Darwinist biologist, studied the *fauna of Danube Delta* and *Black Sea*, was the first to use *dioramas* for museum displays, and founded the *National Museum of Natural History* in Bucharest, bearing his name;

1897-1927: <u>Emil Racovitza</u>, Romanian biologist and speleologist, participated in the Belgian Antarctic Expedition, on the ship *Belgica*, and based on his expertise published *La vie des animaux et plantes dans l'Antarctique* 'The Life of Animals and Plants in Antarctica'; explored over 1400 caves in France, Spain, Algeria, Italy and Slovenia, and founded *speleology* as presented in his *Essay on bio-speleological problems*, and *Speleology: A New Science of the old underworld mysteries*;

1910-20: <u>Otto Fritz Meyerhof</u>, US biochemist, and <u>Archibald Vivian Hill</u>, English physiologist, discovered a fixed relationship between *consumption of oxygen* and *metabolism of lactic acid* in muscle; and investigated the *production of heat in muscle*, respectively;

1911-13: <u>Alfred Henry Sturtevant</u>, US geneticist, produced the first *chromosome map* and mathematical background for genetic mapping experiments on fruit-fly *Drosophila*, and used the phenomenon of genetic linkage to show that *genes are arranged linearly on chromosomes*;

1922-62: <u>Herman Joseph Muller</u>, US geneticist, worked on the physiological and genetic effects of radiation (X-ray mutagenesis), and published works including *Variation due to Change in the Individual Gene*, lying out the basic properties of the heredity molecule; *The Gene as The Basis of Life*; and *Studies in Genetics*;

1924-64: <u>Aleksandr Ivanovich Oparin</u>, Russian biochemist, worked on the origin of life, and published his results as *Proiskhozhdenie Zhizny* 'The Origin of Life', suggesting that life was initiated by slow binding together of molecules to form droplets, which then absorbed other bio-molecules and spontaneously divided; what was followed by publications such as *The Origin of Life on Earth, The History of the Theory of Genesis and Evolution of Life*, and *The Chemical Origin of Life*;

1927-35: <u>Gerhard Johannes Paul Domagk</u>, German biochemist, researching for new dyes and drugs, and aiming at a treatment against streptococci which caused widespread and generally lethal infections, he discovered a *dye becoming of potent benefit when added to a substance called prontosil*, that only worked in living animal, being converted to sulphanilamide in body;

1928: <u>Frederick Griffith</u>, British bacteriologist, discovered the phenomenon of transformation called *Griffith's experiment*, showing that dead bacteria could transfer genetic material to 'transform' other still-living bacteria;

1930-47: <u>Arne Wilhelm Kaurin Tiselius</u>, Swedish chemist, introduced protein analysis by moving boundary *electrophoresis*, which became best criterion of protein purity, isolated the *bushy stunt and cucumber mosaic viruses*, invented *preparative electrophoresis, electro-kinetic filtration*, and other analytical techniques, as well as methods for *chromatographic separation* and identification of amino acids, sugar and other molecules;

1932-33: <u>Max Kleiber</u>, Swiss chemist and biologist, concluded that ¾ power of body weight is most reliable basis for predicting basal metabolic rate of animals, and set the *law of metabolic rate* 'Kleiber's law', that relates metabolic rate to organism's mass;

1935-58: <u>Erwin Schrödinger</u>, Austrian physicist, published works related to biology, anthropology and genetics, such as *Science and the Human Temperament*; *What is Life?*, approaching the problems of genetics from the physical point of view, and *Mind and Matter*, showing the relationship between thermodynamic entropy and the evolution of life;

1940-46: Max Delbrück, German biophysicist, worked on *genetics of phage virus*, a simple organism with protein coat surrounding a coil of DNA, and discovered that *viruses can exchange genetic material to create new types of virus*;

1940-50: Barbara McClintock, US geneticist, operated on chromosomes of maize, proving the *chromosome theory of heredity*, and evidenced how genes can control other genes and can be copied from chromosome to chromosome;

1942: George Wells Beadle, US biochemical geneticist, and Edward Lawrie Tatum, US biochemist, demonstrated the role of genes in biochemical processes by growing bread mould spores on a variety of nutritional media, and suggesting that each spore had one or more blocks in metabolic pathway for particular nutrients, which led to the *one gene, one enzyme* hypothesis, as a single gene codes for synthesis of one protein;

1943-55: Konrad Emil Bloch, German-born US biochemist, revealed the direct metabolic relationship between *cholesterol* and *bile acids*, and then discovered that the mould *Neurospora* required acetate for growth resulting in the recognition of *mevalonic acid* (a key organic compound in biochemistry) as the first-formed building block;

1948-50: Melvin Calvin, US chemist, elucidated the *Thunberg-Wieland cycle* by which some bacteria, unlike animals, synthesize four-carbon kind of sugar, and hence glucose, from acetate, as shown by KE Bloch; and had idea that the *inverse cycle might fix carbon dioxide gas as in photosynthesis*;

1948-67: Severo Ochoa, US geneticist, studied the *energetics of carbon dioxide fixation in photosynthesis*; achieved the first *synthesis of artificial RNA*, solved *amino acid genetic code*, and identified a number of base triplets, as well as studied the *direction of protein synthesis along DNA*, and the *first amino acid in a peptide sequence*;

1949: Rachel Fuller Brown, US biochemist, by introduction of methods in controlling bacterial forms of disease, she isolated the first antifungal antibiotic, called *nystatin*;

1950-75: Edward Lawrie Tatum, US biochemist, and Joshua

Lederberg, US biologist and geneticist, described *transduction in bacteria*, whereby bacterial virus transfers part of its DNA into a host bacterium, leading to development of techniques for *manipulation of genes*, and evidenced the *sexual process of conjunction* in bacteria reproduction;

1951-53: Maurice Hugh Frederick Wilkins, British physicist, and Rosalind Elsie Franklin, English X-ray crystallographer, produced an *X-ray diffraction picture of DNA*, thus contributing to the double helix model of DNA by their *X-ray data of DNA fibres*;

1952: Alfred Day Hershey and Martha Chase, US biologists, effectuated *Hershey-Chase experiment* showing that DNA, rather than protein, represents genetic material of viruses infecting bacteria; and confirmed that *DNA of other organisms fulfils same role*;

1953: Francis Harry Compton Crick, English molecular biologist, and James Dewey Watson, US biologist, worked on the *structure of DNA* (deoxyribonucleic acid), finding that biological molecule contained in cells carries genetic information; elaborated the famous model of a *double-helical molecule*, consisting of two strands of nucleotide bases wound around a common axis in opposite directions, and suggesting a simple method for duplication, i.e. if strands are separated, new partner strands are reconstructed for each based on sequence of the old strand;

1954-58: William Howard Stein and Stanford Moore, US biochemists, developed a column chromatographic method for identification and quantification of *amino acid mixtures* in proteins and physiological tissues, analyzed the *base sequence of RNA*, and studied *novel protease from streptococcus*, showing that its molecular structure differed from that of plant protease papain, thus evidencing a first example of phenomenon called *convergent evolution*;

1955-90: Edmond Henri Fischer and Edwin Gerhard Krebs, US biochemists, studied the *enzyme phosphorylase*, showing that conversions to and from compounds of phosphorus are involved in activating *glycogen phosphorylase*, as a biological regulatory mechanism;

1956: George Emil Palade, Romanian-born US cell biologist, developed a method of separating cell components, known as *cell fractionation*, and identified these components as *mitochondria, endoplasmatic reticulum, Golgi apparatus*, and *ribosomes*, showing that the *protein synthesis occurs on strands of RNA in ribosomes*; thus remarkably contributing to the development of genetics;

1969-72: <u>Gerald Maurice Edelman</u>, US biochemist, and <u>Rodney Robert Porter</u>, English biochemist, investigated a number of *antibody forms in different vertebrates*, proposed the *bilaterally symmetrical four-chain structure* as basis of all immunoglobulins, and evidenced a typical Y-shaped human immunoglobulin *IgG* antibody molecule;

1971-80: <u>Fred Hoyle</u>, English astronomer and mathematician, and <u>Nalin Chandra Wickramasinghe</u>, Sri Lankan-born British mathematician, astronomer and astrobiologist, worked on spectral analysis of interstellar dust, indicating that *large organic molecules and even bacterial spores occur in space*, that is outside the Earth, and pointing out that natural selection and mutation as too weak mechanism to drive evolutionary progress;

1980-85: <u>Robert Huber</u>, German biophysicist, <u>Johann Deisenhofer</u>, US molecular biologist, and <u>Hartmut Michel</u>, German biochemist, determined the *structure of reaction centre* of purple bacterium *Rhodopseudomonas viridis*, confirming and elaborating predictions about the operation of *energy transfer process in photosynthesis*;

1981-82: <u>Ralph Holloway</u> and <u>David Post</u>, US anthropologists, searched 89 primate species identified as fossil hominids, noticed the nonlinearity of the brain-to-body relationship, and introduced the *encephalization coefficient*;

1982: <u>Alan Robert Templeton</u>, US biologist, and <u>John Seward Johnson</u>, US oceanographer, wrote *Life History Evolution under Pleiotropy and k-selection in a Natural Population of Drosophila mercatorum*, Ecological genetics and evolution, Academic Press; showing that in a population of this organism under K-selection the population actually produced a higher frequency of traits typically associated with r-selection;

1983-95: <u>John Maynard Smith</u>, English geneticist and evolutionary biologist, developed a new phase of mathematical understanding of evolutionary processes, in particular application of game theory to behavioural ecology, published *Evolution and the Theory of Games*, researched mutations and recombination in *human mitochondrial DNA*, and wrote influential book *Theory of Evolution*, as well as *Evolutionary Genetics*, and *The Major Transitions in Evolution*;

1984: <u>Michael Stuart Brown</u> and <u>Joseph Leonard Goldstein</u>, US molecular geneticists, elucidated *gene sequence with several mutations*, which codes for low-density lipoproteins as receptor, opening up possibilities of synthesizing drugs to control cholesterol metabolism;

1985-89: <u>Michael Bishop</u>, US molecular biologist and virologist, discovered *oncogenes*, normal cellular genes involved in growth and development of *all mammalian cells*;

1990-2005: <u>Roger David Kornberg</u>, US biochemist, studied molecular basis of *eukaryotic transcription*, as process of making a RNA-copy of part of DNA, followed by movement of RNA out of cell nucleus to ribosomes;

1992-2004: <u>Didier Raoult</u>, <u>Stéphane Audic</u>, <u>Catherine Robert</u>, <u>Chantal Abergel</u>, <u>Patricia Renesto</u>, <u>Hiroyuki Ogata</u>, <u>Bernard La Scola</u>, <u>Marie Suzan</u>, and <u>Jean-Michel Claverie</u>, French researchers, discovered and studied Mimivirus, and published their work *1.2-Megabase Genome Sequence of Mimivirus*, Science, 306;

1994: <u>Ken McNamara</u>, Australian professor of microbiology, and <u>Stanley Awramik</u>, US microbiologist, published their paper entitled *Stromatolites: a key to understanding the early evolution of life*, Science Progress, 77;

1995-2000: <u>Sydney Brenner</u>, British molecular biologist, <u>Howard Robert Horvitz</u>, US biologist, and <u>John Edward Sulston</u>, British biologist, made discoveries concerning *genetic regulation of organ development* and *programmed cell death*;

2000-01: <u>Simon Wilde</u>, Australian, <u>John Valley</u>, US, <u>William Peck</u>, UK, and <u>Colin Graham</u>, US, researchers, studied the existence of detrital zircon in different areas, and published their findings as *Evidence from detrital zircons for existence of continental crust and oceans on the Earth 4.4 Gyr ago*, Nature Geoscience;

2001: <u>Richard Stephen Kent Barnes</u>, British biologist, made thorough researches on invertebrates, and wrote *The Invertebrates: A Synthesis*, Wiley-Blackwell;

2011: <u>Sun Kwok</u> and <u>Yong Zhang</u>, Hong Kong scientists, searching different phases of stellar evolution, proto-planetary and planetary nebulae, and especially cosmic dust, as well as analyzing data from the European Space Agency's Infrared Space Observatory and from NASA's Spitzer Space Telescope, they found out that *organic compounds commonly exist throughout the universe*, as proved by an infrared emission band indicating an amorphous carbonaceous solid with *mixed aromatic and aliphatic structures* naturally and rapidly created by stars; the results and consequences of these findings being published in their *Organic Matter in the Universe*;

2011-12: <u>Scott Sanford</u>, US Space science researcher at NASA Ames, studied the chemical processes that occur when high-energy ultraviolet radiation bombard simple ices like those seen in space, and found out that a surprisingly rich mixture of organics is made, such as *amino acids, nucleobases* and *amphiphiles*;

2013: <u>Steven Benner</u>, US geochemist, interpreted the results of the first *analyses of Martian soil* by *Curiosity rover*, and said 'The evidence seems to be building that we are actually Martians; that life started on Mars and came to Earth on a rock';

2013-14: <u>Marius Albu</u>, Romanian-born British geologist and physicist, based on classical studies and recent discoveries, produced the work *Integrated Course of Life, Soul and Mind*, United p.c. publisher, European Union;

<u>*NASA current missions*</u>: *Kepler* (Mission to search for *habitable planets*), and *Curiosity, the Mars Science Laboratory* (Mission to assess whether Mars ever was, or is still, an environment able for supporting microbial life, using an rover landed on Mars in August 2012).

In order to estimate the time when life occurred on the early Earth, scientists were based on a series of crystallographic, chemic, climatic, and biotic evidences.

The crystallographic analyses indicate that the detrital zircon crystals on the Earth, dated to 4400 million years ago, show evidence of having undergone contact with liquid water, suggesting that the planet already had seas or maybe an early ocean in which the emerging organic molecules could been either of terrestrial or extraterrestrial origin. There are numerous theories, models and hypotheses concerning the origin of life on Earth, which can be grouped in two main concepts, namely cosmic ancestry, and abiogenesis.

Cosmic ancestry explains the origin of life based on *panspermia*, and shows that life already existed in the universe and can be only descend from ancestors, being delivered from space. According to this concept, higher life forms, including intelligent life, descend ultimately from pre-existing life which was at least as advanced as their still surviving viruses. Thus, organic molecules of amino acids were formed extra-terrestrially, arriving on Earth via comets, thereby the identification by spectral analyses of organic molecules in comets and meteorites, such as amino acid glycine detected in material ejected from comets, as well as traces of polycyclic aromatic hydrocarbons and fullerenes (molecules composed entirely of carbon, similar in structure to graphite) detected in nebulae.

Abiogenesis is the natural process by which life occurred from inorganic matter as early as 4200 to 4000 million years ago, whereas it was identified at the Earth's surface between 4000 and 3700 million years ago. This process is based on the fact that the earliest known Earth's life already existed between 3900 and 3500 million years ago, when sufficient crust had solidified from previous molten rocks, and on experiments demonstrating that most amino acids can be racemically (by containing equal amounts of dextrorotatory and levorotatory stereo-isomers) synthesized in conditions thought to be similar to those of the early Earth. Among the variants of abiogenesis, can be mentioned 'primordial soup' theory, and spontaneous generation of life forms from non living molecules.

The main stages in explaining the origin of life are: 1st, the origin of monomers; 2nd, the origin of biological polymers; and 3rd, the evolution from molecules to cells.

Within the water accumulated on the Earth, the dissolved substances, according to their interactions and properties, developed from atoms to molecules and from molecules to complex structures, and then from simple to complex forms of life.

It seems that the first life on Earth was represented by *non-cellular living forms* similar with some viruses, i.e. existing without cellular structure, such as the complex *Mimivirus* discovered in 2003, which could synthesize proteins independent of a host cell. These non-cellular living forms occurred about 4200 million years ago, as a separate domain of life, preceding *proto-organisms with ribosomal RNA genes* which occurred approximately 3850 million years ago as unicellular microorganisms with deoxyribonucleic acid lying freely in cytoplasm and playing an important part in synthesis of proteins. These proto-organisms with ribosomal RNA genes derived from the earliest self-reprocessing RNA molecules, reproduce by binary fission, fragmentation, budding, or spores, and include *primitive forms of archaeons, bacteria*, and *protokaryotes*. Around 3800 million years ago the *prokaryotes* appeared as usually single-celled organisms without a nucleus and membrane-bound organelles, but with circular deoxyribonucleic acid (DNA), carrying genetic information, and no proteins; they reproduce by binary fission or spores, and include *archaea*, autotrophic and filamentous *bacteria*, alongside with first *stromatolites*, assembling not only photosynthesizing cyanobacteria but also other micro-organisms. The existence of photosynthetic bacteria that produce oxygen may be deduced from indirect means, such as by analyses of massive deposits of banded iron-formations indicating the activity of oxygen-producing microorganisms, which were found in many places, including at Isua in West Greenland, with ages up to 3800-4000 million years.

Recent studies of growth ring analysis from fossil stromatolites, corals and bivalves indicate significant changes in number of days in the year, decreasing from 435 about 850 million years ago, to 375 around 370 million years ago, and to 372 about 60 million years ago, respectively. This decrease in number of days per year is due to the rate of decline in Earth's rotation, associated with tidal friction mainly caused by the Moon in its rotation around the Earth. The slight slowing of Earth's rotation about its axis had noticeable consequences not only by shortening geological eras, but also by changing ecological and biological conditions.

Subsequently, photosynthetic life enriched the initial atmosphere with oxygen, but life remained mostly microscopic and small until around 580 million years ago, when complex multi-cellular life arose, and then experienced a rapid diversification into most major phyla during the early Phanerozoic eon. Since that time the biosphere has had a significant effect on the atmosphere, formation of ozone layer, proliferation of oxygen, and creation of soil. Today, excepting non-cellular forms of life (viruses), cellular life consists of *Bacteria*, *Archaea*, and *Eukarya*, the last one including Protista, Fungi, Plantae and Animalia.

Life history theory posits that the schedule and duration of key events in an animal's lifetime are shaped by natural selection to produce the largest possible number of surviving offspring. These events, notably juvenile development, age of sexual maturity, first reproduction, number of offspring and level of parental investment, senescence and death, depend on the physical and ecological environment of the organism. Organisms have evolved a great variety of life histories, from Pacific salmon, which produce thousands of eggs at one time and then die, to human beings, which produce a few offspring over the course of decade. Examples of some major life history characteristics include: age at first reproductive event; reproductive lifespan and aging; as well as number and size of offspring.

The global sum of all ecosystems, as the zone of life on Earth, represents a closed and self-regulating system, called *biosphere*, which extends roughly from 11,000 metres below the ocean's surface to 11,000 metres in the atmosphere. In modern biology, there are five unifying principles: cells are the basic units of life; new species and inherited traits result from evolution; genes are the basic units of heredity; an organism regulates its internal environment to maintain stable and constant conditions; and living organisms consume and transform energy. In the current biological classification, there is in use a hierarchy of eight major taxonomic ranks, namely domain, kingdom, phylum, class, order, family, genus, and species, the number of identified species being of over 6.5 million on land and 2.2 million in ocean. Although the enormous number of organisms classified in these taxonomic ranks, a roughly analysis – based on fossil records, genetic lineages, dating techniques, and environmental evidences – can be displayed by a limited number of 'groups of organisms' representing major biological sequences, which are delimited by times in million years ago, as follows:

4200 \

Non-cellular living forms, similar to the present-day Mimivirus, with no cell-membrane; cells resembling prokaryotes; Proto-organisms with ribosomal RNA genes as unicellular micro-organisms with deoxyribonucleic acid lying freely in cytoplasm which reproduce by binary fission, fragmentation, budding, or spores, and include primitive forms of *archaeons, bacteria, protokaryotes*

\ 3850...2400 ≈ 3125 \

Prokaryotes as usually single-celled organisms without a nucleus and membrane-bound organelles, but with circular deoxyribonucleic acid, carrying genetic information, and no proteins; which reproduce by binary fission or spores, and include archaea, autotrophic and filamentous bacteria, as well as photosynthesizing cyanobacteria; *acritarchs*, identified as fossil structures including egg cases of small metazoans and cysts of chlorophyta (green algae)

\ 2300...2030 ≈ 2165 \

Eukaryotes as organisms with cell nucleus, surrounded by a membrane, always having cytoskeleton and linear DNA associated with proteins, which reproduce by mitosis or meiosis, and comprise fundamental plant, animal, and fungal cells; *stromatolites* built up by cyanobacteria together with large microbial communities

\ 1500...1300 ≈ 1400 \

Multicellular red algae; *dinoflagellates* as unicellular organisms, plant-like flagellates, with two flagella; *Vaucherian algae* with apical growth from filaments forming mats which form coenocytes

\ 900...800 ≈ 850 \

Protozoa as the lowest and simplest of animals, in unicellular forms or colonies multiplying by fission; *Ediacaran bacteria* as the first large, complex multicellular organisms; *fungi* occurred as multi-cellular benthic organisms, having filamentous structure with septa, as thallophytes without chlorophyll and capable of anastomosis, and performing an essential role in decomposition of organic matter and in nutrient cycling and exchange; *sponges, corals, anemones*; *arthropods* as trilobites and crustaceans; *equinoderms, molluscs, trilobites, brachiopods, foraminifers, radiolarians*; *graptolites, cephalopods, chitons* (marine molluscs with shell of movable plates); *jawless fishes* as first vertebrates with true bones

\ 490...450 ≈ 470 \

Conodonts, echinoids; *agnathan fishes, land plants*; *ray-finned fishes, arachnids, scorpions*; *lichens, stoneworts, hexapods, ammonoids, tetrapods*; *insects, sharks, crabs, ferns*; *ratfishes, hagfish*; *amphibians, reptiles*; *beetles, gymnosperms*; *dinosaurs, bivalves,* first *mammals*

\ 230...210 ≈ 220 \

Hermit crabs, starfish; *salamanders, newts*; *Archaeopterix, birds*; *cladotherian, triconodontid, symmetrodont, monotreme mammals*; *blood-sucking insects, marsupials*; *angiosperms*; *bees, ants*

\ 62...54 ≈ 58 \

Large *flightless birds, true primate*; *carnivorous* and *lipotyphlan mammals*; *owls,* diversification of *bird groups*; *whales, rodents, tapirs, rhinoceroses, camels*; *bats, butterflies, moths*; *grasses, eucalypts*; *ostracods, thylacinid marsupials*; *pigs, cats*; *deer, giraffes, hyenas, bears*; *mammoths, bovids, kangaroos*; *termites, horses, snakes*; *hippopotami, zebras, elephants, lions, vultures*; *Australopithecus, Homo habilis, Homo erectus, Homo heidelbergensis, Homo neanderthalensis, Homo sapiens*

\ Present.

Any group of organisms evolves by successive generations following one another at repeated intervals of time in accordance with its average lifespan, which is changing, sometimes drastically because it depends on factors such as environment, nutrition, standard of living, plagues, diseases, calamities, catastrophes, etc. The average lifespan equals or even overtakes the average interval of time between two successive couples of generations. Unlikely the average lifespan, the average interval of time between two successive couples of generations is quite stable, each of its halves corresponding to the average age of birth mother for animals, or birth parent for plants. In other terms, the average interval of time between two successive couples of generations represents the time required for a complete cycle. However, the average lifespan is mainly used to estimate parameters such as: the duration of a couple of generations as complete cycle of life (period) T, the number of complete cycles in unit time (frequency) $f = 1/T$, the duration of a generation (semi-cycle of life) $T/2$, and finally the number of generations $N = (t_\sigma - t)/(T/2)$

from the appearance of each group of organisms to the present t_σ (Greek σημερον 'today').

In summary, every generation of followers is more similar to that of the grandparents than to that of parents, this fact being testified by genetics. Meanwhile, despite the fluctuation of the average lifespan, which depends on environmental conditions, the natural mean duration of a generation is steady or quasi-steady, being regulated by DNA codes. Thus, the successive generations are evolving from one to another in waves propagating from a source generation to a receptor generation, so that the recessive generations repeatedly gain a half of average lifespan, revealing a 'red shift' during their succession. Such a matter-of-fact is entirely consistent with the principle of changing wave frequency, proved by the Austrian physicist Christian Johann Doppler (1842), called *Doppler's effect* of *shift*, related to a source of vibrations moving towards or from an observer, that explains the fall of pitch of a railway whistle when the engine passes, and enables astronomers to measure the radial velocity of stars by the displacement of the spectrum lines.

Although the above times delimiting the major biological sequences are ranging, more for earlier and less for later ones, this sequences can be approximately presented, in million years ago, as follows:

4200 \ (*Archaeons, bacteria, protokaryotes*) \ 3125 \ (*Prokaryotes, acritarchs*) \ 2165 \ (*Eukaryotes, stomatolites*) \ 1400 \ (*Dinoflagellates, algae*) \ 850 \ (*Protozoa, fungi, sponges, trilobites, jawless fishes*) \ 470 \ (*Fishes, amphibians, reptiles, gymnosperms*) \ 220 \ (*Birds, angiosperms, mammals, marsupials*) \ 58 \ (*Primates, anthropoids, hominids, humans*) up to the present time.

Alternatively, referring to the life origin, these limits, also in million years, become: 4200 - 4200 = 0; 4200 - 3125 = 1075; 4200 - 2165 = 2035; 4200 - 1400 = 2800; 4200 - 850 = 3350; 4200 - 470 = 3730; 4200 - 220 = 3980; 4200 - 58 = 4142 respectively. The successive ratios of these transitional times from the life origin are 0/(1075) = 0; (1075)/(2035) ≈ 0.528; (2035)/(2800) ≈ 0.727; (2800)/(3350) ≈ 0.836; (3350)/(3730) ≈ 0.898; (3730)/(3980) ≈ 0.937; (3980)/(4142) ≈ 0.961; and they roughly correspond to the successive ratios given in Background to time, Table *(j)* for $z/4 = j$ sequences, in column $t_{j-1/4}/t_j$ = $tanh(j - 1/4)/tanh(j)$, as: $t_0/t_{0.25}$ = 0; $t_{0.25}/t_{0.5}$ = 0.5299926; $t_{0.5}/t_{0.75}$ = 0.7275729; $t_{0.75}/t_1$ = 0.8339730; $t_1/t_{1.25}$ = 0.8978060; $t_{1.25}/t_{1.5}$ = 0.9371765; $t_{1.5}/t_{1.75}$ = 0.9615166 respectively. Because the time limits of biological sequences have been estimated with certainty increasing from earlier to later ones, the most recent of them would be the most accurate, so that $t_{1.75} \approx 4142$ million years from the origin of life is taken as the base for accurate evaluations of the previous transitional times, which are reconsidered, in million years from the life origin, as

$$t_{0.25} = (t_{0.25}/t_{0.5}) \cdot t_{0.5} \approx (0.5299926) \cdot (2033) \approx 1078;$$
$$t_{0.5} = (t_{0.5}/t_{0.75}) \cdot t_{0.75} \approx (0.7275729) \cdot (2795) \approx 2033;$$
$$t_{0.75} = (t_{0.75}/t_{1}) \cdot t_{1} \approx (0.8339730) \cdot (3351) \approx 2795;$$
$$t_{1} = (t_{1}/t_{1.25}) \cdot t_{1.25} \approx (0.8978060) \cdot (3732) \approx 3351;$$
$$t_{1.25} = (t_{1.25}/t_{1.5}) \cdot t_{1.5} \approx (0.9371765) \cdot (3983) \approx 3732;$$
$$t_{1.5} = (t_{1.5}/t_{1.75}) \cdot t_{1.75} \approx (0.9615166) \cdot (4142) \approx 3983.$$

Also based on the time of start of the last sequence $t_{1.75} \approx 4142$ years from life origin, the fate of life on Earth can be predicted at the final time $t_{\bullet} = t_{1.75}/tanh(1.75) \approx (4142)/(0.9413755) \approx 4400$ million years from the life origin, or $4400 - 4200 \approx 200$ million years from the present into the future. Thus, the life timeline, sequences, and intrinsic characteristics are presented below.

$z/4$ $= j$	Time (million years)		Life sequences with representatives
	from origin $t_j =$ $t_{\bullet} \cdot tanh(j)$	from present $t_j - 4200$	
	4400	+200	
...	...	...	
2	4242	+42	−
1.75	4142	-58	_Primates, anthropoids, hominids, humans
1.5	3983	-217	_Birds, angiosperms, mammals, marsupials
1.25	3732	-468	_Fishes, amphibians, reptiles, gymnosperms
1	3351	-849	_Protozoa, fungi, sponges, trilobites, jawless fish
0.75	2795	-1405	_Dinoflagellates, algae
0.5	2033	-2167	_Eukaryotes, stromatolites
0.25	1078	-3122	_Prokaryotes, acritarchs
0	0	-4200	_Archaeons, bacteria, protokaryotes

$z/4$ $= j$	Time from origin $t_j = t_{\bullet} \cdot tanh(j)$ (million years)	Period $\tau_j =$ $(t_{\bullet}^2 - t_j^2)/(2t_{\bullet})$ (million years)	Frequency $f_j =$ $(2t_{\bullet})/(t_{\bullet}^2 - t_j^2)$ (million years)$^{-1}$	Angular speed $\omega_j =$ $2\pi/\tau_j = 2\pi \cdot f_j$ (million years)$^{-1}$
	4400	0	∞	∞
...	...	...	...	...
2	4242	155.2	0.0064448	0.0404940
1.75	4142	250.4	0.0039930	0.0250890
1.5	3983	397.2	0.0025174	0.0158171
1.25	3732	617.3	0.0016200	0.0101786
1	3351	924.0	0.0010823	0.0068003
0.75	2795	1312.3	0.0007620	0.0047880
0.5	2033	1730.3	0.0005779	0.0036312
0.25	1078	2067.9	0.0004836	0.0030384
0	0	2200.0	0.0004545	0.0028560

According to the timeline above, life subsisted and evolved despite being heavily obstructed by: impacts of asteroids, meteorites and meteors; penetrations of radiation and climate changes; continental drifts, seismic shocks and volcanic eruptions; repulsions or attractions between species and migrations; as well as recently deforestations, industrial activities and environmental contaminations.

Obstruction to life was marked by major mass extinctions, from which only the last five have been identified:

443 million years ago - *first major mass extinction*, with 25% of life families lost;

359-354 million years ago - *second major mass extinction*, with 19% of life families lost;

248 million years ago - *third major mass extinction*, with 54% of life families lost, including the trilobites;

206-200 million years ago - *fourth major mass extinction*, with 23% of life families lost; and

65 million years ago - *fifth major mass extinction*, caused by an asteroid impact in Central America combined with massive volcanic eruptions in India, with 17% of life families lost, including the dinosaurs.

Such global obstructions to life reveal its miraculous capacity not only to survive and recover, but also to re-adapt, diversify, and evolve even faster than before.

6. Anthropology

The science of man in its widest sense, or the study of human origins, societies and culture is named *anthropology* (from Greek άνθρωπος 'man' and λογία 'study'). Usually, anthropology is defined as the study of humankind from past to present, which is drawn and built upon knowledge from social and biological sciences, and the humanities and the natural sciences. As one of its sub-fields, the palaeoanthropology combines the disciplines of palaeontology and physical anthropology, focusing on the study of ancient humans as found in fossil hominid evidence such as petrified bones and footprints.

Biological anthropology and physical anthropology are synonymous terms to describe anthropological research focused on the study of humans and non-human primates in their biological, evolutionary, and demographic viewpoints. It examines the biological and social factors that have affected the evolution of humans and other primates, maintain or change contemporary genetic and physiological variation.

Until the present, anthropology has diversified into numerous branches, such as social, cultural, biological, archaeological, linguistic, musical, visual, economic, applied and development, medical, nutritional, psychological, cognitive, transpersonal, political, legal, digital, ecological, environmental, historical, and urban anthropologies, as well as anthropologies of nature, science and technology, religion, art, kinship, feminism, gender and sexuality, dance and film.

The related human evolution is the process leading up to the appearance of modern humans, and usually covering only the evolutionary history of primates, in particular the genus *Homo*, and the emergence of *Homo sapiens* as a distinct species of hominids (or, great apes). The study of human evolution involves scientific disciplines such as physical anthropology, primatology, archaeology, ethology, linguistics, evolutionary psychology, embryology, and genetics.

Evolutionary history of the primates is traced back to 62...54 ≈ 58 million years ago, when the oldest known primate-like mammal species, called *Plesiadapis*, was widely spreading in North America, Eurasia and Africa during the tropical conditions of the late Palaeocene - early Eocene (Kordos and Begun, 2001), and indicating a lineage from early primates to African apes and humans, including *Dryopithecus*, by a south migration from Europe or Western Asia to Africa. The surviving tropical population of primates, which was most completely evidenced in Eocene-Oligocene fossil beds of the Fayum depression, southwest of Cairo, gave rise to all living species: *lemurs* in Madagascar; *lorises* in Southeast Asia; *galagos* 'bush-babies' in Africa; and *anthropoids* such as *platyrrhine* 'New World monkeys', *catarrhines* 'Old World monkeys', and *great apes*, including humans.

The earliest known catarrhine was *Kamoyapithecus* found in upper Oligocene at Eragaleit in the Northern Kenya Rift Valley, and dated 24 million years ago. Its ancestry is thought to be species related to *Aegyptopithecus*, *Propliopithecus*, and *Parapithecus* evidenced at Fayum as being around 35 million years old. Subsequently, *Saadanius* was described as a close relative of the last common ancestor of the *crown catarrhines* from 29-28 million years ago. Fragments of a fossil *Victoriapithecus*, the earliest Old World Monkeys, were dated 20 million years ago. The series continued with *Proconsul*, *Rangwapithecus*, *Dendropithecus*, *Limnopithecus*, *Nacholapithecus*, *Equatorius*, *Nyanzapithecus*, *Afropithecus*, *Heliopithecus*, and *Kenyapithecus*, all in East Africa, which seem to be around 13 million years old.

The youngest of the Miocene hominoids, *Oreopithecus*, was discovered in Italian coal beds and dated 9 million years ago. The lineage of *gibbons* diverged from Great Apes between 18 and 12 million years ago; and *orangutans* diverged from the other Great Apes around 12 million years ago. Fossil proto-orangutan *Sivapithecus* (India) and *Griphopithecus* (Turkey) were dated about 10 million years ago. Between 8 and 4 million years ago, first *gorillas* and then *chimpanzees* split off from the line leading to humans, so that 98.4% of human DNA is the same as that of chimpanzees. The evolutionary series of hominoids continued with *Sahelanthropus tchadensis*, 7; and *Orrorin tugenensis*, 6 million years ago.

On the hominids' line, human ancestry probably started at the time of *Ardipithecus kadabba*, between 5.40...5.38 ≈ 5.39 and 5.20...5.18 ≈ 5.19 million years ago, i.e. 5.39...5.19 ≈ 5.29 million years ago, as the initial time of human ascendancy. This ascendancy continued at the times, in million years ago, of *Ardipithecus ramidus*, 5.0 - 4.2; *Australopithecus*, 4.0 - 1.8; *Kenyanthropus (platyops)* 3.0 - 2.7; *Paranthropus*, 3.0 - 1.2; *Homo habilis*, 2.4 - 1.4; *Homo rudolfensis* (Kenya) and *Homo georgicus* (Georgia), 1.9 - 1.6; *Homo erectus*, 1.8 - 0.07; *Homo ergaster*, 1.8 - 1.25; *Homo antecessor*, 1.2 - 0.5; *Homo cepranensis*, 0.8; *Homo heidelbergensis*, 0.8 - 0.3; *Homo rhodesiensis*, 0.45 - 0.125; *Homo neanderthalensis*, 0.4 - 0.03; *Homo sapiens* (modern human), 0.2 - present. Notably, other representatives of Homo species were discovered and dated, in million years ago, as *Homo sapiens idaltu* (Ethiopia), 0.16; *Homo floresiensis* (Indonesia), 0.1 - 0.012; and *Denisova hominin* (Siberia), 0.041.

Local populations of Homo neanderthalensis and Homo floresiensis became extinct because either changing environmental conditions, or interbreeding with Homo sapiens, so that nearly all modern non-African humans have between 1% and 4% of their DNA derived from Neanderthals or Floresiens DNA, and up to 6% of their genome common with Denisova hominid.

The attempt to explain hominid evolution is mainly based on the *encephalization*, as the amount of brain mass in relation to an animal's total body mass, and is calculated by the encephalization quotient

$$E = C \cdot S^r,$$

where E is the brain weight, C is the cephalization factor, S is the body weight, and r is the exponential constant considered 0.28 for primates and either 0.56 or 0.66 for mammals in general. This quotient has values of 2.09 for Rhesus monkey, 2.49 for chimpanzee, 5.31 for dolphin, and 7.44 for human.

Human species developed a much larger brain than that of other primates - typically 1330 cm^3 in modern humans, over twice the size of that of a chimpanzee or gorilla. The increase in volume over time has affected areas within the brain unequally - the *temporal lobes*, which contain centres for language processing, have increased disproportionately, as has the *prefrontal cortex*, which has been related to complex decision-making and moderating social behaviour. Encephalization has been tied to an increasing emphasis on meat in the diet, or with the development of cooking, and it has been proposed that intelligence increased as a response to an increased necessity for solving social problems as human society became more complex.

Usually, hominids are characterized by their cranial capacity expressed in cubic centimetres, with average values such as: *Australopithecus afarensis*, 438 cm^3; *Australopithecus africanus*, 452 cm^3; *Paranthropus boisei*, 521 cm^3; *Paranthropus robustus*, 530 cm^3; *Homo habilis*, 612 cm^3; *Homo rudolfensis*, 700 cm^3; *Homo ergaster*, 871 cm^3. Notably, the cranial capacity of modern humans (1300 - 1500 cm^3) is less than that of Neanderthals (1500 - 1800 cm^3), thus contradicting the concept of human evolution. This inconvenience can be overpassed by re-conceptualizing the assessment of brain energy E [$J = kg \cdot m^2 \cdot s^{-2} = N \cdot m = V \cdot C = W \cdot s$] as a product of its *extensive quantity Q* [m^3] and its *intensive quality q* [$J \cdot m^{-3}$], where, as SI units, J

= joule, kg = kilogram, m = metre, s = second, N = newton, V = volt, C = coulomb, and W = watt. The extensive quantity Q simply symbolizes brain volume, while the intensive quality q complexly symbolizes brain efficacy related to cerebral potential of processing and storing information, imagining, intuiting, problem-solving, and so on. Therefore, Homo neanderthalensis' brain energy $E_n = Q_n \cdot q_n$ must be less than Homo sapiens's brain energy $E_s = Q_s \cdot q_s$, i.e.

$$Q_n \cdot q_n < Q_s \cdot q_s, \text{ or } q_n/q_s < Q_s/Q_n,$$

where $Q_n = 1500...1800$ cm$^3 \approx 1650$ cm$^3 = 1.65 \cdot 10^{-3}$m^3, and $Q_s = 1300...1500$ cm$^3 \approx 1400$ cm$^3 = 1.40 \cdot 10^{-3}$m^3, so that

$$q_n/q_s < Q_s/Q_n = (1.40 \cdot 10^{-3}\text{m}^3)/(1.65 \cdot 10^{-3}\text{m}^3) = 0.8485; \text{ or}$$
$$q_n < (0.8485) \cdot q_s.$$

Such an approach enables a deeper understanding of the evolutionary process for primates in general, and for hominids in particular.

The human evolution described above was unveiled, discovered, and reconstituted by travellers, naturalists, physicians, anthropologists, palaeoanthropologists, anatomists, archaeologists, biochemists, palaeontologists, and primatologists, including those mentioned below.

1271-95: Marco Polo, Venetian merchant and traveller, crossed Central Asia and Gobi Desert to China, where served as envoy to Yunnan, northern Burma, Karakorum, Cochin-China, and southern India, and as Governor of Yang Chow helping to subdue the city of Saianfu; then he returned to Venice, and was prisoner at Genoa, where he wrote *Divisament dou Monde* 'Description of the World', an important account of his systematic observations on nature, *anthropology*, and geography of the visited regions, which was presumably recounted to Rustichello da Pisa by Polo when they were both in the Genoese prison, and later published; the readers of this work named him 'the father of modern anthropology';

1735: Carolus Linnaeus, Swedish naturalist and physician, developed the system of botanical nomenclature that was included in his published work *Systema Naturae*, and first used the name *Homo* of the biological genus to which humans belong (from Latin *homo* 'human being') and coined the species binomial *Homo sapiens*;

1798: Immanuel Kant, German philosopher, based on his lectures for students, published the first major treatise on anthropology, which was entitled *Kants Anthropologie aus pragmatischer Sicht* 'Anthropology from a Pragmatic Point of View', and revealed not only his unique contribution to the newly emerging discipline, but also his desire to offer a *practical view of the world and of humanity's place in it*;

1871-73: <u>Charles Robert Darwin</u>, English naturalist and originator of the theory of evolution by natural selection, wrote and published treatises including *The Descent of Man and Selection in relation to Sex*, and *The Expression of the Emotions in Man and Animals*;

1911-40: <u>Franz Boas</u>, US anthropologist, published *The Mind of Primitive Man, Anthropology and Modern Life*, and *Race, Language and Culture*, outlining new and less simple concepts of *culture* and of *race*, as well as arguing *against racial ideology*;

1925: <u>Raymond Arthur Dart</u>, South African anatomist, worked on anthropology and described an ape-like infant part-skull found in Botswana which he considered to be a human ancestor, *Australopithecus africanus*;

1934-50: <u>Louis Seymour Bazett Leakey</u>, British archaeologist, and his wife <u>Mary Leakey</u>, British palaeoanthropologist, wrote the book *Adam's Ancestors: The Evolution of Man and His Culture*; amassed fossils of australopithecines, including the extinct ape *Proconsul*, at Olduvai and Lake Turkana; then extended their research in the Kenyan Rift Valley, and published another book entitled *Excavations at Njoro River Cave*;

1967: <u>Vincent Sarich</u>, US anthropologist, and <u>Allan Wilson</u>, US biochemist, measured the strength of immunological *cross-reactions of blood serum albumin* between pairs of creatures, including humans and African apes, and estimated the divergence time of humans and apes up to 5 million years ago; they published *Immunological time scale for hominid evolution*, Science, 158;

1974-81: <u>Donald Carl Johanson</u>, <u>Maurice Taieb</u> and <u>Yves Coppens</u>, US palaeontologists, discovered and studied the 3.2 million-year-old fossil of a female hominid *Australopithecus afarensis*, known as 'Lucy', that was found in the Afar Triangle region of Hadar, Ethiopia, and later popularized in the book *Lucy: The Beginning of Humankind*, New York: Simon and Schuster;

1981-94: <u>Richard Leakey</u>, Kenyan palaeoanthropologist, wrote and published a series of books such as *The Making of Mankind*, Michael Joseph, London; *Origins Reconsidered*, Little, Brown & Co., London; and *The Origin of Humankind*, Weidenfeld & Nicolson, London;

1992-1995: <u>Tim White</u>, US anthropologist, researching for primate fossils in Africa, discovered *Ardipithecus ramidus* in Ethiopia, dated

4.45-4.35 million years ago, which is considered one of the *oldest human ancestors*;

1993-98: <u>Ian Tattersall</u>, British-born US palaeoanthropologist, published a series of his books including *The Human Odyssey*, Prentice Hall, New York; *The Fossil Trail*, and *Becoming Human*, Oxford University Press;

1994-98: <u>Meave Leakey</u>, London-born Kenyan palaeoanthropologist, and <u>Allan Walker</u>, US anthropologist, discovered *Australopithecus anamensis*, which was made known by their published works including *New four-million-year-old hominid species from Kanapoi and Allia Bay, Kenya*, Nature 376; *Early Hominid Fossils from Africa*, Scientific American, 276 (6); and *New specimens and confirmation of an early ape for Australopithecus anamensis*, Nature, 393;

1996: <u>Robin McKie</u>, English Science Editor of the Observer, and <u>Chris Stringer</u>, English palaeontologist, attentively looking at new discoveries in human history, wrote the book entitled *African Exodus: The Origins of Modern Humanity*;

1996-2013: <u>Virginia Morell</u>, US natural historian and science-writer, published a series of books including *Ancestral Passions: The Leakey Family and the Quest for Humankind's Beginnings*, Touchstone, New York; *Animal Wise: The Thoughts and Emotions of Our Fellow Creatures*, Kindle Edition; and *Blue Nile: Ethiopia's River of Magic and Mystery*, Aventure Press;

1997-2004: <u>Tim White</u>, US anthropologist, together with <u>Gen Suwa</u> and <u>Yohannes Haile-Selassie</u>, African anthropologists, discovered and studied the first known bipedal called *Ardipithecus kadabba*, in Middle Awash Valley, Ethiopia, which was dated from 5.6-5.4 to 5.2 million years ago, as the earliest human ancestor;

2000: <u>Walter Bodmer</u>, English geneticist, and <u>Robin McKie</u>, English Science Editor of the Observer, wrote *The Book of Man: The Quest To Discover Our Genetic Heritage*, Orion Audio Books; and <u>Robin McKie</u>, English Science Editor of the Observer, published his book *Ape●Man: The Story of Human Evolution*, BBC Worldwide Ltd.;

2000-01: <u>Brigitte Senut</u> and <u>Martin Pickford</u>, French palaeoanthropologists, discovered so called *Orrorin tugenensis* in Tugen Hills of Kenya, dated 6.1 - 5.7 million years ago, which may been similar to our early ancestors; and published *First hominid from the Miocene (Lukeino Formation, Kenya)*, Comptes Rendus de l'Académie de Sciences, 332 (2); and <u>Michael Brunet</u>, French palaeontologist, excavated in Chad where he found a 7.2 million year old skull of so-named *Sahelanthropus tchadensis*, considered by him to be a hominin, but then argued to be in the human evolutionary line;

2001: László Kordos, Hungarian geologist, and David Begun, Canadian anthropologist, published their works *Primates from Rudabánya: allocation of specimens to individuals, sex and age categories*, and *A new cranium of Dryopithecus from Rudabánya, Hungary*, Journal of Human Evolution, 40 (1) and 41 (6) respectively;

2005: Frans de Waal, Dutch-born US primatologist, investigated genetic characteristics of chimpanzees and bonobos, and published his results in *Our Inner Ape: the Past and Future of Human Nature*;

2008-10: Michael Tomasello and Esther Herrmann, German evolutionary anthropologists, made comparative studies of apes and humans, and published their work *Ape and Human Cognition: What's the Difference?*, Current Directions in Psychological Science, 19;

2009: Henry Malcom McHenry, US anthropologist, researched human ancestry and wrote the article 'Human Evolution', that was inserted in the book *Evolution: The First Four Billion Years*, Michael Ruse & Joseph Travis, eds.;

2013: *Skull 5* - a 1.8 million years old skull, was found in Dmanisi, Georgia, and considered to be from one of the human ancestors; the fossil combines a long face, massive jaw and large teeth with a small braincase, being analyzed by Swiss researchers, led by the neurobiologist Christoph Zollikofer, and popularized by Live Science contributor Charles Choi who published *Oddball 'Skull 5' Fossil Suggests Early Humans Belong to Same Species*.

Although the data concerning human ancestry are still not yet unanimously agreed, maybe incomplete, and not enough accurately established, some of them seem to be acceptable for use in an attempt to display the humans' timeline and sequences: 1st, the origin of our ancestors can be reasonably dated 5.39...5.19 $\approx$ 5.29 million years ago, as the initial time of human ascendancy; 2nd, the more accurately dated transitions from one sequence to another, in million years ago, seem to be:

Homo antecessor \ 0.88...0.80 $\approx$ 0.84 \ *Homo heidelbergensis* \ 0.64...0.60 $\approx$ 0.62 \ *Homo rhodesiensis* \ 0.46...0.44 $\approx$ 0.45 \ *Homo neanderthalensis* \ 0.32...0.30 $\approx$ 0.31 \ *Homo sapiens* 0.20;

corresponding to 5.29 - 0.84 = 4.45; 5.29 - 0.62 = 4.67; 5.29 - 0.45 = 4.84; 5.29 - 0.31 = 4.98; 5.29 - 0.20 = 5.09 million years from the origin of human ascendancy. Thus, the successive ratios of the last transitional times are $4.45/4.67 \approx 0.953$; $4.67/4.84 \approx 0.965$; $4.84/4.98 \approx 0.972$; $4.98/5.09 \approx 0.978$. Similar values of these ratios can be identified in Background to time, Table *(k)* for $z/8 = k$ sequences, in the following succession: $t_{1.125}/t_{1.25} = 0.9540454$; $t_{1.25}/t_{1.375} = 0.9641485$; $t_{1.375}/t_{1.5} = 0.9720249$; $t_{1.5}/t_{1.625} = 0.9781726$. Taking into account that the emergence of Homo sapiens is most accurately dated as 0.2 million years ago, or $t_{1.625} = 5.29 - 0.20 = 5.09$ years from the origin of human ascendancy, and this time is $t_{1.625} = t_{\bullet} \cdot tanh(1.625) = 5.09$, it follows that the final time can be evaluated as $t_{\bullet} = (5.09)/tanh(1.625) = (5.09)/(0.9253462) = 5.50$ million years from the origin of human ascendancy, or $5.50 - 5.29 = 0.21$ million years from present into the future. Therefore, the above times of transition from one sequence to another can be calculated, in million years from the origin of human ascendancy, as

$$t_{1.125} = t_{\bullet} \cdot tanh(1.125) \approx (5.50) \cdot (0.8093011) \approx 4.45;$$
$$t_{1.25} = t_{\bullet} \cdot tanh(1.25) \approx (5.50) \cdot (0.8482836) \approx 4.67;$$
$$t_{1.375} = t_{\bullet} \cdot tanh(1.375) \approx (5.50) \cdot (0.8798267) \approx 4.84;$$
$$t_{1.5} = t_{\bullet} \cdot tanh(1.5) \approx (5.50) \cdot (0.9051483) \approx 4.98.$$

Continuing to apply the formula $t_k = t_{\bullet} \cdot tanh(k)$ for the other values of argument, the anthropological timeline, sequences, and intrinsic characteristics can be modelled as shown in the tables below.

$z/8 = k$	Time (million years)		Anthropological sequences
	from origin $t_k = t_{\bullet} \cdot tanh(k)$	from present $t_\sigma - 5.29$	
	5.50	+0.21	
...	...	...	
1.75	5.18	-0.11	
1.625	5.09	-0.20	_ Homo sapiens
1.5	4.98	-0.31	_ Homo neanderthalensis
1.375	4.84	-0.45	_ Homo rhodesiensis
1.25	4.67	-0.62	_ Homo heidelbergensis
1.125	4.45	-0.84	_ Homo antecessor
1	4.19	-1.10	_ Homo erectus
0.875	3.87	-1.42	_ Homo ergaster
0.75	3.49	-1.80	_ Homo rudolfensis and georgicus
0.625	3.05	-2.24	_ Homo habilis
0.5	2.54	-2.75	_ Australopithecus garhi
0.375	1.97	-3.32	_ Australopithecus africanus
0.25	1.35	-3.94	_ Australopithecus anamensis and afarensis
0.125	0.68	-4.61	_ Ardipithecus ramidus
0	0	-5.29	_ Ardipithecus kadabba

$z/8 = k$	Time from origin $t_k = t_\bullet \cdot tanh(k)$ (million years)	Period $\tau_k =$ $(t_\bullet{}^2 - t_k{}^2)/(2t_\bullet)$ (million years)	Frequency $f_k =$ $(2t_\bullet)/(t_\bullet{}^2 - t_k{}^2)$ (million years)$^{-1}$	Angular speed $\omega_k =$ $2\pi/\tau_k = 2\pi \cdot f_k$ (million years)$^{-1}$
	5.50	0	∞	∞
...	...	...	...	...
1.75	5.18	0.3106909	3.2186330	20.223267
1.625	5.09	0.3947182	2.5334531	15.918155
1.5	4.98	0.4954182	2.0184968	12.682589
1.375	4.84	0.6204000	1.6118633	10.127636
1.25	4.67	0.7673727	1.3031477	8.1879184
1.125	4.45	0.9497727	1.0528835	6.6154619
1	4.19	1.1539909	0.8665580	5.4447442
0.875	3.87	1.3884636	0.7202205	4.5252790
0.75	3.49	1.6427182	0.6087471	3.8248711
0.625	3.05	1.9043182	0.5251223	3.2994409
0.5	2.54	2.1634909	0.4622159	2.9041884
0.375	1.97	2.3971909	0.4171549	2.6210617
0.25	1.35	2.5843182	0.3869493	2.4312739
0.125	0.68	2.7079636	0.3692812	2.3202621
0	0	2.7500000	0.3636363	2.2847947

7. Humanity

Analytic, critical, or speculative methods for understanding the development of human history and abilities led to foundation of the academic study of human condition defining *humanity* (French *humanité* 'humanity, mankind' coming from Latin *humanitas* 'humanity, human nature'). Besides this meaning, it is also understood as human being collectively, or the total world population; strengths focused on tending and befriending others; total experience of being human; or psychological characteristics common for humans in general.

The humans are primates of the Hominidae family, and the only extant species of the *Homo* genus, who are characterized by having a large brain relative to body size, with a particularly well developed neocortex, prefrontal cortex, and temporal lobes, making them capable of abstract reasoning, language, introspection, problem solving, and culture through social learning. This mental capability combined with the adaptation to bipedal locomotion, that frees the hands for manipulating objects, have allowed humans to make far greater use of tools than any other species.

Humans are the only extant species known to build fires and cook their food, to clothe, create and use numerous technologies and arts; to utilize systems of symbolic communication such as language and art for self-expression; to exchange ideas and organize; as well as to create complex social structures composed of many cooperating and competing groups, from families and kinship networks to states.

Social interactions between humans have established an extremely wide variety of values, social norms and rituals, altogether constituting the basis of human society. Distinctively, humans are noted for their desire to understand and influence the environment, and to explain and manipulate phenomena through science, technology, philosophy, mythology, religion, etc.

The advance of scientific and medical understanding in the last centuries resulted in development of fuel-driven technologies and improved health, causing an exponential increase of human population. Meanwhile, humans have had a dramatic effect on the environment, exemplified by the extinction of a number of species, as human predation and habitat loss, and other negative impacts including pollution, widespread loss of wetlands and other ecosystems, alteration of rivers, and introduction of invasive species.

Originating in Classical Greece and Persia, and studying or trying to understand observable cultural diversity, *human history* has been central in the development of several late 20th century interdisciplinary fields such as cognitive science, global studies, and various ethnic studies. Meanwhile, related anthropological studies were divided into four main fields of approach: biological or physical anthropology, social or cultural anthropology, archaeology, and anthropological linguistics.

Human prehistory is divided into three main ages, variously delimited around the world, but roughly dated as:

Palaeolithic (from Greek παλαιός 'old' and λίθος 'stone') lasting from around 2600000 to 20000 years ago; with subdivisions *Lower Palaeolithic* from 2600000 to 110000-100000 years ago, *Middle Palaeolithic* from 110000-100000 to 40000-30000 years ago, and *Upper Palaeolithic* from 40000-30000 to around 20000 years ago;

Mesolithic (from Greek μεσος 'middle' and λίθος) recorded between around 20000 and 12000-10000 years ago; and

Neolithic (from Greek νεος 'new' and λίθος) lasting from 12000-10000 to 8000-7000 years ago.

The subsequent ages, in years ago, include:
Copper Age (Chalcolithic) from 8000-7000 to 5100-5000; *Bronze Age* from 5100-5000 to about 3000; *Iron Age* from about 3000 to 1500-1300; *Medieval* from 1500-1300 to 150-130 years ago; and *Modern* from 150-130 years ago to present time.

The evolution of humans with their characteristics, capabilities, societies, culture, and ascendancy were studied by a series of historians, anthropologists, ethnologists, psychologists, ethologists, sociologists, explorers, physicians, scholars, philosophers, diplomats, clergymen and journalists, only some of them being mentioned below, in chronological order, with their main contributions.

570-550BC: <u>Anaximander</u>, pre-Socratic Greek philosopher, believed that *humankind had to adapt to environmental conditions*, and put forward the idea that humans had to spend part of their transition 'inside the mouths of big fish' to protect themselves from the Earth's climate until they could come out in open air and lose their 'scales'; he thought that, considering humans' extended infancy, we could not have survived in the primeval world in the same manner we do presently;

443-431BC: <u>Herodotus</u>, Greek historian, widely travelling to Thrace, Persia, Tyre, Egypt, Cyrene, Sicily and Lower Italy, accounted several *ancient Celtic peoples*, and collected historical, *ethnological*, mythological and *archaeological material* for his great *Histories*;

AD90-110: <u>Gaius Cornelius Tacitus</u>, Roman historian, travelled through Germania, where he acquired and accounted several *ancient Germanic peoples*, which were described and characterized in his 12-volume *Historiae* 'Histories';

1017-30: <u>Abu al-Rayhan al-Biruni</u>, Persian-Muslim scholar and polymath, travelled to the Indian subcontinent, recording details on peoples, customs, and religions there; he wrote *Kitab ta'rikh al-Hind* 'The book of the History of India', becoming the 'founder of Indology', and the 'first anthropologist';

1246-47: <u>John of Plano Caprini</u>, Italian clergyman and scholar, as a papal legate, travelled via Kiev, across Dnieper, Don, Volga, and Ural rivers, Caspian Sea, Aral lake, Syr Darya river, and reached the imperial camp called *Sira Orda* 'Yellow Pavilion', where he met the Great Khan of the Mongol Empire, and visited the central area of that empire, collecting precious information about the Mongols and their non-European culture; then he went back to Lyon to meet the Pope, and wrote the remarkable book *Ystoria Mongalorum*;

1735-49: <u>Carolus Linnaeus</u>, Swedish naturalist and physician, published his system of botanical nomenclature in *Systema Naturae*, and introduced binomial nomenclature, giving each organism a Latin generic name with a specific adjective; including the species binomial *Homo sapiens*;

1783: <u>Adam František Kollár</u>, Slovak historian and ethnologist, was credited with coining the term *ethnology* (from Greek *ἔθνος* 'nation'), for which provided its first definition, which was inserted in the two-volume *Historiae Jurisque Publici Regni Ungariae Amoenitates*, published by Typis a Bavmeisterignis, Vindobonae (Vienna); later the ethnology was defined as the branch of anthropology that compares and analyses the characteristics of different peoples and the relationship between them;

1809-22: <u>Jean-Baptiste Lamarck</u>, French naturalist and evolutionist, published his famous two-volume *Philosophie zoologique* 'Zoological Philosophy' in which he postulated that *acquired characters can be inherited by later generations*; and broke with the old notion of immutable species, recognizing that *species needed to adapt to survive environmental changes*, and preparing the way for the now accepted theory of evolution, also applicable to human history;

1819-36: Christian Jörgensen Thomsen, Danish archaeologist and numismatist, on the basis of material used in making weapons and tools, he classified the specimens into three groups representing chronologically successive ages of *Stone*, *Bronze*, and *Iron*, describing them in print in his *Ledetraad til Nordisk Oldkyndighed* 'A Guide to Northern Antiquities';

1843-52: Jens Jacob Asmussen Worsaae, Danish antiquary and archaeologist, published his major works *Danmarks Oldtid* 'The Primeval Antiquities of Denmark', and *The Danes and Norwegians in England*, confirming and developing the Three Age System (Stone, Bronze, and Iron);

1853-55: Joseph Arthur de Gobineau, French Orientalist and diplomat, was author of the essay *The Inequality of Human Races*, formulating 'scientific' *racism*, and being the real inventor of 'superman' and super-morality;

1863-93: Thomas Henry Huxley, English biologist, wrote *Evidence as to Man's Place in Nature, Science and Culture*, and *Evolution and Ethics*;

1865-70: John Lubbock, English politician and biologist, kept the earlier and later subdivisions of the Age System in his works *Prehistoric Times*, introducing the terms *Palaeolithic* 'Old Stone Age' and *Neolithic* 'New Stone Age', as published in *Origin of Civilization*;

1865-81: Edward Burnett Tylor, English anthropologist, wrote a comprehensive work entitled *Researches into the Early History of Mankind and the Development of Civilization*; and then the monumental two-volume *Primitive Culture*, showing that human culture is governed by definite laws of evolutionary development, so that the beliefs and practices of primitive nations may be taken to represent earlier stages in the progress of mankind; he also published and a general introductory called *Anthropology*, suggesting that cultural variation may be due to *racial differences* in mental endowment;

1866-72: Hodder Westropp, Irish archaeologist, studying the human prehistory, he introduced the *Mesolithic* as a technology intermediate between Palaeolithic and Neolithic, and published his book *Prehistoric Phases*;

1890-1915: <u>James George Frazer</u>, Scottish social anthropologist, classicist and folklorist, wrote his major work *The Golden Bough: A Study in Comparative Religion*, from which derived most of the material for comparative studies through extensive reading, not fieldwork; he was considered an antecedent to modern *social anthropology* in Britain;

1894-97: <u>Émile Durkheim</u>, French sociologist, published his methodological writings such as *Les Règles de la méthode sociologique* 'The Rules of Sociological Method', based on 'social facts' which should be treated as 'things' for explaining solely by reference to other social facts, not in terms of any individual person's actions, and greatly influenced *social anthropology*;

1911-40: <u>Franz Boas</u>, US anthropologist, compiled and published *The Mind of Primitive Man*; *Anthropology and Modern Life*; and also *Race, Language and Culture*; in which he outlined new and less simple concepts of *culture* and of *race*, as well as arguing *against racial ideology*;

1921-24: <u>Knud Johan Victor Rasmussen</u>, Danish explorer and ethnologist, in support of the theory that *Inuits* and *North American Indians* both descend from people migrated from Asia, he cross-examined Greenland and Bering Strait, contacting these populations and collecting valuable *ethnological information*;

1922-44: <u>Bronisław Malinowski</u>, Polish-born British anthropologist, was a founder of *modern social anthropology*, publishing *Argonauts of the Western Pacific*; *Crime and Custom in Savage Society*; *Sex and Repression in Savage Society*; two-volume *Coral Gardens and their Magic*; and *A Scientific Theory of Culture*; he pioneered the fieldwork method and proposed the *functionalism in anthropology*;

1922-52: <u>Alfred Reginald Radcliffe-Brown</u>, English social anthropologist, after carrying out field research in the Andaman Islands and Australia, wrote *The Andaman Islanders*, and *The Social Organization of Australian Tribes*; regarded *social anthropology* as the comparative study of 'primitive' societies, whose aim was to establish generalizations about the forms and functioning of social structures, as well as distinguished social anthropology sharply from ethnology that is rather descriptive than a theoretical enterprise; finally he published *Structure and Function in Primitive Society*, containing all the essentials of his theoretical programme;

1925: <u>Marcel Mauss</u>, French sociologist and anthropologist, studied the habits of various communities, and produced the well-known work

Essai sur le don, translated in English as 'The Gift', in which he demonstrated the importance of gift exchange in *primitive social organization*;

1925-47: <u>Vere Gordon Childe</u>, Australian archaeologist, wrote books such as *The Dawn of European Civilization*; *The Most Ancient Near East*; *The Danube in Prehistory*; and most importantly *The Dawn of Europe*, where he affirmed the *Mesolithic*, sufficient data had been collected to determine that it was in fact necessary and was indeed a transition, and intermediary between the previous accepted Palaeolithic and Neolithic;

1928-51: <u>Margaret Mead</u>, US anthropologist, wrote *Coming of Age in Samoa*; *Growing up in New Guinea*; *Male and Female*; and *Growth and Culture*; arguing that personality characteristics, especially as they differ between men and woman, are shaped by cultural conditioning rather than heredity, and advocating for *gender equality* and *sexual liberation*;

1949-72: <u>Claude Lévi-Strauss</u>, French social anthropologist and philosopher, published his remarkable works *Les Structures élémentaires de la parenté* 'The Elementary Structures of Kinship', establishing a new approach for analyzing various collective phenomena such as kinship, ritual and myth; *Anthropologie structurale* 'Structural Anthropology', contributing to the philosophy of structuralism; and the extensive four-volume study *Mythologiques* 'Mythologics', revealing the systematic order behind codes of expression in different cultures, and arguing that myths are not 'justifications' but are, instead, attempts to overcome 'contradictions', thus founding *structural anthropology*;

1949-83: <u>Meyer Fortes</u>, British social anthropologist, carried out fieldwork in west Africa among the Tallensi and Ashanti peoples, and his results were published in the monographs *The Web of Kinship among Tallensi*, and *Time and Social Structure*, as foundations of the *theory of descent*; also *Kinship and the Social Order*, as a complete theoretical statement; and *Rules and the Emergence of Society*, as a critique of socio-biological theories of *human kinship* and *social organization*;

1960-83: <u>Clifford James Geertz</u>, US cultural anthropologist, was author of *The Religion of Java*; *Agricultural Involution*; *Person, Time and Conflict in Bali*; *Islam Observed*; *The Interpretation of Cultures*; and *Local Knowledge*; stating that 'anthropology is perhaps the last of the great nineteenth-century conglomerate disciplines still for the most part organizationally intact';

1969-80: <u>Ulf Hannerz</u>, Swedish social anthropologist, studied urban anthropology, and published books such as *Soulside: Inquiries into Ghetto Culture and Community*, and *Exploring the City: Inquiries toward an Urban Anthropology*;

1978: <u>Louise Audino Tilly</u> and <u>Joan Wallach Scott</u>, US anthropological historians, wrote an influential pioneering study entitled *Women, Work, and Family*, New York: Holt, Rinehart & Winston;

1984: <u>Edwin Fleishman</u> and <u>Laurie Broedling</u>, US psychologists, published their work *Taxonomies of human performance: the description of human tasks*, Academic Press;

1985: <u>Donna Haraway</u>, US anthropologist, wrote the *Cyborg Manifesto*, founding the so-called *cyborg anthropology*, exploring the philosophical and sociological ramifications of this kind of anthropology;

1987-91: <u>Isac Chiva</u>, French anthropological ethnologist, decided to break away from the older anthropology of France focused on a folkloric past rather than on existing contemporary society, and published works such as *Entre livre et musée - Emergence d'une ethnologie de la France*, and *Les revues ethnologiques en Europe: richesses et paradoxes*;

1992: <u>Edwin Fleishman</u> and <u>Maureen Reilly</u>, US psychologists, studied and systematized information from various people, and published their results in a comprehensive work entitled *Handbook of human abilities: definitions, measurements, and job task requirements*, Consulting Psychologists Press; and <u>Aaron Yakovlevich Gurevich</u>, Russian medievalist historian and a pioneer of historical anthropology, wrote *Historical Anthropology of the Middle Ages*, University of Chicago Press;

2006: <u>Joseph Jordania</u>, Australian-Georgian ethnomusicologist and evolutionary musicologist, published his book entitled *Who Asked the First Question? The Origins of Human Choral Singing, Intelligence, Language and Speech*, Logos;

2007: <u>Richard Haier</u> and <u>Rex Jung</u>, US psychologists, argued that human intelligence arises from a distributed and integrated neural network comprising brain regions in the frontal and parietal lobes, and published their work *The Parieto-Frontal Integration Theory (P-FIT) of Intelligence: Converging neuroimaging evidence*, Cambridge University Press;

2010: <u>Jon Cohen</u>, US science journalist, investigated characteristic features of great apes and humans, and wrote the book *Almost Chimpanzee: Searching for What Makes Us Human, in Rainforests, Labs, Sanctuaries, and Zoos*, Macmillan/Time Books.

Anthropological researches revealed a 195000 years old fossil form of 'Omo 1' site in Ethiopia, so that the origin of *Homo sapiens* would be indeed about 200000 years ago. Not too far, there were discovered a 160000 years old skull of 'Homo sapiens idaltu' at Herto site in Middle Awash area, also in Ethiopia; 120000 years old human fossils at Laetoli in Tanzania; and 115000 years old fossils in South Africa.

Some groups of those humans moved to North Africa, and from there, through the Isthmus of Suez to the Near East in west Asia approximately 115000...97000 ≈ 106000 years ago, according to dating by the electron spin resonance method applied for a layer with human remains at Qafzeh, and by the thermo-luminescence method applied for skeletons found at Skuhul, both places in Israel.

Subsequently, humans spread to the Middle East, South Asia, South-east Asia, the Far East; and then, during the last glacial maximum, when sea level was probably more than 150 metres below than today, they crossed by sea on a distance of more than 90 kilometres between Sunda-Schelf (South of the Indonesian Islands) and Sahul Schelf (North Australia), arriving in Australia between 55000 and 40000 years ago, as shown by archaeological evidences found in the Arnhem Land the Malakunanja II rock shelter, and in the upper Swan River, Western Australia, respectively.

Coming also from Near East, groups of humans travelled, across the temporarily lowered sea level at the Straits of Bosporus and Dardanelles, to Europe, where they first arrived at least 40000 years ago, and inhabited places such as Pestera cu Oase 'The Cave with Bones' in present-day Romania from about 36000-34000 years ago, and Combe Capelle in present-day France from about 35000-30000 years ago.

Other human groups populated Asia up to its north-east extremity, and, between 40000 and 16500 years ago, they arrived in Beringia (Eastern Alaska), from where, during the last glacial period around 16000-13000 years ago, penetrated into North America.

Finally, humans colonized the Pacific islands around 4500BC; Fiji, Tonga, and Samoa about 1300BC; and then, between AD300 and

1280, settled in remote islands such as Hawaii, Easter Island, Madagascar, and New Zeeland.

Studies of molecular biology and anatomical adaptations, evidence from the fossil record, as well as research on genetic development and behavioural features all indicate that *Homo sapiens* originated in Africa about 200000 years ago, as the initial time of humanity, experiencing life in the Lower, Middle and Upper Palaeolithic, then in Mesolithic, Neolithic, Copper Age, Bronze Age, Iron Age, and later Medieval and Modern times.

According to evidences from Sub-Saharan Africa, the Near East, Western Europe, the Middle East and the Far East, humanity's timeline can be divided into sequences of durations less accurately dated for earlier than for later ones, so that it is convenient to operate with the latest which are roughly delimited as:

$$Copper\ Age \setminus 3100...2950BC \approx 3025BC \setminus$$
$$Bronze\ Age \setminus 1050...950BC \approx 1000BC \setminus$$
$$Iron\ Age \setminus AD580...660 \approx AD620 \setminus$$
$$Medieval \setminus AD1868...1880 \approx AD1874 \setminus Modern;$$

i.e. 2014 + 3025 = 5039; 2014 - 1000 = 3014; 2014 - 620 = 1394; 2014 - 1874 = 140 years ago; or 200000 - 5039 = 194961; 200000 - 3014 = 196986; 200000 - 1394 = 198606; 200000 - 140 = 199860 years from the origin of humanity. Although being roughly estimated, the successive ratios of last times delimiting sequences are approximated as 194961/196986 ≈ 0.990, 196986/198606 ≈ 0.992, 198606/199860 ≈ 0.994, which are similar to the successive ratios displayed in Background to time, Table *(k)* for $z/8 = k$ sequences, where these ratios are presented as: $t_{1.875}/t_2 = tanh(1.875)/tanh(2) = 0.9896452$; $t_2/t_{2.125} = tanh(2)/tanh(2.125) = 0.9919278$; $t_{2.125}/t_{2.25} = tanh(2.125)/tanh(2.25) = 0.9937084$.

Based on these corresponding ratios and on the most accurately established transitional time $t_{2.25} = 199860$ years from the origin of humanity, other times of transition from one sequence to another can be calculated, in years from the same origin, as follows:

$$t_{1.875} = (0.9896452){\cdot}t_2 = 194959;$$
$$t_2 = (0.9919278){\cdot}t_{2.125} = 196999;$$
$$t_{2.125} = (0.9937084){\cdot}t_{2.25} = 198602.$$

Therefore, the final time of humanity $t_{\bullet} = t_k/tanh(k)$ can be evaluated as $t_{\bullet} = t_{2.25}/tanh(2.25) = (199860)/(0.9780261) = 204350$ million years from the origin of humanity, or 204350 -200000 = 4350 years from the present into future. Continuing to apply the formula $t_k = t_{\bullet}{\cdot}tanh(k)$ for the earlier values of argument k, the transitional times from one sequence to another, as well as the humanity timeline and its sequences are presented in the table below.

| $z/8 = k$ | Time (years) | | Sequences of humanity |
	from origin $t_k = t_{\bullet} \cdot tanh(k)$	from present (BC or AD) t_a - 200000	
	204350	+4350	
...	...	...	
2.375	200844	+844 (AD2858)	$-$ Modern
2.25	199860	-140 (AD1874)	$_-$ Medieval
2.125	198602	-1398 (AD616)	$_-$ Iron Age
2	196999	-3001 (987BC)	$_-$ Bronze Age
1.875	194959	-5041 (3027BC)	$_-$ Copper Age
1.75	192370	-7630 (5616BC)	$_-$ Neolithic
1.625	189095	-10905 (8891BC)	$_-$ Mesolithic (2)
1.5	184967	-15033 (13019BC)	$_-$ Mesolithic (1)
1.375	179793	-20207 (18193BC)	$_-$ Upper Palaeolithic (2)
1.25	173347	-26653	$_-$ Upper Palaeolithic (1)
1.125	165381	-34619	$_-$ Middle Palaeolithic (5)
1	155632	-44368	$_-$ Middle Palaeolithic (4)
0.875	143843	-56157	$_-$ Middle Palaeolithic (3)
0.75	129793	-70207	$_-$ Middle Palaeolithic (2)
0.625	113332	-86668	$_-$ Middle Palaeolithic (1)
0.5	94434	-105566	$_-$ Lower Palaeolithic (n)
0.375	73230	-126770	$_-$ Lower Palaeolithic (n-1)
0.25	50049	-149951	$_-$ Lower Palaeolithic (n-2)
0.125	25412	-174588	$_-$ Lower Palaeolithic (n-3)
0	0	-200000	$-$

During their 200000 years of existence, humans developed capabilities which differentiate them from other primates, such as creativity, consciousness, personality, abstract thinking, spirit, moral judgements, social skills and learning, as well as language, art, music, literature, spirituality, religion, philosophy, self-reflection, science, technology, and so on. Although psychological studies suggested numbers of human abilities ranging from several to 52 (Fleishman and Broedling, 1984; Fleishman and Reilly, 1992), the above data offer the following keys to put them in a suitable order:

Language, art, religion and music appeared in Africa, before humans moved to other continents, from 200000 to around 100000 years ago;

Exploration, inhabitation, knowledge and navigation successively emerged until about 50000 years ago when humans crossed quite wide straits of sea to arrive in Australia;

Weather awareness and *medicine* began to be practiced before about 40000-30000 years ago when old cave paintings from that time indicate primary knowledge in anatomy and the use of trepanation;

Society, construction, science and *industry* appeared until c.9000BC when copper was first discovered and used;

Civilization, legislation, philosophy, aero-astronautics and *high technology* emerged from about 6200-5000BC to the late 19th century, according to historical and scientific evidences.

Among these abilities, *language* emerged (the same time as *Homo sapiens* himself) c.200000; followed by other abilities; and then by *industry* 12000...10000 ≈ c.11000; *civilization* 8000...7000 ≈ c.7500; *legislation* c.5000; *philosophy* c.3000; and later abilities. Comparing their times of origin to the times which mark the sequences of human history, the correspondence between them is established, in years ago, such as: *Language* c.200000 and origin of *Homo sapiens* 200000; ...; *Industry* c.11000 and beginning of *Neolithic* 10905; *Civilization* c.7500 and beginning of *Copper Age* 7630; *Legislation* c.5000 and beginning of *Bronze Age* 5041; *Philosophy* c.3000 and beginning of *Iron Age* 3001; ... Therefore, the timeline, sequences, and intrinsic characteristics of human abilities are:

$z/8 =$ k	Time (years)		**Human abilities**
	from origin $t_k = t_\bullet \cdot tanh(k)$	from present $t_\sigma - 200000$	
	204350	+4350	
...	...	...	...
2.375	200844	+844 (AD2858)	...
2.25	199860	-140 (AD1874)	**High technology**
2.125	198602	-1398 (AD616)	**Aero-astronautics**
2	196999	-3001 (987BC)	**Philosophy**
1.875	194959	-5041 (3027BC)	**Legislation**
1.75	192370	-7630 (5616BC)	**Civilization**
1.625	189095	-10905	**Industry**
1.5	184967	-15033	**Science**
1.375	179793	-20207	**Construction**
1.25	173347	-26653	*Society* - **Sociology**
1.125	165381	-34619	**Medicine**
1	155632	-44368	*Weather awareness* - **Meteorology**
0.875	143843	-56157	**Navigation**
0.75	129793	-70207	**Knowledge**
0.625	113332	-86668	**Inhabitation**
0.5	94434	-105566	**Exploration**
0.375	73230	-126770	*Music* - **Musicology**
0.25	50049	-149951	*Religion* - **Theology**
0.125	25412	-174588	**Art**
0	0	-200000	*Language* - **Linguistics**

$z/8 =$ k	Time from origin $t_k =$ $t_\bullet \cdot tanh(k)$ (years)	Period $\tau_k =$ $(t_\bullet{}^2 - t_k{}^2)/(2t_\bullet)$ (years)	Frequency $f_k =$ $(2t_\bullet)/(t_\bullet{}^2 - t_k{}^2)$ (years)$^{-1}$	Angular speed $\omega_k =$ $2\pi/\tau_k = 2\pi \cdot f_k$ (years)$^{-1}$
	204350	0	∞	∞
...	...	...	...	...
2.375	200844	3475.9	0.0002877	0.0018076
2.25	199860	4440.7	0.0002252	0.0014149
2.125	198602	5667.2	0.0001765	0.0011087
2	196999	7218.8	0.0001385	0.0008704
1.875	194959	9175.2	0.0001090	0.0006848
1.75	192370	11628.8	0.0000860	0.0005403
1.625	189095	14685.6	0.0000681	0.0004278
1.5	184967	18463.7	0.0000542	0.0003403
1.375	179793	23081.5	0.0000433	0.0002722
1.25	173347	28651.2	0.0000349	0.0002193
1.125	165381	35253.4	0.0000284	0.0001782
1	155632	42910.7	0.0000233	0.0001464
0.875	143843	51549.1	0.0000194	0.0001219
0.75	129793	60956.0	0.0000164	0.0001031
0.625	113332	70748.2	0.0000141	0.0000888
0.5	94434	80355.1	0.0000124	0.0000782
0.375	73230	89053.8	0.0000112	0.0000706
0.25	50049	96046.0	0.0000104	0.0000654
0.125	25412	100594.9	0.0000099	0.0000625
0	0	102175.0	0.0000098	0.0000615

8. Linguistics

The scientific study of language in its widest sense, in every aspect and in all its varieties is named *linguistics* (from Latin *lingua* 'tongue, speech, language'). Among the subfields of linguistics can be mentioned:

1) Study of language structure, or *grammar*, focusing on the system of rules followed by its speakers, which includes the study of morphology (the formation and composition of words), syntax (the formation and composition of phrases and sentences from words), and phonology (sound systems);

2) Study of language, concerning how languages employ logic structures and real-world references to convey, process, assign meaning, and how they resolve ambiguity, which comprises the study of semantics (how meaning is inferred from words and concepts), and pragmatics (how meaning is inferred from context);

3) Study of the broader context in which language is influenced by social, cultural, historical and political factors, including evolutionary linguistics (related to the origins and growth of languages), historical linguistics (related to language change), sociolinguistics (referring to the relation between linguistic variation and social structures), psycholinguistics (exploring the representation and function of language processing in brain), language acquisition (related to the acquirement of language by children and adults), and discourse analysis (involving the structure of text or conversations).

Linguistics emerged from a series of observations, studies, and analyses on the differentiation, changes, rules, and relationship of various languages, as well as the comparison to each another, decipherment of former or extinct languages, research for cerebral centres of language, and language programming, which have been carried out, in the last twenty six centuries, by prominent personalities, such those mentioned below, in chronological order, with their contributions.

c.600BC: Pānini, Indian grammarian, was author of *Ashtadhyayi* 'Eight Lectures', a grammar of Sanskrit comprising 4,000 aphoristic statements which provide the rules of word formation and, to a lesser extent, sentence structure; representing the basis of all later Sanskrit grammars;

485-480BC: <u>Confucius</u>, Chinese philosopher, emphasized the moral commitment implicit in a name, *zhengming*, as a contribution to Chinese linguistics;

405-400BC: <u>Socrates</u>, Greek philosopher, worked on linguistics, and introduced *dialectics as a new text genre*;

360-350BC: <u>Plato</u>, Greek philosopher, argued that words denote concepts which are eternal and exist in the world of ideas, as exposed in his *Cratylus dialogue*, introduced the word *etymology* to describe the history of a word's meaning, and used the word *grammar* in its original meaning as *Τέχνη Γραμματική* 'The Art of Writing';

335-323BC: <u>Aristotle</u>, Greek philosopher and scientist, supported the conventional origins of meaning, defined the *logic of speech* and the argument, and worked on *rhetoric* and *poetics*;

260-240BC: <u>Xunzi</u> (Master Xun), Chinese Confucian philosopher, wrote elaborately argued essays, including the 'proper use of terms' from *zhengming*, and adopted a conventional view for the *origin of sound-to-meaning mapping*, although the objects signified by the term remain real;

c.250BC: <u>Kātyāyana</u>, Ancient Indian Sanskrit grammarian and Vedic priest, elaborated the *Varttika*, based on Pānini's grammar, which became a core part of the *Vyākarana* 'Grammar' canon;

c.140BC: <u>Patañjali the Grammarian</u>, Indian linguist, wrote a *substantial commentary* on Pānini's *Ashtadhyayi*, by which he contributing to the study of grammar in his country;

c.100BC: <u>Dionysius Thrax</u>, Greek grammarian, produced the *Techne Grammatike* 'Art of Letters', as the basis of all European works on grammar;

AD110-140: <u>Shuōwén Jiězi</u>, Han Dynasty Chinese linguist, analyzed the seal script characters, and introduced 540 *xiaozhuan* radicals, as the first *Chinese dictionary*, which was considered a major conceptual innovation in understanding the Chinese writing system;

c.150: <u>Apollonius Dyskolos</u>, Alexandrian grammarian, first reduced Greek syntax to a system, and wrote the treatise *On Syntax*, and shorter works on *pronouns*, *conjunctions*, and *adverbs*;

c.350: <u>Aelius Donatus</u>, Roman grammarian, produced treatises on Latin grammar entitled *Ars grammatica*, as the only textbook used in schools of the Middle Ages, so that *Donat* in western Europe came to mean a 'grammar book';

c.600: <u>Bhartrihari</u>, Hindu poet and philosopher, studying the Sanskrit language and its rules, he produced a *Sanskrit grammar*;

700-730: <u>'Abd Allah ibn Abi Ishaq al-Hadrami</u>, Arabic grammarian, compiled a *prescriptive grammar* by referring to usage of the Bedouins, whose language was seen as especially pure, his work being considered influential upon later grammarians;

785-795: <u>Sibawayh</u> (Uthman ibn Qunbar Al-Bisri), Persian Muslim linguist and grammarian of the Arabic language, made a detailed description of Arabic in his monumental work *al-Kitāb fī an-nahw* 'The Book of Grammar';

1318-20: <u>Dante Alighieri</u>, Italian poet, wrote the unfinished work *De Vulgari Eloquentia* 'On the Eloquence of Vernacular', discussing the origin of language, the divisions of languages, and the dialects of Italian in particular, extending the scope of linguistic enquiry *from Latin/Greek to include the languages* of his time;

1630-31: <u>John Amos Comenius</u>, Czech educationist, pioneered new *language teaching methods*, and published *Janua Linguarum Reserata*;

1772-87: <u>William Jones</u>, English jurist and Orientalist, published *Persian Grammar*, and pointed out striking resemblance of Sanskrit to Latin and Greek languages, becoming a pioneer of *comparative linguistics*;

1773-92: <u>James Burnett Monboddo</u>, Scottish judge and anthropologist, published six-volume *Of the Origin and Progress of Language*, as a learned but idiosyncratic, but his theory of human affinity with monkeys anticipated the theory of evolution, and the *modern science of anthropology*;

1810-20: <u>Wilhelm von Humboldt</u>, German politician and philologist, was the first to study the *Basque language* scientifically, worked on the *languages of the East* and the *South Sea Islands*, wrote *Über die Verschiedenheit des menschlichen Sprachbaues und ihren Einfluß auf die geistige Entwickelung des Menschengeschlechts* 'On the Variety of the Structure of Human Language and its Influence upon the Mental Development of the Human Race', and formulated 3959 rules of *Sanskrit morphology*;

1816-56: <u>Franz Bopp</u>, German philologist, produced the Indo-European grammar *Über das Conjugationssystem der Sanskritsprache* 'On the System of Conjugation in Sanskrit', tracing common origin of grammatical forms of Indo-European languages; and the greatest six-volume *A Comparative Grammar of Sanskrit, Zend, Greek, Latin, Lithuanian, Old Slavonic, Gothic and German*, later including *Old Armenian*;

1819: <u>Jacob Ludwig Carl Grimm</u>, German philologist, wrote *Deutsche Grammatik*, considered 'first great scientific linguistic work of the world', and formulated *Grimm's Law*, which states that in Proto-Indo-European voiceless stops become voiceless fricatives, voiced stops become voiceless stops, and voiced aspirates become voiced stops or fricatives (depending on context);

1822-32: <u>Jean François Champollion</u>, French linguist and Egyptologist, using a copy of the *Rosetta stone*, and his acquired knowledge of the Coptic language, he was the first to decipher the Egyptian hieroglyphics, showing that the Egyptian writing system was a *combination of phonetic and ideographic signs*; and founding *Egyptology*; his published works include *Panthéon égyptien, collection des personnages mythologiques de l'ancienne Égypte*; *Précis du système hiéroglyphique des anciens Égyptiens*; two-volume *Monuments de l'Égypte et de la Nubie*; *Grammaire égyptienne*; and *Dictionnaire égyptien en écriture hiéroglyphique*; by which the studies of early Egyptian history and culture have been properly promoted;

1836-85: <u>Ivar Andreas Aasen</u>, Norwegian philologist and lexicographer, carried out important studies for replacing Dano-Norwegian *Riksmål* by *Landsmål*, later known as *Nynorsk*, based on western Norwegian dialects; and published *Grammar of the Norwegian Dialects*, and *Dictionary of the Norwegian Dialects*;

1853-68: <u>August Schleicher</u>, German philologist, wrote fundamental works for the studies of Indo-European languages, including *Die ersten Spaltungen des indogermanischen Urvolkes*; *Handbuch der litauischen Sprache*, the first scientific compendium of Lithuanian language; *Kurzer Abriss der Geschichte der italienischen Sprachen*, relating to the Italian language; *Compendium der vergleichenden Grammatik der indogermanischen Sprachen*, relating to the German language; and the great *A Compendium of the Comparative Grammar of the Indo-European, Sanskrit, Greek, and Latin Languages*, attempting to reconstruct the Proto-Indo-European language;

1861: <u>Paul Broca</u>, French surgeon and anthropologist, was the first to locate the *motor speech centre* in the brain, since known as the convolution of Broca or *Broca's gyrus*, and introduced the *motor aphasia*, the meaning of aphasia being the loss of speech by a severe defect in understanding it;

1874: <u>Carl Wernicke</u>, German neurologist and psychiatrist, published *Der Aphasische Symtomencomplex* 'The Aphasic Syndrome', showing that *sensory aphasia* typically originates in an area of the left temporal lobe, since known as *Wernicke's area*;

1877: <u>Karl Verner</u>, Danish linguist, formulated *Verner's Law*, according to which Proto-Germanic fricatives become voiced if next conditions are met: they are not initial, but preceded and followed by voiced, and accent is not on immediately preceding syllable;

1878: <u>Ferdinand de Saussure</u>, Swiss linguist, published *Mémoire sur le système primitif des voyelles dans les langues indo-européenes* 'Memoir on the Primitive Vowels in the Indo-European Languages', constituting the first serious attempt to determine the nature of language as the object of which linguistics is the study, and showing the importance of *synchronic analysis*, although this focus has shifted and the term 'philology' has been later used for the study of a language's grammar, history, and literary tradition;

1893-1905: <u>Lazarus Ludwig Zamenhof</u>, Polish oculist and philologist, invented *Esperanto* 'One who hopes' as an international language to promote world peace, and published *Fundamento de Esperanto* 'Basis of Esperanto';

1897-1925: <u>Antoine Meillet</u>, French philologist, produced standard works on *Old Slavonic, Greek, Armenian, Old Persian*, etc., as well as on *comparative Indo-European grammar*, and *linguistic theory*; such as *Research on the Use of the Genitive-Accusative in Old-Slavonic*; *Esquisse d'une grammaire comparée de l'arménien classique*; *Caractères généraux des langues germaniques*; and *La méthode comparative en linguistique historique* 'The comparative method in historical linguistics';

1898: <u>Ma Jianzhong</u>, late Qing Dynasty Chinese official and scholar, was author of *Mashi Wentong* 'Basic principles for writing clearly and coherently by Mister Ma', a textbook of *Chinese grammar*, in modern sense, based on the Latin 'prescriptive' model;

1902-21: <u>Jules Gilliéron</u>, Swiss linguist, published *Atlas linguistiques de la France, La Généalogie des mots qui désignent l'abeille* 'The Etymology of Words Relating to the Bee', and *Pathologie et thérapeutique verbales* 'The Pathology and Treatment of Words';

1914-33: <u>Leonard Bloomfield</u>, US linguist, had a major part in making *linguistics an independent scientific discipline*, by publishing

Introduction to the Study of Language, and the major work *Language* on linguistic theory, pioneering *structuralism*;

1939-43: Louis Trolle Hjelmslev, Danish linguist, devised a system of linguistic analysis known as *glossematics*, which was based on the study of the distribution of, and the relationships between, the smallest meaningful units of a language, called *glossemes*; and wrote *Prolegomena to a Theory of Language*;

1960-65: William Stokoe, US scholar, published *Sign Language Structure*, and co-authored *A Dictionary of American Sign Language on Linguistic Principles*, proving that American Sign Language fits the criteria for a natural language;

1960-87: Noam Chomsky, US linguist, contributed at linguistic theory by works such as *Aspects of the Theory of Syntax*; *Cartesian Linguistics*; *The Sound Pattern of English*; *Language and Mind*; *Reflections on Language*; *The Logical Structure of Linguistic Theory*; and *Language and Problems of Knowledge*; by which he established the *generative theory* of linguistics;

1967: Jacques Derrida, French philosopher, published the influential work *La Voix et le phénomène* 'Speech and Phenomena', and also *L'Écriture et la différence* 'Writing and Difference', by which he founded *semiotics*;

1967-2006: Johan Frederik Staal, Dutch-born US scholar of Greek and Indian logic and philosophy, and Sanskrit grammar, studied the methods of linguistic theory, and the applications of modern mathematical logic to linguistics, stating that 'Pānini is the Indian Euclid'; and published *Word Order in Sanskrit and Universal Grammar*; *A Reader on the Sanskrit Grammarians*; *Universals. Studies in Indian Logic and Linguistics*; and *Artificial Languages across Sciences and Civilizations*;

1996: Hans Henrich Hock and Brian Joseph, US linguists, published *Language History, Language Change, and Language Relationship: An introduction to Historical and Comparative Linguistics*, a comprehensive study of languages;

1996-2000: Mario Alinei, Italian linguist, published two-volume *Origini delle Lingue d'Europa* 'The Origins of Language of Europe', formulating *Palaeolithic Continuity Theory* with proposal that Indo-Europeans arrived in Europe tens of thousands of years ago, and that by the end of Ice Age already differentiated into local language speakers occupying territories within or close to their now-traditional homelands; and hypothesizing that a homogeneous early Indo-

European people appeared in Europe about 6000 years ago; he was founder and editor of *Quaderni di semantica*, a journal of theoretical and applied semantics;

2000: <u>Venceslas Kruta</u>, French linguist, wrote *Les Celtes – Histoire et dictionnaire. Des origines à la romanisation et au christianisme*, an important book for Celtic culture and language;

2001-04: <u>Andrei Alexandrescu</u>, Romanian C++ programmer and author, working on *policy-based design* implemented via *template metaprogamming*, developed an initial *computer programming language (C)* into the *C plus plus (C++)* *programming languages*; and wrote the books *Modern C++ Design: Generic Programming and Design Patterns Applied*; and *C++ Coding Standards: 101 Rules, Guidelines, and Best Practices*, both published by Addison-Wesley;

2010: <u>Marius Albu</u> and <u>Ioana Gauntlett</u>, Romanian-born British researchers of old Celtic language, compiled the work *Celtic Names in Western and Eastern European Languages - Evidences for Cultural Diffusion*, The Edwin Mellen Press, Lewiston-Queenston-Lampeter.

Looking backwards, and comparing with non-human primates, humans are characterized by significant adaptations, such as bipedalism, increased brain size, lengthened ontology (gestation and infancy) and decreased sexual dimorphism. One important morphological change included the evolution of a power and precision grip.

Speech organs evolved in the first instance not for speech, but for more basic bodily functions such as feeding and breathing. Our species' unprecedented use of the *tongue, lips,* and *vocal organs* as an instrument of communication developed in association with *larynx* (voice box), *respiratory control*, and involving a special shaped lingual bone, named *hyoid bone* (Latin *os hyoideum*, from Greek χιοειδες 'upsilon shaped') and situated in the anterior midline of neck between the chin and the thyroid cartilage. Nonhuman primates have broadly similar organs, but with different neural control; and Neanderthals had also the hyoid bone, as evidenced by recently discovered remains of their skeletons.

The *phonatory process*, or voicing, occurs when air is expelled from the lungs through *glottis*, creating a pressure drop across the larynx. When this drop becomes sufficiently large, the vocal folds start to oscillate. The minimum pressure drop required to achieve phonation is called *phonation threshold pressure*, which for humans with normal vocal folds is approximately 2-3 centimetres column of water. The sound produced by the larynx is a harmonic series, i.e. consists of a fundamental tone (called the fundamental frequency, as the main acoustic cue for percept pitch) accompanied by harmonic overtones, which are multiples of the fundamental frequency. According to source-filter theory, the resulting sound excites the resonance chamber that is the *vocal tract* to produce individual speech sounds. The *voice* consists of sound made by a human using the *vocal folds* for talking, singing, laughing, crying, screaming, etc. Habitual speech fundamental frequency ranges between 75-150 hertz for man and 150-300 hertz for woman. The mechanism for generating the human voice can be subdivided into three parts: the lungs, the vocal folds within the larynx, and articulators.

Language is defined as: human speech; variety of speech or body of words and idioms; mode of expression; diction; manner of expressing thought or feeling; artificial system of signs and symbols with rules for forming intelligible communications; as well as national branch of one of religious and military orders. In the human brain, language is derived and controlled from specialized centres of the cerebral cortex, in the left hemisphere for 97% of right-handed people and 19% of left-handed people, and in both left and right hemispheres for 68% of them. These centres are situated in five language areas of the brain, namely *Angular Gyrus*, *Supra-marginal Gyrus*, *Paul Broca's area*, *Carl Wernicke's area*, and *Primary Auditory Cortex*, which can be found in both hemispheres, contributing to processing and understanding language: the left hemisphere processing the linguistic meaning of prosody (rhythm, stress, and intonation of connected speech), while the right hemisphere processes the emotions conveyed by prosody.

The centres of main areas for language are interconnected within a *total area* varying during a life time between 0.01 and 0.04 square metres, where the neurons are properly active between 0.01% and 0.10%, and therefore the *active area* for language ranges between $1 \cdot 10^{-6}$ and $4 \cdot 10^{-5}$ square metres. This active area multiplied by the *thickness* of cerebral cortex that varies between $1.5 \cdot 10^{-3}$ and $4.5 \cdot 10^{-3}$ metres, results in an linguistic *active volume* that ranges between $(1 \cdot 10^{-6}) \cdot (1.5 \cdot 10^{-3}) = 1.5 \cdot 10^{-9}$ and $(4 \cdot 10^{-5}) \cdot (4.5 \cdot 10^{-3}) = 1.8 \cdot 10^{-7}$ cubic metres.

The linguistic signals are propagating in the neocortex with a *speed* decreasing during the life time approximately from 150 to 23 metres per second. As the mental active volume ranging from $9.0 \cdot 10^{-9}$ to $1.12 \cdot 10^{-6}$ cubic metres requires a power varying from 25 to 35 watts; it follows that the linguistic active volume, ranging from $1.5 \cdot 10^{-9}$ to

$1.8 \cdot 10^{-7}$ cubic metres, requires a *linguistic power* varying from 4.17 to 5.63 watts, corresponding to a *linguistic energy* varying from $2.3 \cdot 10^8$ to $1.35 \cdot 10^{10}$ joules.

Language is the most precious treasure of every culture, preserving and transmitting traces of its historical background. For instance, present-day English includes elements from pre-Celtic, Celtic, Latin, German, Scandinavian, and French, as the result of historical and cultural development. As media's influence is spreading, there is an increasing tendency to uniform our cultures, but meanwhile to threaten many languages with extinction. Like their speakers, languages were emerging, differentiating, flourishing, and spreading, while some of them were diminishing, surviving, and even extinguishing.

In 1998-2000, the *number of languages* currently used was of 209 in Europe, 2034 in Asia, 1995 in Africa, 1341 in Pacific areas, and 949 in Americas; *official languages* being prioritized according to associated populations given in millions such as English 1400, Chinese 1070, Hindi 700, Spanish 280, Russian 270, French 220, Arabic 170, Portuguese 160, Malay 160, Bengali 150, Japanese 120, etc.

It is believed that all spoken languages emerged from the following *proto-languages*: in <u>Africa</u> *Proto-Afroasiatic* with Proto-Semitic and Proto-Berber branches; in <u>Near East</u> *Proto-Anatolian, Proto-Northwest Caucasian* with Proto-Abazgi and Proto-Circassian branches, as well as *Proto-Kartvelian* and *Proto-Semitic* branches; in <u>Eurasia</u> *Proto-Basque, Proto-Dravidian,* and *Proto-Indo-European* with Proto-Greek, Proto-Armenian, Proto-Indo-Iranian, Proto-Baltic-Slavic, Proto-Celtic, Proto-Germanic, and Proto-Romance branches; in <u>North Asia</u> *Proto-Turkic, Proto-Mongolic, Proto-Uralic,* and *Proto-Chukotko-Kamchatkan*; in <u>Pacific Rim</u> *Proto-Pama-Nyungan, Proto-Austronesian, Proto-Tai-Kadai, Proto-Tibeto-Burman, Proto-Hmong-Mien,* and *Proto-Mon-Khmer*; and in <u>Americas</u> *Proto-Eskimo-Aleut, Proto-Algonquian, Proto-Iroquoian, Proto-Uto-Aztecan, Proto-Mayan,* and *Proto-Oto-Manguean.*

From their homelands in Anatolia, 7500-5500BC, and in Pontic steppes, 5500-4500BC, the Proto-Indo-European languages developed into *Indo-European languages*, representing today one of the most spoken and spread family of languages comprising 439 languages and dialects, from which 221 belong to the Indo-Aryan sub-branch. They include major current language of Europe, the Iranian plateau, and South Asia. With written attestations appearing since the Bronze Age in the forms of Anatolian languages and Mycenaean Greek, the Indo-European family is significant to the field of historical linguistics as possessing the longest record history after the Afro-Asiatic family. Indo-European languages are spoken by almost three billion native speakers, and have ten major branches which are presented in the chronological order of their earliest surviving written attestations:

1) *Anatolian*, with isolated terms in Old Assyrian sources from the 19th century BC and Hittite texts from the 16th century BC, becoming extinct by Late Antiquity;

2) *Hellenic*, with records in Mycenaean Greek from 1450 to 1350BC and in Homeric texts of the 8th century BC;

3) *Indo-Iranian*, descending from Proto-Indo-Iranian by the 3rd millennium BC, and comprising: Iranian attested from around 1000BC in form of Avestan, and epigraphically from 520BC in form of Old Persian in Behistun inscription; Indo-Aryan attested from the late 15th to the early 14th centuries BC in Mitani texts showing traces of Indo-Aryan, and also epigraphically from the 3rd century BC in form of Prakrit in Edicts of Ashoka, and intact records of Rigveda coming from oral tradition dating from the mid-2nd millennium BC in form of Vedic Sanskrit, Dardic, and Nuristani;

4) *Italic*, including Latin and its descendants (Romance languages: Italian, French, Spanish, Portuguese, Romanian, Raeto-Romanic), attested from the 7th century BC;

5) *Celtic*, descending from Proto-Celtic, with Gaulish inscriptions from the 6th century BC, Celtiberian from the 2nd century BC, Old Irish manuscript tradition from the 8th century AD, and inscriptions in Old Welsh from the same period;

6) *Germanic*, descending from Proto-Germanic, with earliest testimonies in runic inscriptions from the 2nd century AD, earliest coherent text in Gothic from the 4th century AD, and Old English manuscript tradition from the 8th century AD;

7) *Armenian*, with alphabet writings from the beginning of the 5th century AD;

8) *Tocharian*, extant in two dialects (Turfanian and Kuchean) from the 6th to the 9th centuries AD, and probably extinct by the 10th century;

9) *Balto-Slavic*, with Slavic, coming from Proto-Slavic, from the 9th century AD, and earliest texts in Old Church Slavonic; and Baltic, attested from the 14th century AD, with archaic features attributed to Proto-Indo-European; and

10) *Albanian*, coming from Proto-Albanian likely emerging from Palaeo-Balkan predecessors attested from the 14th century AD.

A language in general and an Indo-European one in particular combines words which are usually divided into eight parts of speech called *nouns* (n), *pronouns* (pn), *adjectives* (aj), *verbs* (v), *adverbs* (av), *prepositions* (pp), *conjunctions* (cj), and *interjections* (ij), according to their functions, and optionally grouped in *sentences*, *phrases* and *clauses* (spc) such as

$$[(pp)\cdot(n / pn)\cdot(aj)] * [(v)\cdot(av)] + [(cj), (ij)] = (spc),$$

where the signs {/, ·, +, *, =} mean {'instead', 'added', attached', 'operation', 'equal'} respectively.

In technical terms, each part of speech can be interpreted like: noun 'substance', pronoun 'spare', adjective 'aspect', verb 'power', adverb 'gear', preposition 'setting', conjunction 'connection', and interjection 'suddenness'.

Although Neanderthals had a simple kind of speech, human language emerged from very beginning of Homo sapiens, i.e. about 200000 years ago, later evolved by writings and scripts, literature and printing, and will have the same fate as their speakers. Emergence, spread, differentiation, and derivation of languages and linguistic laws, together with related scripts and writings, printings and literatures, literary works and movements, promoters and authors are briefly displayed below.

200000-175000 years ago: Establishment of a *basic vocabulary* as an initial *common root of languages*, originating in Africa;

150000-100000 years ago: *Development of language* by African ancestors of all of us;

100000-30000BC: *Differentiation of languages* due to human movement to other continents;

30000-10000BC: *Language changes* in the new colonized territories;

10000-7800BC: Appearance of *Pre-boreal languages* (corresponding to transition from Mesolithic to Early Neolithic);

6000-5300BC: Emergence of *Protolanguages*, including Proto-(west African, Nilo-Saharan and Malayo-Polynesian) at latitudes less than 15°North; Proto-(Afro-Asiatic, Dravidian, and Tibeto-Burman) at latitudes of 15°-35°North; Proto-(Hamitic, Semitic, Sumerian, Indo-Iranian, and Sino-Tibetan) at latitudes of 23°-40°North; Proto-(Iberian-Basque, Indo-European, Ural-Altaic, and Manchu-Ainu) at latitudes of 45°-55°North; and Proto-(Finno-Uralic/Ugric, and Eskimo-Aleut) at latitudes of more than 55°North;

3600-3500BC: Invention of *pictographic writing* at Uruk, and first evidence of cuneiform writing in Mesopotamia, and of *hieroglyphic writing* in Egypt;

3300-2800BC: Emergence of *Bronze Age languages* (corresponding to transition from Late Neolithic to Bronze Age);

3200-3100BC: *Sumerian cuneiform writing* was used for first literary language and literature, and also development of *pictographic writing* in Sumer, South Mesopotamia;

3100-3000BC: Emergence of *Historic languages*;

3000BC: Earliest examples of the *Indus script*;

2500BC: Use of *Aboriginal languages*, such as Nootka and Salish, in West Canada;

2400BC: Emergence of *pictographic writing* in India;

2150-2000BC: *Epic of Gilgamesh*, written on clay tablets, as the greatest surviving work of Sumerian literature, and describing events related to <u>Gilgamesh</u> (c.2500BC), the 5th king of Uruk;

2000-1500BC: Origins of *Veda*, sacred Hindu scriptures, hymns written in Old Sanskrit, later developed as *Rig-Veda*, *Yajur-Veda*, *Sâma-Veda*, and *Atharva-Veda*, in India;

2000-1450BC: *Main families of languages*, including Indo-European family, were establishing and spreading in the world;

1700BC: Use of *Linear A script* in the Aegean area;

1700-1046BC: *History of China* - a written text from the Shang Dynasty times;

1550BC: *Book of the Dead*, an Egyptian funerary text, written on papyrus scrolls (now in British Museum);

1500BC: Origin of the earliest *Indian literature*; use of *Linear B script* in the Aegean area; *Hurrian* and *Hittite writings* in Anatolia; and *first alphabets* in Syria and Palestine;

1250BC: Use of *hieroglyphic Hittite* in Anatolia;

1200BC: Invention of *Phoenician script* in East Mediterranean;

1200-800BC: Emergence of *Later spoken languages* (corresponding to transition from Bronze Age to Iron Age), including Indo-European languages (*Celtic* in West Europe; *Germanic* in North Europe; *Slavic* in North central Europe; *Italic* in South Europe; *Illyrian* and *Thracian* in South-east Europe; *Greek* in South Europe and West Asia Minor; *Anatolian* in Asia Minor; *Iranian* in North Pontic-Caucasus-Caspian,

West central Asia and North Arabian Sea; *Armenian* in South-west Caucasus; *Indo-Aryan* in North-west India; and *Tocharian* in central Asia);

1100-900BC: Origins of *Earlier literatures*; and the invention of early *Chinese writing*;

800BC: Flourishing of *Sanskrit literature*, the religious thought and caste system in India;

780-750BC: Emergence of earliest *Greek alphabetic inscriptions* in Europe;

c.750BC: *Beginning of Greek epic poetry* – Homer elaborated great poems, such as the *Iliad*, dealing with episodes in Trojan War; and *Odyssey*, dealing with Odysseus's adventures on his return from Troy, which are commonly attributed to him;

600BC: Development of *Etruscan script* in central Italy;

c.500BC: Appearance of *Latin script* in Italian peninsula;

498-440BC: *Greek lyric poetry* – Pindar produced many works, from them being still extant *Pythian Ode X*, odes for tyrants of Syracuse and Macedonia, as well as for free cities of Greece; *hymns* to gods, paeans, dithyrambs, mimic dancing songs, convivial songs, dirges, but unfortunately only fragments are extant; and the entirely preserved poem *Epinikia* 'Triumphal Odes';

484-406: *Greek tragic drama* – Aeschylus wrote about 60 plays, from which only several are extant, namely *The Persians*, *Seven against Thebes*, *Prometheus Bound*, *Suppliants*, and *Orestia*; and Euripides produced *Medea, Andromache, Supplices, Troades, The Women of Troy, Phoenissae, Orestes, Bacchae* and *Iphigenia in Aulidensis*;

445-405BC: *Greek tragedy* – Sophocles was author of *Ajax, Antigone, Oedipus Tyrannus, Trachiniai* 'Women of Trachis', *Electra, Philoctetes*, and *Oedipus Coloneus*;

424-388BC: *Greek comic drama* – Aristophanes wrote *Hippeis* 'Knights', *Nephelai* 'Clouds', *Sphekes* 'Wasps', *Eirene* 'Peace', *Ornithes* 'Birds', *Lysistrata* 'Destroyer of Armies', *Thesmophoriazusae* 'The Woman attending the Thesmophoria', *Batrachoi* 'Frogs', and *Ecclesiazusae* 'Women in Parliament';

330BC: *Hellenism* – defined by Greek idiom, language, and culture, which increased during and after the Macedonian expansion;

285-230BC: *Greek Sceptic literature* – Timon of Phlius was author of *satyr-plays, comedies, tragedies, epic poems*, and the famous series of

Silloi as satirical mock-heroic poems parodying and insulting most of the earlier Greek philosophers;

250-240BC: Introduction of *Brahmi* script in India;

250-230BC: <u>Manetho</u>, Egyptian historian and priest, whose works dealt with Egyptian matters but written solely in Greek language; from his works the best known and studied is the three-volume *Aegyptiaca* 'History of Egypt', valuable for its chronology of the reigns of pharaohs; and others including *Against Herodotus*, *The Sacred Book*, *On Antiquity and Religion*, *On Festivals*, *On the Preparation of Kyphi*, and *Digest of Physics*;

166-160BC: *Roman drama* – <u>Publius Terentius</u> (Terence) wrote comic dramas, such as *Andria*, *Hecyra* 'Mother-in-Law', *Heauton Timoroumenos* 'Self-Tormentor', *Eunuchus*, *Phormio*, and *Adelphi* 'Brothers';

150BC: Use of *Mayan script* in Mesoamerica;

80-43BC: *Climax of Roman oratory* – <u>Marcus Tullius Cicero</u> elaborated speeches *Pro Roscio Amerino*, *Pro Lege Manilia*, *Pro Sestio*, *Pro Milone*, *Philippics*, as well as essays *De Senectute*, *De Amicitia*, and *De Oficiis*, which remained style models;

70-27BC: *Roman prose* – <u>Marcus Terentius Varro</u> produced works including *Saturae Menippeae* 'Menippean Satires', *Antiquitates Rerum Humanorum et Divinarum* 'Antiquities in Matters Human and Divine', *De Lingua Latina* 'On the Latin Language', *De Re Rustica* 'Country Affairs', and the encyclopaedic *Disciplinarum Libri IX*;

35BC- AD17: *Climax of Roman literature* – <u>Quintus Horace</u> wrote *Satires*, *Epodes*, four *Odes*, and three epistles including *Ars Poetica*, with profound influence on poetry and literary criticism in the 17th-18th centuries Europe; <u>Publius Virgil</u> produced great *Aeneid*, representing a national epic based on the story of Aeneas the Trojan, legendary founder of the Roman nation, and of the Julian family; <u>Sextus Propertius</u> wrote books of *poems*, first of them being devoted to his central figure of inspiration, mistress Cynthia; and <u>Publius Naso Ovid</u> was author of collection of mythological tales, named *Metamorphoses*, in Rome; as well as elegies, called *Tristia* 'Sorrows', and *Epistolae ex Ponto* 'Letters from Pontus Euxinus' which were written during his exile at Tomis (now Constanta, Romania);

AD50-100: *Imperial Roman literature* – <u>Lucius Annaeus Seneca</u> wrote three treatises *Consolationes*, and then *Epistolae morales ad Lucilum*, *Apocolocyntosis divi Claudii* 'The Pumpkinification of the

Divine Claudius', *On the Shortness of Life*, and *Tenne Tragedies*; Tiberius Catius Asconius Silius Italicus was author of the longest surviving Latin poem, entitled *Punica*, an epic in 17 books on Second Punic War; and Marcus Fabius Quintilianus produced *Institutio Oratoria* 'Education of an Orator', a complete system of rhetoric in 12 books, remarkable for its sound critical judgements, purity of taste, admirable form, and perfect familiarity with literature of oratory;

140-160: *Greek literature* – Athenaeus wrote *Deipnosophistae* 'Banquet of the Learned', a collection of anecdotes and excerpts from ancient authors, arranged as a scholarly dinner-table conversation; and *Roman literature and rhetoric* – Lucius Apuleius was author of *Metamorphoses* 'The Golden Ass', a tale of adventure containing elements of magic, satire and romance, and also of *Apologia* 'Apology', an eloquent speech in defence;

155-170: *Greek satire and rhetoric* – Lucian wrote in the elegant Attic style works including *Deorum Dialogi* 'Dialogues of the Gods', *Mortuorum Dialogi* 'Dialogues of the Dead', *Charon, Symposium, Halieus, Biōn Prasis, Drapetae*, and *Vera Historia*;

355-400: *Late Latin and Greek literature* – Flavius Claudius Julian, Roman emperor, wrote *Epistles, Orations, Caesars*, and *Misopogon* which are still extant, but his chief work *Kata Christianon* was lost; Gregory of Nyssa, Greek Christian theologian, produced his main works *Twelve Books against Eunomius*, a treatise on Trinity, great *Catechetical Oration*, and also ascetic treatises, sermons and epistles; and Marcus Aurelius Clemens Prudentius, Latin Christian poet, was author of poems such as *Cathemerinon Liber, Peristephanon, Apotheosis, Hamartigenia, Psychomachia, Contra Symmachum*, and *Diptychon*;

381-407: *Latin religious literature due to Syrians* – St John Chrysostom, Doctor of the Church, wrote *Homilies, Commentaries* on whole Bible, *Epistles, Treatises* on Providence and Priesthood, and *Liturgies*;

c.550: *Pre-Islamic Arabic literature* – Antar ('Antarah Ibn Shaddad al-'Absi) was author of odes and poems, which became known as one of *Seven Golden Odes* of Arabic literature, his best or chief poem being contained in the *Mu'allaqat*; whence the 10th century 'Romance of Antar' derived as a model of Bedouin heroism and chivalry;

590-650: Beginning of *Later literatures* in the Old World;

740-759: *Classical Japanese poetry* – Akahito Yamabe no produced his great anthology of *Japanese poetry* of his time; and, together with

Hitomaro, wrote the comprehensive *Man'yōshū* 'Collection of a Myriad Leaves';

860-868: *Cyrillic alphabet* – St Cyril, the Apostle of Slaves, after his mission among Khazars and Moravians, invented the *Cyrillic alphabet*, basically using Greek written signs (from letters) to represent Slavonic sounds, that later was spread by Orthodox Church through parts of eastern and central Europe;

948-965: *Icelandic poetry* – Egill Skallagrímsson compiled the eulogy *Höfuðlausn* 'Head Ransom', the great lament *Sonatorrek* 'On the Loss of Sons', and the verse-sequence *Arinbjarnarkviða* 'The Lay of Arinbjörn';

995-1010: *Persian poetry* – Firdausi (Hakim Abul-Qasim Firdawsi Tusi) produced the masterpiece *Shah Náma* 'Book of Kings', mainly composed of mythological and fanciful incidents, based on events from Persian annals; short pieces as *kasidas* and *ghazals*; and also *Yusuf u Zulaykha*, a story of Joseph and Potiphar's wife;

1137-48: *Byzantine literature* – Princess Anna Comnena wrote *Alexiad*, a text-book where she described the life of her father, Emperor Alexius I Comnenus, and containing a valuable account of the First Crusade;

1170-90: *French early literature* – Chrétien de Troyes collected romances dealing with *Arthurian* legends and subjects, whereby the character *Lancelot*; his known romances include *Yvain et Lancelot*; *Érec et Énide*; *Cligès*; and the unfinished *Perceval, ou le Conte du Graal* 'Percival, or the Story of the Holy Grail';

1185-1783: *Serbian and Croatian early literature* – Gligorije the Pupil, Serbian writer, produced *The Gospel Book*, a masterpiece of calligraphy and illustrations (now entered onto the UNESCO *Memory of the World* List), which also explains the origin of the Cyrillic script and its letters; Peter Zoranić, Croatian novelist, published the first Croatian novel entitled *Planine*; and Dositej Obradović, Serbian writer and polyglot, was author of a captivating autobiography entitled *The Life and Adventures of Dimitrije Obradović*;

1190-1220: *German early literature* – Austrian Walther von der Vogelweide produced stories and poems, such as *Ich saz ûf eine steine* 'Sitting on a Stone', and collections of sayings *Reichssprüche* 'Sayings of Holy Empire', *Papstsprüche* 'Sayings about Popes', and *Kaisersprüche* 'Sayings about the Emperors'; and German Wolfram von Eschenbach wrote lyrics, including *Taglieder*, love songs, short epic *Willehalm*, and most notable the poem *Parzival*, with its main theme related to the popular Grail's history, which later derived in E Wagner's libretto of his opera 'Parsifal';

1307-58: *Italian literary language and early literature* – Dante Alighieri created great *Divina Commedia*, narrating a journey through Hell and Purgatory to Paradise, by which Italian was established as literary language, and in addition wrote *De Monarchia* (in Latin) expounding Dante's theory of divinely intended government of the world by a universal pope, and left unfinished *De Vulgari Eloquentia* discussing the origin and division of languages; Francesco Petrarch wrote the epic poem *Africa*, historical prose *De Viris Illustribus*, dialogues *De Contemptu Mundi*, treatises *De Otio Religiosorum*, and *De Vita Solitaria* which influenced G Boccaccio, as well as his later *Opera Omnia*; altogether with unprecedented influence across Europe; and Giovanni Boccaccio produced outstanding work *Decameron* with medieval subject matter and classical form, mythological *De genealogia deorum gentilium* 'The Genealogies of the Gentile Gods', and treatises such as *De claris mulieribus* 'Famous Women', and *De Montibus* 'On Mountains';

1369-87: *English literary language* – Geoffrey Chaucer produced first his work *Book of the Duchess*, and then best known *The Canterbury Tales* by which the southern English dialect was established as the English literary language; as well as *The Parliament of Fowls*; *The House of Fame*; *Troilus and Cressida*; and *The Legend of Good Woman*;

1395-1410: *Scottish chronicle* – Andrew of Wyntoun wrote *The Orygynale Cronykil of Scotland*, as a specimen of old Scots, covering a period from the creation until 1406 and outlining the geography and history of ancient and mediaeval Scotland;

1400-1640: *Renaissance* – the revival of letters and art as a transition from the mediaeval to the modern world, in Europe, under the influence of Greek and Latin literatures;

1430-60: *English chronicle* – John Capgrave wrote *Nova legenda Angliae*; *De illustribus Henricis*; *Vita Humfredi Ducis Glocestriae*; and *A Chronicle of England from the Creation to 1417*;

1440-45: *Invention of printing press*, of movable type, by German printer Johannes Gensfleisch Gutenberg in Strasbourg and Mainz, who printed *Fragment of the Last Judgement*, and editions of the Aelius Donatus' Latin school grammar;

1440-71: *Italian literature* – Laurentius Valla wrote humanist works including *De Donatione Constantini Magni* 'On the Donation of

Constantine the Great', and *De Elegantia Latinae Linguae* 'Elegances of the Latin Language';

1450-70: Culmination of *Humanism* as literary culture and classical studies that puts human interests and mind paramount, rejecting supranatural or belief, and applying *pragmatism, critical thinking* and *evidence* (rationalism, empiricism) over established doctrine or faith (fideism), in all sciences, including linguistics;

1463-70: *French poetry* – François Villon produced works mainly consisting of *Petit Testament*, and *Grand Testament*, with 40 and 172 eight-syllabic octaves respectively;

1474-75: *First printed English book*, namely *Recuyell of the Historyes of Troye*, was performed by printer William Caxton, joined with Flemish calligrapher Colard Mansion;

1506-1787: *Italian literature* – Ludovico Ariosto was author of Renaissance style-comedies, satires, sonnets, and poems, including the great poem *Orlando Furioso* 'Orlando Enraged' that developed the epic tale of Roland (Orlando); Marco Girolamo Vida produced poems on silk culture and chess, as well as Latin orations and dialogues, such as *Christias*, and *De Arte Poetica* 'On the Art of Poetry'; Torquato Tasso wrote Renaissance style-verses and philosophical dialogues, pastoral play *Aminta*, tragedy *Il Re Torrismondo* 'King Torismondo', epic poem *Gerusalemme Liberata* 'The Recovery of Jerusalem' describing the First Crusade, and *Gerusalemme Conquistata* 'Jerusalem Conquered'; and Carlo Goldoni was author of plays such as *La Locandiera* 'The Mistress of the Inn', *I rusteghi* 'The Boors', and *Le baruffe chiozzote* 'The Squabbles of Chioggia';

1512-1743: *Romanian early literature* – St Neagoe Basarab, Wallachian Prince, wrote *The Teachings of Neagoe Basarab to his Son Theodosie*, with subjects such as philosophy, diplomacy, morals and ethics; Grigore Ureche, Moldavian chronicler, produced *The Chronicles of the Land of Moldavia*, showing the common Romanian origin of Moldavians, Wallachians and Romanians of Transylvania; Miron Costin, Moldavian political figure and chronicler, wrote *The Chronicles of Moldova From the Prince Aron's Rule* (to 1660), and the *Polish Verse History of Moldavia and Wallachia*; Dimitrie Barila (St Dosoftei, or Dositheiu), Moldavian metropolitan, scholar and poet, published the first *volume of poetry in Romanian*, a verse translation of the *Psalms*; Constantin Cantacuzino, Wallachian nobleman, was author of *The Political and Geographical History of the Romanian*

Countries; <u>Nicolae Costin</u>, Moldavian statesman and chronicler, became known by his historical writings including *The Chronicle of the Principality of Moldavia from the World's Building to 1601*, and *The Chronicle of Nicolae Mavrocordat's Rule*; <u>Antim Ivireanul</u>, Georgian-born Wallachian metropolitan, author and typographist, published the popular book *The Flower of the Gifts*, and *The Sermons*; and <u>Ion Neculce</u>, Moldavian chronicler, wrote his main work *The Chronicles of the Land of Moldavia (from the Prince Dabija's Rule to the second rule of Constantin Mavrocordat)*;

1515-72: <u>*Portuguese literature and literary language*</u> – <u>Gill Vicente</u> wrote plays and farces, such as *Inês Pereira*, *Juiz da Beira*, and his best *Autos das barcas*, entitled *Inferno*, *Purgatório* and *Glória*; and <u>Luis de Camoëns</u> produced great Renaissance style-epic poem *Os Lusiadas* 'The Lusiads', and also plays, sonnets and lyrics, founding the literary language of his people;

1532-1818: <u>*French literature*</u> – <u>François Rabelais</u> published Renaissance style-books *Pantagruel*, *Gargantua*, *Tiers Livre* 'Third Book', *Quart Livre* 'Fourth Book', as well as scraps and notes for a fifth book called *L'Isle sonante*; <u>Michel Eyquem de Montaigne</u> wrote sceptical philosophical *Apologie de Raymon Sebond* 'Apologia for Raymond Sebond', and provided a major contribution to literary history introducing *essays* as a new literary genre; <u>Pierre Corneille</u> produced comedy *Mélite* 'Melite', plays *L'Aveugle de Smyrne* 'The Blind Man of Smyrna', and *La Grande pastorale* 'The Great Pastoral', great *Médée* 'Medea', the well-known *Le Cid*, tragedies *Horace* 'Horatius', *Cinna* 'Cinna's Conspiracy', and *Polyeucte* 'Polyeuctes', comedy *Le Menteur* 'The Mistaken Beau', as well as *Pulchérie*, and *Suréna*; <u>Madeleine de Scudéry</u> was author of romance *Ibrahim ou l'illustre Bassa* 'Ibrahim, or The Illustrious Bassa', famous *Artamène, ou le Grand Cyrus* 'Artamenes, or the Grand Cyrus', and then *Clélie* 'Clelia', and *Mathilde d'Anguilon* 'Mathilda of Aquilar'; <u>Jean Baptiste Poquelin Molière</u> produced comic works *Les Précieuses ridicules* 'The Conceited Young Ladies', masterpieces *Tartuffe*; *Le Misanthrope* 'The Misanthrope'; *Amphitryon*; *Le Bourgeois gentilhomme* 'The Citizen turned Gentleman'; and *George Dandin*; then *L'Avare* 'The Miser', *Les Fourberies de Scapin* 'The Cheats of Scapin', and *Malade imaginaire* 'The Imaginary Invalid'; <u>Jean Racine</u> wrote plays *La Thébaïde ou Les Frères ennemis* 'The Fatal Legacy', *Alexandre le grand* 'Alexander the Great', then *Andromaque* 'Andromache', *Les Plaideurs* 'The Litigants', *Britannicus*, *Bérénice* 'Titus and Berenice', *Bajazet* 'The Sultaness',

Mithridate 'Mithridates', masterpiece *Iphigénie* 'Achilles, or Iphigenia in Aulis', marvellous *Phèdre* 'Phaedre and Hippolytys', as well as plays *Ester* and *Athalie*; <u>Jean de La Fontaine</u> was author of the ballad *Les Rieurs du Beau-Richard*, elegy *Pleures*, and *Nymphes de Vaux*, as well as fables, including *Le coq et la perle* 'The Cock and the Pearl', *La colombe et la fourmi* 'The Dave and the Ant', *Le lion malade et le renard* 'The Fox and the Sick Lion', *Le cheval et l'âne* 'The Horse and the Donkey', *Le vieux chat et le jeune souris* 'The Old Cat and the Young Mouse', *Le rat de ville et le rat des champs* 'The Town Mouse and the Country Mouse', *Le loup et l'agneau* 'The Wolf and the Crane', and *Le loup devenu berger* 'The Wolf who Played Shepherd'; <u>Nicolas Boileau</u> wrote works including *L'Art poétique* 'The Art of Poetry', comic *Lutrin* 'Lectern', and *Dialogue des héros de roman* 'A Conversation between the Heroes of Novels'; <u>Prosper Jolyot de Crébillon</u> produced tragedies *Idoménée*, *Atrée et Thyeste*, *Électre*, and masterpiece *Rhadamiste et Zénobie*, as well as *Catalina*; <u>François Marie Arouet Voltaire</u> created great works, such as dramas *Œdipe* 'Oedipus', *Mahomet* 'Mohamet the Imposter', *Princesse de Navarre*, and *Irène*; poetry *La Ligue ou Henri le Grand* 'Henriade', and *Poème sur le désastre de Lisbonne* 'Poem on the Lisbon Earthquake'; philosophical tales *Lettres écrites de Londres sur les Anglais* 'Letters Concerning the English Nation', and *Candide* 'Candid'; as well as philosophical and historical writings *Traité de métaphysique* 'Treatise on Metaphysics', *Siècle de Louis Quatorze* 'The Age of Louis XIV', *Les Mœurs et l'esprit des nations* 'The General History and State of Europe', and *Dictionnaire philosophique portatif* 'The Philosophical Dictionary for the Pocket'; <u>Pierre Augustin Caron de Beaumarchais</u> produced plays *Eugénie* 'The School for Rakes', and *Les Deux Amis* 'The Two Friends'; reputed satire *Mémoires du Sieur Beaumarchais par lui-même* 'Autobiography'; and famous comedies *Le Barbier de Séville* 'The Barber of Seville', and *La Folle journée ou le mariage de Figaro* 'The Follies of a Day, or The Marriage of Figaro'; and <u>Madame de Staël</u> wrote the celebrated *Lettres sur Rousseau*; *Réflexions sur la paix intérieur* 'Reflexions on Civil Peace'; *Influence des passions*; *Littérature et ses rapports avec les institutions sociales* 'The Influence of Literature upon Society'; then the novel *Delphine*, and romance *Corinne*; as well as *De l'Allemagne* 'Germany';

1579-1781: <u>*English literature*</u> – <u>Edmund Spenser</u> was author of works such as *Shepheard's Calender*; *The Faerie Queene*; *Mother Hubberd's Tale*; *Colin Clout's Come Home Again*; *Complaints*; *The Early Tears of the Muses*; sonnet sequence *Amoretti*; supreme marriage poem *Epithalamion*; then *Four Hymns*, and *Prothalamion*; as well as prose *View of the Present State of Ireland*; Thomas Lodge published *Defence of Poetry*; *An Alarum against Usurers*; *The Detectable Historie of Forbonius and Priscilla*; *Scillaes Metamorphosis*; *Rosalinde*, supplying W Shakespeare with many incidents in 'As You Like It'; also a collection of poems, namely *Phillis*, and *A Fig for Momus*; Thomas Watson wrote lyric works, such as sonnets *Hecatompathia or Passionate Century of Love*, and *The Tears of Fancie*, which were probably studied by W Shakespeare; William Warner became known by writings such as *Pan his Syrinx Pipe*, and a long metrical history in 14-syllable verse named *Albion's England*; Christopher Marlowe produced noticeable writings, including *Tamburlaine the Great*, a renewed style of English tragedy; *The Tragical History of Dr Faustus*, a series of detached scenes; *The Jew of Malta*, an uneven work; *Edward II*, a mature play; as well as *The Tragedy of Dido*, and *Hero and Leander*, both unfinished; William Shakespeare created great Renaissance works, including early plays *The Two Gentlemen of Verona*, *Henry VI*, *Titus Andronicus*, *The Taming of the Shrew*, *The Comedy of Errors*, *Love's Labour's Lost*, *Romeo and Juliet*; histories *Richard III*, *Richard II*, *King John*, *Henry IV*, *Henry V*; later comedies *A Midsummer Night's Dream*, *The Merchant of Venice*, *The Merry Wives of Windsor*, *Much Ado About Nothing*, *As You Like It*, *Twelfth Night*, *Troilus and Cressida*, *Measure for Measure*, *All's Well That Ends Well*; Roman plays *Julius Caesar*, *Antony and Cleopatra*, *Coriolanus*; later tragedies *Hamlet*, *Othello*, *Timon of Athens*, *King Lear*, *Macbeth*; late plays *Pericles*, *The Winter's Tale*, *Cymbeline*, *The Tempest*, *Henry VIII*; as well as non-dramatic works *Venus and Adonis*, *The Rape of Lucrece*, *'The Phoenix and the Turtle'*; and finally *Sonnets*, and *'A Lover's Complaint'*; George Chapman produced *The Blind Beggar of Alexandria*, *The Gentleman Usher*, *Tragedie of Charle, Duke of Byron*, *The Widow's Tears*, *Caesar and Pompey*, *Euthymiae and Raptus*, *Petrarch's Seven Penitentiall Psalmes*, *The Divine Poem of Musaeus*, and *The Georgicks of Hesiod*; John Marston was author of the poem *The Metamorphosis of Pygmalion's Image and Certain Satires*; tragedies *Antonio and Mellida*, and *Antonio's Revenge*; comedy *The Malcontent*; as well as plays *The Dutch Courtesan*; *Parasitaster, or the Fawn*; *Sophonisba*; and *What You Will*; John Donne wrote passionate and erotic poems *Songs and Sonnets*; *Satires*; *Elegies*; verse *Anniversaire*; religious temper in lyrical form *Divine Poems*; prose works *Pseudo-Martyr*, and *Biothanatos*; as well as two sonnet sequences *La Corona*, and *Holy Sonnets*; Thomas Dekker wrote comedies *The Shoemaker's Holiday, or the Gentle Craft*; *The Pleasant Comedy of Old Fortunatus*; and *The Roaring Girl*; together with T Middleton, powerful drama *The Honest Whore*; with J Webster, plays *Famous History of Sir Thomas Wyat*; *Westward Ho!*; and *Northward Ho!*; and with F and W Rowley, powerful tragedy *The Witch of Edmonton*; as well as pamphlets *The Wonderful Year*, and *The Bellman of London*; Benjamin Jonson became known for his masterpieces *Volpone*, a satire on senile sensuality and greedy legacy hunters; *The Silent Woman*, a farcical comedy involving a heartless hoax; *The Alchemist*, with a plot and strict adherence to the unities; and *Bartholomew Fair*, unveiling author's anti-Puritan prejudices; Thomas Heywood wrote poetry, such as *Nine Bookes of Various History concerning Women*; plays, including domestic tragedy *A Woman Kilde with Kindnesse*; and *The English Traveller*; together with W Rowley, *Fortune by Land and Sea*; then *The Rape of Lucrece*, notable for its songs; *A Challenge for Beautie*, expressing tenderness; and *The Royal King and Loyall Subject*, stressing the doctrine of passive obedience to kingly authority; John Fletcher produced plays *The Faithful Shepherdess*; *The Humorous Lieutenant*; and *Rule a Wife and Have a Wife*; together with F Beumont, *Philaster*, *A King and No King*; and *The Maid's Tragedy*; as well as, together with W Shakespeare, *Two Noble Kinsmen*, and *Henry VIII*; John Webster wrote two best known tragedies *The White Devil*, and *The Duchess of Malfi*; Francis Quarles was author of verse, including *A Feast of Wormes*; *Argalus and Parthenia*; *Divine Poems*; *The Historie of Samson*; *Divine Fancies*; and *Emblems*; as well as prose such as *Enchyridion*, and *The Profest Royalist*; George Wither produced a book of five pastorals, *The Shepherds Hunting*; love elegy *Fidelia*; satirical *Wither's Motto*; his main poem *Fair Virtue, or the Mistress of Philarete*; then *Hymns and Songs of the Church*; *Psalms of David translated*; *Emblems*; and *Hallelujah*; John Milton was author of *The Doctrine and Discipline of Divorce*; *Poems*; *Areopagitica*; *A Speech for the Liberty of Unlicensed Printing*; *Eikonoklastes*; two *Defensiones*; *Paradise Lost*; *Paradise Regained*; and *Samson Agonistes*; Izaak Walton produced celebrated *The Compleat Angler, or the Contemplative Man's Recreation*, and several biographies of contemporary personalities; John Dryden wrote *Heroic Stanzas*, and *Astrea Redux*; verse plays *The Indian Emperor*, and *Aurungzebe*; comedy for stage *The Rival Ladies*; *Essay of Dramatic Poesy*; *Defence of the Epilogue*; *The Hind and the Panther*; and *Discourse Concerning the Origin and Progress of Satire*; William Wycherley produced comedies *Love in a Wood, or St James's Park*, and *The*

Gentleman Dancing-master, as well as plays *The Country Wife*, and *The Plain Dealer*; John Vanbrugh wrote *The Relapse, The Provok'd Wife, The Confederacy*, and unfinished *The Provok'd Husband*; Jonathan Swift was author of satires *A Tale of a Tub*, and *Gulliver's Travels*; verses *Journal to Stella, The Grand Question Debated*, and *Verses on His Own Death*; pamphlet *On the Conduct of the Allies*; poem *Cadenus and Vanessa*; verse satire *On Poetry; A Rhapsody*; ironical writings *Directions to Servants*, and *A Complete Collection of Genteel and Ingenious Conversation*; as well as *History of the Four Last Years of the Queen*; Daniel Defoe wrote books *Robinson Crusoe; Journal of the Plague Year; Moll Flanders; Roxana; Tour through the Whole Island of Great Britain; The Great Law of Subordination Considered*; and *Augusta Triumphans, or the Way to make London the Most Flourishing City in the Universe*; Edward Young produced tragedies *Busiris, The Revenge*, and *The Brothers*; satire *The Love of Fame, the Universal Passion*; poem *The Instalment*; and *The Complaint, or Night Thoughts on Life, Death and Immortality*; Samuel Richardson was author of novels *Pamela; Clarissa, Or the History of a Young Lady*; and *Sir Charles Grandison*; Henry Fielding wrote *An Apology for the Life of Mrs Shamela Andrews; The Adventures of Joseph Andrews and his Friend; Mr Abraham Adams; Miscellanies; The History of Tom Jones, a Foundling*; and *Amelia*; and Samuel Johnson (Dr Johnson) produced didactic poem *The Vanity of Human Wishes; Dictionary of the English Language*; moral fable *Rasselas: The Prince of Abyssinia*; and monumental *Lives of the Most Eminent English Poets*;

1585-1680: *Spanish literary language and literature* – Miguel de Cervantes Saavedra wrote Renaissance works, such as the pastoral romance *La Galatea*; plays *La Numancia*, and *El trato de Angel*; and immediately popular great writing *Don Quixote*; as well as tales, and a poem, founding the Spanish literary language; Lope de Vega produced works such as poem *Angelica*, miscellaneous *Rimas*, romance *Peregrino en su Patria* 'The Pilgrim of Casteele', religious pastoral *Pastores de Belén*, miscellanies *Filomana* and *Circe*, and epic *Rimas de Tomé de Burguillos*; Luis de Góngora was author of long poems such as *Soledades* 'The Solitudes', *Polifemo* 'Polyphemus and Galatea', and *Piramo y Tisbe* 'Pyramus and Thisbe', written in an affected style later called 'gongorism'; and Pedro Calderón de la Barca wrote famous dramas *La vida es sueño* 'Life's a Dream', and *El alcalde de Zalamea* 'The Mayor of Zalamea'; also writings *autos sacramentales* divided into biblical, classical, ethical, 'cloak and sword plays', and dramas of passion, the first of them being *El divino*

Orfeo 'The Divine Orpheus';

1654-59: *Dutch poetry and drama* – Joost van den Vondel produced poetry and dramatic works, such as *Lucifer*, and *Jephtha*, greatly influencing German poetical revival after the devastating Thirty Years War;

1759-73: *Irish literature* – Oliver Goldsmith wrote *Enquiry into the Present State of Polite Learning in Europe*; novel *The Vicar of Wakefield*; poetry *The Deserted Village*; and play *She Stoops to Conquer*;

1767-1912: *German literature* – Johann Wolfgang von Goethe was author of plays *Die Laune des Verliebten* 'The Beloved's Whim', and *Die Mitschuldigen* 'The Accomplices'; masterpiece drama *Götz von Berlichingen*; novels *Die Lieden des jungen Werthers* 'The Sorrows of Young Werther', and *Wilhelm Meisters Theatralische Sendung* 'Wilhelm Meister's Theatrical Mission'; verse dramas *Iphigenie auf Tauris*, *Egmont*, and *Torquato Tasso*; epic idyll *Hermann und Dorothea*; drama *Die natürliche Tochter* 'The Natural Daughter'; and novel *Die Wahlverwandtschaften* 'The Elective Affinities'; Johann Gottfried Herder wrote collection of folksongs *Stimmen der Völker in Liedern* 'Voices of the Peoples in Songs', treatise *Vom Geist der Ebraïschen Poesie* 'The Spirit of Hebrew Poetry', a version of *Cid*, and especially *Ideen zur Geschichte der Menschheit* 'Outlines of a Philosophy of the History of Man'; Johann Christoph Friedrich Schiller produced plays *Die Räuber* 'The Robbers', *Fiesko* 'Fiesco', *Kabale und Liebe* 'Cabal and Love', and *Don Carlos*; stories *Verbrecher aus verlorener Ehre* 'The Dishonoured Irreclaimable', and *Der Geisterseher* 'The Gost-Seer'; poems *An die Freude* 'Ode to Joy', and *Die Künstler* 'The Artists'; famous *Über naïve und sentimentalische Dichtung* 'On Simple and Sentimental Poetry'; celebrated *Xenien* 'Epigrams'; ballads *Der Taucher* 'The Diver', *Der Ring des Polykrates* 'The Ring of Polykrates', *Die Kraniche des Ibykus* 'The Cranes of Ibycus', and especially *Das Lied von der Glocke* 'Song of the Bell'; as well as the dramatic trilogy *Wallenstein*, the greatest historical drama in German language; then psychological study *Maria Stuart*, and half-legend *Wilhelm Tell*; Friedrich Hölderlin wrote poems, philosophical novel *Hyperion*, elegy *Menon's Laments for Diotima*, and works *Brot und Wein* 'Bread and Wine', and *Der Rhein* 'The Rhine'; Heinrich von Kleist was author of the fine tale *Michael Kohlhaas*, and popular play *Prinz Friedrich von Homburg* 'The Prince of Homburg'; Brothers Jacob Ludwig Carl Grimm and Wilhelm Carl Grimm founded science of comparative folklore, and wrote *Kinder- und Hausmärchen* 'Children and House Tales' as a foundation of science of comparative folklore; *Deutsche Sagen* 'German Sagas'; *Deutsche Grammatik* 'German Grammar'; *Geschichte der deutsche Sprache* 'History of German Language'; and *Deutsche Wörterbuch* 'German Dictionary'; <u>Ernst Theodor Wilhelm Hoffmann</u> produced short and long tales, and music critic, such as *Die Serapionsbrüder* 'The Serapion Brothers', *Elixiere des Teufels* 'The Devil's Elixirs', and *Lebensansichten des Katers Murr* 'Opinions of the Tomcat Murr'; <u>Heinrich Heine</u> wrote poetry, prose, and other works, including the well-known *Das Buch der Lieder* 'Book of Songs'; and <u>Gerhart Johann Robert Hauptmann</u> produced play *Vor Sonnenaufgang* 'Before Dawn'; then *Die Weber* 'The Weavers', introducing a theatrical phenomenon of 'collective hero'; *Florian Geyer*, marking transition to a mixture of fantasy and naturalism; *Die Versunkene Glocke* 'The Sunken Bell', and *Rose Bernd*; plays *Der Biberpeltz* 'The Beaver Coat', and *Der rote Hahn* 'The Conflagration'; as well as novels *Der Narr in Christo: Emanuel Quint* 'The Fool in Christ: Emanuel Quint', and *Atlantis*;

1771-1832: <u>*Scottish literature*</u> – <u>Robert Fergusson</u> wrote poems in Scots and English, including *The Daft Days*; *Auld Reekie*; *Elegy on the Death of Scots Music*; *Hallow Fair*; *To the Tron Kirk Bell*; *Leith Races*; and *The Rising of the Session*; <u>Robert Burns</u> was author of *Poems, Chiefly in the Scottish Dialect*; and songs such as *John Anderson My Jo*; *Age Fond Kiss*; *Ye Jacobi's by Name*; *The Banks Odon*; and *Auld Lang Sine*; and <u>Walter Scott</u> produced *The Minstrelsy of the Scottish Border*; *Waverley*; *Ivanhoe*; *The Talisman*; *Rob Roy*; *Life of Napoleon*; and *Tales of a Grandfather*;

1795-1904: <u>*English literature*</u> – <u>Robert Southey</u> became known by his *poems*, epic poem *Joan of Arc*, well-known biographies of *Nelson, Wesley*, and *Bunyan*, as well as *A Vision of Judgement, Naval History*, and *The Doctor*; <u>William Wordsworth</u> wrote *Lyrical Ballads*; poems *Michael*; *Ruth*; *Lucy*; *The Solitary Reaper*; *The Recluse*; *The Prelude*; and overall *The Excursion*; *Ecclesiastical Sonnets*; and *Memorials*; <u>George Gordon Noel Byron</u> produced satires, poems and oriental pieces, including *English Bards and Scotch Reviewers*; *Childe Harold's Pilgrimage*; as well as *Glamour*; *Lara*; *Siege of Corinth*; *Bipod*; *A Vision of Judgement*; and *Don Juan*; <u>Jane Austen</u> wrote novels and literary works, such as *Sense and Sensibility*; *Pride and Prejudice*; *Mansfield Park*; *Emma*; and *Persuasion*; <u>Percy Bysshe Shelley</u> created lyrical poetry and prose pieces, including *Queen Mob*; *The Assassins*; *The Revolt of Islam*; and masterpiece *Prometheus*

Unbounded; John Keats wrote sonnets, poems, romantic works and splendid romances such as *The Examiner*; *Edition*; *The Eve of St Agnes*; *On a Grecian Urn*; *To Autumn*; and *On Melancholy*; Charles John Huffman Dickens was author of great works including *Oliver Twist*; *Nicholas Nickel*; *David Copperfield*; and *Great Expectations*; William Makepeace Thackeray, wrote pieces such as *The Paris Sketchbook*; great *Vanity Fair*; *Henry Edmond*; and *The Newcomes*; Emily Jane Brontë wrote powerful novel *Wuthering Heights*, a tale of love and revenge in style of Greek tragedy; and her sister Charlotte Brontë produced masterpiece *Jane Eyre*, and other novels, memoires, and stories; William Morris became known by his poetry including *The Story of Sigurd the Volsung and the Fall of the Nibelungs*; *The Dream of John Ball*; *The Roots of the Mountains*; *The Wood beyond the World*; and *The Water of the Wondrous Isles*; and Thomas Hardy published novels, poems, and dramas such as *The Return of the Native*; *Wessex Poems*; and *The Dynasts*;

1801-1914: *French literature* – René de Chateaubriand produced works including *Atala*; *Le Génie du Christianisme* 'The Beauties of Christianity'; *Itinéraire de Paris à Jerusalem* 'Travels in Greece, Palestine, Egypt and Barbary'; and celebrated *Mémoires d'outre-tombe* 'Memoirs from Beyond the Grave'; Alphonse Marie Louis de Lamartine wrote poems and historical literature, including *Méditations*; *Harmonies poétiques et religieuses* 'Political and Religious Harmonies'; *Souvenirs d'Orient* 'Recollections of a Pilgrimage to the Holy Land'; *La Chute d'un ange* 'The Fall of an Angel'; *Histoire des Girondins* 'History of the Girondins'; and *Histoire de la Restauration*; Victor Marie Hugo produced odes and ballads, dramas, histories, and overall novels *Notre Dame de Paris*, and *Les Misérables*; Henry Marie Beyle Stendhal wrote masterpieces *Le Rouge et le noir* 'Red and Black', and *La Chartreuse of Parma* 'The Charterhouse of Parma'; Honoré de Balzac published novels unveiling a complete picture of modern civilization in his *Comédie humaine*, and other known works such as *Le Père Goriot*, and *Eugénie Grandet*; Théophile Gautier wrote poems including *Comédie de la mort* 'The Comedy of Death', as well as novels *Émaux et camées* 'Enamels and Cameos', and *Mademoiselle de Maupin*; Alexandre Dumas (Dumas père) produced novels and plays including the well-known *Le Comte de Monte Cristo* 'The Count of Monte Cristo', *Les Trois mousquetaires* 'The Three Musketeers', *Vingt ans après* 'Twenty Years After', *La Reine Margot* 'Queen Margot', and *La Tulipe noire* 'The Black Tulip'; Eugène Labiche wrote the novel *La Clef des champs* 'Key of the Fields', comedies, farces, and vaudevilles; <u>Charles Pierre Baudelaire</u> published his collection of poems *Les Fleurs du mal* 'The Flowers of Evil', *Les Paradis artificiels* 'The Artificial Paradises', and *Petit Poèmes en prose* 'Little Poems in Prose'; <u>Gustave Flaubert</u> published works including *Madame de Bovary*; *La Tentation de St Antoine*; and *Trois contes*; <u>Prosper Mérimée</u> wrote novels including *Carmen*; *Arsène Guillot*; and *L'Abbé Aubain*; Brothers <u>Edmond de Goncourt</u> and <u>Jules de Goncourt</u> published novels such as *Les Hommes de lettres* 'The Men of Letters', *Germinie Lacerteux*, and great *Madame Gervaisais*, as well as founded Académie Goncourt to foster fiction with *annual Prix Goncourt*; <u>Alphonse Daudet</u> produced theatrical pieces, notably *L'Arlésienne* 'A Woman from Arles', sketches, and short stories including *Tartarin de Tarascon*, and naturalistic novels such as *Sapho*, and *L'Immortel* 'The Immortal One'; <u>Jules Verne</u> became worldwide known for his novels such as *Cinq semaines en ballon* 'Five Weeks in a Balloon', *Voyage au centre de la terre* 'A Journey to the Centre of the Earth', *Vingt mille lieues sous les mers* 'Twenty Thousand Leagues Under the Sea', and *Le Tour du monde en quatrevingts jours* 'Around the World in Eighty Days'; <u>Sully-Prudhomme</u> wrote in Parnassian style pieces including *Stances et poèmes* 'Stanzas and Poems', *Impressions de la guere* 'Impressions of War', and *La Révolte des fleurs* 'The Flowers' Revolution'; <u>Émile Zola</u> produced great novels such as *Les Rougon-Macquart*; *Nana*; *La Bête Humaine* 'The Beast in Man'; as well as *La Débâcle* 'The Downfall'; and *Le Docteur Pascal* 'Doctor Pascal'; <u>Paul Verlaine</u>, was author of pieces including *Romances sans paroles* 'Romances Without Words', *Sagesse* 'Wisdom', *Poètes maudits* 'Accused Poets', and *Louis Leclerc*; <u>Stéphane Mallarmé</u> was one of leaders of *Symbolist school*, and became well-known by his poem *L'Après-midi d'un faune* 'A Faun's Afternoon', and *Les Dieux antiques* 'The Ancient Gods'; <u>Anatole France</u> produced stories, novels such as *Le Crime de Sylvestre Bonnard*, critical studies including the Parnassian *Le Livre de mon ami* 'My Friend's Book', satirical and sceptical works including *Les Opinions de Jérôme Coignard*, *Île des pingouins* 'Isle of Penguins', and *La Révolte des anges* 'The Angels' Revolt'; and <u>Guy de Maupassant</u> became known by novels such as *Boule de suif* 'Ball of Tallow', *La Peur* 'The Fear', and full-length novels such as *Une Vie* 'A Woman's Life', and *Bel-Ami*;

1807-27: *Irish literature* –<u>Thomas Moore</u> produced *Irish Melodies*; *The Twopenny Postbag*; *Lalla Rookh*; *The Loves of the Angels*; and novel *The Epicurean*; <u>William Butler Yeats</u> published works including *Mosada: A Dramatic Poem*; *Fairy and Folk Tales of the*

Irish Peasantry; *The Celtic Twilight*; *The Collected Works in Prose and Verse*; and *The Wild Swans at Coole*; <u>George Bernard Shaw</u> became well-known by works such as *The Quintessence of Ibsenism*; *The Perfect Wagnerite*; *Mrs Warren's Profession*; *Three Plays for Puritans*; *Man and Superman*; *The Doctor's Dilemma*; *Androcles and the Lion*; *Common Sense About the War*; *Heartbreak House*; *Back to Methuselah*; *The Intelligent Woman's Guide to Socialism and Capitalism*; *The Black Girl in search of God*; *Too True to Be Good*; and *The Simpleton of the Unexpected Isles*; <u>Bram Stoker</u> wrote *The Jewel of the Seven Stars*; *The Lady of the Shroud*; *The Lair of the White Worm*; and overall classic vampire story *Dracula*, by which he became a celebrity; and <u>James Augustine Aloysius Joyce</u> wrote *Chamber Music*; *Dubliners*; and seminal novel *Ulysses*;

1808-1916: <u>*US literature*</u> – <u>Washington Irving</u> pioneered American literature, writing satirical essays, boisterous works, tales and historical studies, such as *Salmagundi*; *A History of New York, by Diedrich Knickerbocker*; *Tales of a Traveller*; *The History of the Life and Voyages of Christopher Columbus*; *The Conquest of Granada*; *Tour on the Prairie*; and *The Adventures of Captain Bonneville*; <u>James Fenimore Cooper</u> produced stories of sea and of Native American Indians, including *The Spy*; *The Pilot*; *The Last of the Mohicans*; *The Prairie*; *The Red Rover*; *The Bravo*; *The Pathfinder*; *The Deerslayer*; *The Two Admirals*; *Wing-and-Wing*; and *Satanstoe*; <u>Edgar Allan Poe</u> wrote the volumes of verse *Tamerlane and other Poems*, and *Al Aaraaf*; then *Narrative of Arthur Gordon Pym*; *Tales of the Grotesque and Arabesque*; and *The Raven and Other Poems*, thus pioneering the modern detective story; <u>Nathaniel Hawthorne</u> published novels, stories, sketches, and studies, including best-known *The Scarlet Letter*; *Wonder Book*; and *Tanglewood Tales*; <u>Harriet Beecher Stowe</u> produced anti-slavery novels *Uncle Tom's Cabin*, and *Dred: A Tale of the Dismal Swamp*; <u>Herman Melville</u> wrote masterpiece *Moby-Dick*, novel *Pierre*, and short stories such as *The Piazza Tales*; <u>Walter Whitman</u> published poems including *Leaves of Grass*; *Drum Taps*; and *Sequel to Drum Taps*; <u>Mark Twain</u> produced works such as *The Prince and the Pauper*; *A Connecticut Yankee in King Arthur's Court*; and especially well-known *Tom Sawyer*, and *Huckleberry Finn*; and <u>Henry James</u> wrote works including *Washington Square*; *Portrait of a Lady*; *The Bostonians*; *The Tragic Muse*; *The American Scene*; and *The Altar of the Dead*;

1820-1923: <u>*Russian literature*</u> – <u>Alexander Sergeyevich Pushkin</u> created noticeable poems, novels, splendid romance *Eugene Onegin*, and drama *Boris Godunov*; <u>Nikolai Vasilevich Gogol</u> published his major work *Vechera na khutore bliz Dikanki* 'Evenings on a Farm near Dikanka', and great novel *Myortvye dushi* 'Dead Souls'; <u>Mikhail Yuriyevich Lermontov</u> wrote impressive novel *Geroy nashevo vremeni* 'A Hero of Our Times', and romantic verse play *Maskarad* 'Masquerade'; <u>Ivan Sergeyevich Turgenev</u> was author of plays and novels, including *A Month in the Country*; *Zapiski okhotnika* 'A Sportsman's Sketches'; *Ottsy I dety* 'Fathers and Sons'; and powerful piece *Stepnoy Korol'Lir* 'A Lear of the Stepes'; <u>Leo Nikolayevich Tolstoy</u> produced works such as the great *Voinya i mir* 'War and Peace', and *Anna Karenina*; as well as *Plody prosveshcheniya* 'The Fruits of Enlightenment', and *Voskreseniye* 'Ressurection'; <u>Nikolai Alekseyevich Nekrasov</u> wrote *poems* depicting social wrongs of peasantry, and unfinished narrative epic *Komu na Rusi zhit khorosho?* 'Who Can Be Happy and Free in Russia?'; <u>Fyodor Mikhailovich Dostoevsky</u> produced great novels such as *Crime and Punishment*; *The Idiot*; *The Devils*; and *Brothers Karamazov*; making him widely recognized as a deep thinker; <u>Anton Pavlovich Chekhov</u> wrote *Pëstrye Rasskazy* 'Motley Stories', *Medved* 'The Bear', *Predlozheniye* 'A Marriage Proposal', *Dyadya Vanya* 'Uncle Vanya', *Vishnyovy Sad* 'The Cherry Orchard', and *Tri Sestry* 'The Three Sisters'; and <u>Maxim Gorky</u>, was author of story *Chelkash*, transitional Romanticism-Realism *Foma Gordeyev*, better-known play *Na dne* 'The Lower Depths', as well as autobiographical trilogy *Detstvo* 'My Childhood', *Vlyudakh* 'In the World', and *Moi universitety* 'My University';

1835-55: <u>*Danish literature*</u> – <u>Hans Christian Andersen</u> published great stories, poetry, travel books, novels, plays, and pamphlets of fairy tales, such as *The Emperor's New Clothes*; *The Snow Queen*; and *The Little Mermaid*; which spread across Europe;

1844-1906: <u>*Hungarian literature*</u> – <u>Sándor Petöfi</u> produced works including the popular *János vitéz* 'Janos the Hero', and the novel *A hóhér kötele* 'The Hangman's Rope'; and <u>Endre Ady</u> broke away from prevailing conservative poetry, according to Symbolist procedures, and produced the incisive collection *Uj versek* 'New Verses';

1850-1900: *Modern languages* – emerged in all inhabited continents and islands;

1850-1937: <u>*Norwegian literature*</u> – <u>Henrik Ibsen</u> was author of dramas including *Samfundets støtter* 'Pillars of Society', *Fruen fra havet* 'The Lady from the Sea', and *Bygmester Solness* 'The Master Builder'; and <u>Knut Hamsun</u> wrote novels *Sult* 'Hunger', *Mysterier*

'Mysteries', lyrical *Pan*, and masterpiece *Markens grøde* 'Growth of the Soil';

1862-64: <u>*Icelandic literature*</u> – <u>Jón Árnason</u> assembled folk-tales and fairy-tales in his *Íslenskar þjóðsögur og œvintýri* 'Icelandic Legends';

1870-1938: <u>*Romanian literary language and literature*</u> – <u>Mihail Eminescu</u> perfected the lyric verse, as in *Călin (Pages from a Fairy Tale)*; *Oh Mother*; *The Satires* 'Epistle-satires'; *From that Star*; and especially the long poem *Luceafărul* 'The Evening Star'; as well as prose including *The Tear Drop Prince*; *Empty Genius*; *Wretched Dionis*; and *Caesara*; expressing the Latin structure and vocabulary of literary Romanian language; <u>Ion Luca Caragiale</u> produced plays, short-stories, and dramas such as *Mr Leonida*; *A Stormy Night*; and *A Lost Letter*; then *The False Accusation*; *An Easter Torch*; and *Kir Ianulea*; <u>Barbu Stefanescu Delavrancea</u> published historical writings and pieces including *Etymologicum magnum Romaniae*, trilogy *The Sunset, The Snowstorm* and *Venus*, as well as *Sic Cogito*, and *Hagi-Tudose*; <u>George Cosbuc</u> wrote *Ballads and Pastorals*; *We Demand Land*; and *Life's Struggle*; *The Mother*; *Zamfira's Wedding*; *El Zorab*; stories assembled in *Verses and Prose*; as well as collections *Threads of Spun-yarn*; and *Ballades and Idylls*; <u>Gheorghe Toparceanu</u> published volumes *Ballads; Merry and Sad*; *Original Parodies*; and *Bitter Almonds*; celebrated poems *The Ballad of a Tiny Cricket*; *Fall Rhapsodies*; *The Bullet Train*; and *The Crow*; as well as *Memoires from the Battle of Turtucaia*; *Letters with No Address*; and *Pirin-Planina*; and <u>Liviu Rebreanu</u> produced short stories *The Hooligans*; *Confession*; and *Resentfulness*; modern novels *Ion* 'John'; *The Little King*; *The Revolt*; and *The Gorilla*; psychological novels *Forest of the Hanged*; *Adam and Eve*; and *Ciuleandra*; as well as plays *The Quadrille*; *The Envelope*; and *The Apostles*;

1872-1918: <u>*Swedish literature*</u> – <u>August Strindberg</u> was author of dramas, novels, and plays such as *Mäster Olof*; *Gustav Vasa*; *Erik XIV*; *Hemsöborna* 'The People of Hemsö'; and trilogy *Till Damaskus* 'To Damascus'; and <u>Selma Ottiliaa Lovisa Lagerlöf</u> published *Gösta Berlings saga* 'The Story of Gösta Berling'; *Antikrists Mirakler* 'The Miracles of Anti-Christ'; children's classic *Nils Holgerssons underbara resa genom Sverige* 'The Wonderful Adventures of Nils'; and trilogy *The Rings of the Lowenskolds*;

1878-1940: <u>*Indian philosophic literature*</u> – <u>Rabindranath Tagore</u> wrote poetry and prose such as *A Poet's Tale*; *Karuna*; *The Tragedy of Rudachandra*; major work *Binodini*; *The Crescent Moon*; *Chitra*; and *My Reminiscences*;

1879-1965: _Italian literature_ – Gabriele d'Annunzio wrote _Primo vere_ 'In Early Spring', trilogy comprising novels _Il Piacere_ 'The Child of Pleasure', _L'Innocente_ 'The Intruder', and _Il Trionfo della morte_ 'The Triumph of Death'; tragedies _La Gioconda_, and _Francesca da Rimini_; as well as the great play _La figlia di Jorio_ 'The Daughter of Jorio'; Luigi Pirandello produced powerful and realistic _Il Fu Mattia Pascal_ 'The Late Mattia Pascal', _Si Gira_ 'Shoot!', _Sei personaggi in cerca d'autore_ 'Six Characters in Search of an Author', and _Come Tu Mi Vuoi_ 'As You Desire Me'; and Salvatore Quasimodo was author of works such as _Ed è Subito Sera_ 'And Suddenly It is Evening', _La Vita non è sogno_ 'Life is Not a Dream', _La Terra impareggiabile_ 'The Matchless Earth', and the collection _Selected Poems_;

1884-1919: _Polish literature_ – Henryk Sienkiewicz produced classical novels such as the trilogy _Ogniem I mieczem_ 'With Fire and Sword', _Potop_ 'The Deluge', _Pan Wolodyjowski_ 'Pan Michael', and the well-known _Quo Vadis?_; and Władysław Stanisław Reymont wrote novels including _Ziemia obiecana_ 'The Promised Land', his masterpiece _Chłopi_ 'The Peasants', and the historical trilogy _Rok 1794_ 'The Year 1794';

1886-1936: _British literature_ – English Rudyard Kipling created successful collections of verse _Barrack Room Ballads_, and _The Seven Seas_; classic animal stories _Jungle Books_; as well as _Rewards and Fairies_, and _Debits and Credits_; Scottish Arthur Conan Doyle produced historical romances; novel _Rodney Stone_; famous modern detective _A Study in Scarlet_; well-known serial _The Adventures of Sherlock Holmes_; and books _The Sign of Four_, and _The Hound of the Baskervilles_; Irish Oscar Fingal Wilde was author of the classic children's stories _The Happy Prince and Other Tales_; novel _The Picture of Dorian Gray_; dramatic-like _Lady Windermere's Fan_; _A Woman of No Importance_; and _An Ideal Husband_; as well as the masterpiece _The Importance of Being Earnest_; and Irish Edith Somerville wrote novels such as _An Irish Cousin_; _The Real Charlotte_; and _Some Experiences of an Irish R.M._;

1889-1910: _Belgian literature_ – Maurice Maeterlinck published the volume of poetry _Les Serres chaudes_ 'The Greenhouses'; prose plays _La Princesse Maleine_ 'The Princess Maleine'; _Pelléas et Mélisande_; _Joyzelle_; _Marie-Magdeleine_; and popular writing _La Vie des abeilles_ 'The Life of the Bee';

1895-1951: _French literature_ – Paul Valéry wrote poetry and poems _La Jeune parque_ 'The Young Fate', and _Charmes ou poems_; aesthetic studies such as *Eupalinos ou l'architecte* 'Eupalinos, or the Architect', and *L'Âme et la danse* 'Dance and the Soul'; and also the short play *Le Solitaire* 'The Solitary Man'; <u>André Paul Guillaume Gide</u> produced novels such as *Les Caves du Vatican* 'The Vatican Swindle', *La Symphonie pastorale* 'Two Symphonies', and *Les Faux-monnayeurs* 'The Counterfeiters'; <u>Jules Romains</u> was author of poems *La Vie unanime* 'The Unanimous Life', *Manuel de deification* 'A Treatise on Deification'; novels *Mort de quelqu'un* 'The Death of a Nobody', and *Les Copains* 'The Friends'; successful play *Knock, ou le triomphe de la médicine* 'Doctor Knock'; and great cycle *Les Hommes de bonne volonté* 'Men of Good Will'; <u>Roger Martin du Gard</u>, produced eight-novel series *Les Thibault* 'The Thibaults'; <u>Romain Rolland</u> published *Beethoven*; *Jean-Christophe*; *Au dessus de la mêlée* 'Above the Fray'; and *L'Âme enchantée* 'The Enchanted Spirit'; <u>Guillaume Apollinaire</u> wrote poetry including *L'Enchanteur pourrissant* 'The Decaying Magician', and *Calligrammes*; play *Les Mamelles de Tirésias* 'The Breasts of Tiresias' coining the term 'surrealist'; and the Modernist manifesto *L'Esprit nouveau et les poètes* 'The New Spirit and the Poets'; <u>Marcel Proust</u> created the major *À la recherche du temps perdu* 'Remembrance of Time Past' including a series of autobiographical novels such as *Du côté de chez Swann* 'Swann's Way', *À l'ombre des jeunes filles en fleur* 'Within a Budding Grove', *Le Côté de Guermantes* 'The Guermantes' Way', and *Sodome et Gomorrhe* 'The Cities of the Plain'; and <u>Max Jacob</u>, wrote the collection of prose-poems *Le Cornet de dés* 'The Dice Cup', and mystical *L'Homme de cristal* 'The Crystal Man';

1895-1962: *English literature* – <u>Herbert George Wells</u> wrote *The Time Machine*; *The Invisible Man*; *The War of the Worlds*; *The First Men in the Moon*; *Mr Britling Sees It Through*; *The Outline of History*; *The World of William Clissold*; *The Shape of Things to Come*; and *Mind at the End of its Tether*; <u>John Galsworthy</u> produced collection of short stories *From the Four Winds*; novels *The Island Pharisees*, and *The Patrician*; plays *Strife*; *Justice*; *The Skin Game*; and *Loyalties*; as well as two cycles of celebrated series *Forsyte Saga*; <u>Virginia Woolf</u> wrote *The Voyage Out*; *Night and Day*; *Mrs Dalloway*; *To the Lighthouse*; *Orlando*; *The Waves*; *The Years*; *Three Guineas*; *Between the Acts*; and the outstanding *Jacob's Room*; and <u>Aldous Leonard Huxley</u> was author of *Crome Yellow*; *Antic Hay*; *Those Barren Leaves*; *Point Counter Point*; *Proper Studies*; the famous *Brave New World*; *Eyeless in Gaza*; *After Many a Summer*; *Time must have a Stop*; study *The Devils of Loudun*; *The Doors of Perception*; *Heaven and Hell*; as well as the optimistic Utopian

Island;

1900-42: *Austrian literature* – <u>Rainer Maria Rilke</u> wrote *Vom lieben Gott und Anderes* 'Stories of God'; *Das Stundenbuch* 'Poems for the Book of Hours'; *Auguste Rodin*; *Neue Gedichte* 'New Poems'; *Die Aufzeichnungen des Malte Laurids Brigge* 'Journal of My Other Self'; and overall the major *Die Sonnette an Orpheus* 'Sonnets to Orpheus', and *Duineser Elegien* 'Duino Elegies'; <u>Franz Kafka</u> produced influential novels *Prozess* 'The Trial', *Das Schloss* 'The Castle', and *Amerika*; and <u>Stefan Zweig</u> was author of short stories such as *Kaleidoskop* 'Kaleidoscope', and especially novels *Der Zwang* 'Passion and Pain', and *Ungeduld des Herzens* 'Beware of Pity', all notable for their deep psychological insights;

1903-62: *US literature* – <u>Jack London</u> wrote novels such as *The Call of the Wild*; *The Sea-Wolf*; *White Fang*; *The Iron Heel*; *Martin Eden*; and *John Barleycorn*; <u>William Carlos Williams</u> published poems and novels including *Poems*; *The Tempers*; *Sour grapes*; *Spring and All*; *The Great American Novel*; *In the American Grain*; *White Mule*; *In the Money*; *The Build-Up*; the masterpiece *Paterson*; and *The Collected Later Poems*; *The Collected Earlier Poems*; and *Pictures from Breughel*; and <u>Robert Lee Frost</u> was author of *A Boy's Will*; *North of Boston*; and volumes *West-Running Brook*; *A Witness Tree*; *Steeple Bush*; and *In the Clearing*;

1904-55: *Swiss literature* – <u>Hermann Hesse</u> published novel *Peter Camenzind*; prose *Rosshalde*; *Knulp*; *Demian*; *Narziss und Goldmund* 'Death and the Lover'; *Steppenwolf*; *Das Glasperlenspiel* 'The Glass Bead Game'; and poetry *Die Gedichte* 'Hours in the Garden and Other Poems', as well as *Beschwörungen* 'Affirmations';

1907-43: *Welsh literature* – <u>William Henry Davies</u> published *A Soul's Destroyer*; *The Autobiography of a Super-tramp*; autobiographical *Beggars*; *The True Traveller*; *A Poet's Pilgrimage*; *Later Days*; and *Adventures of Johnny Walker*;

1909-14: *Futurism* – a literary and artistic movement initiated by <u>Emilio Filippo Tommaso Marinetti</u>, Italian poet, who published, in the newspaper *Le Figaro*, his *Manifesti del Futurismo* 'Futurist Manifesto', glorifying the war, machine age-speed, and a sort of 'dynamism' as a revolt against tradition;

1916-20: *Dadaism* – a movement in literature and art promoted by <u>Hugo Ball</u>, German artist, <u>Hans Arp</u>, Alsatian sculptor and poet, and <u>Tristan Tzara</u>, Romanian poet, as opposing war and cultural values, but espousing anarchic individualism and freedom from artistic convention;

1918-25: *Chinese literature* – <u>Lu Xun</u> wrote short-stories such as *Diary of a Madman*, successful *The True Story of Ah Q*, and cycles *Cry*, and *Hesitation*;

1918-90: *Soviet literature* – <u>Vladimir Vladimirovich Mayakovsky</u> published the play *Misteriya-Buff* 'Mystery-Bouffe'; the long poem *150,000,000*; political *Pro eto* 'About This'; satirical plays *Klop* 'The Bedbug', and *Banya* 'The Bath-House'; and finally *Vladimir Ilych Lenin*, and *Khorosho!* 'Good!'; <u>Ilya Grigorevich Ehrenburg</u> wrote *The Extraordinary Adventures of Julio Jurenito*; *The Fall of Paris*; *The Storm*; *The Thaw*; and *People, Years, Life*; <u>Boris Leonidovich Pasternak</u> produced *Devyat'sot pyaty god* 'The Year 1905'; *Vtoroye rozhdeniye* 'Second Birth'; *Detstvo Lyuvers* 'The Childhood of Luvers'; *Provest'* 'The Last Summer'; and the famous *Doktor Zhivago* 'Doctor Zhivago'; and <u>Mikhail Aleksandrovich Sholokhov</u> was author of the masterpiece *Tikhy Don* 'And Quiet Flows the Don', or 'The Don Flows Home to the Sea'; and *Podnyataya tselina* 'Virgin Soil Up-turned and Harvest on the Don'; <u>Aleksandr Isayevich Solzhenitsyn</u> became known by his novel *Odin den'Ivana Denisovicha* 'One Day in the Life Ivan Denisovich'; which was followed by other novels titled *Rakovy Korpus* 'The Cancer Ward', and *V Kruge pervom* 'The First Circle'; then the factual account of Stalinist terror *Arkhipelag Gulag* 'The Gulag Archipelago'; and also *Bodalsya telyonok s dubom* 'The Oak and the Calf', and *Kak Nam Obustroit' Rossiyu?* 'Rebuilding Russia';

1918-83: *French literature* – <u>André Maurois</u>, published *Les Silences du Colonel Bramble* 'The Silences of Colonel Bramble', *Les Discours du Docteur O'Grady* 'The Speeches of Dr O'Grady', and biographical *Ariel*; then *Disraeli*; *Voltaire*; *À la recherche de Marcel Proust* 'In Search of Marcel Proust'; and *La vie de Sir Alexander Fleming* 'The Life of Sir Alexander Fleming'; <u>François Mauriac</u> produced *Le Baiser au lépreux* 'The Kiss to the Leper'; *Génitrix*; *Thérèse Desqueyroux*; *Le nœud de vipères* 'Vipers' Tangle'; and *Asmodée*; <u>André Malraux,</u> wrote *Les Conquérants* 'The Conquerors', *La Condition humaine* 'Man's Fate', *L'Espoir* 'Man's Hope', *La Psychologie de l'art* 'The Psychology of Art', and *Les Voix du silence* 'The Voices of Silence'; <u>Georges Joseph Christian Simenon</u> created famous novels *M. Gallet décède* 'The Death of Monsieur Gallet', *Le Pendu de Saint-Pholien* 'The Crime of Inspector Maigret', and *Les Mémoires de Maigret* 'Maigret's Memoirs'; <u>Jean-Paul Sartre</u> published *La Nausée* 'Nausea'; *Le Mur*; *Les Mouches* 'The Flies'; *Les Mains sales* 'Crime

Passionnel'; *L'Être et le néant* 'Being and Nothingness'; and *Les Chemins de la liberté* 'The Paths of Freedom'; <u>Albert Camus</u> was author of *L'Étranger*; *Le Myth of Sisyphus*; the masterpiece *La Peste*; and ironic *La Chute*; <u>Jean Genet</u> produced *Notre-Dame des fleurs* 'Our Lady of the Flowers', *Miracle de la rose* 'Miracle of the Rose', *Pompes funèbres* 'Funeral Rites', *Les Condamnés á mort* 'Those Condemned to Death', and *Chants Secrets* 'Secret Songs'; and <u>Françoise Sagan</u> wrote *Bonjour tristesse*; *Un Certain Sourire* 'A Certain Smile'; *Dans un mois, dans un an* 'Those Without Shadows'; *Aimez-vous Brahms* 'Goodbye Again'; *La Chamade*; and *Le Rendez-vous manqué* 'The Missed Rendezvous'; also plays *Château en Suède* 'Castle in Sweden', and *Un Piano dans l'herbe* 'A Piano on the Grass'; as well as novels *La Femme fardée* 'The Painted Lady', and *Un Orange immobile* 'The Still Storm';

1918-95: *German literature* – <u>Bertolt Friedrich Brecht</u>, published *Trommeln in der Nacht* 'Drums in the Night', *Mann ist Mann* 'A Man's a Man', *Dreigroschenoper* 'The Three-penny Opera', *Mutter Courage und ihre Kinder* 'Mother Courage and her Children', and *Furcht und Elend des dritten Reiches* 'Fear and Loathing under the Third Reich'; and <u>Thomas Mann</u> produced *Betrachtungen eines Unpolitischen* 'Reflections of a Non-political Man', *Der Zauberberg* 'The Magic Mountain', *Achtung Europa! Deutsche Hörer!* 'Listen Germany: Twenty-Five Messages to the German People over the BBC'; the great *Doktor Faustus*; and the comic *Bekenntnisse des Hochstaplers Felix Krull* 'Confessions of Felix Krull, Confidence Man'; <u>Lion Feuchtwanger</u> was author of works including *Die hässliche Herzogin* 'The Ugly Duchess', *Jud Süss* 'Jew Süss', *Erfolg* 'Success'; and part-biographies of F Goya, and JJ Rousseau; <u>Heinrich Böll</u> published *Der Zug war pünktlich* 'The Train was on Time'; trilogy *Und sagte kein einziges Wort* 'Acquainted with the Night', *Haus ohne Hüter* 'The Unguarded House', and *Das Brot der frühen Jahre* 'The Bread of our Early Years' depicting life in Germany during and after Nazi regime; then later novels including *Gruppenbild mit Dame* 'Group Portrait with Lady', and *Die vorlorene Ehre der Katherina Blum* 'The Lost Honour of Katherina Blum'; and <u>Günter Wilhelm Grass</u> wrote novel *Die Blechtrommel* 'The Tin Drum', and important books such as *Katz und Maus* 'Cat and Mouse', *Hundejahre* 'Dog Years', *Örtlich betäubt* 'Local Anaesthetic', *Der Butt* 'The Flounder', *Das Treffen in Telgte* 'The Meeting at Telgte', *Die Ratten* 'The Rats', *Unkenrufe* 'The Call of the Toad', and *Ein weites Feld* 'A Broad Field';

1920-92: *US literature* – Thomas Stearns Eliot was author of *The Sacred Wood*; *Homage to Dryden*; *The Use of Poetry and the Use of Criticism*; *Elizabethan Essays*; *On Poetry and Poets*; *After Strange Gods*; *Essays Ancient and Modern*; *The Idea of a Christian Society*; *Notes towards the Definition of Culture*; *For Lancelot Andrewes*; then *The Rock*; *Murder in the Cathedral*; *The Cocktail Party*; *The Confidential Clerk*; and *The Elder Statesman*; Ernest Millar Hemingway, produced *Three Stories and Ten Poems*; *Our Time*; *The Sun Also Rises*; *Men Without Women*; *A Farewell to Arms*; *Death in the Afternoon*; *Green Hills in Africa*; *For Whom the Bell Tolls*; *Across the River and Into the Trees*; and also *The Old Man and the Sea*; Eugene Gladstone O'Neill published *The Great God Brown*; *Marco Millions*; *Strange Interlude*; *Lazarus Laughed*; *Mourning Becomes Electra*; *Ah, Wilderness*; *Days Without End*; *The Iceman Cometh*; *A Moon for the Misbegotten*; *Long Day's Journey into Night*; *A Touch of the Poet*; and *Hughie*; William Faulkner wrote *The Sound and the Fury*; *Sartoris*; *As I Lay Dying*; *Sanctuary*; *Light in August*; *Absalom, Absalom!*; *Hamlet*; *Intruder in the Dust*; *A Fable*; *The Town*; and *The Mansion*; Richard Ghormley Eberhart was author of *A Bravery of Earth*; *Reading the Spirit*; *The Quarry*; *Fields of Grace*; *Selected Poems 1930-1965*; and *Collected Poems 1930-1976*; John Ernest Steinbeck, published *Tortilla Flat*; *In Dubious Battle*; *Of Mice and Men*; *The Moon is Down*; *The Pearl*; *Burning Bright*; *East of Eden*; *Winter of our Discontent*; also humorous *Cannery Row*; and *The Short Reign of Pippin IV*; Kenneth Lee Pike wrote *Phonetics*; *Phonemics*; *Tone Languages*; and *Language in Relation to a Unified Theory of the Structure of Human Behaviour*; and Toni Morrison (Chloe Anthony) produced the story *The Bluest Eye*; novels *Sula*; *Song of Solomon*; *Tar Baby*; *Beloved*; *Jazz*; and *Paradise*; as well as the study *Playing in the Dark: Whiteness and the Literary Imagination*;

1921-77: *Czech poetry* – Jaroslav Seifert produced collections *Město v slzáck* 'City of Tears', and *Samá láska* 'All Love'; as well as patriotic *Přilba Llíny* 'A Helmet of Earth', and *Morový sloup* 'The Prague Column';

1922-85: *Japanese literature* – Yasunari Kawabata wrote short stories *Tales to hold in the Palm of your Hand*; and novels *Izu no odoriko* 'The Izu Dancer', *Yukiguni* 'Snow Country', *Sembazuru* 'Thousand Cranes', and *Yama no oto* 'The Sound of the Mountain'; Abe Kobo, produced novels *Daiyon Kampyoki* 'Inter Ice Age Four', *Suna no onna* 'The Woman in the Dunes', and *Mikkai* 'Secret Rendezvous'; and Oë Kenzaburo published *Shisha no Ogori* 'The Arrogance of the

Dead'; *Nip the Buds, Shoot the Kids*; *Kojinteki na taiken*; *A Personal Matter*; and *Man'en gannen no futtuboru* 'The Silent Cry', as well as short novels collected as *Teach Us to Outgrow Our Madness*; and *The Crazy Iris and Other Stories of the Atomic Aftermath*;

1923-80: *French literature due to Romanians* – Panait Istrati became known by works such as *The Thistles of the Bărăgan*, and *Kira Kyralina*, as well as *Vers l'autre flamme, confession pour vaincus* directed against USSR's Stalinist regime; and Eugen Ionesco introduced a new style of drama called the *Theatre of the Absurd*, comprising *La Cantatrice chauve* 'The Bold Prima Donna'; *Les Chaises* 'The Chairs'; *Amédée*; *Le Tableau* 'The Picture'; *Rhinocéros*; *Jeux de massacre* 'Wipe-Out Game'; *Macbett*; *Voyages chez les morts ou Thème et variations* 'Journey Among the Dead'; and the novel *Le Solitaire* 'The Hermit';

1924-66: *Chilean poetry* – Pablo Neruda published *Veinte poemas de amor y una canción desesperada* 'Twenty Love Poems and a Song of Despair', *Residencia en la tierra* 'Residence on Earth', *Alturas de Macchu Picchu* 'The Heights of Macchu Picchu', *Odas elementales* 'Elementary Odes', and comprehensive *Canto General* 'Poems from Canto General';

1927-48: *Indian literature* – Mahatma Gandhi (Great Soul) produced autobiographical *The Story of My Experiment with Truth*, and works comprised in the 90-volume *The Collected Works of Mahatma Gandhi*, with a great influence for peace as a message not only for India but for entire world;

1931-65: *Greek literature* – George Seferis published a collection of poetry entitled *Strophe* 'Turning Point' with an immediate success, *Mythistorima* 'Myth History' containing some of first free-verse Greek poems, and also collections including the three-volume *Hemeologhia Katastromatos* 'Logbook';

1931-68: *Icelandic prose* – Halldór Kiljan Laxness wrote epic novels like *Salka Valka*; *Sjálfstætt fólk* 'Independent People'; *Heimsljós* 'World Light'; *Íslandsklukkan* 'Iceland's Bell'; *Atómstöðin* 'The Atom Station'; *Gerpla* 'The Happy Warriors'; *Brekkukotsannáll* 'The Fish Can Sing'; *Paradisarheimt* 'Paradise Reclaimed'; and *Kristnihald undir Jökli* 'Christianity at Glacier';

1931-2004: *English literature* – John Betjeman published the collection of verse *Mount Zion; or, In Touch with the Infinite*; the book *Ghastly Good Taste*; and other collections including *Continual Dew: A Little Book of Bourgeois Verse*; *Old Lights for New Chancels*;

New Bats in Old Belfries; *A Few Late Chrysanthemums*; and *Collected Poems*; also *Cornwall*; and the volumes *A Nip in the Air*, and *High and Low*; Roald Dahl wrote collections *Someone Like You*; *Kiss Kiss*; and *Switch Bitch*; novel *My Uncle Oswald*; children's stories *Charlie and the Chocolate Factory*; *Charlie and the Great Glass Elevator*; *James and the Giant Peach*; *Fantastic Mr Fox*; *The Enormous Crocodile*; *The BFG*; *Matilda*; and *Esio Trot*; as well as screen-plays *You Only Live Twice*, and *Chitty Chitty Bang Bang*; Catherine Ann Cookson, inspired mainly by her deprived youth in South Tyneside, North-east England, she was a prolific writer with more than 70 books published, notably tragedy and romance, including *Mallen trilogy*, and *Tilly Trotter series*; she became well-known for her filmed *The Fifteen Streets*; the books *A Grand Man*, as a basis of the film 'Jacqueline', and *Rooney*, also adapted for screen; then *Katie Mulholland*, adapted into a stage musical; as well as the filmed *The Black Valvet Gown*; Ted Hughes wrote *The Hawk in the Rain*; *Wodwo*; *Crow*; *Cave Birds*; *Season Songs*; *Gaudete*; *Moortown*; *Remains of Elmet*; *River*; *Wolf Watching*; *Tales from Ovid*; *The Iron Man*; and *The Iron Woman*; as well as books for children, collection *Rain-Charm for the Duchy and other Laureate Poems*, and essays such as *Winter Pollen: Occasional Prose*; Harold Pinter produced *The Birthday Party*, and filmed work *The Caretaker*; which were followed by television plays *The Lover*, and *The Collections*; radio and stage *The Dwarfs*, and *The Homecoming*; film-scripts *The Servant*, and *The Pumpkin*; plays *No Man's Land*; *Betrayal*; *Party Time*; and *Moon light*; short pieces *Other Voices* displayed at the National Theatre in London; political themes *One for the Road*; *Mountain Language*; and *A New World Order*; and finally film-scripts *The French Lieutenant's Woman*; *The Handmaid's Tale*; and *The Comfort of Strangers*; Phyllis Dorothy James was author of detective stories such as *Cover Her Face*; series *A Mind to Murder*; *The Black Tower*; and *Death of an Expert Witness*; which were followed by *A Taste for Death*; *Devices and Desires*; *The Skull Beneath the Skin*; and finally *The Children of Men*; *Original Sin*; and *A Certain Justice*; Tom Stoppard wrote radio plays including *The Dissolution of Dominc Boot*; works *Rosencrantz and Guildenstern are Dead*; *Jumpers*; *Travesties*; *The Real Inspector Hound*; *Professional Foul*; *The Real Thing*; *Arcadia*; *Indian Ink*; and *The Invention of Love*; also *The Russia House*, and especially *Shakespeare in Love*; and Lynda La Plante produced works such as *Widows*; *Prime Suspect*; *Civvies*; *Framed*; *Comics*; *She's Out*; *The Governor*; *Supply and Demand*; novels *Bella Mafia*, and *Seekers*; *Trial and Retribution*; *Killer Net*; *Profiler*; and *Above Suspicion*;

many of them being adapted for television;

1932-2000: *Chinese literature* – Ba Jin wrote a major trilogy, comprising *Jia* 'The Family', *Chun* 'Spring', and *Qiu* 'Autumn', in which he attacked the traditional family system, and became immensely popular with younger generation; and Gao Xingjian was author of plays including *Bus Stop*, and *The Other Shore*; as well as the novel *Soul Mountain*, searching for roots in a shattered Communist China;

1933-47: *Spanish literature* – Federico García Lorca wrote plays *Bodas de Sangre* 'Blood Wedding'; *Yerma*; and *La Casa de Bernarda Alba* 'The House of Bernarda Alba'; gypsy songs *Canciones*, and *Romancero Gitano* 'Gypsy Ballads'; and elegiac poems *Llanto por la muerte de Ignacio Sánchez Mejías* 'Lament for the Death of a Bullfighter and Other Poems';

1938-93: *Irish prose* – Samuel Barclay Beckett published novels *Murphy*, and *Watt* in English; trilogy *Molloy*, *Malone Meurt*, and *L'Innommable* in French; plays *En attendant Godot*, and *Fin de partie* also in French; then *Happy Days*; *Not I*; and *Ill Seen Ill Said*; and finally the short piece *Breath*; and Dame Iris Murdoch produced *The Sandcastle*; *The Bell*; *A Several Head*; *The Sea*; *Nuns and Soldiers*; *The Good Apprentice*; *The book and the Brotherhood*; *The Message to the Planet*; and *The Green Knight*;

1949-95: *South African prose* – Nadine Gordimer wrote the collections *Face to Face*, and *The Soft Voice of the Serpent*; the novel *The Lying Days*; the books *Occasion for Loving*; *The Late Bourgeois World*; *A Guest of Honour*; *The Conservationist*; *Burger's Daughter*; *July's People*; and *A Sport of Nature*; as well as *None to Accompany Me*; and *Writing and Being*;

1951-88: *Scottish literature* – Dame Muriel Sarah Spark published short story *The Seraph and the Zambesi*; then *The Comforters*; *Memento Mori*; *The Ballad of Peckham Rye*; and *The Bachelors*; the novel *The Prime of Miss Jean Brodie*; and *The Girls of Slender Means*; *The Mandelbaum Gate*; *The Abbess of Crewe*; *Loitering with Intent*; *The Only Problem*; and *A Far Cry from Kensington*;

1955-91: *Romanian literature* – Marin Preda produced his masterpiece *Moromeții* 'The Moromites'; and other works including the novel *Cel mai iubit dintre pământeni* 'The Most Loved from Countrymen'; Nichita Stanescu wrote sensible works including *The Book of Re-Reading*; *Epica Magna*; *The Imperfect Works/Creations*; and *Breathings/Respirations*; and Ioan Petru Culianu published several outstanding books including *Expérience de l'Extase*; and *Eros et Magie à la Renaissance*; as well as *Dictionaire des Religions*, and *Out-of this World*;

1960-61: *French West Indian prose* – Frantz Omar Fanon, Martinique-born French psychiatrist, philosopher and writer, was author of a study on Algerian revolution, entitled *Les Damnés de la terre* 'The Wretched of the Earth', which became source of inspiration and manifesto for liberation struggles throughout the Third World;

1965-2001: *Northern Irish literature* – Seamus Justin Heaney produced *Eleven Poems*; the collections *Death of a Naturalist*; *Wintering Out*; *North*; *Bog Poems*; *Stations*; *Preoccupations*; *The Spirit Level*; and *Electric Light*; as well as prose poems *Field Work*; *Station Island*; *The Haw Lantern*; and *Seeing Things*;

1965-2008: *Moldovan literature* – Grigore Vieru published the book *Poetry for Readers of All Ages*; poems *Ars Poetica*; *Friday's Star*; *I Do Not Hate You, Death*; *Autobiographical*; *Testament (To Mihai Eminescu)*; *In Your Language*; *Bliss*; and *A Letter from Bessarabia*; and also songs including *Melancholia*, becoming representative of new literary movement in Moldova;

1985-2010: *German literature due to Transylvanian Saxons* – Herta Müller was focused on German minority in Communist Romania under repressive Nicolae Ceausescu's regime and also on modern history of Germans in Banat and Transylvania, and produced acclaimed novel *Atemschaukel* 'The Hunger Angel' depicting deportation of Romania's German minority to Stalinist Soviet Gulags during the Soviet occupation; and other works including *Drückender Tango* 'Oppressive Tango', *Der Teufel sitzt im Spiegel* 'The Devil is Sitting in the Mirror', and *In der Falle* 'In a Trap'.

Trying to assemble the above data, it seems to be a very difficult task to find time sequences in linguistic development. However, there are two clues for such a task:
1st, the language emerged in the same time as their speakers about 200000 years ago (see chapter 7), and will have the same fate as them at $t_\bullet$ = 204350 years from the origin of language, or 204350 - 200000 ≈ 4350 years later than present time, so that the timeline is common for both humans and their languages, i.e. the language sequences are the same with quarter-cyclical human sequences displayed in the table at the end of chapter 7; and
2nd, the last experienced language sequences are approximately delimited by approximate times as follows:

10000...7800BC ≈ 8900BC \ *Pre-boreal languages*
\ 6000...5300BC ≈ 5650BC \ *Protolanguages*
\ 3100...3000BC ≈ 3050BC \ *Historic languages*
\ 1100...900BC ≈ 1000BC \ *Early literatures*
\ AD590...650 ≈ AD620 \ *Late literatures*
\ AD1850...1900 ≈ AD1875 \ *Modern literatures*.

The transitional times in years ago are therefore: 10914 (corresponding to transition from Mesolithic to Neolithic); 7664 (corresponding to transition from Neolithic to Copper Age); 5064 (corresponding to transition from Copper Age to Bronze Age); 3014 (corresponding to transition from Bronze Age to Iron Age); 1394 (corresponding to transition from Iron Age to Medieval); 139 (corresponding to transition from Medieval to Modern) respectively. Referring to the origin of language about 200000 years ago, these times become 200000 - 10914 ≈ 189086; 200000 - 7664 ≈ 192336; 200000 - 5064 ≈ 194936; 200000 - 3014 ≈ 196986; 200000 - 1394 ≈ 198606; 200000 - 139 ≈ 199861 years ago; which divided by the final time $t_{\bullet}$ = 204350 from the origin of language result in the approximate ratios 189086/204350 ≈ 0.925; 192336/204350 ≈ 0.941; 194936/204350 ≈ 0.954; 196986/204350 ≈ 0.964; 198606/204350 ≈ 0.972; 199861/204350 ≈ 0.978, corresponding to the accurate values given in Background to time, Table *(k)* for $z/8 = k$ sequences, as:

$$t_{1.625}/t_{\bullet} = tanh(1.625) = 0.9253462;$$
$$t_{1.75}/t_{\bullet} = tanh(1.75) = 0.9413755;$$
$$t_{1.875}/t_{\bullet} = tanh(1.875) = 0.9540453;$$
$$t_{2}/t_{\bullet} = tanh(2) = 0.9640276;$$
$$t_{2.125}/t_{\bullet} = tanh(2.125) = 0.9718727;$$
$$t_{2.25}/t_{\bullet} = tanh(2.25) = 0.9780261;$$

whence the transitional times, in years from the origin of language:

$$t_{1.625} = t_{\bullet} \cdot tanh(1.625) = 189095;$$
$$t_{1.75} = t_{\bullet} \cdot tanh(1.75) = 192370;$$
$$t_{1.875} = t_{\bullet} \cdot tanh(1.875) = 194959;$$
$$t_{2} = t_{\bullet} \cdot tanh(2) = 196999;$$
$$t_{2.125} = t_{\bullet} \cdot tanh(2.125) = 198602;$$
$$t_{2.25} = t_{\bullet} \cdot tanh(2.25) = 199860.$$

Therefore, linguistic development with its timeline, sequences, and intrinsic characteristics is presented as follows:

z/8 = k	Time (years)		Linguistic sequences
	from origin $t_k =$ $t_\bullet \cdot tanh(k)$	from present t_k - 200000	
	204350	+4350	
...	...	...	
2.375	200844	+844 (AD2858)	_Modern literatures
2.25	199860	-140 (AD1874)	_Later literatures
2.125	198602	-1398 (AD616)	_Earlier literatures
2	196999	-3001 (987BC)	_Historic languages
1.875	194959	-5041 (3027BC)	_Protolanguages
1.75	192370	-7630 (5616BC)	_Pre-boreal languages
1.625	189095	-10905	_Language changes in new environs
1.5	184967	-15033	_ Simple verses
1.375	179793	-20207	_ Long stories
1.25	173347	-26653	_ Short stories
1.125	165381	-34619	_ Conversations, communications
1	155632	-44368	_ Messages and dialogues
0.875	143843	-56157	_ Relative clauses, differ. languages
0.75	129793	-70207	_ Subordinate clauses
0.625	113332	-86668	_ Co-ordinate clauses
0.5	94434	-105566	_ Clauses in sentences; African lang.
0.375	73230	-126770	_ Phrases in sentences
0.25	50049	-149951	_ Complex sentences
0.125	25412	-174588	_ Simple sentences, basic vocabulary
0	0	-200000	_

z/8 = k	Time from origin $t_k = t_\bullet \cdot tanh(k)$ (years)	Period $\tau_k =$ $(t_\bullet^2 - t_k^2)/(2t_\bullet)$ (years)	Frequency $f_k =$ $(2t_\bullet)/(t_\bullet^2 - t_k^2)$ (years)$^{-1}$	Angular speed $\omega_k =$ $2\pi/\tau_k = 2\pi \cdot f_k$ (years)$^{-1}$
	204350	0	∞	∞
...	...	...	...	...
2.375	200844	3475.9	0.0002877	0.0018076
2.25	199860	4440.7	0.0002252	0.0014149
2.125	198602	5667.2	0.0001765	0.0011087
2	196999	7218.8	0.0001385	0.0008704
1.875	194959	9175.2	0.0001090	0.0006848
1.75	192370	11628.8	0.0000860	0.0005403
1.625	189095	14685.6	0.0000681	0.0004278
1.5	184967	18463.7	0.0000542	0.0003403
1.375	179793	23081.5	0.0000433	0.0002722
1.25	173347	28651.2	0.0000349	0.0002193
1.125	165381	35253.4	0.0000284	0.0001782
1	155632	42910.7	0.0000233	0.0001464
0.875	143843	51549.1	0.0000194	0.0001219
0.75	129793	60956.0	0.0000164	0.0001031
0.625	113332	70748.2	0.0000141	0.0000888

0.5	94434	80355.1	0.0000124	0.0000782
0.375	73230	89053.8	0.0000112	0.0000706
0.25	50049	96046.0	0.0000104	0.0000654
0.125	25412	100594.9	0.0000099	0.0000625
0	0	102175.0	0.0000098	0.0000615

9. Art

The expression or application of human creative skill and imagination, typically in a visual form, is called *art* (from Latin *ars, artis* 'skill, method, technique'). From its widely ranging forms of expression, painting, sculpture and architecture have been dominant over millennia; printmaking, photography, film and other visual media were quite recently included; while literature, music and dance are approached in other chapters of the present work.

Art is a diverse range of human activities and the products of those activities. It originated by drawing simple lines on soft natural materials around 180000-170000 years ago, and developed by use of pigments in the Rift Valley and South Africa dated 150000 years ago, by simple paints such as those probably held in containers dated 100,000 years ago, and by oldest known art objects in the world, namely a series of tiny, drilled snail shells about 75,000 years old, found in a South African cave.

Palaeolithic sculptures, cave painting, rock painting and petro-glyphs dated between 40,000 and 17000 years ago have been also found in a some caves of Western Europe (Spain and France).

Many traditions in art have foundations in the art of the great ancient civilizations: *Ancient Egypt, Mesopotamia, Persia, India, China, Ancient Greece, Rome,* as well as *Inca, Olmec* and *Maya.*

In *Byzantine* and *Medieval art* of the Western Middle Ages, much art focused on the expression of Biblical and religious truths, and used styles that showed the higher glory of a heavenly world, such as architecture of churches and great cathedrals, use of gold in the background of paintings, or glass in mosaics and windows, which also presented figures in idealized, patterned (flat) forms.

Renaissance art had a greatly increased emphasis on the realistic depiction of the material world, and the place of humans in it, reflected in the corporeality of the human body, and development of a systematic method of graphical perspective to depict recession in a three-dimensional picture space.

In the East, *Islamic art*'s rejection of iconography led to emphasis on geometric patterns, calligraphy, and architecture. *Chinese styles* varied greatly from era to era, and each one is traditionally named after the ruling dynasty. So, for example, Tang Dynasty paintings are monochromatic and sparse, emphasizing idealized landscapes, but Ming Dynasty paintings are busy and colourful, and focused on telling stories via setting and composition. *Japanese* named their styles after imperial dynasties too, and also saw much interplay between the styles of calligraphy and painting. Woodblock printing became important in Japan after the 17ᵗʰ century.

In the last decades, *Computer art* was developed using computers to produce or display artworks. Such art can be an image, animation, video, CD-ROM, DVD-ROM, videogame, web site, algorithm, performance or gallery installation, for example an artist may combine traditional painting with algorithm art and other digital techniques.

The quality of an artistic work w_j emerges from its intrinsic yield y_i multiplied by a code x^i_j that values the yield, i.e.

$$w_j = x^i_j \cdot y_i,$$

where the distinctive notations i and j indicate the artist's and valuators' attributes respectively. Thus the resultant value of artistic work W can be interpreted as the sum

$$W = \Sigma_j \, w_j = \Sigma_j \, x^i_j \cdot y_i,$$

which becomes more or less valuable in time. Noticeably, the more valuable, the more persistent an artistic work would be.

Old forms of art, artistic styles, schools, movements, computer art, artists, architects, sculptors, goldsmiths, bronze-casters, medallists, engravers, lithographers, potters, glassmakers, painters, watercolourists, decorators, designers, planners, film directors, producers, filmmakers and computer artists, together with their times and works are displayed below.

100000-75000 years ago: *Blombos cave*, South Africa – containers that may have been used to hold paints, the oldest unmistakable artwork represented by a piece of red ochre, adorned with *scratched patterns of lines*, as well as a series of *tiny, drilled snail shells*;

40800 years ago: *El Castillo cave*, Spain – a *red dot*, contemporary with the arrival of first modern humans in Europe, showing that rock art, cave paintings and engravings characterized the early humans as *Homo aestheticus*;

40000-18000 years ago – *Indigenous Australian art* represented by *dot painting* of animals, lakes and dreamtime; *rock art* in Western Australia's Pilbara region and the Olary district of South Australia, depicting extinct megafauna such as *Genyornis*; *bark painting* and *aerial landscape art*, a kind of map like, bird's eye view of the desert landscape, on rock, sand or body painting; *rock engraving* at Murujuga in Western Australia, New South Wales, and Panaramitee in Central Australia; and also *carving* and *sculpture*;

35600-18500 years ago: *Altamira cave*, Spain – images created by using charcoal and ochre, as well as haematite, which are examples of *drawings* and *polychrome rock paintings of wild animals and human hands*;

17300 years ago: *Lascaux caves*, France – cave paintings with nearly 2000 *figures of animals* (mainly equines and stags), *human figures*, and *abstract signs*;

9500-8400BC: *Göbekli Tepe decorations*, Turkey – *reliefs* depicting lions, bulls, boars, foxes, gazelles, donkeys, snakes and other reptiles, insects, arachnids, birds including vultures, as well as *humanoid figures*, *human arms*, and debris of *pottery shards*;

7000BC: *Bhimbetka rock shelters*, India – stone age painting compositions portraying *bulls, buffaloes, deer, antelopes*, a *peacock*, a *tiger*, and scenes of *communal life*, executed more in red and white, and less in green and yellow;

3700BC: *Temple of Ġgantija's reliefs*, Malta – impressive old *wall low-relief*, and *spirally carved reliefs*;

3500-2500BC: <u>*Sumerian art*</u> – *Inanna*, or *Ishtar*, a female head from Uruk, who was an Earth and later a horned Moon goddess (Iraq Museum of Bagdad); *Lady of Warka* from Erech, a female head carved in white marble with simplicity and subtlety (Iraq Museum of Bagdad); *White Temple and Ziggurat*, in Uruk, of a size 22 per 17 metres; *Statuettes from the Temple of Abu*, Tel Asmar, representing mainly males stand or sit with hands clasped in an attitude of prayer, often naked above the waist and wear a woollen skirt curiously woven in a pattern that suggests overlapping petals (now in Iraq Museum of Bagdad, and Oriental Institute in University of Chicago); *Wooden harp*, detailed with gold and mosaic inlay, picturing mythological scenes on the soundbox, and surmounted by a black-bearded golden head of a bull (University Museum in Philadelphia); and *Ram in a Thichet* shaped ritual offering stand from Ur, made of silver, lapis lazuli, and mussel shells, rearing on his hind legs to eat from a tree of gold (British Museum in London);

3300-1500BC: <u>*Indus Valley and Vedic sculpture*</u> – including *Priest-king, Dancing Girl, Male Torso, seals* such as 'Unicorn', jars, figurines and necklaces;

3100-850BC: <u>*Ancient Egyptian art*</u> – *Narmer's Palette*, a small dark green schist stone carved into a shield-shaped ceremonial palette depicting Pharaoh Narmer, the first ruler of unified Egypt and founder of the First Dynasty; *Great Sphinx of Giza*, an impressive structure with dimensions of 20 by 6 by 73.5 metres; Imhotep, the earliest known architect of the Third Dynasty, who served under King Djoser and built *Djoser's Pyramid Step* at the complex of Saqqara necropolis; *Khufu's Statue*, a small ivory statue of Pharaoh Khufu, of Cheops, discovered next to the Great Pyramid; *Menkaura (Mycerinus) and Queen Khamerernebty II*, a greywacke statue (Museum of Fine Art, Boston); Senenmut, great architect and steward during the 18th Dynasty, who designed and supervised the construction of the 'Holy of Holies' *Djeser-Djeseru mortuary temple*, the Pharaoh Hatshepsut's tomb complex, a colonnaded structure of perfect harmony; *Thutmose III Statue*, a basalt artistic masterpiece, immortalizing the sixth Pharaoh of the 18th Dynasty (Luxor Museum, Egypt); *The Golden Death Mask of Tutankhamun* (Egyptian Museum, Cairo); Bek, or Bak, royal sculptor of the 18th Dynasty, oversaw construction of the *great temple statues* of the Pharaoh Akhenaten; Ineni, another architect of the 18th Dynasty, was responsible for major structure projects, expanded the Temple of Karnack and oversaw the construction of *Amenhotep I's tomb and mortuary temple*; *Bust of Nefertiti*, a limestone piece, layered with painted stucco, immortalizing Akhenaten's wife, which was created by artist sculptor Thutmose at the climax of the Amarna Period art (Egyptian Museum of Berlin); *Osiris* (on a lapis lazuli pillar) flanked by *Horus* and *Isis* (Louvre, Paris); also pottery such as vases, amulets, images of deities, animals, and *art pottery covered by enamel*, as well as *papyri with painted figures*;

2500BC: *Mesoamerican pyramid works of art* – pieces examplified by *jade beads shaped like howler monkeys, crocodiles and gourds*, as well as *ceramic pots, ritual axes, iron-pyrite mirrors* and a *red-painted stucco mask*;

2150-1600BC: *Late-Sumerian and Babylonian art* – largest *Ziggurat of Ur*, Mesopotamia; *temples* and *palaces* built by King Gudea of Lagash, who placed portraits of himself offering goods to gods; and achievements of Babylonian artists who first experimented with *foreshortening* (a means of suggesting depth by representing a figure or object at an angle);

2000-1400BC: *Minoan art*, Crete – the palaces of *Knossos, Phaistos*, and *Malia*, with original architectural and structural features, such as light-wells and polythyra, painted plaster, marble revetment and *wall-paintings* adorning the rooms and passages, *frescos* including the *Prince of the Lilies, Bull Hunt* and *dolphins*, as well as exceptionally high quality works of *pottery, vessels* and *figurines*;

1700-1200BC: *Hittite Art* – *Lion Gate* with tall lions guarding the main entrance to capital Hattusha, Anatolia, and recurring theme that *wild beasts of fantastic monsters* have been placed at gates/entrances to protect cities, tombs, palaces and temples from evil;

1600-1100BC: *Mycenaean art*, Greece – architecture, sculpture and painting, such as fortresses and palaces of *Mycenae*, *Tiryns* and *Pylos*, yielding a wealth of artefacts and fragmentary frescoes; pottery, bronze, faience and *ivory vessels* with decorations of mythic, warrior or animal motifs; *terracotta figurines* painted with stripes or zigzags; *silver and clay figures*; *frescoes* depicting female figures, scenes of hunting, bull leaping, battles and processions; as well as a lavish grave *gold and silver objects, jewellery, weapons* and *pottery*;

1550-150BC: *Classical Doric, Ionic and Corinthian Styles* – great works due to Greek and later Roman *architects*, *sculptors* and *painters*;

1200BC-AD1535: *Inca art*, Peru – remarquable works such as *Chavin temples* in Chavin de Huantar; *Nazca drawings of animals* at Nazca and Palpa in north Peru; *Lambayeque gold figures* and *fine masonry* at Machu Picchu, in the Urubamba Valley; and also ceramic *Storage Jar* at Urpu;

1100-200BC: *Olmec art* – sculptures, in Mexico, including ceramic-cinnabar-red ochre *'Baby Figure'*, ceramic-pigment *Hunchback*, and greenstone *Bench Figure*;

900-612BC: *Assyrian art* – *palaces* resembling to fortified citadels with gates guarded by *monstrous lamassu* (a protective deity often depicted with a bull or lion's body, eagle's wings, and human's head); *painted reliefs* depicting the king in battle and hunting lions which decorated the walls of vast ceremonial halls; limestone *Lamassu at Dur Sharrukin*, in Iraq, winged, a human-headed bull from the citadel of <u>Sargon II</u>, and multiple lamassus around the citadel to ward of king's enemies;

612-330BC: *Late-Babylonian and Persian Art* – *Ishtar Gate*, Babylon, in Iraq, created under <u>Nebuchadnezzar II</u>, with *glazed bricks*, each brick moulded and glazed separately, patterns of *lions* and *other animals*; and a huge *palace complex* built in Persepolis by <u>Darius I</u> and <u>Xerxes</u>;

560-330BC: *Ancient Greek art* – <u>Chersiphron</u> and his son <u>Metagenes</u> were architects of the *Temple of Aremis* at Ephesus; <u>Polygnotus</u> was the first to indicate *perspective* and *landscapes* in his works, such as *wall murals*, in Athens, Delphi, and Plataea; <u>Apollodorus</u> introduced the technique of *chiaroscuro* (light and shade); <u>Ictinus</u> and <u>Callicrates</u> designed the peripteral octostyle Doric temple of *Parthenon* in Athens; <u>Mnesikles</u> designed the monumental Doric-Ionic building of *Propylaea* in Athens; <u>Phidias</u> supervised the construction and decoration of both the *Propylaea* and *Parthenon*, the second one holding the sculpture of *Athena*, on Acropolis in Athens, and the statue of *Zeus* at Olympia as the 4th Wonder of Ancient World; <u>Praxiteles</u> was author of usually marble great works, almost all of them perished, though his statue *Hermes Carrying the Infant Dionysus* was found at Olympia, and several other statues are known from Roman copies, for example *Aphrodite of Cnidos* (Vatican Museum, Rome); <u>Skopas</u> was known by his sculptures for the *Temple of Athena* at Tegea, and by works for part of the *Mausoleum of Halicarnassus*; <u>Leochares</u> decorated much of the *Mausoleum of Halicarnassus*, the 5th Wonder of Ancient World, produced sculptures of *Demeter* (British Museum, London), *Philip II of Macedon* and *Alexander the Great*; and <u>Lysippus of Sicyon</u> made bronze statues of *Apoxyomenos* 'Man using a Strigil' (with a Roman marble copy in Vatican Museum, Rome) and a portrait bust of *Alexander the Great*;

350BC-AD550: <u>*Ancient Roman art*</u> – works of architecture, sculpture and painting, which emerged under influences from Etruscan and then Greek arts, became Roman art itself by *Triumphal painting*, and developed by a series of unknown or anonymous artists; this ancient art was promoted, patronized, encouraged, or remembered by some consuls, but especially by Roman emperors such as <u>Octavian Augustus</u>, <u>Titus Flavius Vespasian</u>, <u>Flavius Sabinus Titus</u>, <u>Marcus Ulpius Trajan</u>, <u>Publius Aelius Hadrian</u>, <u>Antoninus Pius</u>, <u>Marcus Aurelius</u>, <u>Gaius Aurelius Valerius Diocletian</u>, <u>Constantine I the Great</u>, <u>Tetrarchs</u>, and <u>Justinian I the Great</u>, whose successive achievements included *'rebuilding of Rome from brick to marble'*; *Colosseum* of Rome; *Arch of Titus*; *Trajan's Column* designed by architect <u>Apollodorus of Damascus</u>; *Athenaeum, Pantheon* and *Mausoleum* (now part of Castle St Angelo) in Rome, magnificent *villa at Tibur* and the city of *Adrianopolis*; *Column of Antoninus Pius*; *Marcus Aurelius' Column*; *Baths of Diocletian, Arch of Constantine*; porphyry *The Four Tetrarchs* (moved from Constantinople to Venice); and *Justinian's famous mosaics* of Ravenna respectively; moreover, Roman art was widely represented by monuments such as *Tropaeum Traiani* (now in Romania), as well as by *amphitheatres, arches, mausoleums*, carved *sarcophagi*, colossal and normal-sized *sculptures* and *busts*, as well as *mural paintings* all over the empire, including those preserved under the volcanic ash and debris ejected by Vesuvius over Herculaneum and Pompeii, in Italy, vase *pottery* and *silverware* in Britannia, and later Christian catacombs and churches with their *sarcophagi*, *mosaics* and *paints*;

100BC-AD1000: <u>*Mayan art*</u> – abundant works including ceramic *Cylindrical Vessel* in Guatemala; limestone *Spouted Vessel* in Mexico-Guatemala; cinnabar *Deity Figure* in Honduras; ceramic *Censer with Seated Figure* in Guatemala; ceramic *Carved Bowl* and jade *Head Pendant* in Mexico-Guatemala; pigment *Costumed Figure* in Mexico; ceramic *Cylindrical Vessel with Throne Scene* in Guatemala; ceramic *Censer Support* and limestone *Head of a Rain God* in Mexico;

AD330-1450: <u>*Byzantine architecture and art*</u>, Near East, Southern and Eastern Europe – works of Byzantine architects, mosaicists and painters, who built and decorated great structures such as: basilica *Sant'Apollinaire Nuovo* in Ravenna; *Church of St Sophia* (*Hagia Sophia*, with famous mosaics including *Christ Pantocrator*) and *Hagia Irene* in Istanbul; *Great Palace of Constantinople*; *Hagios Demetrios* (with outstanding *mosaics*) in Thessaloniki, *St Catherine Monastery* on Mount Sinai, *Jvari Monastery* in Georgia, three Armenian *Churches of Echmiadzin*, *Hasios Lukas* in Greece, *Daphni Monastery* near Athens, *Cappella Palatina* in Sicily; *St Mark's Basilica* and *Torcello Cathedral* in Venice; *Church of the Holy Saviour at Chora*, with admired *mosaics*, and *St Mary Pammakaristos* in Istanbul;

490-1400: <u>*Early French architecture*</u> – Structures such as Pre-Romanesque *St Gereon* in Cologne, and Abbey *Saint-Germain des-Prés* in Paris; Romanesque *Abbey of Cluny*, *Angoulême Cathedral*, *Notre-Dame of Domfront* in Normandy, *Saint-Étienne* in Caen, and *Trinité Church* of Caen; Gothic *Abbey Church of St Denis* in northern Paris, *Sens Cathedral* in Bourgogne, *Notre-Dame of Laon* in Picardy, *Notre-Dame de Paris*, *Lyon Cathedral*, *Toul Cathedral* in Lorraine, *Cathedral of Notre-Dame de Chartres*, *Amiens Cathedral*, and *Bourges Cathedral*; Pre-Renaissance *Palace of Aachen* in Aix-la-Chapelle, and the grand *Palais des Papes* in Avignon;

550-1530: <u>*Aztec art*</u>, Mexico – Examples include the stone *Coiled Serpent*, *Deity as Standard Bearer*, *Earth Monster Relief*, and *Female Figure*; basalt *Head of a Water Deity*; sandstone *Standard Bearer*; ceramic *Water Deity* (Chalchihuitlicue); and *Temple Mayor* (Great Temple) in Tenochtitlan;

629-1630: <u>*Early Islamic architecture and art*</u>, Near East, Middle East,

North Africa and Iberian Peninsula, India and China – Mosques with minarets; palaces with halls, pointed arches, courts, arabesques, fountains, gardens and water supply systems; towns with fortresses and medina marketplaces; as well as calligraphy, painting, glass, ceramics and textiles; including structures such as *Al-Masjid al-Nabawi of Medina* in Saudi Arabia, *Great Mosque of Kairouan* in Tunisia, *Dome of the Rock* in Jerusalem, *Great Mosque of Damascus* in Syria, *Great Mosque of Xi'an* in China, *Great Mosque of Samarra* in Iraq, *Mosque of Ahamad Ibn Tulun* in Cairo (Egypt), *Koutoubia Mosque* in Marrakesh (Morocco), *Bibi-Heybat Mosque* in Baku (Azerbaijan), *Soltaniyeh Dome* in Zanjan (Iran), the necropolis *Shah-e Zendah* in Samarkand (Uzbekistan), *Jama Masjid Mosque of Herat* in Afghanistan, *Selimiye Mosque* in Erdine (Turkey), the monument and mosque *Charminar* 'Four Towers' at Old City in Heyderabad (India), *Lotfallah Mosque* and *Shah Mosque* in Isfahan (Iran); and also *Moorish architecture and art* in the Iberian Peninsula;

660-1714: <u>*Early British architecture*</u> – Successive Anglo-Saxon, Norman, Gothic, Vernacular, Tudor, Elizabethan, Jacobean and Stuart styles led to various kinds of architectural structures, such as: <u>English</u> *Rochester, Ripon, Oxford, Lichfield, Chester, Lincoln, Worcester, St Albans, Durham, Canterbury, Carlisle, Norwich, Ely, Bristol, Wells, Salisbury, Peterborough* and *Glouchester Cathedrals*; *Westminster Abbey, Rochester Castle, Ightham Mote* and *Alfriston Clery House*; <u>Welsh</u> *Llanfair Kilgeddin, Penrhos, Llangynwyd, Cardiff, Dinefwr, Llanrhystud, Abergavenny, Dinas Brân, Mold, Pembroke, Caernarfon, Conwy, Harlech, Beaumaris* and *Gwydir Castles*; *Bangor, St David's* and *St Asaph Cathedrals*; <u>Scottish</u> *Dunblane, Glasgow, Elgin, St Machar's, Dunkeld* and *St Giles Cathedrals*; *Holyrood Palace* in Edinburgh, *Church of St Rule, Dunfermline Abbey, Kings College Chapel* in Aberdeen; *Prevost Skene's House* in Aberdeen, and *Heriot's Hospital* in Edinburgh;

737-1450: <u>*Early Iberian architecture*</u> – Christian and Moorish structures in <u>Spain</u> such as the Asturian church *San Julián de los Prados* in Oviedo, palace pavilions *Santa Maria del Naranco* and *San Miguel de Lillo* on Naranco Mountain; Andalusian *Great Mosque* of Córdoba and *Palace of the Aljaferia* in Zaragoza; Almohad synagogue of *Santa María la Blanca* in Toledo; Nasrid palaces of *Alhambra*, and *Generalife* in Granada; Mudéjar *Casa Pilatos* in Seville, and *Santa Clara Monastery* in Tordesillas; Romanesque *Cathedral of Santiago de Compostela* in Galicia; Gothic *Cathedral of Ávila* and *Cathedrals of Burgos, León* and *Toledo*; Architectural structures in <u>Portugal</u>

including Pre-Romanesque *Saint Frutuoso Chapel* near Braga; Moorish *Silves Castle, Paderne Castle* in Algrave, *Sintra Castle* near Lisbon, and *Mértola Mosque* in Alentego; Romanesque *Braga Cathedral, Monastery of Rates, Cathedral of Coimbra, Lisbon Cathedral* and *Lisbon Castle*; Gothic *Monastery of Alcobaça, Monastery of Santa Clara-a-Velha* in Coimbra, *Monastery of Batalha, Guarda Cathedral* and *Sintra Royal Palace*;

750-1150: *Romanesque art*, Western Europe – churches and cathedrals with massive walls, rounded arches and small windows, such as *Durham Cathedral* in England;

840-1530: *Early Italian architecture* – Romanesque structures including the *Leaning Tower of Pisa, Basilica Sant'Ambrogio* in Milan, *Modena's Duomo, St Mark's Basilica* in Venice, *Orvieto's Duomo* in Umbria, and *Siena's Duomo* in Tuscany; Gothic works including the *Cathedral of Milan*, and *San Petronio Basilica* in Bologna; and the transition to Renaissance by works such as *Florence Cathedral, Basilica of San Lorenzo* in Florence, *Basilica of Sant'Andrea* in Mantua, the iconic *Tempieto* at San Pietro in Montecitoria, Rome, and rebuilding of *Saint Peter's Basilica* in Rome;

850-1600: *Orthodox architecture and art*, South-eastern and Eastern Europe, Near East and Siberia – under Byzantine influence, constructions such as *Church of St Apostles Peter and Paul* at Novi Pazar, and marble *Studenica Monastery* in Serbia; *Boyana Church*, and *St Nicholas Mirlikiyski Church* in Bulgaria; *St Sophia Cathedral*, and *St Cyril's Monastery* in Kyiv, Ukraine; *Novgorod Cathedral, St Margaret's Chapel Church of All Saints no Kulichkakh, Cathedral of the Annunciation* and *Cathedral of the Archangel Michael* in Kremlin, *Cathedral of the Metropolitan Peter of Moscow*, and *Cathedral of St Basil* in Moscow, Russia; *Tismana Monastery, Dealu Monastery*, and *Curtea de Arges Cathedral* in Wallachia, Romania; as well as *Voronet, Arbore, Humor, Moldovita*, and *Sucevita Monasteries of Bukovina* with their famous *exterior wall-frescoes* featuring portraits of saints and prophets, scenes from Jesus' life, images of angels and demons, and of heaven and hell; all of them yielding *icon paintings* in Byzantine style, in Moldavia, Romania;

900-1300: *Toltec art* – carved massive, block-like sculptures including the *free-standing columns* at Tula in Mexico; standing *human/deity statues* and *heads*, gold, jade and turquoise objects of art, andesit-dacit paint *Eagle Relief* in Mexico, as well as architectural structures including *large palaces* and *pyramids* with friezes running around them;

950-1410: *Early German architecture* – Works including Pre-Romanesque *St Michael's Church* in Hildesheim; Romanesque *Imperial Cathedral Basilica of the Assumption and St Stephen* in Speyer, *Worms* and *Mainz Cathedrals*, *Limburg* and *Lübeck Cathedrals*, *Brunswick* and *Bamberg Cathedrals*, and *Wartburg Castle*; Gothic *Liebfrauenkirche* (*Church of Our Lady*) in Trier, magnificent *Freiburg Cathedral* and *Cologne Cathedral*, *St Mary's Church* in Lübeck, the *Town Hall* and *St Nicholas Church* at Stralsund, and *Bremen Town Hall*;

990-1230: *Chinese painting* – monumental, spontaneous and impressionistic styles represented by silk hanging scroll of the famous Fan Kuan's *Travelling in Streams and Mountains* (Palace Museum, Taiwan) in monumental style; traditional painting such as Wen Tong's masterpieces of *Bamboo paintings* in spontaneous style; outstanding painting of Su Dongpo's *Wen ren hua* 'literati painting', remembering his statement 'To paint the bamboo, it is necessary to have it entirely within yourself'; and delicate and almost impressionistic spatial views such as Xia Gui's *Landscapes*;

1000-1500: *Early Scandinavian architecture* – Works in Sweden including Romanesque *Lund Cathedral*, *Sigtuna Monastery*, *Husaby Church*, and *Alvastra Monastery*; Gothic *Cathedral of Västerås* (*Strängnäs) Cathedral*, and *Uppsala Cathedral*; in Norway such as Stave churches, for example *Stave in Lom*; Romanesque *Churches at Ringsaker* and *Kviteseid*; Gothic *Stavanger Cathedral*, and *Nidaros Cathedral*; in Denmark Romanesque *St Bendt's Church* in Ringsted, *Church of Our Lady* in Kalundborg, and *Lund Cathedral* in Scania (southern Sweden); Gothic *St Canute's Cathedral* in Odense, *St Peter's Church* in Næstved, and *Glimmingehus Castle* in Scania;

1000-1712: *Early Ukrainian and Russian architecture* – Ukrainian Rus' style including *St Sophia of Kiev*, *Church of the Saviour* in Berestove, and *St Cyril's Monastery* in Kiev; as well as Vernacular and Baroque styles at *St Andrew's Church* in Kiev; Russian church style including *Saint Sophia Cathedral* in Novgorod, *Golden Gate* in Vladimir, *Church of the Intercession* on the Neri, *Dormition Cathedral* in Vladimir, *St Andronik Monastery* in Moscow, *Deposition Church* of the Moscow Kremlin, *Holy Spirit Church* of the Holy Trinity Lavra, and *Cathedral of Novodevichy Convent* in the Kremlin; Renaissance Russian style at *Saint Basil's Cathedral* in Moscow; Muscovite style such as *Church of the Ascension* in Kolomenskoe near Moscow, *Kazan Cathedral* and *Nativity Church at Putinki* in Moscow, *Palace of the Patriarch on Moscow,*

Kolomenskoye summer residence of the Tsars, Church of St John the Baptist in Yaroslavl, *Cathedral of the Resurrection* in New Jerusalem Monastery at Istria near Moscow, and *Church of the Intercession* in Kizhi on Lake Onega;

1100-1400: *Early Polish architecture* – Structures such as Romanesque *Cathedral of the Blessed Virgin Mary of Masovia* in Płock, *Saint Martin Collegiate Church* in Opatow, *Church of St Jacob* in Sandomierz, *Archcathedral Basilica* in Poznań, *St Mary's Basilica* in Kraków, and *Kraków Academy* (*Jagiellonian University*); Gothic *Cathedral Basilica of the Assumption of the Blessed Virgin Mary and St Adalbert* in Gniezno, *Wawel Cathedral* in Kraków, the old towns *Kazimierz* and *Wiślica*;

1150-1550: *Gothic art*, Western Europe – buildings with pointed arches, flying buttress, and curtain walls with extensive stained-glass windows, exemplified by *Sens Cathedral*, and *Chartres Cathedral* in France; *Salisbury Cathedral, Wells Cathedral*, and *Kings College Chapel* in England; as well as *St Stephan's Cathedral*, and *Abbey Church of Heiligenkreuz* in Vienna, Austria;

1165-1853: *Early Great palaces* – Norman-Georgian *Windsor Castle* in Berkshire, England, a royal residence, on an area of 45,000 m^2; Moorish *Alhambra* located on a plateau dominating the city of Granada in southern Spain, a fortress-palace-garden built by the Nasrid sultans, which covers an area of 142,000 m^2; Chinese *Forbidden City* in central Beijing, the world's largest palace complex, on an area of 150,000 m^2; Renaissance *Apostolic Palace* in Vatican City, the current Papal Palace and Vatican Museums of Rome, on an area of 162,000 m^2; Islamic *Topkapi Palace* in Istanbul, the former residence of Ottoman sultans, on an area of 700,000 m^2; Russo-Byzantine *Grand Kremlin Palace* on Borovitsky Hill in Moscow, the official residence of the President of Russia, on an area of 24,000 m^2; Gothic-Renaissance *Louvre Palace* in Paris, the former residence of the Kings of France, now housing the Louvre Museum, on an area of 60,600 m^2; Renaissance *Quirinal Palace* in Rome, the ancient Pope's palace and present palace of the President of Italy, on an area of 110,500 m^2; Renaissance *Palace of Versailles* near Paris, the later residence of the Kings of France, on an area of 67,000 m^2; Tibetan *Potala Palace* on Marpo Ri hill above the Lhasa valley, the residence of the Dalai Lama, on an area of over 130,000 m^2; Baroque *Royal Palace of Stockholm*, residence of the Kings of Sweden, on an area of 61,200 m^2; Edwardian-Victorian *Buckingham Palace* in the City of Westminster, London, the official residence of the British monarch, on an area of 77,000 m²; Baroque *Winter Palace* in Saint Petersburg, the official residence of the Russian monarchs, now housing the Hermitage Museum, on an area of 60,000 m²; and Baroque *Royal Palace of Madrid*, the official residence of the Spanish Royal Family, on an area of 135,000 m²;

1270-1380: *Pre-Renaissance art*, Italy – Arnolfo di Cambio's portrait statue of *Charles of Anjou*, tomb of *Cardinal de Braye* at Orvieto, bronze statue of *St Peter* in Rome, and design of *Florence Cathedral*; Giovanni Cimabué's works in Lower Church of *San Francesco* at Assisi, and mosaic figure of *Saint John* in the *apse of Pisa Cathedral*; Giotto di Bondone's *mosaics* for Florence Baptistery, *frescoes* in Arena Chapel of Padua and Peruzzi Chapel of S Croce Church in Florence, *Navicella mosaic* of St Peter's in Rome, *Ognissanti Madonna* in Florence, and *façade decorations* for Florence Cathedral; Duccio di Buoninsegna's masterpieces *Maestà* for the altar of Siena Cathedral, *Rucellai Madonna* in S Maria Novella at Florence, and *Annunciation, Christ Healing the Blind Man*, and *Transfiguration* (National Gallery, London); and Andrea da Firenze's *monumental fresco cycle* for Spanish Chapel of Dominican Church S Maria Novella in Florence, and *frescoes Life of St Ranieri* at Camposanto in Pisa;

1300-1710: *Early Romanian architecture* – in Byzantine style at the Wallachian *Princely Church* in *The Court of Arges* and the *Cozia Monastery*; Gothic style at the Transylvanian *Black Church* in Brasov, *Bran Castle* and *Hunyad Castle*; Moldavian Style at the painted *Monasteries of Bucovina* (*Arbore, Humor, Moldovita, Probota, Sucevita* and *Voronet*, all of them in UNESCO world Heritage), *Neamt Monastery*, and *The Three Hierarchs Monastery* in Iasi; Brancovan style at the *Monastery of Horezu*, and the *Princely Palace of Mogosoaia*;

1360-1425: *Russian icon painting* – under the influence of Byzantine art and especially of Greek painter Theophanes, the *Novgorod School* painted icons such as *Crucifixion* (Louvre, Paris); and Andrei Rublev produced great *icon paintings* in Byzantine style, excelling with the famous *Old Testament Trinity* 'Hospitality of Abraham' (now in Tretyakov Gallery, Moscow), as three graceful angels;

1400-1640: *Renaissance Art* – artistic movement started in Italy and flourished in Western Europe, *imitating and developing ancient Greek and Latin art*, and represented such as:

ITALY - painter Gentile da Fabriano with *Adoration of the Magi*

(Uffizi Gallery, Florence); architect and sculptor <u>Filippo Brunelleschi</u> with panels for *Baptistery doors* of Florence Cathedral (now in Bargello Museum), and design of *Dome* belonging to Florence Cathedral; goldsmith and sculptor <u>Lorenzo Ghilberti</u> with *bronze doors* for Florence Baptistery (now in Bargello Museum, Florence), and *bronze figures* of Sts Matthew, Stephen, and John (on the exterior of Or San Michele, Florence); sculptor <u>Donatello di Niccolo</u> with figures of saints on exterior of *Or San Michele* in Florence, and of prophets at *Campanile* in Florence; painter and medallist <u>Antonio Pisanello</u> with surviving frescoes *Annunciation* (Saint Fermo, Verona), and *St George and the Princess of Trebizond* (S Anastasia, Verona); sculptor <u>Luca Della Robbia</u> with *Cantoria, Bronze Door* for Florence Cathedral, marble *Tomb of the Bishop of Fiesole*, and *terracotta figures*; painter <u>Piero della Francesca</u> with very precise and geometric attitude towards composition, as displayed in his series of frescoes illustrating *The Legend of the Holy Cross* (Choir of San Francesco, Arezzo), *Flagellation* (Urbino), and unfinished *Nativity* (National Gallery, London); painter <u>Paolo Uccello</u> with *Battle of San Romano* (now in London, Paris, and Florence), *The Rout of San Romano* and *St George and the Dragon* (both in National Gallery, London), as well as *The Hunt in the Forest* (Ashmolean Museum, Oxford); painter <u>Andrea Mantegna</u> with *Saint Zeno Altarpiece, Saint Sebastian* in Padua, tempera pictures *The Triumph of Caesar* (Hampton Court Palace, England), decoration of *ceiling of Camera degli Sposi* in Mantua, and a collection of *Classical statuary*; painter <u>Antonello da Messina</u> with *Salvator Mundi, Self-portrait* (National Gallery, London), and fragments of Venetian *San Cassiano altarpiece* (Vienna); sculptor <u>Antonio Rossellino</u> with *Madonna and the Laughing Child* (Victoria and Albert Museum, London), marble tomb of *Cardinal of Portugal* (San Miniato al Monte, Florence), marble portrait bust of Florentine *Matteo Palmieri*, and marble *Madonna della Latte* (on Nori monument in Santa Croce, Florence); bronze sculptor and painter <u>Andrea del Verrocchio</u> with bronze statue *David* in Florence, equestrian monument to *B Colleoni* in Venice, and painting *Baptism* in Florence; painter <u>Sandro Botticelli</u> with mythological *Primavera, The Birth of Venus* (Uffizi Gallery, Florence), scheme of decoration of *Sistine Chapel* (Vatican, Rome), severe and emotional *Mystic Nativity* (National Gallery, London), and illustrations for A Dante's *Divina Comedia*; painter <u>Luigi Vivarini</u> with *Madonna and Six Saints* in Academy of Venice; painter <u>Giovanni Bellini</u> with *Madonnas* (Italy), and *drawings* (Louvre, Paris, and British Museum); painter, sculptor and architect <u>Leonardo da</u>

Vinci with paintings *Baptism of Christ, Adoration of the Magi* (Uffizi Gallery, Florence), famous *Last Supper* (Refectory of Santa Maria delle Grazie), portrait *La Belle Ferronnière* (Louvre, Paris), cartoons *Madonna and Child with St Anne* (Royal Academy, London), and *The Battle of Anghiari* (Palazzo delle Signoria, Florence), as well as celebrated easel picture *Mona Lisa* (Louvre, Paris); painter Filippino Lippi with *Frescoes of Brancacci Chapel* in Carmine (Florence), *frescoes of Strozzi Chapel* in S Maria Novella and of *Caraffa Chapel* in S Maria sopra Minerva in Rome, outstanding easel pictures *The Virgin and Saints* (National Gallery, London), *The Adoration of the Magi* (Uffizi, Florence), and *The Vision of St Bernard* (Badia, Florence); painter Domenico Ghirlandaio with fresco *St Francis Raising the Dead Child* including reminiscent female figures such as *Portrait of a Young Woman* as a fine example of Renaissance beauty (Calouste Gulbenkian Museum, Lisbon), altarpiece *Adoration of the Shepherds* (Florentine Academy) for the church of S Trinità in choir of S Maria Novella, fresco *Christ Calling Peter and Andrew* (Sistine Chapel, Rome), easel pictures *Adoration of the Magi* (Innocenti, Florence), and *Visitation of the Virgin* (Louvre, Paris), as well as mosaic *Annunciation* (Cathedral of Florence); architect Donato Bramante with design of *S Maria delle Grazie* in Milan, new *Basilica of St Peter's, Belvedere courtyard, Tempietto di S Pietro* in Montorio, as well as *Palazzo dei Tribunali*, and *Palazzo Caprini* in Rome; sculptor and painter Michelangelo di Lodovico Buonarroti with famous sculptures *Battle of the Centaurs* and *Madonna of the Steps* in Florence, marbles *Cupid* in Bologna, *Bacchus* (Museum of Florence), and *Pieta* (St Peter's, Rome), colossal marble block *David* in Florence, *Holy Family of the Tribune* and *Madonna* (National Gallery, London), sublime statue of *Moses* in Rome, and almost superhuman *Decorations of the Sistine Chapel* (Vatican, Rome); painter Barbarelli Giorgione with *The Tempest, The Family of Giorgione* (Venice), *The Three Philosophers* (Vienna), and *Sleeping Venus* (Dresden Gallery); painter Raphael Sanzio with fresco series *The School of Athens*, portraits of *Julius II, Virgin of the Popolo* (Vatican, Rome), *Altoviti* (Munich), *Leo X* and *Inghirami* (Florence), frescoes of *Camera dell'Incendio* (Vatican chambers), *Sistine Madonna, St Cecilia* of Bologna, and *Ezechiel* of the Pitti (Florence), as well as *Spasimo, Holy Family*, and *St Michael*; painter Vecellio Titian with *Madonna of the Pesaro Family* (Venice), *Feast of Venus, Presentation of the Virgin, Charles V at the Battle of Mühlberg, Ecce Homo, Diana and Actaeon, Diana and Callisto, Perseus and Andromeda, The Fall of Man, The Entombment, Christ Crowned with Thorns, Madonna*

Suckling the Child, and *Pietà*; painter <u>Antonio Allegri da Correggio</u> with *Ecce Homo* (National Gallery, London), *The Night* (Dresden Gallery), *Jupiter and Antilope* (Louvre, Paris), *Education of Cupid* (National Gallery, London), *Danae* (Borghese Gallery, Rome), and *Leda* (Berlin Museum); sculptor <u>Baccio Bandinelli</u> with *Hercules and Cacus* (near Vecchio Palace, Florence), *Adam and Eve* (National Museum, Florence), and *bas-reliefs* in Florence Cathedral; architect <u>Andrea Palladio</u> with *San Georgio Maggiore* in Venice, and writing *Quattro Libri dell'Architettura* 'The Four Books of Architecture'; painter <u>Paolo Caliari Veronese</u> with *The Triumph of Venice* (on ceiling of the Ducal Palace), *The Marriage Feast at Cana* (Louvre, Paris), *The Adoration of the Magi* (National Gallery, London), and *Feast in the House of Levi* (Venice); architect <u>Giacomo Barozzi da Vignola</u> with design of *Villa di Papa Giulio* and *Church of the Il Gesu* in Rome, as well as of *Palazzo Farnese* in Piacenza; and architect and sculptor <u>Francesco Borromini</u> with *S Carlo alle Quattro Fontane, S Ivo della Sapienza, S Andrea delle Fratte*, and *Oratorio of S Philippo Neri*;

FLANDERS - <u>Jan van Eyck</u> with *Man in a Red Turban* (National Gallery, London), and panel altarpiece *The Adoration of the Holy Lamb* (Ghent); painter <u>Rogier van der Weyden</u> with famous *Last Judgement* painted for Chancellor Rolin; painter <u>Hugo van der Goes</u> with *The Adoration of the Shepherds* (Uffizi, Florence), and magnificent *Portineri Altar-piece for S Maria Nouva*; artist <u>Pieter Brueghel the Elder</u> with *Blind Leading the Blind, The Peasant Wedding*, and *The Peasant Dance* (Vienna), as well as *The Adoration of the Kings*, and *The Death of the Virgin* (National Gallery, London); sculptor and architect <u>Giovanni Bologna</u> with *Flying Mercury, Rape of the Sabines*, and *Hercules and the Centaur*; painter <u>Peter Paul Rubens</u> with masterpiece triptych *Descent from the Cross* (Antwerp Cathedral), *decoration of Luxembourg Palace* (Paris), *portraits of Philip IV of Spain, Peace and War* (National Gallery, London), sketches for *Apotheosis of James VI and I* (Whitehall), and *The Crucifixion of St Peter* (Cologne); painter <u>Anthony Van Dyck</u> with full-length portrait of *James VI and I* (Windsor), *Prince of Orange and his Family* and *Ferdinand of Austria*, portraits of *Charles I* and *Queen Henrietta Maria and the two royal children* (Windsor), and magnificent *Le Roi à la chasse*; portrait and genre painter <u>Frans Hals</u> with *Portrait of a Man, Paulus von Berestyn, Catherine, Jacob Pietersz Olycan* and *Aletta Hanemans, The Laughing Cavalier* (Wallace collection, London), *Banquet of the Company of St Adrian, Gypsy Girl* (Louvre, Paris), *Malle Babbe* (Berlin), *Jolly Toper*

(Amsterdam), *Pieter van den Broecke* (Kenwood, London), as well as *Man in a Slouch Hat,* and *The Seated Man* (Kassel); and painter Rembrandt Harmensz van Rijn with *Rembrandt's Mother as the Prophetess Hannah* (Rijksmuseum, Amsterdam), *The Anatomy Lesson of Doctor Tulp* (The Hague), *The Entombment of Christ* (Hunterian Gallery, Glasgow), *Belshazzar Feast* and *Christ and the Woman taken in Adultery* (National Gallery, London), *The Military Company of Captain Frans Banning Cocq* (Rijksmuseum, Amsterdam), *Susanna surprised by the Elders* (Gemäldegalerie, Berlin), *Jacob blessing the Sons of Joseph* (Kassel), and *The Return of the Prodigal Son* (Hermitage, St Petersburg);

GERMANY - painter Konrad Witz with *Christ Walking on the Water,* remarkable because it is set on Lake Geneva, and is the earliest known landscape in European art; painter and engraver Albrecht Dürer with the *Dresden triptych* and *Baumgartner altarpiece* in Munich, *The Prodigal Son, Adam and Eve* (Madrid) and *Adoration of the Trinity* (Vienna), drawings *Triumphal Car* and *Triumphal Arch,* copperplates *Small Passion, Knight, Death and the Devil, St Jerome in his Study,* and *Melancholia*; artist and architect Matthias Grünewald with great *Isenheim altarpiece* (Colmar Museum, Alsace), design of *waterworks* for Magdeburg and dramatic *Crucifixion* (Kunsthalle, Karlsruhe); painter and engraver Hans Baldung with *Die Frau und der Tod* (Basle), and *Die Eitelkeit* 'Vanity' (Vienna);

FRANCE - sculptor Jean Goujon with *Diana reclining by a Stag,* and reliefs for *Fountain of the Innocents* (Louvre, Paris), as well as monument to *Duke of Brézée* in Rouen Cathedral; sculptor Germain Pilon with *The Three Graces* (Louvre, Paris), *statues of Henry II* and *Catherine de Médicis* (Saint Denis), as well as *Virgin of the Sorrows* (St Paul de Louis, Paris), marble *Christ Risen,* and bronze *Cardinal René de Birague* (Louvre); artist Georges de La Tour with paintings *St Joseph the Carpenter* (Louvre, Paris), and *The Lamentation over St Sebastian* (Berlin); etcher and engraver Jacques Callot with etchings of *Siege of La Rochelle,* and engravings including *Misères et malheurs de la guerre* 'Miseries and Misfortunes of War' and *Gypsies*; and Nicolas Poussin with *The Adoration of the Golden Calf, Cephalus and Aurora* and *Bacchanalian Festival* (National Gallery, London), *The Inspiration of the Poet, The Arcadian Shepherds* and *Landscape with the Burial of Phocion* (Louvre, Paris), as well as *The Rape of the Sabine Women* (Metropolitan Museum of Art, New York);

SPAIN - metal engraver Juan de Arfe with impressive and artistic *altarpieces,* now exposed in cathedrals at Avial, Burgos and Seville;

175

and painter <u>Diego de Silva y Velázquez</u> with great works including *Old Woman Cooking Eggs* (National Galleries of Scotland), baroque *Surrender of Breda*, *Infanta Margarita* and *Infanta Maria Theresa*, jester *Don Juan de Austria*, impressionistic *Views from Villa Medici*, and late masterpieces *Las Mañinas* 'Maids of Honour', *Las Hilanderas* 'The Tapestry Weavers', and famous *Venus and Cupid*;

ENGLAND - goldsmith and miniaturist <u>Nicholas Hilliard</u> with *An Unknown Youth Learning Against a Tree Among Roses* (Victoria and Albert Museum, London), and masterpiece *The Pelican portrait of Queen Elizabeth*; and architect and stage designer <u>Inigo Jones</u> with designs of *Queen's House* at Greenwich, rebuilding of *Banqueting House* at Whitehall, *nave*, *transepts* and *Corinthian portico* of Old St Paul's in London, *Marlborough Chapel* and *Double Cube room* at Wilton, as works marking the foundation of classical English architecture;

1570-90: *Mannerist art* – style of painting and architecture characterized by *distortion of human figure*, and *bright colours*, exemplified by works of Italian painter <u>Agnolo Bronzino</u> with *Venus, Folly, Cupid and Time* (National Gallery, London), and *Christ in Limbo* (Uffizi, Florence); and German painter <u>Hans Holbein the Younger</u> with portraits of *Erasmus* (National Gallery, London) and *Sir Thomas More* (Frick Collection, New York), portrait group *The Ambassadors* (National Gallery, London), charming *Lady with a Squirrel and a Starling* (National Gallery, London), and great *Henry VIII granting a Charter to the Barber-Surgeons*;

1620-1850: *Baroque art*, in Southern and Western Europe – based on Classical models with *exuberant and extravagant decoration* of large public buildings, and represented as follows:

ITALY - sculptor, architect and painter <u>Gian Lorenzo Bernini</u> produced *Life-size marble statues* (Borghese Gallery), famous *Baldacchino for St Peter's* (Rome), sculptural and architectural *work in Vatican* and *Cornaro Chapel* (Santa Maria della Vittoria Church) with sculpture depicting *The Ecstasy of St Theresa*; painter <u>Michelangelo Merisi da Caravaggio</u> was author of the famous *Life of Saint Matthew, Conversion of Saint Paul*, and *Crucifixion of Saint Peter* (Rome); goldsmith, sculptor and engraver <u>Benvenuto Cellini</u> became known by the gold salt-cellar *Neptune and Triton* (Vienna), and bronze *Perseus with the Head of Medusa*; painter <u>Jacopo Robusti Tintoretto</u> created *The Last Judgement, The Golden Calf, The Marriage of Canna, The Removal of the Body of St Mark, The Origin of the Milky Way, Paradiso, Entombment*, and *The Last Supper*; and architect <u>Guarino Guarini</u> designed *San Lorenzo, Capella della SS Sindone* and *Palazzo Carignano* in Turin, as well as palaces for Bavaria and Baden, and published the influential *Architectura Civile*;

SPAIN - painter <u>Domenico Theotocopoulos El Greco</u> became well-known by his *View of Mount Sinai and the Monastery of St Catherine* (Historical Museum of Crete), famous *Burial of Count Orgaz* (Toledo), *Crucifixion* and *Resurrection* (Prado, Madrid), as well as *View of Toledo* (New York and National Gallery, London);

ENGLAND - painter of miniatures <u>Richard Gibson</u> was a page to Charles I, and then made several portraits of *Cromwell*, and was himself painted by <u>Peter Lely</u> who also painted *Windsor Beauties*, and a series of English *Admirals* at Greenwich, marking the contrast to his other portraits which sometimes have a hasty superficial appearance;

1640-90: *French art* – an artistic development due to architect <u>François Mansart</u> with major works such as the *north wing of Château de Blois*, churches *Santa Marie de la Visitation*, and *Val-de-Grâce*, as well as new deigns of *Hôtel de la Vrillière*, and *Château de Maisons*; historical painter <u>Charles Le Brun</u> who founded *French school of painting* and Academy of Painting and Sculpture, was director of *Gobelins tapestry works*, and produced *decorations of Versailles*; and sculptor, painter and architect <u>Pierre Puget</u> who worked on *ceilings* of Barberini Palace (Rome) and Pitti Palace (Florence), and created sculptures such as *Hercules, Milo of Croton, Alexander the Great* and *Diogenes* (Louvre, Paris);

1656-65: *Dutch painting* – <u>Jan Vermeer</u> produced *The Allegory of Faith, The Procuress* (Dresden), *Christ in the House of Martha and Mary, View of Delft, Girl with a Pearl Earring* (The Hague), *Woman with a Water Jug* (New York), *Woman reading a Letter* (Amsterdam), and *Allegory of Painting* (Vienna);

1660-80: *Italian art* – <u>Carlo Dolci</u> painted *Madonnas, St Cecilia, Herodias with the Head of John the Baptist* (Dresden), and *Magdalene* (Uffizi, Florence);

1663-1715: *English art* – architect <u>Christopher Wren</u> designed the *Chapel* at Pembroke College (Cambridge), *Sheldonian Theatre* (Oxford), *Library* of Trinity College (Cambridge), *New St Paul's Cathedral* (London), *Royal Exchange Greenwich Observatory*, and *Ashmolean Museum* (Oxford); and sculptor and woodcarver <u>Grinling Gibbons</u> decorated *King's rooms* at Windsor and the *Choir* at St Paul's Cathedral (London), made *festoons* and *Cherubs's heads* (Chatsworth, Burghley, Hampton Court, and Blenheim), as well as

177

statues of *Charles II* at Chelsea Hospital, and *James VII and II* in Trafalgar Square (London);

1682-1706: *Swedish architecture* – Nicodemus Tessin completed the *Drottningholm Palace* and its royal church in Stockholm, designed the *Steninge Castle* and the *Tessin Palace* (now Governor's Palace), also *Trinity Church* (Karlskrona), *Fredrik's Church* (Karlshamm), and made important works on the *Amelienborg Castle* (Copenhagen), *Louvre* (Paris), *Apollo Temple* (Versailles), and overall the *Royal Palace* (Stockholm);

1700-1850: *Rococo art* – an architecture extending Baroque style, with *greater extravagance of design motifs*, and using *newly lightness of detail and elements*, such as shells, flowers and trees, which is exemplified by works of French painter Jean Antoine Watteau with mythological *Embarquement pour l'îsle de Cythère*, and *Fêtes galantes*; French painter François Boucher with *The Rising* and *The Setting of the Sun*; and English painter and engraver William Hogarth with *A Midnight Modern Conversation, A Rake's Progress, The Pool of Bethesda, The Good Samaritan, Marriage à la mode, Captain Coram* (Foundling Hospital, London), and *The Shrimp Girl* (National Gallery, London);

1730-70: *Italian decorative painting* – Giovanni Battista Tiepolo produced *An Allegory with Venus and Time* (National Gallery, London), part of a ceiling in Contarini Palace (Venice), created movement and energy including *Antony and Cleopatra* at Labia Palace, as well as *Allegory of the Power of Eloquence, St Aloysius Gonzaga in Glory, The Adoration of the Magi* and *The Martyrdom of St Agatha* (all of them in Courtauld Institute, London);

1733-89: *English art* – painter and engraver William Hogarth with *Southwark Fair* and *A Midnight Modern Conversation, A Rake's Progress* (Soane's Museum, London), conventional *The Pool of Bethesda* and *The Good Samaritan*, masterpiece series *Marriage à la mode* (National Gallery, London), series *Industry and Idleness, Gin Lane* and *Beer Street*, as well as informal *Captain Coram* (Foundling Hospital, London), *Self-Portrait* and *The Shrimp Girl* (National Gallery, London); landscape and portrait painter Thomas Gainsborough with *The Charterhouse, Mr and Mrs Andrews, Earl Nugent*, masterpieces *Lord and Lady Howe, Mrs Portman* (Tate Gallery, London) and *Blue Boy* (Huntington Collection, Pasadena), great landscapes *The Harvest Wagon* (Barber Institute, Birmingham) and *The Watering Place* (Tate Gallery), then *The King's Daughters*, character study *Mr Truman*, luxuriant *Mrs Graham* (Edinburgh),

George III and *Queen Charlotte* (Windsor Castle), as well as *Mrs Siddons*, landscapes *Cottage Door* (Pasadena), *The Morning Walk*, and rococo *Cattle Crossing a Bridge*; and painter <u>Joshua Reynolds</u> with works including *Commodore Keppel* (National Maritime Museum, London), *Dr Samuel Johnson* (National Portrait Gallery, London), *Tragic Muse* for Sarah Siddons, and pictures of children such as *The Strawberry Girl*, and *Simplicity*;

1750-1900: *<u>Neoclassical architecture</u>* – a style focused on severe production of *large-scale rebuilding* of cities in England, Scotland, France and USA, exemplified by works of Scottish architects <u>Robert Adam</u> and <u>James Adam</u> who designed *Home House* in London's Portland Square, *Lansdowne House*, *Derby House*, *Register House* in Edinburgh, *Old Quad* of Edinburgh University, and *Oval staircase* of Culzean Castle in Ayrshire; French financier and town-planner <u>Georges Eugène Haussmann</u> who redesigned the city of Paris by *widening streets*, *laying out boulevards* and *parks*, *building bridges*, and thus distinctively modernizing the whole city; and British-born US civil engineer <u>Benjamin Henry Latrobe</u> who designed *parts of the Capitol*, and the *White House* in Washington, as well as *waterworks* in Philadelphia and New Orleans;

1755-1805: *<u>French art</u>* – genre and portrait painter <u>Jean Baptiste Greuze</u> created paintings of Italian subjects, and remarkable studies of girls such as *The Broken Pitcher* (Louvre, Paris), and *Girl with Doves* (Wallace Collection, London); Rococo and Baroque sculptor <u>Étienne Maurice Falconet</u> produced figures of *Venus*, *Bathers* (Louvre, Paris), and bronze *equestrian monument to Peter I the Great* (St Petersburg); painter and engraver <u>Jean Honoré Fragonard</u> made genre pictures of contemporary life, notably *The Progress of Love*, landscapes foreshadowing Impressionism, *La Bacchante endormie* 'The Sleeping Baccante', and *La Chemise enlevée* (Louvre, Paris); architect <u>Claude Nicolas Ledoux</u> designed *Château at Louveciennes*, *Theatre at Besançon*, *Saltworks at Arc-et-Senans*, and *tax buildings* around Paris; and Baroque sculptor <u>Jean Antoine Houdon</u> became known by sculpture of *St Bruno* (Santa Maria degli Angeli, Rome), monument to *G Washington* (Virginia, USA), busts of *Diderot* and *Voltaire* (Théâtre Français, Paris; and Victoria and Albert Museum, London), *Napoleon*, *Catherine the Great*, and *JJ Rousseau* (Louvre, Paris);

1763-69: *<u>English pottery</u>* – a revival in pottery due to <u>Josiah Wedgwood</u> who obtained patent of a cream-coloured ware, called *Queen's ware*, produced *unglazed blue Jasper ware*, with its raised designs in white, and also black basalt ware, called *Etruria*, expressing the *Wedgwood ware*;

1775-1851: *English art* – sculptor and draughtsman <u>John Flaxman</u> with monument to *Earl of Mansfield*, statue of *J Kemble* (Westminster Abbey, London), sculptures to poets *W Collins* (Chichester Cathedral), *T Chatterton* (Bristol), and to *Lord Nelson* (St Paul's Cathedral, London), as well as statue of *R Burns* (National Portrait Gallery, Edinburgh); painter and engraver <u>William Blake</u> with *The Canterbury Pilgrims* (Canterbury Cathedral), *The Spiritual Form of Pitt guiding Behemoth* (National Gallery, London), *Jacob's Dream*, and *The Last Judgement*; architect <u>John Nash</u> with designs of the *new Regent's Park, Regent Street, Trafalgar Square, St James's Park, Marble Arch*, and *Carlton House Terrace*, as well as recreation of *Buckingham Palace* and *Brighton Pavilion* in oriental style; painter <u>Joseph Mallord William Turner</u> with *The Shipwreck, Frosty Morning*, and *Crossing the Brook*, engravings of series *Liber Studiorum*, and also *The Fighting Téméraire, Rain*, and *Steam and Speed* (National Gallery, London); watercolourist <u>Peter de Wint</u> with *The Cricketers, The Holy Harvest, Nottingham, Richmond Hill* and *Cows in Water* (mainly in Lincoln Art Gallery), as well as oils *A Cornfield*, and *A Woody Landscape* (Victoria and Albert Museum, London); architect <u>Charles Barry</u> with designs of *Travellers' Club, Manchester Athenaeum, Reform Club*, and the *new Palace of Westminster*; and architect <u>Augustus Welby Northmore Pugin</u> with *drawings* for Houses of Parliament, designs of several Roman Catholic churches, including the *Catholic Cathedral* in Birmingham and *St Oswald's* in Liverpool, and his books *Contrasts between the Architecture of the 15th and 19th Centuries, True Principles of Christian Architecture*, and *Chancel Screens*;

1778-1953: *Spanish art* – artist <u>Francisco de Goya</u> performed *Blind Guitarist*, portraits *The Family of Charles IV, Maja nude* and *Maja clothed* (Prado, Madrid), etchings *Los Caprichos*, and sardonic series *The Disasters of War*; architect <u>Antoni Gaudí</u> designed unconventional buildings *Palacio Güell, Parque Güell, Casa Batlló, Casa Milá*, and celebrated *Sagrada Familia* in Barcelona; painter and pioneer of Cubism <u>Pablo Picasso</u> created traditional *Gypsy Girl on the Beach*, Neo-Impressionist *Longchamp* and *The Blue Room*, portraits of *G Stein* and *Stalin*, analytical Cubist *Demoiselles d'Avignon*, series *MaJolie*, grotesque *Three Dancers*, and immense canvas *Guernica*; and painter <u>Salvador Dalí</u> produced *The Persistence of Memory*, known as *Limp Watches* (Museum of Modern Art, New York), and *Christ of St John of the Cross* (St Mungo Museum, Glasgow);

1782-1954: _Italian art_ – sculptor <u>Antonio Canova</u> with _Theseus_ (Victoria and Albert Museum, London), colossal statue of _Napoleon I_ (Paris), statue of _Pauline Bonaparte_ (Borghese Gallery, Rome), and sculpture _The Three Graces_ (Edinburgh and London); painter and sculptor <u>Amedeo Modigliani</u> with _elongated stone heads in African style_, richly-coloured _portraits_, and very frank _nudes_; artist <u>Giorgio de Chirico</u> with dreamlike pictures of deserted squares, such as _Nostalgia of the Infinite_ (Museum of Modern Art, New York), and influence on Surrealists by _metaphysical painting_ which included semi-abstract geometric figures and stylized horses; and sculptor and painter <u>Marino Marini</u> with theme _horse and rider_ in many versions, combination of different techniques, such as _Dancer_, and _portraits of I Stravinsky, M Chagall_, and _H Miller_;

1784-1985: _French art_ – Neoclassical painter <u>Jacques Louis David</u> with _Oath of the Horatii, Death of Socrates, Brutus Condemning his Son, Death of Marat, The Rape of the Sabines, Napoleon Crossing the Alps_, and _Le Sacre de Napoléon Ier par le Pape Pie VII_ 'The Coronation of Napoleon I by Pope Pius VII'; painter <u>Jean Auguste Dominique Ingres</u> with _Achilles Receiving the Ambassadors of Agamemnon_ (École des Beux-Arts), famous nudes _La Grande Odalisque, The Valpinçon Bather_ and _La Source_ (all in Louvre, Paris), historical _Paolo and Francesca_ (Chantilly) and _The Oath to Louis XIII_ for Montauban Cathedral, then _Apotheosis of Homer_ (Louvre ceiling) and _The Martyrdom of St Symphorian_ (Autun Cathedral), also successful _Stratonice_ (Chantilly), _Vierge à l'hostie_ (Louvre), and _Odalisque à l'esclave_; painter <u>Eugène Delacroix</u> with _Massacre at Chios_ (Louvre, Paris), historical _The Execution of Faliero_ (Wallace collection) and _Liberty Guiding the People_ (Louvre, Paris), oriental flavoured _Algerian Woman_, and _panels_ for library of the Deputies' Chamber in Paris; painter and founder of Realism <u>Gustave Courbet</u> with _Peasants of Flagzey_ (Musée des Beaux-Arts, Besançon), _Burial at Ornans_ (Musée d'Orsay, Paris), and famous canvas _Studio of the Painter: an Allegory of Realism_ (Louvre, Paris); sculptor <u>Albert-Ernest de Carrier-Belleuse</u> with _Femme à la Rose, Michel Angelo, Raphael Sanzio, Dante, Candelabra, Melody, La Liseuse, Minerve_ and _Diana_, as well as equestrian statues of _Joan of Arc_ (Orléans) and _Michael the Brave_ (Bucharest); painter <u>Édouard Manet</u> with _Spanish Guitar Player_ (awarded at 1861 Salon), _Déjeuner sur l'herbe_ (rejected at 1863 Salon), _Olympia_, and _Un Bar aux Folies-Bergère_; Romantic sculptor <u>Auguste Rodin</u> with _L'Homme au nez cassé_ 'The Man with the Broken Nose', _Saint Jean Baptiste_ 'St John the Baptist', _Porte de l'enfer_ 'The Gate of Hell', _Le Baiser_ 'The Kiss',

Le Penseur 'The Thinker', *Les Bourgeois de Calais* 'The Burghers of Calais', also portrait busts of *Bastien-Lepage, P de Chavannes, V Hugo* and *GB Shaw* (Musée Rodin, Paris; Rodin Museum, Philadelphia; and Victoria and Albert Museum, London); artist <u>Pierre Auguste Renoir</u> with sunlight leaf-filtering picture *Moulin de la Galette* (Louvre, Paris), *The Umbrellas* (National Gallery, London), series paintings *Bathers*, and *The Judgement of Paris*; painter <u>Paul Cézanne</u> with *Aix: Paysage rocheux* 'Rocky Landscape in Aix' and *Le Jardinier* 'The Gardner' (Tate Gallery, London), *L'Homme au chapeau de paille* 'Man in a Straw Hat' (Metropolitan Museum, New York), *The Card Players* (Musée d'Orsay, Paris), and *La Vieille au chapelet* 'The Old Woman with Beads' (National Gallery, London); designer and glass maker <u>Émile Gallé</u> with design of the glass called *Gallé*, reflecting experiments with materials and techniques, generally in the style of Art Nouveau, and novel designs of *artistic furniture*; painter <u>Claude Monet</u> with *Impression: soleil levant* (presented at first Impressionist Exhibition), *Haystacks, Rouen Cathedral,* and *Waterlilies*; artist <u>Georges Pierre Seurat</u> with Pointillist canvases *Une Baignade, Asnières, Un Dimanche d'été à la Grande-Jatte* 'Sunday Afternoon on the Island of La Grande-Jatte' (Chicago), *Les Poseuses,* and *Le Cirque*; painter and lithographer <u>Henri Toulouse-Lautrec</u> with portrait of *Van Gogh* (Amsterdam), *Jean Avril dansant*, drawing of *Oscar Wilde, Le Lit* and *La Toilette* (Museé d'Orsay), *Monsieur Boileau at the Café*, paintings of *Moulin Rouge* and *The Two Friends* (Tate Collection), *La Clownesse Cha-u-Kao* (Musée d'Orsay), as well as *The Bar, At the Races,* and *Tête-à-tête supper*; Post-Impressionist painter <u>Paul Gauguin</u> with *Still Life with Three Puppies* (Museum of Modern Art, New York), *The Vision after the Sermon* (National Gallery, Edinburgh), *D'où venons-nous?* 'Whence do we come?', *Que sommes-nous?* 'What are we?', and *Où allons-nous?* 'Where are we going?' (Boston); architect <u>Hector Germain Guimard</u> with *Castel Béranger* apartment block, and well-known *Paris Métro entrances*; sculptor <u>Aristide Maillol</u> with female nudes *Three Graces, Mediterranean, Crouching Woman* (Museum of Modern Art, New York), and sculpture *The Mountain*; architect <u>Tony Garnier</u> with Utopian ideal *Une Cité industrielle*, combining functionalism with political and social reform, and sophisticated reinforced concrete buildings at Lyons including *Grange Blanche hospital* and *Stadium*, as well as *Hôtel de Ville* at Boulogne-Billancourt; painter <u>Henri Matisse</u> with high-pitch coloured *Woman with the Hat*, stained glass for *Dominican Chapelle du Rosaire at Vence* (South France), and paintings *Bonheur de vivre, L'Escargot* and *La Lieuse* 'Woman

Reading'; sculptor <u>Henri Gaudier-Brzeska</u> with abstract style both in carving and drawing, exemplified by *Red Stone Dancer* (Tate Collection, London), *Crouching Figure* (Minneapolis), and bronze *Two Men With a Bowl* (National Museum of Wales, Cardiff); Alsatian sculptor <u>Hans Arp</u> with abstract reliefs in wood *Madame Torso in a Wavy Hat* (Berne) and *Forest* (Penrose Collection, London), three-dimensional works *Landmark* and *Ptolemy* (Paris), monumental *Wood relief* for Harvard (USA), and *Mural* for UNESCO building (Paris); architect <u>Charles Édouard Jeanneret Le Corbusier</u> with designs of *Unité d'habitation* (Marseilles), *Chandigarh* (Punjab), and buildings on stilts 'pilotis' first applied for *Swiss Pavilion* in Cité Universitaire (Paris); Russian-born painter <u>Wassily Kandinsky</u> with publication of the famous book *Über das Geistige in der Kunst* 'On the Spiritual in Art', introduction of *Abstract painting*, and paintings predominantly geometric, in line with Suprematist and Constructivist works; Polish Jewish-born sculptor <u>Jacques Lipchitz</u> with bronze figure and animal compositions *Mother and Child* (Cleveland Museum of Art, Ohio), *Benediction* (Private collection, New York), and *Spirit of Enterprise* (Philadelphia bronze sketch in Tate Gallery, London); and sculptor <u>César Baldaccini</u> with *Petit déjeuner sur l'herbe* 'Breakfast Picnic', *L'Homme de Saint-Denis* (Tate Modern, London), *Yellow Buick* (Museum of Modern Art, New York), and *The Champions*, a permanent memorial to four damaged and burnt-out racing cars;

1800-35: *Japanese painting* – <u>Katsushika Hokusai</u> produced coloured woodcut print *The Great Wave off Kangawa*, 15-volume *Mangwa* 'Random Sketches' depicting most facets of Japanese life, and *impressive landscapes* including *36 Views of Mount Fuji*, which greatly influenced the French Impressionists;

1848-1980: *Romanian sculpture* – <u>Karl Storck</u> was author of sculptures *Princess Balasa, Mihail Cantacuzino, Ana Davila, Carol Davila, Minerva Crowning the Arts and Science* and allegoric pieces; busts of *Theodor Aman, Alexandru Ioan Cuza, Mihail Kogalniceanu, C.A. Rosetti, Elena Cuza*, and *Grigore III Ghica of Moldavia*; *iconostasis* of Viforata Monastery; as well as *facade* and *interior* of Sutu Palace; <u>Frederik Storck</u> created statues of *Mihai Eminescu* and *Doctor Constantin Codrescu*, bust of *Spiru Haret*, bronze *Spleen*, and also the piece *Melancholy*; sculptor <u>Constantin Brancusi</u> with *The Table of Silence, The Kiss Gate* and *The Column of Infinite*, all of them in Targu Jiu (Romania), abstract sculptures two block-like versions of *The Kiss*, highly-polished egg-shaped carvings *Sleeping Muse* (Pompidou Centre, Paris), African-influenced *The Prodigal Son*, and several versions of *Mademoiselle Pogany*, *Bird in Space*, and *The Sea-Lions*; Dimitrie Paciurea is remembered for his symbolic style exhibited in works such as the celebrated *The Giant* 'The Man Emerging From Rock', as well as *The Primitive Man*, *The Sphinx*, *The Girl with the Pot*, the *Chimaerae (of the Sky, of the Earth, of the Water, with Wings, of the Night)*, *God of War*, *Head of a Fawn*, and impressionistic *Pan*; and Ion Jalea became well-known by allegoric compositions *The Fall of the Angels* and *The Fall of Lucifer*; statues *The Fight of Hercules Against the Centaur*, *Archer Resting* (with a bronze copy at the Court of Justice, Luxembourg), *Spiru Haret*, *The Hammer Thrower*, and *George Enesco*; statuary group *The Mother and the Schoolboy*; equestrian statues *Mircea the Elder* and *Decebal*; busts of *Mihai Eminescu* and *Octavian Goga*; and the *reliefs* of the *Union Obelisk*;

1850-1900: *Neo-Gothic architecture* – a Gothic revival in Europe and USA, which became obvious in churches and buildings, noticeably English *Houses of Parliament*;

1858-1961: *Romanian painting* – Nicolae Grigorescu with celebrated *The Romanian Ring Dance*, *The Wagon with Oxen*, *Girl with Hope-chest*, *The Battle of Smardan*, and *The Return from Fair*; painter of Armenian descent Theodor Aman with *The Union of the Principalities*, portrait of *Michael the Brave*, and *Székelys bring the Head of Cardinal Andrew Báthory to Michael the Brave after the Battle of Shelimber*; painter Ion Andreescu with *Beech Forest*, *Street from Barbizon during Summer Time*, *The Red Scarf*, *Peasant Woman with Green Kerchief*, *Still Life*, and *The Winter*; painter, engraver and lithographer Nicolae Tonitza with *The Man of a New World*, *The Forester's Daughter*, *Portrait of a Child*, *The Garden in Valeni*, and *Nude*; and expressionist and abstract oil painter Ion Tuculescu with *Self-Portrait on an Autumn Leave*, *Butterfly Dance*, *Charming Sea*, *The Peacocks of the Looks*, *Daphne Monastery*, *Moods of the Field Traces*, *Looks of the Colour*, *Landscape from Paros Island*, and *Landscape at Mangalia*;

1880-1910: *Art Nouveau* – an architectural style countering Neo-Gothic with sinuous flowing shapes for buildings and interior design, like *Glasgow Art School* in Scotland, and *Church of the Holy Family* at Barcelona in Spain;

1881-88: *Dutch painting* – Vincent Willem Van Gogh produced peasant domestic *The Potato Eaters* (Van Gogh Museum, Amsterdam), *Boots* (Museum of Art, Baltimore), *Sunflowers*, *The Bridge*, *The Chair and the Pipe*, *Ravine*, and *Cornfields with Flight of*

Birds;

1897-1927: *Scottish architecture* – architect, designer and watercolourist Charles Rennie Mackintosh with *Glasgow School of Art, Cranston tearooms*, houses like *Hill House* in Helensburgh, and interior design, textiles, furniture, metalwork, and watercolours;

1900-50: *Chinese art* – under influence of early Qing individualists, artist Qi Baishi painted *birds, flowers, fruit, landscapes*, and subjects from daily life, exposing a spontaneous, calligraphic, and even humorous style, and unveiling a keen observation of nature;

1903-45: *English architecture* – Giles Gilbert Scott designed *Anglican Cathedral* in Liverpool, the new *nave* at Downside Abbey, new buildings at Clare College, a new *Bodleian Library* at Oxford, and a new *Cambridge University Library*, as well as made the plans for the new *Waterloo Bridge*;

1905-20: Foundation of *Expressionist school Die Brücke* 'The Bridge' by German painter Erich Heckel who produced excellent works of lithography and wood-cut, including *Self-portrait*;

After 1905: *Modernist architecture* – a style *using spare line and form*, emphasizing rationalism, and eliminating ornament, based on technological advance in glass, steel, concrete, and construction;

1907-58: *British sculpture* – Russian-Polish Jew-born Jacob Epstein with marble *Genesis* (Granada Television) and *Ecce Homo*, bronze portrait heads of *J Conrad, A Einstein, TS Eliot* and *GB Shaw*, two bronze *Madonna and Child* (New York, and London), and aluminium *St Michael and the Devil* (Coventry Cathedral);

1907-64: *US art* – Slovenian-born sculptor Ivan Meštrović designed the *National Temple* at Kossovo, colossal *Monument to the Unknown Soldier* in Belgrade, and made portrait busts including that of *T Beecham*; architect Frank Lloyd Wright designed the earthquake-proof *Imperial Hotel* (Tokyo), *Falling Water* (Pennsylvania), *Johnson Wax* (Wisconsin), *Florida Southern College*, and *Guggenheim Museum of Art* (New York); pioneer of kinetic art Alexander Calder with works relying upon air currents to set them as *rotating and casting intricate, ever-changing shadows* (New York and Paris); German-born architect Ludwig Mies van der Rohe with designs of high-rise flats for *Weissenhof Exhibition* (Germany), *German Pavilion* for Barcelona International Exposition, and two *glass apartment towers on Lake Shore Drive* (Chicago); and sculptor David Roland Smith with bronze relief plaques *Medals of Dishonour, Hudson River Landscape* (Whitney Museum, New York City), as well as *Subi XVIII* and *Subi XIX* (New York);

1925-44: *Russian cinematography* – film director <u>Sergei Mikhailovich Eisenstein</u> produced propaganda films on Russian Revolution with *The Battleship Potemkin*, on October Revolution with *Ten Days That Shook The World*, and then patriotic epic *Alexander Nevski*, *The Magic Seed*, and masterpiece *Ivan the Terrible* with its sequel *The Boyars Plot*;

1926-61: *Belgian Surrealist painting* – <u>René François Ghislain Magritte</u> with *The Menaced Assassin* (Museum of Modern Art, New York) and *The Red Model* (Pompidou Centre, Paris), as well as murals *La Fée Ignorante* (Palais des Beaux-Arts, Charleroi), and *Les Barricades Mystérieuses* (Palais des Congrès, Brussels);

1928-64: *US cinematography* – artist and film producer <u>Walter Disney</u> produced the films *Plane Crazy* and *Steamboat Willie*, creating famous character *Mickey Mouse*, cartoon films *Snow White and the Seven Dwarfs*, *Pinocchio*, *Dumbo* and *Fantasia*, swashbuckling films for young people such as *Treasure Island*, *Robin Hood*, *The Swiss Family Robinson*, and also family films exemplified by *Mary Poppins*;

1930-47: *Swiss art* – sculptor and painter <u>Alberto Giacometti</u> built Surrealist constructions of a symbolic kind, including 'thin man' *bronzes*, long spidery *statuettes*, suggesting transience, change and decay, such as *Man Pointing*; and artist <u>Paul Klee</u> made abstract pictures with supreme technical skill in many media and subtle colouring, such as *Twittering Machine* (Museum of Modern Art, New York);

1931-2000: *English art* – sculptress <u>Dame Barbara Hepworth</u> with wood *Forms in Echelon* and *Group II (People Waiting)*, wood and stone *Two Forms with White (Greek)*, marble *Three Monoliths*, and bronze *Four Square (Walk Through)*; sculptor <u>Henry Spencer Moore</u> with *Madonna and Child* for St Matthew's Church in Northampton, *Decorative frieze* on Time-Life Building in London, as well as *Massive reclining figures* for UNESCO Building in Paris and for Lincoln Center in New York; painter <u>Richard Hamilton</u> with *Hommage à Chrysler Corp*, collage picture *Just What is it that Makes Today's Homes so Different, so Appealing?* as an introduction to Pop Art, *Study of Hugh Gaitskell as a Famous Monster of Film Land*, and *The Citizen*; sculptor <u>Reginald Cotterell Butler</u> with a *steel and bronze working model*, and then *linear sculpture* presented in constructions in *wrought iron*; painter <u>Patrick Heron</u> with *Lemon and Ultramarine*,

Cadmium with Violet, Scarlet, Emerald, as well as *Lemon and Venetian*; architect <u>Denys Louis Lasdun</u> with designs of *Royal College of Physicians* (London), *University of East Anglia* (Norwich), *National Theatre* (London), *European Investment Bank* (Luxembourg), and *Institute of Education* (London); sculptor <u>Anthony Caro</u> with abstract style based on large pieces of metal welded together and painted in primary colours such as *Early One Morning*, figurative bronzes *Triangle Workshop* in New York State, and co-design of London's *Millennium Bridge*; architect <u>Richard George Rogers</u> with designs of *Pompidou Centre* in Paris, *Lloyds of London*, and *European Court of Human Rights* in Strasbourg; and architect <u>Norman Robert Foster</u> with designs of *Hong Kong* and *Shanghai Banks*, *Century Tower* in Tokyo, a *new terminal at Stansted Airport* in England, *Communications Tower* for 1992 Barcelona Olympics, *ITN headquarters*, and *airport at Chek Lap Kok* in Hong Kong, redesign of *Reichstag* in Berlin, and completion of *Millennium Bridge* in London;

1935-90: <u>*Spanish art*</u> – Lanzarote-born <u>César Manrique</u> produced pictorial and architectural works of encouraging a unique way of understanding *human impact* on its territory, and established a model and a philosophy which allowed *Lanzarote Island* to retain a significant part of its essence, and monuments of a unique *cactus garden*, a house containing paintings in *Fundacion César Manrique*, and *public sculptures* all over Lanzarote;

1940-72: <u>*English cinematography*</u> – filmmaker <u>Alfred Joseph Hitchcock</u> became master of suspense and innovative camerawork, producing films including *Rebecca, Spellbound, Notorious, Rope, Dial M for Murder, Rear Window, Psycho, The Birds*, and *Frenzy*;

1950-91: <u>*Japanese cinematography*</u> – film director <u>Akira Kurosawa</u> with his *Rashomon, The Seven Samurai, The Throne of Blood, Dersu Uzala, Kagemushi* 'Shadow Warrior', *Ran* 'Chaos', and *Rhapsody in August*;

1953-86: <u>*Italian cinematography*</u> – film director <u>Federico Fellini</u> produced *I Vitelloni* 'The Young and the Passionate', *La Strada* 'The Road', *Le Notte di Cabiria* 'The Nights of Cabiria', *La Dolce Vita* 'The Sweet Life', *Otto e Mezzo, Giulietta degli Spiriti* 'Juliet of the Spirits', *Satyricon, Fellini's Roma, Amarcord* 'I Remember', *Casanova, Orchestra Rehearsal, Città delle Donne* 'City of Women', *The Ship Sails On*, and *Ginger and Fred*;

1975-2001: <u>*US cinematography*</u> – filmmaker <u>Steven Spielberg</u> with

187

successful films *Jaws, Close Encounters of the Third Kind, Raiders of the Lost Ark, Indiana Jones and the Temple of Doom, The Color Purple, Empire of the Sun, Hook, Jurassic Park, Men in Black, Saving Private Ryan,* and *A.I. Artificial Intelligence*;

After 1980: *Post-Modernist art* – a movement in USA, UK and Japan, which rejected Modern style functionalism in favour of *mixture of styles and motifs*, using irony, parody, and illusion as distinct from Modernist ideals of truth; and *High Tech art* – a constructional style deriving from Modern movement and expressing highly developed structures and technical innovations, e.g. *Hong Kong* and *Shanghai Banks* in Hong Kong, and *Lloyd Building* in London;

1986-95: *British art* – sculptor <u>Richard Deacon</u> with *Listening to Reason, Double Talk* and *The Back of my Hand; Struck Dumb* and *Kiss and Tell*; as well as public commissions *Between the Eyes* (Toronto), *Let's Not be Stupid* (Warwick University), and *Never Mind* (Antwerp); and painter and installation artist <u>Damien Hirst</u> with *The Physical Impossibility of Death in the Mind of Someone Living* (Saatchi Collection, London), installation *The Asthmatic Escaped* (Tate Gallery, London), *Mother and Child, Divided*, as well as a touring exhibition entitled *Some Went Mad, Some Ran Away* featuring a white lamb in formaldehyde called *Away From the Flock* which was vandalized with black ink at the Serpentine Gallery, London;

1990-2013: *Adobe Systems*, the graphic software, made drawing painting and image manipulation software popular; then the offset and inkjet printing was developed into a robot-artwork, called *Zanelle*, using brush strokes and artist grade paints; followed by multiple other robotic painters currently mass-produced.

The above data indicate some noticeable sequences in art development, which are selected and delimited by transitional times, in years, as:

Pre-classic ancient art \ 2000...1700BC ≈ 1850BC \
Classic ancient art \ 250BC...AD100 ≈ 75BC \
Religious art \ AD1260...1400 ≈ AD1330 \
Revived art \ AD1590...1600 ≈ AD1595 \ *Late styles of art*;

or converted in years ago 3864; 2089; 684; 419. On the other hand, taking into account the specifications at the end of chapter 7, art originated 174588 years ago, so that the times of changing its sequences, in years from the origin of art, are: *Pre-classic ancient art* \ 174588 - 3864 ≈ 170724 \ *Classic ancient art* \ 174588 - 2089 ≈

172499 \ *Religious art* \ 174588 - 684 ≈ 173904 \ *Revived art* \ 174588 - 419 ≈ 174169 \ *Later styles of art.*

According to the table in chapter 7, art will cease 4350 years after the present, as its final time $t_{\bullet}$ ≈ 174588 + 4350 = 178938 years; and then the ratios of above times of changing art sequences to the final time are 170724/178938 ≈ 0.954, 172499/178938 ≈ 0.964, 173904/178938 ≈ 0.972, and 174169/178938 ≈ 0.973. Similar values of these ratios are given in Background to time, Table *(m)* for $z/8 = k$ sequences, in the same succession, as $t_{1.875}/t_{\bullet}$ = 0.9540453; $t_2/t_{\bullet}$ = 0.9640276; $t_{2.125}/t_{\bullet}$ = 0.9718727; $t_{2.25}/t_{\bullet}$ = 0.9780261 respectively.

Using these values of ratios, the transitional times, in years from the origin of art, can be more accurately evaluated as:

$$t_{1.875} ≈ (0.9540453)\cdot 178938 ≈ 170715;$$
$$t_2 ≈ (0.9640276)\cdot 178938 ≈ 172501;$$
$$t_{2.125} ≈ (0.9718727)\cdot 178938 ≈ 173905;$$
$$t_{2.25} ≈ (0.9780261)\cdot 178938 ≈ 175006.$$

Applying the formula $t_k = t_{\bullet}\cdot tanh(k)$ for calculating the earlier transitional times, the art development with its timeline, sequences, and intrinsic characteristics is finally displayed in the next two tables.

$z/8 = $ k	Time (years)		Art sequences
	from origin $t_k =$ $t_{\bullet}\cdot tanh(k)$	from present t_k - 174588	
	178938	+4350	
...	...	...	
2.375	175868	+1280 (AD3294)	–
2.25	175006	-418 (AD1596)	_Late styles of art
2.125	173905	-683 (AD1331)	_Revived art
2	172501	-2087 (73BC)	_Religious art
1.875	170715	-3873 (1859BC)	_Classic ancient art
1.75	168448	-6140 (4126BC)	_Pre-classic ancient art
1.625	165580	-9008 (6994BC)	_Temple spiral carved reliefs
1.5	161965	-12623	_Stone carved drawings
1.375	157434	-17154	_Ivory and jade decorative objects
1.25	151790	-22798	_Figures of animals and humans
1.125	144815	29773	_Painted human faces
1	136278	-38310	_ Figures of human hands, animals
0.875	125955	48633	_Sculptures and cave painting
0.75	113652	-60936	_Petro-glyphs and rock painting
0.625	99239	-75349	_ Limestone and clay figurines
0.5	82690	-91898	_Drilled snail shells
0.375	64124	-110464	_Decorative bones and amulets
0.25	43825	-130763	_Drawings of curves and spirals
0.125	22251	-152337	_Red ochre engraved with lines
0	0	-174588	_Simple parallel and crossed lines

$z/8 = $ k	Time from origin $t_k = t_\bullet \cdot tanh(k)$ (years)	Period $\tau_k = $ $(t_\bullet^2 - t_k^2)/(2t_\bullet)$ (years)	Frequency $f_k = $ $(2t_\bullet)/(t_\bullet^2 - t_k^2)$ (years)$^{-1}$	Angular speed $\omega_k = $ $2\pi/\tau_k = 2\pi \cdot f_k$ (years)$^{-1}$
	178938	0	∞	∞
...	...	...	...	...
2.375	175868	3043.7	0.0003286	0.0020643
2.25	175006	3888.8	0.0002571	0.0016157
2.125	173905	4962.2	0.0002015	0.0012662
2	172501	6321.2	0.0001582	0.0009940
1.875	170715	8034.1	0.0001245	0.0007821
1.75	168448	10182.5	0.0000982	0.0006171
1.625	165580	12859.4	0.0000778	0.0004886
1.5	161965	16168.0	0.0000619	0.0003886
1.375	157434	20211.9	0.0000495	0.0003109
1.25	151790	25088.6	0.0000399	0.0002504
1.125	144815	30869.4	0.0000324	0.0002035
1	136278	37574.8	0.0000266	0.0001672
0.875	125955	45138.9	0.0000222	0.0001392
0.75	113652	53376.1	0.0000187	0.0001177
0.625	99239	61950.0	0.0000161	0.0001014
0.5	82690	70362.8	0.0000142	0.0000893
0.375	64124	77979.3	0.0000128	0.0000806
0.25	43825	84102.3	0.0000119	0.0000747
0.125	22251	88085.5	0.0000114	0.0000713
0	0	89469.0	0.0000112	0.0000702

10. Theology

The field of systematic and rational study and analysis concerning concepts of God and its attributes and relations to the universe, as well as the study of divine things or religious truth is called *theology* (from Greek *Θεός* 'God' and *λόγος* 'study of'), which also includes the nature of religious truths, or the profession acquired by completing specialized training in religious studies, usually at a school, university or seminary of divinity.

The earliest evidence of religious practice dates back to the Lower Palaeolithic. Archaeologists refer to apparent intentional burials indicating religious ideas from around 300000 years ago, that is about 100000 years before the emergence of modern Homo sapiens, such as the earliest undisputed intentional *Neanderthal burial* found at Krapina in Croatia and dated 130000 years ago.

As modern humans emerged about 200000 years ago, their religious beliefs, ideas and practices emerged probably 50000 years later, and followed a course pointed out by archaeological discoveries, decipherment of scripts and texts, historical studies of rituals and religions, pantheons and deities, diversifications and movements, ascendency or decline of religions, religious-military conflicts, founders and promoters, martyrs and propagators, historians and archaeologists, philosophers and physicians, theorists and experts, writers and journalists, politicians and biologists, scientists and even mathematicians; from which some are displayed as follows:

100000 years ago: *Undisputed human burial* in *Skhul cave* at *Qafzeh*, Israel, which contains *skeletal remains stained with red ochre*; indicating that religion have been previously practiced in Africa;

42000 years ago: *Ritual burial* of a man at *Lake Mungo*, Australia, containing a *body sprinkled with copious amounts of red ochre*, indicating that the aborigines brought religious rituals from Africa, via Asia;

33000-30000 years ago: *Religious belief* evidenced by cave paintings of *Chauvet cave*, on the valley of Ardeche near Pont-d'Arc in southern France;

32000 years ago: *Upper Palaeolithic burial* at *Dolní Věstonice* on Dyje River, Czech Republic, which holds the earliest known undisputed remains of a *female shaman*;

32000-28000 years ago: *Lion-Human* from *Hohlenstein-Stadel*, Germany, also indicates early *religious ideas and practices*;

30000-28000 years ago: *Upper Palaeolithic burials* of *Sungir*, at Vladimir in Russia, 200 kilometres east of Moscow, show an *elaborate ritual*;

17300 years ago: *Shamanistic belief* was indicated at *Lascaux caves* near Montignac in Dordogne, France, where anthropomorphic painting consists of strange beasts such as *half-human and half-bird*, or *half-human and half-lion*, as forms of deities;

9130-7370BC: *Early Neolithic sanctuary* was discovered at *Göbekli Tepe* (Turkish 'Potbelly Hill'), in South-eastern Anatolia, as the oldest human-made *place of worship*, that is 15-metre tall by 300-metre diameter;

7500-5700BC: <u>*Cult of Great Goddess*</u> was unveiled at *Çatalhöyük* (Turkish 'Fork mound') on Konya Plain in southern Anatolia, as a spiritual centre with figurines of females, carved in marble, blue and brown limestone, or schist, calcite, basalt, alabaster, and clay; each of them representing a *female deity*;

4000-2110BC: <u>*Sumerian pantheon*</u> included *Anu* 'father of the gods and lord of heaven', *Enlil* 'ruler of the gods, represented by the air or wind', *Enki* or *Ea* 'god of the earth and water', and *Ishtar* or *Inanna* 'goddess of love and fertility';

3200BC: <u>*Early religious writings*</u> were recorded in the Near East, unveiling human *religious experiences and ideas*, and opening the way for history of religions;

3150-1000BC: <u>*Egyptian pantheon and deities*</u> were represented by *Ra* 'the Sun and a creator deity', *Shu* 'embodiment of wind or air', *Tefnut* 'goddess of rain, dew, clouds and wet weather', *Thoth* 'god of the Moon, drawing, writing, geometry, wisdom, medicine, music, astronomy and magic', *Geb* 'god of the Earth and first ruler of Egypt', *Nut* 'goddess of heaven and the sky', *Isis* 'goddess of magical power and healing', *Osiris* 'god of the underworld', *Nephthys* 'goddess of death, holder of the rattle', *Seth* 'god of storms, evil, desert', *Horus* 'falcon-headed god of the sky, war and protection', *Hapi* 'god embodied by the Nile', *Aken* 'ferryman to the underworld', *Anubis* 'dog or jackal god of embalming and tomb-caretaker who watched over the dead', *Apis* 'bull-god of fertility', *Mut* 'mother of the primordial waters of cosmos', *Aten* 'the Sun disk', and *But* 'cow goddess of the cosmos and the essence of soul';

2400-2300BC: *The Pyramid Texts* from ancient Egypt were deciphered as the oldest known *religious texts* in the world;

2240-2220BC: Narām-Sin, ruler of Akkad, was the first leader with religious attributes to be known by name deciphered on his extant *Stela*, discovered in Susa (Iran), and associated with several *epic tales*, suggesting that a sort of religion was obeyed in his empire;

2100BC: *Imperial cult* in southern Mesopotamia was imposed by Ur-Nammu, ruler of Sumer, resulting in the *Ziggurat of Ur* dedicated to the deity called *Nanna*;

2100-2000BC: Abraham (Abram), Old Testament father of the Hebrew people, was accredited to have a *Convent* with the Lord 'Yahweh', becoming ancestor of three great monotheistic religions - Judaism, Christianity, and Islam;

1750-200BC: *Hindu pantheon* included *Brahma the Creator*; *Shiva the Destroyer*; *Vishnu the Preserver*; *Buddha*, an incarnation of Vishnu; and *The Bull Nandi*, a fertility symbol;

1600-336BC: *Ancient Greek deities* were represented by *Chaos* 'the void', *Tartarus* 'the abyss', *Gaia* 'the Earth', *Eros* 'god of love, procreator and sexual desire', *Erebus* 'darkness', *Nyx* 'night'; *Typhon* 'storms', *Uranus* 'the sky', *Ourea* 'mountains', *Pontus* 'sea', *Aether* 'heaven', *Hemera* 'day'; *Oceanus* 'god of the ocean', *Chronos* 'god of time', *Aeolus* 'god of air and winds', *Rhea* 'goddess of nature', *Themis* 'goddess of divine order, law and custom', *Helios* 'god of the Sun', *Selene* 'goddess of the Moon', *Crios* 'crab protecting the sea nymphs'; *Zeus* 'king of the gods and ruler of mankind', *Hera* 'queen of the gods and goddess of the family', *Poseidon* 'god of the seas', *Hades* 'god of the underworld', *Apollo* 'the youthful god of the Sun and music', *Artemis* 'goddess of the hunt', *Persephone* 'queen of the underworld and goddess of vegetation', *Ares* 'god of war', *Nike* 'goddess of victory', *Aphrodite* 'goddess of love and beauty', *Dionysus* 'god of wine', *Hebe* 'goddess of youth', *Hephaestus* 'god of metallurgy', *Athena* 'goddess of wisdom and arts', *Hermes* 'god of trade', *Demeter* 'goddess of the harvest', *Hestia* 'goddess of the hearth', and *Asclepius* 'god of health and medicine';

1379-1362BC: Akhenaten and Nefertiti, Egyptian king and queen, changes the traditionally religious beliefs, broke up with the worship of old gods, and introduced a purified and *monotheistic solar cult* with the sole god *Aten* 'the Sun disk'; this form of religion being called *Atenism*;

c.1300BC: Moses (Môsheh), Old Testament Hebrew prophet and

lawgiver, received the *Divine Law* and the *Ten commandments* from God, by which the will of the Lord was revealed in *Torah* 'Law of Moses', or *Pentateuch*, comprising first five books of the Old Testament, namely Genesis, Exodus, Leviticus, Numbers, and Deuteronomy;

1300-1100BC: Rishis were authors of *Riks* or *mantra*, by intuitive perception, composing the *hymns of the Rig Veda*, as the foundation for the *historical Vedic religion*;

1200-200BC: *Chinese deities* included the *Four Fiends* - Hundun 'chaos', Taotie 'gluttony', Táowù 'ignorance', and Qióngqi 'deviousness'; *Presiding deities* - Shangdi 'deity of the Jade Emperor and Tian', Yu Di 'the Jade Emperor', Tian 'heaven', Nüwa and Fuxi 'ancestors of all humankind, often represented as half-snakes and half-humans', and Pangu 'first sentient being and creator making the heavens and the earth'; *Yinglong* 'dragon as god of rain'; and the *Four Symbols of the Chinese constellation* - Azure Dragon 'east', Black Tortoise 'north', White Tiger 'west' and Vermillion Bird 'south';

950-850BC: *Classical Hinduism*, so called from *Hindu* (Persian *Sindhu* 'Indus river'), was based on the earliest *Upanishads*, and known as *Vedanta* 'conclusion of the Veda';

900-200BC: *Axial age* was coined by German philosopher Karl Theodor Jaspers (1932), that was applied to the age when 'the spiritual foundations of humanity were laid simultaneously and independently...And these are the foundations upon which humanity still subsists today', which later was summarized by US historian Peter Watson (2002) as the age of foundation of many of humanity's most influential philosophical and religious traditions, such as *monotheism* represented by *Hinduism, Jainism* and *Buddhism* in India; *Zoroastrianism* in Persia; *Taoism* and *Confucianism* in China; and *Platonism* in Greece;

830-790BC: Parshva, the 23rd Tirthankara, was the earliest Jain leader who is accepted as a historic figure, and religious founder of *Jainism* in India;

710-680BC: Numa Pompilius, Second Roman king, based on the Etruscan tradition, he codified and organized early *Roman religion*;

600-570BC: Nebuchadnezzar II, Neo-Babylonian Emperor, built the *Etemenanki*, by which *Marduk* was established as the patron deity of Babylon;

550BC: Laozi (Lao Tzu), Chinese philosopher and sage, founded

Taoism, also called *Daoism*, a philosophy with central notion of *Dao* 'unchanging principle of universe' related to living in harmony with balance between *yin* 'feminine' and *yang* 'masculine', as well as between universal elements; and Zoroaster (Zarathustra), Persian religious leader and prophet, completed the scriptures of *Gathas*, by which *Zoroastrianism* was founded, as a dualistic religion appeared in the earliest part of the sacred book *Avesta*, including some forms of later *Mithraism* that extended even through the Roman Empire;

550-530BC: Mahavira, the final Tirthankara, was the last to contribute to Jainism, by completing it in the *late form propagated in Indian subcontinent*;

550BC-AD330: *Ancient Roman deities* were represented by *Janus* 'god of beginnings and transitions', *Jupiter* 'king of the gods, and god of sky and thunder', *Saturn* 'god of generation, dissolution, plenty, wealth, agriculture, periodic renewal and liberation', *Genius* 'instance of a general divine nature, present in every person, place or thing', *Mercury* 'patron god of financial gain, commerce, eloquence, communication, travel, boundaries, luck, trickery and guide of souls to the underworld', *Apollo* 'god of light and the Sun, truth and prophesy, healing, plague, music and poetry', *Mars* 'god of war and guardian of agriculture', *Vulcan* 'god of fire and volcanoes', *Neptune* 'god of water, horses and the sea', *Sol* 'solar deity', *Orcus* 'god of the underworld and punisher of broken oaths', *Liber* 'god of viticulture and wine, fertility and freedom', *Tellus* or *Terra Mater* 'goddess of the Earth', *Ceres* 'goddess of agriculture, grain crops, fertility and motherly relationship', *Juno* 'protectress-goddess and special counsellor of the state, and goddess of marriage', *Luna* 'divine embodiment of the Moon', *Diana* 'goddess of hunt, the moon and birthing', *Minerva* 'goddess of wisdom and sponsor of arts, trade and defence', *Venus* 'goddess of love, beauty, sex, fertility and prosperity', and *Vesta* 'virgin goddess of the hearth, home and family';

540-500BC: Pythagoras, Greek philosopher, mystic and mathematician, founded *Pythagoreanism*, as the belief in immortality and transmigration of soul, moral asceticism and purification; which was developed and propagated by his former slave Zalmoxis (Gebeleïzis), influential sage in Thracian, Celtic and Greek cultures, who preached to Thracians for developing customs such as: prohibition against setting down in written from proper store of knowledge, celebration of birth with mourning and of death with joy, obsession with war and high-spiritedness to battle, and beliefs closed to later assertions of Stoics;

530-485BC: <u>Buddha</u> 'the enlightened one', Indian Prince Gautama Siddhartha, after six years of extreme self-mortification seeing in contemplative life a perfect way to self-enlightenment, he taught *Buddhism* in Nepal and India in the next 40 years of his life;

497-480BC: <u>Confucius</u>, Chinese philosopher and sage, spent a dozen years as an itinerant sage followed by a group of disciples, wandering from court to court and seeking a sympathetic patron, then spent his final years teaching, and, after his death, his disciples compiled the volume of *Analects* recording his sayings and doings, by which *Confucianism* was founded;

450-50BC: *Celtic deities* included *Abellio* 'god of apple trees', *Alaunus* 'god of healing and prophecy', *Ambisagrus* 'god of thunder and lightning', *Andraste* 'goddess of victory', *Ankou* 'a god of death', *Arausio* 'a god of water'; *Belatucadros* 'a god of war', *Belenus* 'a god of healing', *Belisama* 'goddess of lakes, rivers, fire, crafts and light', *Borvo* 'a god of mineral and hot springs', *Borrum* 'a god of the winds', *Camulus* 'a god of war and sky', *Cissonius* 'a god of trade', *Coventina* 'goddess of wells and springs', *Damara* 'a fertility goddess', *Dis Pater* 'a god of the underworld', *Epona* 'fertility goddess and protectress of horses, donkeys and mules', *Erecura* 'earth goddess', *Fagus* 'a god of beech trees', *Lenus* 'a healing god', *Lugus* 'god of creation and learning', *Nantosuelta* 'goddess of nature, the earth, fire and fertility', *Sucellos* 'a god of love and time', *Sulis* 'a solar nourishing and life-giving goddess', *Taranis* 'a god of thunder', and *Vindonnus* 'a hunting and healing god';

400-360BC: <u>Plato</u>, Greek philosopher, in his major work *Republic*, Book *ii*, chapter *18*, introduced the term *theology* (Greek θεόλόγια with the meaning of 'discourse on God');

273BC: <u>Aśoka</u> (Ashoka), Mauryan ruler of India, organized Buddhism as the *state religion*, acting in favour of disseminating *dharma* 'moral principles' and inscribing his *pronouncements* on rocks and pillars placed across his state;

200BC-AD600: *Germanic (Old Norse) deities* were represented by *Baldr* 'god of shining day', *Dellingr* 'god of dayspring, or shining one', *Eir* 'goddess of peace and clemency, or help and mercy', *Frigg* 'goddess of love', *Heimdallr* 'world-brighter god', *Hel* 'goddess who covers up, or hides something', *Hermóðr* 'war-spirit god', *Höðr* 'warrior god', *Meili* 'the lovely god', *Nótt* 'goddess of night', *Odin* 'god of frenzy', *Sigyn* 'victorious girl-friend goddess', *Sjöfn* 'goddess of love', *Snotra* 'the clever goddess', *Sól* 'goddess of the Sun', *Thor* 'god of thunder', and *Vár* 'beloved goddess';

AD27-30: <u>Jesus Christ</u>, Central figure of Christian faith, had two *missionary journeys* through Galilee, as well as in Tyre and Sidon, was submitted to *Crucifixion*, from which the cross remains a Christian symbol, experienced *Resurrection*, and then *Ascension*, being assisted by his apostles; thus marking the beginning of *Christianity*;

30-35: <u>St Stephen</u>, New Testament figure and First Christian martyr, was appointed by Jesus Christ's apostles to *manage finances and alms of early church*, and finally was submitted to trial for blasphemy by the supreme Jewish court, and *martyrdom* by stoning to death;

30-62: *Christian apostles* - <u>St Bartholomew</u>, abandoned the Hebrew Gospel in India, and did *missionary work* in Ethiopia, and then in Mesopotamia where took place his *martyrdom*; <u>St Peter</u>, named *Peter* as 'Stone' of founding the new religion, did *missionary activity* in Pontus, Cappadocia, Galatia, Asia and Bithynia, wrote the *Epistles*, and ended by *martyrdom* in Rome; <u>St Andrew</u>, patron saint of Scotland, Russia, Romania and Greece, preached the *Gospel* in Asia Minor and Scythia, was crucified in Achaia (Greece), and from the 14th century his crucifixion was associated with a X-shaped cross; and <u>St Paul</u>, Christian missionary and martyr, Apostle of Gentiles, did *missionary journeys* to Cyprus, Pamphylia, Lycaonia, Galatia, Phrygia, Macedonia, Achaia (mainly in Corinth, Greece), and wrote his *Epistles*;

40-100: *New Testament evangelists* - <u>St Mathew</u>, <u>St Mark</u>, <u>St Luke</u> and <u>St John</u> were authors of the first, second, third, and fourth *Gospels* of the New Testament, based on real facts or imaginary visions; St Mark also introduced Christianity to the Egyptians, in Alexandria, which in the next century was taken over by the Coptic people, South of Egypt;

177-200: <u>St Irenaeus</u>, Greek theologian and Christian father of Greek (Orthodox) Church, was a successful missionary bishop, opposed to *Gnosticism*, especially to Valentinians, and wrote invaluable *Against Heresies*, as well as activated for maintenance of contact between Eastern and Western sections of church;

180-201: <u>St Clement of Alexandria</u>, Greek Church Father, wrote *Who is the Rich Man that is Saved*, and also the trilogy *The Missionary*, *The Tutor*, and *The Miscellanies*;

c.200: <u>Nāgārjuna</u>, Indian Buddhist and monk-philosopher, was founder of the *Madhyamaka* 'Middle Path school of Buddhism';

211-232: <u>Origen</u>, Christian scholar, theologian and early Greek father of Church, wrote *Hexapla*, textual criticism of the Scriptures, and *Eight Books against Celsus*, constituting great early Christian apologies;

319-335: <u>Arius</u>, Libyan theologian, founded the heresy called *Arianism*, maintaining that in doctrine of Trinity the Son was not co-equal or co-eternal with the Father, but only first and highest of all finite beings, created out of nothing by an act of God's free will; he made known his ideas by the theological work *Thaleida*;

320-330: <u>Arnobius the Elder</u>, Numidian teacher of rhetoric, vigorously defended Christianity by writing *Adversus Nationes*, which was translated in 'Ante-Nicene Library';

324-325: <u>Constantine I the Great</u>, Roman emperor, authorized *Christianity as the state religion*, and initiated the great Church Council of Nicaea, which sided against the Arians, and adopted the *Nicene Creed*;

341-348: <u>Ulfilas</u>, Cappadocian prelate, was consecrated as a *missionary bishop to Visigoths*, converted the *Goths to Christianity*, and translated the Bible from Greek into a Germanic language;

360-370: <u>St Athanasius</u>, Greek Christian theologian and prelate, made polemical works of Trinity, Incarnation, and divinity of Holy Spirit, containing *Athanasian Creed* 'Athanasian beliefs';

397-426: <u>St Augustine of Hippo</u>, Numidian Christian and Latin Doctor of Church, wrote *Confessions*, a classic of world literature and a spiritual autobiography; *De Civitate Dei* 'The City of God', in 22 books, as an influential and important vindication of Christianity; and *De Trinitate* 'The Trinity', a weighty exposition of Trinity's doctrine;

398-404: <u>St John Chrysostom</u>, Syrian churchman and one of the Doctors of the Church, wrote *Homilies*, *Commentaries* on the Bible, *Epistles*, *Treatises* on Providence, Priesthood, and others, as well as *Liturgies*;

400-420: <u>Pelagius</u>, British monk and heretic, had a view that salvation could be achieved by the exercise of one's basically good moral nature, thus founding *Pelagianism*, then was condemned by the Councils, excommunicated and banished from Rome;

405-435: <u>St John Cassian</u>, Romanian monk and theologian, instituted several monasteries, including *Abbey of St Victor* at Massilia (now Marseille), a model for many others in France and Spain, as well as wrote *Collationes*, on Desert Fathers, and a book on *monasticism*;

420-450: Eutyches, Greek religious Archimandrite at Constantinople, founded *Eutychianism*, holding that after the incarnation, human nature became merged in the divine, and that Jesus Christ had therefore but one nature; then he was condemned by a synod at Constantinople, but later was restored; and Nestorius, Syrian ecclesiastic, defended the Presbyter Anastasius in denying that the Virgin Mary could be truly called the Mother of God, emphasizing the distinction of the divine and human natures, thus founding *Nestorianism*; subsequently he was deposed by a general council in Ephesus, banished to Petra in Arabia, and died after imprisonment;

432-454: St Patrick, Christian apostle and patron saint of Ireland, went to France, where he became a monk, was consecrated bishop and sent by Pope Celestine as a *missionary to Ireland*, where he converted many Celtic chiefs, spent the rest of his life in missionary work, and wrote the spiritual autobiography *Confession*;

520-540: Bodhidharma, Indian monk, travelled to China, and then founded *Ch'an* or *Zen Buddhism*, arguing that merit leading to salvation could not be accumulated through good deeds, and taught meditation as the means of return to Buddha's spiritual precepts;

560-600: St David (Dewi), Patron saint of Wales, had the presidency over two Welsh synods, at Brefi and 'Lucus Victoriae', instituting *Christianity in Wales*, and, according to *Annales Cambriae*, died as Bishop of Moni Judeorum (Menevia);

590-600: Gregory I the Great, Pope and saint, a Doctor of Church, achieved a complete overhaul of public services and ritual in *Roman Church* and systematized its sacred chants, from which arose *Gregorian chant*, entrusted a mission to convert English, and reconciled the Gothic kingdom of Spain with Rome, as well as wrote *homilies* on Ezekiel, Job, and on Gospels, as well as *Regulae Pastoralis liber* 'Book of Rules for Pastors', *Sacramentarium*, and *Antiphonarium*;

622: Muhammad, Arab prophet, founded *Islam*, a monotheistic religion based on *teachings of Prophet Muhammad* embodied by Muhammad as submitted to will of Allah; the beginning of the Muslim era was marked by his migration, named *Hegira*, following an invitation from an agricultural community of Yathrib to the north, whence Islam was then widely propagated;

710-754: St John of Damascus, Greek theologian of Eastern Church, wrote *Fount of Wisdom*, an encyclopaedia of Christian theology, *treatises* against superstitions, Jacobite and Monophysite heretics, *homilies*, as well as *Barlaam and Joasaph*, a version of Buddha's life;

860-870: St Cyril and St Methodius (Apostles of the Slaves), Greek Christian missionaries, worked among the Tartar Khazars, Bulgarians of Thrace and Moesia, and in Moravia, evangelized the Slavs, and invented the *Cyrillic alphabet*;

925-935: Abu al-Hasan al-Ash'ari, Islamic theologian and philosopher, founded the school of tenets of faith that bears his name *Ash'ari*, defended the idea of God's omnipotence and reaffirmed traditional interpretations of religious authority within Islam, then developed the major thesis written in one of his books entitled *Maqalat al-eslamiyin*;

990-1010: Ælfric (Grammaticus), Anglo-Saxon churchman and writer, produced two books of 80 *Homiles*, as a paraphrase of first seven books of Bible, as well as the books *Live of the Saints*, and Latin *Colloquium* between a master, his pupil, and various craftsmen, giving a vivid picture of social conditions in England of his time;

1000-1500: *Inuit deities* included *Agloolik* 'evil god of the sea who can hurt boats by biting them', *Akna* 'mother goddess of fertility', *Anguta* 'god as gatherer of the dead and carrier into the underworld', *Igaluk* 'the moon god', *Nanook* 'the master-god of polar bears', *Pinga* 'goddess of the hunt, fertility and medicine', *Qailertetang* 'female deity as weather spirit, taking care of animals, fishers and hunters', *Sedna* 'the mistress of sea animals', and *Sila* 'personification of the air';

1080-90: St Wulfstan, Anglo-Saxon prelate, preached against slave trading, helped to compile the *Domesday Book*, a manuscript recording the great Norman survey of the country, and wrote at least part of the *Anglo-Saxon Chronicle*;

1095-1390: *The Crusades* were the medieval Christian military expeditions to recover the *Holy Land* from the Muslims, which took place in chronological order as: *1ˢᵗ Crusade*, French-Norman-German-Italian knights (1095-1099); *2ⁿᵈ Crusade*, French-South German armies (1147-1149); *3ʳᵈ Crusade*, German-French-English knights (1187-1192); *4ᵗʰ Crusade*, French-English-German armies (1202-1204); an intermediate *Children's Crusade* (1212); *5ᵗʰ Crusade*, German-Flemish-Frisian-Hungarian forces (1217-1221); *6ᵗʰ Crusade*, German-English armies (1228-1229); *7ᵗʰ Crusade*, French army (1248-1254); *8ᵗʰ Crusade*, French knights (1270); *9ᵗʰ Prince Edward's Crusade*, English-French forces (1271-1272); and finally *Alexandrian Crusade*, Cypriot forces (1365); and *Mahdian crusade*, French-

Genoese soldiers (1390);

1162-70: <u>Thomas Becket</u>, English saint and martyr, Archbishop of Canterbury, opposed to King Henry II to subordinate the church to the state, shielded from oppression of nobility, and *separated the Church of England from the king's authority*;

1197-1533: *Inca deities* were represented by *Apocatequil* 'god of lightning', *Apu* 'a god of mountains', *Catequil* 'a god of thunder and lightning', *Copacati* 'a lake goddess', *Inti* 'the sun god', *Kuka Mama* 'a goddess of health and joy', *Mama Allpa* 'a fertility goddess', *Mama Qucha* 'the sea and fish goddess', *Mama Sara* 'the goddess of grain', *Pacha Kamaq* 'a chthonic creator god', *Supay* 'god of death', and *Wiraqucha* 'the god of everything';

1248-1521: *Aztec Pantheon* included *Ahuiateteo* 'gods of excess', *Atlacoya* 'goddess of drought', *Chalchiuhtlicue* ' goddess of water, lakes, rivers, seas, streams and storms', *Chicomecoatl* 'goddess of agriculture', *Chimalma* 'goddess of fertility, patroness of life and death, guide of the rebirth', *Cihuacoatl* 'goddess of childbirth and picker of souls', *Cinteteo* 'gods of the maize', *Huitzilopochtli* 'god of will and the Sun, patron of war and fire, and lord of the South', *Ixtlilton* 'god of medicine', *Metztli* 'goddess of the moon', *Miquizlitecuhtli* 'god of death', *Quetzalcoatl-Tlaloc* 'deity of water, fertility and storm', *Teotlale* 'god of the deserts', *Tezcatlipoca* 'omnipotent deity of rulers, sorcerers and warriors', *Tlaloc* 'god of thunder, rain and earthquakes', *Tlaltecuhtli* 'god of the earth', *Toltecatl* 'a god of wine', *Tonatiuh* 'god of sun', *Tzapotlatena* ' goddess of nature', *Xiuhtotontli* 'gods of fires', and *Yacatecuhtli* 'god of commerce and bartering';

1255-80: <u>Albertus Magnus</u> (Doctor Universalis), German philosopher and cleric, wrote notable works such as *Summa theologiae*, and *Summa de creaturis*, influenced his pupil Thomas Aquinas, and became patron saint of Roman Catholic Church;

c.1260: <u>St Thomas Aquinas</u>, Italian scholastic philosopher and theologian, tried to combine and *reconcile Aristotle's scientific rationalism with Christian doctrines* of faith and revelation, and had an immense influence on theological thought;

1350-80: <u>Nicodim</u>, Wallachian orthodox metropolitan, founded monasteries such as *Vodiţa*, and *Tismana*, reproduced *Evangheliarul* 'The Evangeliar', and organized the state orthodox religion in Wallachia;

1376-80: <u>John Wycliffe</u>, English religious reformer, wrote a book entitled *De Domino Divino*, an attack of the constitution of Church, declaring that it would be better without pope and prelates; began the *first English translation of Bible*; and pointed out the importance of studying English Medieval Church;

1408-15: <u>John Huss</u>, Bohemian religious reformer, produced *De Ecclesia* 'On the Church', which was followed by author's call before the *General Council at Constance* (today Konstanz, at the border between Germany and Switzerland, on *Lake Constance*, called *Bodensee* in German), and his burn at stake initiated the *Hussite Wars* lasting until the mid-15th century;

1489-97: <u>Girolamo Savonarola</u>, Italian religious and political reformer, wrote sermons, theological treatises, and *The Triumph of the Cross*, a chief apology of orthodox Catholicism;

1500-23: <u>Desiderius Erasmus</u>, Dutch humanist and scholar, was author of *Adagia* 'Adages'; *Enchiridion Militis Christiani* 'Handbook of a Christian Soldier'; famous *Encomium Moriae* 'In Praise of Folly'; and *De Libero Arbitrio*, attacking M Luther;

1519-25: <u>Huldreich Zwingli</u>, Swiss reformer, preached the gospel boldly, and produced *Reformed doctrines in 67 theses*, and *Commentarius de vera et falsa religion*;

1520-52: <u>Martin Luther</u>, German religious reformer, wrote *Christian Nobles of Germany*; followed by the treatise *On the Babylonian Captivity of the Church of God*, attacking doctrinal system of Church of Rome; also published *De Servo Arbitrio*, opposing to D Erasmus; as well as *Table-talk*, *Letters*, and *Sermons*, which marked the culmination of *German Reformation* by ascendancy of <u>*Lutheranism*</u> as a form of Protestant Christianity, which was stated by *Augsburg Confession*;

1521-34: <u>Henry VIII</u>, King of England, published a *Defence of Sacraments*, declaring that the *king was sole supreme head of Church of England*; ordered to the *dissolution of monasteries*; and separated English Church from the Catholic Church of Rome;

1531-53: <u>Michael Servetus</u>, Spanish theologian and physician, wrote *De Trinitatis Erroribus*, and *Christianismi Restitutio*, denying the Trinity and the divinity of Jesus, then managed to escape of Inquisition, but was burnt for heresy;

1534-40: <u>St Ignatius Loyola</u>, Spanish soldier, founded the <u>*Jesuits*</u> as a Christian religious order, and wrote *Spiritual Exercises*, a vital work in training of Jesuits;

1536-61: <u>John Calvin</u>, French theologian, wrote *Christianae Religionis Institutio* 'Institutes of the Christian Religion', rendering a double service to Protestantism, namely systematization of its doctrine and organization of its ecclesiastical discipline; and also *Readings on the Prophet Daniel*;

1536-63: <u>Nicolaus Olahus</u>, Romanian-Hungarian Catholic prelate, historiographer and statesman, defended Roman Catholicism against the Protestant rising, as he did in his writings *Breviarium Ecclesiae Strigoniensis*, and *Ordo et Ritus Ecclesiae Strigoniensis*;

1541-47: <u>Johannes Honterus</u>, Transylvanian Saxon religious reformer, initiated Lutheran principles in Braşov (now in Romania), writing *Reformatio ecclesiae Coronensis ac totius Barcensis provinciae*, and founded the 'Ecclesia Dei Nationis Saxonum' in Transylvania;

1549-53: <u>Thomas Cranmer</u>, English prelate and Archbishop of Canterbury, compiled *Edward VI's Book of Common Prayer*, which converted the Mass into Communion; composed *42 articles of religion*, later called the '39 Articles'; and rephrased the *Prayer Book*;

1558-59: <u>John Knox</u>, Scottish Protestant reformer, wrote *First Blast of the Trumpet against the Monstrous Regiment of Women*, and established the *Church of Scotland*;

1572: <u>Catherine de Médicis</u>, Queen and Regent of France, and her son <u>Charles IX</u>, King of France, instigated and authorized the infamous slaughter of Huguenots (Calvinist Protestants), known as *St Bartholomew's Day Massacre*, followed by a wave of Roman Catholic mob violence against Huguenots led by the admiral Gaspard de Coligny, with an estimated death toll of 10000-20000;

1598: <u>Henry IV</u>, First Bourbon King of France, supervised the *Edict of Nantes*, guaranteeing rights of Huguenot minority, and ending more than 40 years of religious wars in France;

1630-38: <u>Cornelius Otto Jansen</u>, Dutch Roman Catholic theologian, founded the <u>*Jansenist movement*</u> *of reform*, and completed the great work *Augustinus*, trying to prove that teaching of St Augustine against Pelagians, free will and predestination were directly opposed to teaching of the Jesuit schools;

1646-74: <u>George Fox</u>, English religious leader, preached a gospel of brotherly love, calling his society the 'Friends of Truth'; founded the *Society of Friends*, called *Quakers*; and travelled to Barbados, Jamaica, America, Holland and Germany, preaching and writing by means of his *Journal*;

1650-54: <u>James Ussher</u>, Irish prelate, was Calvinistic in theology and moderate in his ideas of Church government, and wrote *Annales Veteris et Novi Testamenti*, where Creation was fixed at 4004BC;

1653-63: <u>John Eliot</u>, English missionary, preached to the Native Americans, writings of *A Primer or Catechism, in the Massachusetts Indian Language*, and *The Christian Commonwealth*; he also translated the Bible into the Native American language, becoming known as 'The Apostle to the Indians';

1656-84: <u>John Bunyan</u>, English writer and preacher, wrote *Some Gospel Truths Opened*; *Profitable Meditations*; *I Will Pray with the Spirit*; *Christian Behaviour*; *The Holy City*; *The Resurrection of the Dead*; *Grace Abounding*; *The Pilgrim's Progress*; *Life and Death of Mr Badman*; and also *Holy War*;

1679-1715: <u>Gilbert Burnet</u>, Scottish churchman and Anglican historian, was author of *History of the Reformation*; *Exposition of the thirty-nine Articles*; and *History of My Own Time*;

1717-29: <u>William Law</u>, English churchman and writer, published *Three Letters*; *Remarks on Mandeville's Fable of the Bees*; and overall *Serious Call to a Devout and Holy Life*;

1739-78: <u>John Wesley</u>, English Christian evangelist, led the small dedicated group Holy Club and Oxford Methodists, founding <u>Methodism</u> as an evangelical movement, and published the *Methodist Magazine*;

1753-69: <u>George Whitefield</u>, English evangelist, propagated *Methodism*, compiled a *hymn book*, preached more than 18000 formal sermons, of which seventy-eight have been published, and made evangelistic journeys in England, Scotland, Wales, and America;

1758-76: <u>Ann Lee</u> (Mother Ann), American mystic and religious leader, joined the 'Shaking Quakers', or <u>Shakers</u>, and proclaimed a new gospel drawn from her divine revelations, founding the parent Shaker settlement at Niskayuna, near Albany, New York;

1825-35: <u>Joseph White</u>, British theological writer, published *Evidence against Catholicism* and notable sonnet *Night and Death*, was tutor in Archbishop Whately's family in Dublin, and then fled to Liverpool on adopting Unitarian views;

1830-43: <u>Joseph Smith</u>, US religious leader, compiled the *Book of Mormon*, containing postulated history of America to the 5th century AD, supposedly written by a prophet named *Mormon*, and thus he established 'the new Church of Jesus Christ of Latter-day Saints' and founded _Mormonism_ 'Latter Day Saint movement' with headquarters initially at Kirtland, Ohio, and then in Illinois;

1862-93: Herbert Spencer, English revolutionary philosopher, advocated Social Darwinism, published _System of Synthetic Philosophy_, and proposed the concept of _animism_ as attribution of a soul to natural objects and phenomena;

1865-70: John Lubbock, English politician and biologist, published _Prehistoric Times_, and _Origin of Civilization_, used the term 'fetishism', and was a member of the 'X Club' which conspired to _replace the ecclesiastical establishment_ with a scientific one;

1865-71: Edward Burnett Tylor, English anthropologist, publish a major study entitled _Researches into the early History of Mankind and the Development of Civilization_, and also _Primitive Culture_, showing that human culture, in its religious aspect, is governed by laws of evolution, such that the beliefs and practices of primitive nations may be taken to represent earlier stages in the progress of mankind;

1866-1908: Mary Baker Eddy, US founder of the Christian Science Church, developed the spiritual and metaphysical system 'Christian Science', published her beliefs in _Science and Health with Key to the Scriptures_, as well as founded _Church of Christ, Scientist_ at Boston, as well as _Christian Science Journal_, and the _Christian Science Monitor_;

1870-80: Wilhelm Mannhardt, German scholar and folklorist, worked on Baltic mythology, championed _solar theory_, and suggested that religion began in 'naturalism' as mythological explanation of natural events;

1878-1910: Max Müller, German-born British philologist and Orientalist, edited the 51-volume _Sacred Books of the East_, and theorized that religion began in _hedonism_;

1916-35: Mordecai Menahem Kaplan, US rabbi and philosopher, founded the Jewish Center in New York City, and the Society for the Advancement of Judaism; originated the _Reconstructionist movement in Judaism_, celebrating the richness of Jewish culture but questioning some of its religious traditions; wrote the book _Judaism as a Civilization_; and founded his journal _The Reconstructionist_;

1938-67: Karl Barth, Swiss theologian, published works including _Knowledge of God and the Service of God_, and monumental _Church Dogmatics_;

1949-86: Mircea Eliade, Romanian historian and philosopher of comparative religion, published numerous books and papers, including *The Myth of the Eternal Return*; *Patterns in Comparative Religion*; *Yoga: Immorality and Freedom*; *The Sacred and the Profane*; *A History of Religious Ideas*; as well as co-ordinated and edited 16-volume *The Encyclopaedia of Religion*;

1963-67: <u>James Mellaart</u>, British archaeologist, was remarked by his discovery of the Neolithic settlement of Çeatalhöyük, Turkey, where he discovered the well-known *Mother goddess*, and published *Çeatalhöyük, A Neolithic Town in Anatolia*;

1977-92: <u>Geshe Kelsang Gyatso</u>, Tibetan Buddhist teacher and monk, founded the *New Kadampa Tradition of Buddism*, seeking to preserve and promote essence of Buddhist teaching in a form suited to modern world and way of life; he published the books *Clear light of Bliss*; *The Joyful Path of Good Fortune*; and *Introduction to Buddhism*.

2012: *Predominant religions*, millions of adherents - *Christianity*, 1800 -200; *Islam*, 1570; *Hinduism*, 1083 - 1101; *Chinese traditional* (including *Taoism* and *Confucianism*), 394; *Buddhism*, 376 - 1200; *Primal-Indigeneous*, 300; *Folk*-worldwide, 250; *Atheist*, 150; *African Traditional* and *Diasporic*, 100; *Sikhism* (Indian), 28; *Shinto* (Japanese), 27 - 65; *Spiritism*, 15; *Judaism*, 14; *Cheondoism* (Korean), 12.5; *Jainism*, 8 - 14; *Baha'i* (Abrahamic), 7 - 7.5; *Korean Shamanism*, 5 - 15; *Kao Đài* (Vietnamese), 5 - 9; *Tenrikyo* (Japanese), 5; *Seicho-no-le* (Japanese), 5; *Hoa Hao* (Vietnamese), 2 - 4; *Neo-Paganism*, 1; *Unitarian Universalism*, 0.8; *Rastafarianism*, 0.6; *Zoroastrianism* (Parsi), 0.145.

According to the table presented at the end of chapter 7, religion emerged 50049 after humans' origin, i.e. 200000 -50049 $\approx$ 149951 years ago, as its initial time, and will last until about 204350 years from humans' origin, as its final time at $t_\bullet = 204350 - 50049 = 154301$ years from religion origin, or 154301 - 149951 = 4350 years after the present time. A brief examination of the above development indicates that the late course of cults and religions was marked by successive sequences of predominant cult/religion to another, which are delimited by times as:

Cult of Great Goddess \ 6300...4000BC $\approx$ 5150BC \
Early Polytheistic cults and religious texts \ 4000...1370BC $\approx$ 2685BC \
\ *Polytheistic and Monotheistic religions* \ 1370...90BC $\approx$ 730BC \
Polytheistic Greek-Roman religions \ 90BC...AD1710 $\approx$ AD810 \
Monotheistic Christianity and Islam;

or 7164; 4699; 2744; and 1204 years ago respectively. Referring to the origin of religion 149951 years ago, these religious changes took place at the times: 149951 - 7164 = 142787; 149951 - 4699 = 145252; 149951 - 2744 = 147207; 149951 - 1204 = 148747; which divided by the final time $t_\bullet$ = 154301 years from its origin result in 142787/154301 ≈ 0.925; 145252/154301 ≈ 0.941; 147207/154301 ≈ 0.954; 148747/154301 ≈ 0.964. In the same succession, similar figures can be found in Background to time, Table *(k)* for *z/8 = k* sequences, as

$$t_{1.625}/t_\bullet = tanh(1.625) \approx 0.9253462; \quad t_{1.75}/t_\bullet = tanh(1.75) \approx 0.9413755;$$
$$t_{1.875}/t_\bullet = tanh(1.875) \approx 0.9540453; \quad t_2/t_\bullet = tanh(2) \approx 0.9640276;$$

whence the accurate values of these transitional times, in years from the origin of religion, are:

$$t_{1.625} = t_\bullet \cdot tanh(1.625) \approx 142782;$$
$$t_{1.75} = t_\bullet \cdot tanh(1.75) \approx 145255;$$
$$t_{1.875} = t_\bullet \cdot tanh(1.875) \approx 147210;$$
$$t_2 = t_\bullet \cdot tanh(2) \approx 148750.$$

Continuing the calculation for the earlier transitional times, the timeline, sequences, and intrinsic characteristics of religion in its development can be displayed as follows:

z/8 = k	Time (years)		Sequences of religion
	From origin $t_k = t_\bullet \cdot tanh(k)$	from present $t_k - 149951$	
	154301	+4350	
...	...	...	
2.125	149961	+10 (AD2024)	–
2	148750	-1201 (AD813)	_Monotheistic Christian + Islamic
1.875	147210	-2741 (727BC)	_Polytheistic Greek and Roman
1.75	145255	-4696 (2682BC)	_Polytheistic and monotheistic
1.625	142782	-7169 (5155BC)	_Early polytheistic, religious texts
1.5	139665	-10286 (8272BC)	_Cult of Great Goddess
1.375	135758	-14193	_Neolithic sanctuary
1.25	130891	-19060	_Half-human and half-bird/lion
1.125	124876	-25075	_Religious practices
1	117515	-32436	_Female shamans
0.875	108613	-41338	_Other religious beliefs
0.75	98004	-51947	_Shamanic beliefs
0.625	85575	-64376	_Complex religious beliefs
0.5	71305	-78646	_Ritual burials out of Africa
0.375	55295	-94656	_Changes of religious beliefs
0.25	37791	-112160	_Earliest human burials
0.125	19188	-130763	_ Complex religious ideas
0	0	-149951	_ Simple religious ideas

$z/8 =$ k	Time from origin $t_k = t_\bullet \cdot tanh(k)$ (years)	Period $\tau_k =$ $(t_\bullet^2 - t_k^2)/(2t_\bullet)$ (years)	Frequency $f_k =$ $(2t_\bullet)/(t_\bullet^2 - t_k^2)$ (years)$^{-1}$	Angular speed $\omega_k =$ $2\pi/\tau_k = 2\pi \cdot f_k$ (years)$^{-1}$
	154301	0	∞	∞
...	...	...	...	...
2.125	149961	4279.0	0.0002337	0.0014684
2	148750	5451.2	0.0001834	0.0011526
1.875	147210	6928.1	0.0001443	0.0009069
1.75	145255	8780.8	0.0001139	0.0007156
1.625	142782	11089.0	0.0000902	0.0005666
1.5	139665	13941.9	0.0000717	0.0004507
1.375	135758	17428.8	0.0000574	0.0003605
1.25	130891	21634.2	0.0000462	0.0002904
1.125	124876	26619.3	0.0000376	0.0002360
1	117515	32401.0	0.0000309	0.0001939
0.875	108613	38924.0	0.0000257	0.0001614
0.75	98004	46027.0	0.0000217	0.0001365
0.625	85575	53420.6	0.0000187	0.0001176
0.5	71305	60674.9	0.0000165	0.0001036
0.375	55295	67242.8	0.0000149	0.0000934
0.25	37791	72522.7	0.0000138	0.0000866
0.125	19188	75957.4	0.0000132	0.0000827
0	0	77150.5	0.0000130	0.0000814

11. Musicology

The study of music as a branch of knowledge or field of research, distinct from composition or performance, is called *musicology* (from Greek μουσική 'music' and λογία 'study of'). In the broad sense, musicology includes musical disciplines related to humanities and sciences, as well as all manifestations of music in worldwide cultures, and then also includes all systematic musicology related to psychology, biology and computing. Such a broad meaning implies that the parent disciplines of musicology include history, cultural and gender studies, philosophy, aesthetics and semiotics, ethnology and cultural anthropology, archaeology and prehistory, psychology and sociology, physiology and neuroscience, acoustics and psychoacoustic, also computer or information sciences and mathematics. Among its sub-disciplines, there are historical musicology, ethnomusicology, popular music studies, music theory, analysis and composition, music psychology and cognition, and performance practice and research. The history of music was closely associated with the history of dance, and therefore music includes dance as a form of its expression.

Alongside with language, music has a long evolution, being found in every known culture, past and present, varying widely both in time and space. The development of music among humans occurred against the backdrop of natural sounds. It was probably influenced by birdsong and sounds other animals communicate.

Music has been in existence from more than 100000 years ago, the first music being invented in Africa, and then evolved to become a fundamental constituent of human life. The first musical 'instrument' was the human voice itself, which can make a vast array of sounds, from singing, humming and whistling to clicking, coughing and yawning.

The oldest flute 'Divje Babe flute' was discovered in 1995) within the Slovenian cave *Divje Babe I*, from 43000 years ago, which was made from a fragment of the femur of a young cave bear. Other flutes were discovered in 2012, at *Geißenklösterle cave*, near Blaubeuren, Southern Germany, as musical instruments sculptured from bird bone and mammoth ivory, and dated as 43000-42000 years old. Also in Germany, at *Hohle Fels cave*, near Ulm, was found a five-holed flute with a V-shaped mouthpiece, made from a vulture wing bone, and dated 35000 years ago. Several *gudi* 'bone flutes' were discovered at *Jiahu*, Henan province, China, and dated about 6000BC. The oldest known wooden pipes were found near *Greystones*, Ireland, made from yew wood of between 30 and 50 centimetres long, and dated 4000 years ago.

Prehistoric music, once more commonly called 'primitive music', refers to all music produced in preliterate cultures. Traditional Native American and Australian Aboriginal music could be considered prehistoric, although the term is commonly used in relation to music in the Near-Middle East and Europe before the development of writing there. Afterward, ancient music followed as monophonic, but recent archaeological evidence indicates that this view is no longer true.

The *oldest known song* in cuneiform, 4,000 years old from Ur, indicated a composition in harmonies of thirds (like that of around AD1500 English 'gymel', a technique of temporarily dividing up one voice part, usually an upper one, into two parts of equal range, but singing different music), and was written using a Pythagorean tuning of the diatonic scale. The oldest surviving example of a *complete musical composition* in ancient Greek musical notation was deciphered in an epitaph on a marble stele called *Seikilos column* near Aidin, Turkey, and dated between 200BC and AD100, more probably the 1st century AD. Ancient bagpipes and double pipes, as well as a review of ancient drawings on vases and walls unveiled that these musical instruments were widely used in ancient Greek world. Instruments, including the seven-holed flute and various types of stringed instruments have been recovered from the *Indus valley civilization* archaeological sites. Indian classical music, called *marga*, can be found from the scriptures of the *Hindu* tradition, the Vedas; in one of the four Vedas, namely Samaveda, music has been described at length.

The *early music* era may also comprise contemporary but traditional or folk music, including Asian and Persian music; music of India, Mesopotamia and Egypt; Jewish, Greek and Roman music; and Muslim music. Early European classical tradition developed after the Roman Empire collapsed in AD476 until the end of Baroque era by the mid-18th century. What unified these cultures in the Middle Ages was the Roman Catholic Church, along with Greek Orthodox Church, as the focal points of musical development for the first thousand years of that period.

Musical instruments can be divided by type into: strings (plucked or bowed), woodwind, brass, percussion, and keyboard instruments. An *orchestra* has instruments from four families: bowed string instruments (e.g. violin), woodwind (e.g. flute), brass (e.g. trumpet), and percussion (e.g. drums).

Some music researchers, collectors and pioneers, musicologists and musicians, composers and conductors, musical instruments and schools, opera houses and producers of electronic music, together with related choreographers, dancers and impresarios who initiated, developed and perfected choreography, as well as authors of studies on music and dance are mentioned below, with their times and contributions.

100000-60000 years ago: Apart of human voice itself, the earliest musical instruments were *wooden* or *bone trumpets* and *drums* used in ritual and ceremonial events;

45000-35000 years ago: *Bone* and *ivory flutes* have been discovered across the Old World;

7000BC: Paintings of *communal dancing* and *music* were found in *Bhimbetka rock shelters*, Raisen, Madhya Pradesh state of India;

5000-4500BC: <u>Jamshid</u> (Jamšid), legendary king of Greater Iran, was considered by ancient Persians as the first cultivator of vine for producing wine, and meanwhile the inventor of *Persian music*;

3300BC: *Dancing to the clapping of bands* has been exhibited on the *tomb of* <u>Ur-ari-en-Ptah</u>, the 6th Dynasty of Ancient Egypt (now in British Museum, London);

2600-2500BC: *Sumerian/Akkadian lyres, harps, double flute, sistra* and *cymbals* were discovered in archaeological research of the ancient city of Ur;

1000-500BC: *Greek, Etruscan* and *Roman lyres, wind instruments, harps, pipes, organs* and *clappers* were found in many archaeological sites in Europe, the Near East, and North Africa;

543BC: *Ceremonial music and dancing* mentioned in *Mahavamsa*, a Sinhalese ancient chronicle, in which it is stated that when King <u>Vijaya</u> landed in Sri Lanka, he heard sounds of music and dancing from a wedding ceremony there;

530-500BC: <u>Pythagoras</u>, Greek philosopher, mystic and mathematician, discovered and analyzed *chief musical intervals*, and indicated that basic consonant intervals are produced by simple ratios, such as 2:1 for octave, 3:2 for fifth, and 4:3 for fourth;

490-480BC: *Dance of Bacchantes* was painted by the ancient Greek vase-painter <u>Makron</u>, who worked, in the red-figure technique, for the potter Hieron;

335-323BC: <u>Aristotle</u>, Greek philosopher and scientist, wrote *Problems*, where, in the Book XIX, 12, described musical techniques of time, altogether indicating *polyphony*;

c.250BC: *Dancing maenad*, an ancient Greek terracotta statuette, was found in Taranto, Italy;

150-100BC: *Veiled dancer*, an ancient terracotta figurine, was discovered in the Aeolian city *Myrna* (now in Louvre, Paris);

AD120-130: <u>Mesomedes of Crete</u>, Greek lyric poet and composer, produced poetic-musical works such as *Citharoedic Hymn* 'Suidas', *Hymn to Nemesis*, *Hymn to the Muses*, and *Hymn to the Sun*, still extant in manuscripts, as well as other hymns now lost;

270-290: <u>Ktesbios</u>, Greek engineer in Alexandria, first built an *organ* called 'hydraulis', stabilizing the wind pressure by use of water, and preceding other keyboard instruments; later bellows replaced the hydraulic mechanism, creating the pneumatic organ, whereby the increasing volume of sound;

c.520: <u>Anicius Manlius Severinus Boethius</u>, Roman philosopher, politician and musician, apart from his appreciated work on philosophy, wrote *De institutione musica* 'The Principles of Music', printed at Venice in 1491-92, where he divided music into three types, namely *musica mundana* 'music of the universe', *musica humana* 'music of humans', and *musica instrumentalis* 'instrumental music';

590-600: <u>Gregory I the Great</u>, Pope and saint, a Doctor of Church, produced *Gregorian chant*, collected and reorganized *music suitable for church*, in use for centuries by monks and nuns who sang, memorized, and wrote down melodies;

990: <u>Wulfstan</u>, Anglo-Saxon prelate, installed in Winchester Cathedral the oldest-known *organ*;

1025-50: <u>Hermannus Contractus</u>, German composer, preceding the dvelopment of European music, composed sensitive *choral music*, and *plainchant*;

1130-70: <u>Hildegard of Bingen</u>, German Benedictine abbess, mystical philosopher, healer and musician, expressed her apocalyptic visions in literary writings, and compiled a body of *religious music*, including *liturgical music*;

1323-46: <u>Guillaume de Machaut</u>, French composer, cleric and poet, created works such as *Messe de Notre Dame*; *Le Livre du voir-dit* or *Voir Dit* 'Tale of Truth'; and also ballades, rondeaux, and motets;

1375-80: Francesco Landini, blind Florentine composer, organist, singer, poet and instrument maker, developed a commercial type of *portable organ*;

1390-1410: Hermann Monk from Salzburg, German minnesinger, incorporated *folk styles* collected from the Alpine regions in his compositions, and made forays into *polyphony*;

1420-40: Oswald von Wolkenstein, German musician, travelled across Europe learning about *classical traditions*, and brought back some *techniques and styles* to his homeland;

1420-52: Gilles de Bins dit Binchois, Franco-Flemish composer and organist, created secular songs of rondeaux; church music including a *Te Deum*; and compositions *Magnificat*; *Credo*; and *Gloria*;

1430-1562: *Pipe organs* were made for *St Andreas Kirche Ostönnen* (Westphalia, Germany), *Basilica of Valère* (Sion, Switzerland), *S Petronio Basilica* (Bologna, Italy), *Duomo Vecchio* 'Old Cathedral' (Brescia, Italy), and Cathedral *Sé de Évora* (Évora, Portugal);

1440-70: Guillaume Dufay, French composer, was author of eight complete masses, motets, and songs, pioneering *cantus firmus mass*, and greatly influencing the Renaissance composers;

1445-59: John Bedyngham, English composer, produced motets *Manus Dei*; *Salva Jesu*; and *Vide dire*; and also *O Rosa Bella*; *Beata es Virgo Maria*; and *Mi very joy*, of rhythmical complexities;

1493-1502: Heinrich Isaac, Flemish composer, published three-volume *Choralis Constantinus*, including *36 masses*, many *secular songs*, and motets;

1495-1511: Sebastian Virdung, German priest, wrote the oldest-known treatise on instrumental music, *Musica getutscht und ausgezogen*, referring to keyboard instruments, lute, and recorder;

1520: Bernardo Pisano, Italian composer, was one of the first authors of madrigals, and wrote *Musica di messer Bernardo Pisano sopra le canzone del Petrarcha*, as the first secular music collection ever printed containing only the works of a single composer;

1528-52: Pierre Attaignant, first French music printer, brought out in Paris more than fifty *collections of chansons*, about 1500 pieces altogether;

1529: Martin Agricola, German composer of Renaissance music and music theorist, worked on the theory of music, and compiled *Musica instrumentalis deudsch* 'The German Instrumental Music';

1551-90: <u>Giovanni Pierluigi da Palestrina</u>, Italian composer, left after him over 100 *masses*; and a large number of *motets*, *hymns*, and other liturgical pieces, as well as *madrigals*;

1564-1745: <u>Andrea Amati</u>, <u>Nicolo Amati</u>, <u>Giuseppe Guarnieri</u>, and <u>Antonio Stradivarius</u>, Italian greatest violin makers of Cremona, founded and developed the standard *violin*, *violas*, and *violoncellos*, with musical qualities never reached after them, revealing the precision, purity and sensibility in performing highest level of interpretation, and stimulating generations of composers to create adequate musical pieces;

1570-80: <u>Andrea Gabrieli</u>, Italian composer, produced *masses* and other choral works, as well as *organ pieces* including toccatas and ricercaras; foreshadowing the *fugue*;

1585-1605: <u>Tomás Luis de Victoria</u>, Spanish composer, was author of *Officium Hebdonadae Sanctae*, books of motets and masses; and masterly *Requiem Mass*, composed at death of Empress Maria;

1587-1642: <u>Claudio Monteverdi</u>, Italian composer, wrote eight books of *madrigals*, containing original harmonies; operas *Orfeo*; *Il Ritorno d'Ulisse in patria* 'Ulysses's Return to his Native Land'; and *L'Incoronazione di Poppea* 'The Coronation of Poppea'; as well as great church music such in his *Mass*, and *Vespes* of Virgin;

1593-1614: <u>John Wilbye</u>, English composer, published two books of madrigals, including best-known madrigals *Adieu, sweet Amaryllis*; *All Pleasure is of this Condition*; *Down in a Valley*; *Draw on, Sweet Night*; *Flora gave me Fairest Flowers*; *Lady, your Words do Spite Me*; *Softly, softly*; *Stay, Coridon*; *Sweet Honey-Sucking Bees*; and *Weep, Weep mine Eyes*;

1597-1603: <u>Thomas Morley</u>, English composer and organist, became known by his works *A Plaine and Easie Introduction to Practicall Musicke*; popular pieces *Now is the month of maying*; *My bonny lass she smileth*; and *It was a lover and his lass*; also compiled collection *The Triumphes of Oriana*, in honour of Queen Elizabeth I;

1598-1613: <u>Adriano Banchieri</u>, Italian composer, organist and theorist, produced *L'organo suonarino*, containing first precise rules for accompanying from a figured bass; *Moderna practica musicale*; and first comic opera *La pazzia senile*;

1600-35: <u>Giles Farnaby</u>, English composer, was author of *madrigals*, and settings of *psalms* in verse paraphrases for East Psalter, and especially *keyboard music*;

1601-20: <u>Thomas Weelkes</u>, English composer and organist, wrote six-part madrigal *As Vesta was from Latmos hill descending*; then *Ayres or Phantasticke Spirites*; *O Care, wilt though despatch me*; *Thule, the period of cosmography*; *Like two proud armies*; anthems including *Hosanna to the Son of David*, and three-part song *Death hath Deprived me of my Dearest Friend*;

1604-13: <u>Robert Johnson</u>, English composer and lutenist, wrote *Full Fathom Five* and *Where the Bee Sucks* for W Shakespeare's 'The Tempest', and music for J Webster's 'The Duchess of Malfi';

1605-07: <u>Tobias Hume</u>, English composer and army officer, published *First Part of Ayres* for lyra-viol, and *Captaine Humes Poeticall Musicke* as the largest repertory of solo lyra-viol music;

1606-38: <u>Agostino Agazzari</u>, Italian composer and music theorist, produced theoretical works *Del sonare sopra il basso con tutti li stromenti e dell'usu loro nel concerto*; and *La musica ecclesiastica dove si contiene la vera diffinitione della musica come scienza, non più veduta, e sua nobilità*; as well as church music, madrigals, and pastoral drama *Eumelio*;

1607-10: <u>Gregor Aichinger</u>, German organist and composer, was author of fine religious choral works, such as *Cantiones ecclesiasticae*, containing Latin motets; and the set *Magnificat*;

1609-13: <u>John Bull</u>, English composer and virginalist, created virginals, such as *Walsingham*; *God the father*; and *God the son*; contributed to *Parthenia*, and *God Save the King* bearing no resemblance to national anthem, and another untitled piece as a possible source of this melody;

1613-50: <u>Severo Bonini</u>, Italian organist and composer, wrote the treatise *Prima parte de'discorsi e regole sovra la musica* 'First part of discourses and rules about music', and produced Rinuccini's 'Lamento d'Adrianna';

1617-19: <u>Giovanni Francesco Anerio</u>, Italian composer and priest, was author of masses, motets, many madrigals including *Selva armonica*, and the short oratorios *Teatro armonico spiritual*;

1638-44: <u>Heinrich Albert</u>, German organist, poet and composer, became known for his 8 volumes of sacred and secular *Arien*, pioneering basso continuo; as well as opera *Comödien-Musik*; and songs *Ännchen von Tharau*; *Gott des Himmels und der Erden*; and *Ich bin ja, Herr, in deiner Macht*;

1639-66: <u>Pietro Francesco Cavalli</u>, Italian composer, produced operas *Le nozze di Teti e di Peleo*; *Xerse*; *Didone*; *La virtù de'strali d'amore*; *Egisto*; *Ormindo*; *Doriclea*; *Giasone*; *Oristeo*; *Calisto*; *Eritrea*; *Orione*; *Erismena*; *Statira*; *Principessa di Persia*; *Ercole amante*; *Scipione affricano*; *Mutio Scevola*; and *Pompeo magno*;

1665-74: <u>Cristoph Bernhard</u>, German singer, composer and theorist, wrote *Geistliche Harmonien*, containing 20 sacred concertos; compositions showing mastery of counterpoint, notably the hymn *Prudentia prudentiana*; as well as *Tractus compositionis augmentatus*, which classifies music in three styles, namely *relationship of words and music, place of performance*, and *types of dissonance*;

1682-1703: <u>John Blow</u>, English composer and organist, was author of over 100 anthems, 13 services, secular songs, and also masque *Venus and Adonis*, and *Meditationes on a Theme by John Blow*;

1685-95: <u>Henry Purcell</u>, English composer, created the anthem *My Heart is Inditing*; opera *Dido and Aeneas*; spoken dialogues with musical items *Dioclesian*; *King Arthur*; *The Fairy Queen*; *The Tempest*; and songs such as *The Indian Queen*;

1690-1710: <u>Arcangelo Corelli</u> (Il divino), Italian composer, produced *Concerti grossi*, as well as *solo* and *trio sonatas for violin*, marking an epoch in chamber music, and greatly influencing JS Bach and contemporary string technique;

1695-1726: <u>Reinhard Keiser</u>, German composer, was author of the operas *Mahumet II*; *Der hochmütige, gestürtzte und wieder erhabene Croesus*; *Die Römische Unruhe*; *Die verdammte Staat-Sucht*; *Fredegunda*; and *Der Lächerliche Prinz Jodelet*;

1697-1735: <u>André Campra</u>, French composer, became known for his works *Opéra-ballet L'Europe galante*; *Le carnaval de Venise*; *Iphigénie en Tauride*; *Télémaque*; *Alcine*; *Idomenée*; *Camille, reine des volsques*; *Achille et Deidamie*; and also for his church music such as *Requiem, Cantates françoises*, and *Mass*;

1706-17: <u>Godfrey Finger</u>, Moravian composer, produced operas, theatre music, concertos, sonatas, and especially *Suite in G minor*, and *Farewell for the Death of Purcell*;

1708-48: <u>Johann Sebastian Bach</u>, German composer, created great works such as six *Brandenburg Concertos*; *Little Organ Book*; *The Well-tempered Clavier*; *St John Passion*; *St Matthew Passion*; *Mass in B Minor*; *Christmas Oratorio*; *The Musical Offering*; and *The Art of*

Fugue;

1709: <u>Bartolommeo Cristofori</u>, Italian maker of musical instruments and curator for the Medici family, invented the first *piano*, called by him 'keyboard instrument that can be played soft and loud';

1712-25: <u>Antonio Lucio Vivaldi</u>, Venetian violinist and composer, produced 12 concertos *L'Estro Armonico* 'Harmonic Inspiration', and moreover the very popular *Le quattro stagioni* 'The Four Seasons' as an early example of programme music;

1723-28: <u>Nicola Haym</u>, Italian cellist, composer and librettist, was author of *Ottone, Flavio*; *Giulio Cesare*; *Tamerlano*; *Rodelinda*; *Siroe*; and *Tolomeo*; as well as *Complete Sonatas*;

1726-40: <u>Maria Anna de Camargo</u>, French dancer, with debut in J Balon's 'Les Caractères de la dance', introduced the *shortening of traditional ballet skirt*, allowing complicated steps to be seen;

1730-1830: *Golden Age of the Harp* was marked by a series of harp makers, including <u>Jacob Hochbrucker</u>, German, and <u>Jean-Henri Naderman</u>, French, single-action pedal harps makers; <u>Beat Wolf</u>, Swiss maker and restorer of 18th century single-action harps; <u>Georges Cousineau</u> and <u>Sébastien Érard</u>, French producers of *double-action pedal harps*, and of *grand piano* action by repetition lever in the 'double escapement' actions allowing notes to be repeated more easily than in single actions, thus preceding those used in modern grands; <u>John Richards</u> of Llanrwst and <u>Basset Jones</u> of Cardiff, *Welsh triple harp* makers;

1738-55: <u>Francesco Araia</u>, Italian composer, was author of operas including *Artaserse*, and *Mitridate,* also *Cephalus and Procris* as first opera performed in Russia with sung in Russian;

1739-47: <u>George Friderich Handel</u>, German-English composer, produced oratories *Saul*; *Israel in Egypt*; *Messiah*; *Samson, Joseph and His Brethren*; *Semele*; *Judas Maccabeus*; *Solomon*; and *Jephtha*;

1741: *La lezione di danza* 'The Dancing Lesson', oil on canvas, (now in Gallerie dell'Accademia, Venetia) was produced by Venetian painter <u>Pietro Longhi</u>;

1742-55: <u>Johann Stamitz</u>, Bohemian violinist and composer, established composition style later known as *Mannheim school*, and created 74 symphonies, concertos for various instruments, chamber music, and a mass; developed sonata form; introduced *sharp contrasts into symphonic movements*; and wrote fine *concerto music*;

1751-60: <u>Francesco Geminiani</u>, Italian violinist and composer, was author of *The Art of Playing on the Violin*; *Art of Accompaniment*; *Art of Playing the Guitar or Citra*; and created the ballet *La foresta incantata*;

1754-79: <u>Jean-Georges Noverre</u>, French dancer and choreographer, published the important work *Lettres sur la dance* 'Letters on Dance', being inventor of the notion of *ballet d'action* that influenced modern ballet;

1762-78: <u>Christoph Willibald Gluck</u>, German composer, became known by *Orfeo ed Euridice* 'Orpheus and Eurydice', *Alceste*, and moreover *Iphigénie en Tauride*;

1767-91: <u>Wolfgang Amadeus Mozart</u>, Austrian composition genius, created the opera buffa *La finta semplice* 'The Feigned Simpleton'; singspiel *Bastien and Bastienne*; symphonies and quartets; *La finta giardiniera* 'The Feigned Gardener Girl'; violin concertos; *Haffner Serenade*; *Paris Symphony*; *Coronation Mass*; opera series *Idomeneo, rè di Creta* 'Idomeneo, King of Crete'; opera *Die Entführung aus dem Serail* 'The Abduction from the Seraglio'; *Linz*, and *Prague* Symphonies; *Le Nozze di Figaro* 'The Marriage of Figaro', *Don Giovanni*, and *Così fan tutte* 'Women are all Like That'; serenade *Eine kleine Nachtmusik*; operas *Die Zauberflöte* 'The Magic Flute', and *La Clemenza di Tito*; and at last unfinished *Requiem* 'which seems to concentrate the grief of entire world';

1770-94: <u>William Billings</u>, American-born composer, printed the songbooks *The New-England Psalm-Singer*; *The Suffolk Harmony*; and *The Continental Harmony*;

1771-89: <u>Charles Burney</u>, English musicologist, wrote *Present State of Music in France and Italy*, and four-volume *General History of Music*, the bias of the last one towards the then popular Italian style, to the neglect of Bach and his contemporaries, led to a fall in his influence;

1774-1820: <u>Antonio Salieri</u>, Italian composer, taught L van Beethoven, and FP Schubert, and produced 40 *operas*, an *oratorio*, and *masses*;

1776-78: <u>Giuseppe Piermarini</u>, Italian great neoclassical architect, being called by Empress Maria Teresa of Austria, he designed and built the famous <u>*Teatro alla Scala*</u>, Milan, one of the best opera houses in the world, with high acoustic qualities for listening to operatic compositions, and for balletic performances; La Scala became open with A Salieri's opera 'L'Europa riconosciuta' to a libretto by Mattia Verazi;

1780-1800: <u>Luigi Boccherini</u>, Italian composer, was author of *chamber music, String quintet in E*; including most popular *minuet*, as well as *cello concertos*, and *sonatas*;

1791-1808: <u>Joseph Haydn</u>, Austrian composer, become well-known by *The Creation*, and *The Seasons*, 104 symphonies, about 50 concertos, 84 string quartets, 24 stage works, 12 masses, orchestral divertismenti, keyboard sonatas, also chamber, choral, instrumental, and vocal pieces;

1795-1827: <u>Ludwig van Beethoven</u>, German composer and pianist, was recognized as a genial composer, his reputation being due to concert overtures; opera *Fidelio*; five piano concertos, and two for violin; *Diabelli Variations*; 32 piano sonatas; 17 string quartets; *Missa solemnis*; *Symphony Eroica* (No.3), and *Choral Symphony* (No.9) with *Ode to Joy* 'expressing the gladness of life'; also *Moonlight Sonata* 'with its drops of light';

1813-63: <u>Gioacchino Antonio Rossini</u>, Italian operatic composer, was author of *Tancredi*; *L'Italiana in Algeri* 'The Italian Girl in Algiers'; the masterpiece *Il Barbiere de Seviglia* 'The Barber of Seville'; *La Cenerentola* 'Cinderella'; *La Gazza Ladra* 'The Thieving Magpie'; *Armide*; *Mosè in Egitto* 'Moses in Egypt'; *La Donna del Lago* 'The Lady of the Lake'; great *Semiramide*; *Stabat Mater*; and *Petite messe solennelle* 'Little Solemn Mass';

1816-18: <u>Joseph Mohr</u>, Austrian priest, and <u>Franz Xaver Gruber</u>, Austrian schoolmaster and organist, wrote the lyrics and composed the music of the famous carol *Silent Night* (German *Stille Nacht, heilige Nacht*) respectively;

1817: <u>Muzio Clementi</u>, Italian pianist and composer, produced *Gradus ad Parnassum* 'Steps to Parnassus', on which subsequent piano methods have been based;

1818-43: <u>Gaetano Donizetti</u>, Italian composer, became known for the opera *Enrico di Borgogna*; then *Anna Bolena*; *L'Elisir d'amore*; *Lucrezia Borgia*; *Lucia di Lammermoor*; *La Fille du régiment* 'The Daughter of the Regiment'; *La Favorita* 'The Favourite'; and *Don Pasquale*;

1819-45: <u>Jacques Halévy</u>, French composer, won Grand Prix de Rome with cantata *Herminie*, and produced operas including *La Juive*, ballets, cantatas, and songs;

1820-28: <u>Franz Peter Schubert</u>, Austrian composer, achieved important musical works such as the song cycle *Winterreise* 'Winter Journey', and *Die schöne Müllerin* 'Fair Maid of the Mill'; group of songs *Schwanengesang* 'Swan Song'; oratorio *Lazarus*; and operas *Alfonso und Estrella*, and *Fierabras*;

1820-57: Carlo Blasis, Italian dancer, choreographer and teacher, was author of *Noted treatises on the codification of ballet technique*, as the most important ballet teaching of the 19th century;

1826-47: Frédéric François Chopin, Polish composer and pianist, produced great compositions for piano including *two concertos*, and *orchestral works*, which are characterized by great volatility of mood, and rhythmic fluidity; among his best-appreciated compositions can be mentioned *Op.2 Variations*; *Ballade No.2, Op.38*; *Polonaises*, and *Scherzo No.3, Op.39*;

1827-31: Vincenzo Bellini, Italian operatic composer, became known for his *Il Pirata* 'The Pirate'; *I Capuleti ed I Montecchi* 'The Capulets and the Montagues'; *La Sonnambula* 'The Sleepwalking Girl'; *Norma*; and *I Puritani* 'The Puritans';

1830-62: Louis Hector Berlioz, French romantic composer, founded modern orchestration, and wrote *Symphonie fantastique*; dramatic cantatas such as *La Damnation de Faust*; sacred music; and operas *Les Troyens*, and *Béatrice et Bénédict*;

1831-47: Theobald Böhm, German flautist and inventor, intending to make a flute which would be acoustically perfect, he devised a key mechanism improved as a *model on which modern flute is based*;

1836-42: Mikhail Ivanovich Glinka, Russian composer, was author of opera *A Life for the Tsar*, earlier named 'Ivan Susanin', and of national *Russlan and Ludmilla*;

1840-54: Robert Alexander Schumann, German composer, became well-known by compositions mainly for pianoforte with intense poetry and romanticism, such as *Abegg Theme and Variations*; *Papillons*; *Davidsbündlertänze*; *Carnaval*; *Fantasiestücke*; *Études symphoniques* 'Symphonic Studies'; *Kinderszenen* 'Scenes from Childhood'; *Kreisleriana*; *Novelleten*; *Waldscenen* 'Forest Scenes'; and *Albumblätter* 'Album Leaves';

1842-93: Giuseppe Fortunino Francesco Verdi, Italian operatic composer, had major successes with *Nabuco*; *I Masnadieri* 'The Robbers'; *Macbeth*; *Luisa Miller*; *Rigoleto*; *Il Trovatore*; and *La Traviata*; then *Simon Boccanegra*; and *Un Ballo in Maschera* 'A Masked Ball'; as well as *La Forza del Destino* 'The Force of

Destiny'; *Don Carlos*; *Aïda*; *Otello*; and *Falstaff*;

1843-82: <u>Richard Wagner</u>, German composer, changed the course of music by compositions *Der fliegende Holländer* 'The Flying Dutchman'; *Tannhäuser*; *Siegfrieds Tod* 'The Death of Siegfried'; *Götterdämmerung* 'Twilight of the Gods'; *Lohengrin*; the great *Ring* cycle, *Das Rheingold* 'The Rhinegold', *Die Walküre* 'The Valkyrie'; *Tristan und Isolde*; non-tragic drama *Die Meistersinger* 'The Mastersingers'; and opera *Parsifal*;

1848-80: <u>Franz Liszt</u>, Hungarian composer and pianist, was author of 12 symphonic poems, masses, two symphonies, a large number of piano pieces, and moreover the *Hungarian Rhapsodies*;

1852-1914: <u>Camille Saint-Saëns</u>, French composer and music critic, produced *Ode à Sainte Cécile* 'Ode to St Cecilia'; five symphonies; 13 *operas* including best-known *Samson et Dalila*; four *symphonic poems* including *Danse macabre*; *concertos for piano*; *Carnaval des animaux* 'Carnival of the Animals' for two pianos and orchestra; church music including *Messe solennelle* 'Solemn Mass'; chamber music and songs; as well as works including *Harmonie et mélodie* 'Harmony and Melody', *Portraits et souvenirs* 'Portraits and Souvenirs', and *Au courant de la vie* 'During a Lifetime';

1853-65: <u>Louis Moreau Gottschalk</u>, US pianist and composer, was author of piano and piano-orchestral compositions, many of which exploring Creole, African and Spanish idioms, such as the symphonic work *A Night in the Tropics*;

1855-75: <u>Georges Alexandre-César-Léopold Bizet</u>, French composer and pianist of the romantic era, created musical works such as *Symphony No.1 in C minor*; *Roma symphony*; opera in three acts *Les Pêcheurs de Perles*; *La Jolie Fille de Perth*, introducing Scottish colour or mood; piano piece *Jeux d'enfants* and the incidental music for *L'Arlésienne*; culminating with the masterpiece *Carmen*, which became one of the most popular and frequently performed works in the entire opera repertory;

1858-80: <u>Jacques Offenbach</u>, German composer, developed modern *opéra bouffe* by his *Orphée aux enfers* 'Orpheus in the Underworld'; *La Belle Hélène*; *Barbebleue* 'Bluebeard'; and *La Vie Parisienne* 'Parisian Life'; and produced grand opera *Les Contes d'Hoffmann* 'The Tales of Hoffmann';

1866-79: <u>Bedřich Smetana</u>, Czech composer and founder of Czech national music, produced nine operas, notably *Prodaná nevěsta* 'The Bartered Bride', and *Dalibor*; chamber and orchestral works including a string quartet *From my Life*; and the symphonic poems *Má Vlast* 'My Country';

1866-1907: <u>Edvard Hagerup Grieg</u>, Norwegian composer, conductor and pianist, was author of incidental music *Sigurd Jorsalfar*, and *Peer Gynt*; orchestral music *In Autumn*; *Lyric Suite*; *Elegiac Melodies*; *Holberg Suite*; and *Norwegian Melodies*; piano *Lyric Pieces*; *Shepherd's Boy*; *Holberg Suite*; *Norwegian Folk Tunes*; and many songs;

1867-74: <u>Johann Strauss the Younger</u>, Austrian violinist, conductor and composer, became famous by waltzes *An der Schönen blauen Donau* 'The Beautiful Blue Danube', and *Geschichten aus dem Wienerwald* 'Tales from the Vienna Woods'; and also by beautiful polkas, marches, and several operettas including *Die Fledermaus* 'The Bat';

1868-79: <u>Modest Petrovich Mussorgsky</u>, Russian composer, produced operas *Boris Godunov*; *Mlada*; *Khovanshchina*; and *Sorochintsy Fair*; orchestral *Night on the Bare Mountain*; piano pieces *Souvenir d'Enfance*; *Intermezzo*; *Memories of Childhood*; and *Pictures at an Exhibition*; songs *The Nursery*; *Sunless*; *Songs and Dances of Death*; and the famous *Song of the Flea*;

1869-80: <u>Johannes Brahms</u>, German composer, created major works for orchestra, including two serenades, four symphonies, *Piano concertos No.1 in D minor* and *No.2 in B-flat major*, a *Violin Concerto*, a *Double Concerto for violin and cello*, *German Requiem*; *Variations on a Theme of Haydn*; the *Academic Festival Overture*, and the *Tragic Overture*;

1872-92: <u>Pyotr Ilych Tchaikovsky</u>, Russian composer, became renowned by *six symphonies*, of which the last three are best-known; two piano concertos; a violin concerto; tone poems such as *Romeo and Juliet*, and *Italian Capriccio*; ballet music including *Swan Lake*, *The Sleeping Beauty*, and *The Nutcracker*; as well as operas *Eugene Onegin*, *The Queen of Spades*, and *Mazeppa*;

1875-1900: <u>Ion Ivanovici</u>, Romanian bandmaster and composer, created more than 350 dance pieces, marches and waltzes, including famous waltz *The Waves of the Danube*, by which the author became known as 'Johann Strauss of Romanians';

1877-1904: <u>Antonin Leopold Dvořák</u>, Czech composer, produced works including *Klänge aus Mähren*; *Slavonic Dances*; *Stabat Mater*; symphonies; chamber and piano music; as well as operas *Rusalka*, and *Armida*;

1879-83: <u>Ciprian Porumbescu</u>, Romanian composer, was author of operetta *New Moon*; the anthem *Three Colours*; and also *Song for the 1ˢᵗ of May*; *Ballad for Violin and Orchestra*; *Serenada*; *Latin Nation Song*; *Spring Song*; *On Our Flag*; *Putna Monastery's Altar*; and *On the Prut's shores*;

1881-90: <u>César Auguste Franck</u>, French composer, became known for the tone poem *Le Chasseur maudit* 'The Accursed Hunter'; *Variations symphoniques* for piano and orchestra; *Violin Sonata*; *Symphony in D minor*; and his best *String Quartet*;

1882-1907: <u>Nikolai Andreyevich Rimsky-Korsakov</u>, Russian composer, produced orchestral works *Capriccio Espagnol*; *Easter Festival*; and *Scheherazade*; operas *The Snow Maiden*; *Legend of Tsar Saltan*; *The Invisible City of Kitesh*; *Skazaniye o neridimom grade Kitezhe*; and *The Golden Cockerel*;

1884-1905: <u>Claude Debussy</u>, French composer, created the cantata *L'Enfant prodigue*; mature work *Prélude à l'aprés-midi d'un faune*; admired operatic setting of M Maeterlinck's *Pelléas et Mélisande*; then outstanding piano pieces *Images*, and *Préludes*; as well as extended his 'musical Impressionism' to orchestral music in *La Mer*;

1886-1916: <u>Charles Villiers Stanford</u>, Irish composer, was author of *Voyage of Maeldune*; oratorios *The Three Holy Children*, and *Eden*; operas *The Veiled Prophet of Khorassan*; *Savonarola*; *The Canterbury Pilgrims*; *Shamus O'Brien*; *Much Ado about Nothing*; and *The Critic*;

1889-1948: <u>Richard Strauss</u>, German composer, became well-known by symphonic poems *Don Juan*; *Tod und Verklärung* 'Death and Transfiguration'; *Till Eulenspiegels lustige Streiche* 'Till Eulenspiegel's Merry Pranks'; *Also sprach Zarathustra* 'Thus Spake Zarathustra'; *Don Quixote*; and *Ein Heldenleben* 'A Hero's Life'; operas *Guntram*; *Salome*; *Elektra*; and *Capriccio*; as well as *Metamorphosen*, and valedictory *Vier letzte Lieder* 'Four Last Songs';

1891-1906: <u>Edward Elgar</u>, English composer, was author of *Enigma Variations*; as well as of oratorios *The Dream of Gerontius*; *The Apostles*; and *The Kingdom*;

1892-1926: <u>Jean Julius Christian Sibelius</u>, Finnish composer, produced symphonic poems *Kullervo*, and *Swan of Tuonela*; symphonies and symphonic poems *En Saga*, *Finlandia*, and *Tapiola*, and the famous *Valse Triste* as 'an obsessive grief';

1893-1923: <u>Giacomo Puccini</u>, Italian operatic composer, became well-known for his *Manon Lescaut*; *La Bohème* 'Bohemian Life';

Tosca; *Madama Butterfly*; and *Turandot*;

1898-1925: <u>Sergei Pavlovich Diaghilev</u>, Russian ballet impresario, performed *Boris Godunov* in Paris, founded *Ballets Russes de Diaghilev*, and worked with great dancers and composers;

1899-1928: <u>Maurice Ravel</u>, French composer, was author of piano pieces *Sonatine* 'Little Sonata', *Miroirs* 'Mirrors', *Ma Mère l'Oye* 'Mother Goose', and *Gaspard de la nuit* 'Gaspard of the Night'; comic opera *L'Heure espagnole* 'The Spanish hour'; opera *L'Enfant et les sortilèges* 'The Child and His Spells'; and the famous *Boléro*, inspired by sounds of a factory near his house in Paris;

1899-1947: <u>Arnold Franz Walter Schoenberg</u>, Austrian-born German and US composer and conductor, produced *Verklärte Nacht* 'Transfigured Night', and *Pelleas and Melisande*; *Gurrelieder*; *Erwartung* 'Expectation'; *Pierrot Lunaire*; *Die Jacobsleiter*, and *Variations for Orchestra*; opera *Von Heute auf Morgen* 'From One Day to the Next'; stage work *Moses und Aaron*; Hebrew setting *Kol Nidre*; *Ode to Napoleon*; and *A Survivor from Warsaw*;

1900-40: <u>George Enesco</u>, Romanian composer, was a successful virtuoso and teacher of violin, among his pupils being Yehudi Menuhin, and author of *Romanian Rhapsodies*, opera *Oedipus*, three symphonies, as well as orchestral and chamber music;

1901-11: <u>Gustav Mahler</u>, Czechoslovakian-born Austrian composer, became known for his *nine symphonies*, song-symphony *Das Lied von der Erde*, and songs with orchestral accompaniment such as *Kindertotenlieder* 'Songs on the Death of Children';

1905-22: <u>Manuel de Falla</u>, Spanish composer, created operas *La Vida Breve* 'The short Life', and *El Retablo de Maese Pedro* 'Master Peter's Pupet Show'; famous ballet *El Sombrero de Tres Picos* 'The Three-Cornered Hat', and another ballet *El Amor Brujo* 'Love, the Magician'; as well as piano and orchestra suite *Noches en los Jardines de España* 'Nights in the Gardens of Spain';

1905-29: <u>Franz Lehár</u>, Hungarian composer, was author of operettas, including better known *The Merry Widow*; *The Count of Luxembourg*; *Frederica*; and *The Land of Smiles*;

1909-23: <u>Michel Fokine</u>, US dancer and choreographer, taught and choreographed for both theatre and ballet, and created a *more expressive ballet* than before;

1912-31: <u>Albert Roussel</u>, French composer, was author of choral *Évocations*, opera *Padmâvati*, and ballets such as best-known *Le*

Festin de l'araignée 'The Spider's Feast', and *Bacchus et Ariane*;

1914-66: Igor Fyodorovich Stravinsky, Russian-born US composer, produced ballets *Pulcinella*; *Apollo Musagetes*; *The Card Game*; *Orpheus*; and *Agon*; opera-oratorio *Oedipus Rex*; magnificent choral *Symphony of Psalms*; symphonies of wind instruments, and *Symphony in C major*; opera *The Rake's Progress*; piece for voice and clarinets *Elegy for J.F.K.*; and *Requiem Canticles*;

1916-28: Ottorino Respighi, Italian composer, musicologist and violinist, was author of nine operas; symphonic poems *Fontane di Roma* 'Fountains of Rome', *Pini di Roma* 'Pines of Rome', *Gli uccelli* 'The Birds', and *Feste Romane* 'Roman Festivals'; as well as the ballet *La Boutique fanatasque* 'The Fantastic Toyshop';

1916-39: Sergei Sergeyevich Prokofiev, Russian composer, produced piano sonatas; operas *The Gambler*, and *The Fiery Angel*; ballets *The Prodigal Son*; *Romeo and Juliet*; and *Cinderella*; film scores *Lieutenant Kijé*, and *Alexander Nevsky*; opera *War and Peace*; and cantatas *We are Seven*, and *Heil to Stalin*;

1921-38: Bronislava Nijinska, Russian ballet dancer and choreographer, created masterpiece ballets *Les Noces* 'The Wedding', and *Les Biches* 'The Does', and choreographed for many companies in Europe and the USA;

1924-35: George Gershwin, US composer, was author of musical comedy *Lady Be Good*; concert *Rhapsody in Blue*; *Concerto in F*; *An American in Paris*; series of hit musicals including *Of Thee I Sing*; numerous classics of US popular song such as *Someone to Watch over Me*; *Embraceable You*; and *I Got Rhythm*; as well as black opera *Porgy and Bess*;

1925-34: Sergei Vasilevich Rachmaninov, Russian composer and pianist, produced operas, orchestral pieces and songs, best known four concertos, popular *Prelude in C Sharp Minor*, and *Rhapsody on a Theme of Paganini* for piano and orchestra;

1925-39: Joaquin Rodrigo, Spanish composer, was author of orchestral work *Cinco Piezas Infantiles* 'Five Pieces from Childhood'; most frequently played *Concierto de Aranjuez* 'Aranjuez Concerto'; also concertos for violin, harp, and flute;

1930-39: Béla Bartók, Hungarian composer, produced opera *Duke Bluebeard's Castle*, ballets *The Wooden Prince* and *The Miraculous Mandarin*, as well as *Concerto for Orchestra* and *Sonata for 2 pianos and percussion*;

1930-74: <u>Dmitri Dmitriyevich Shostakovich</u>, Russian composer, was known for the ballets *Zolotoy vek* 'The Age of Gold', *Bolt* 'The Bolt', and *Svetytoly ruchey* 'Bright Stream'; also by operas *Nos* 'The Nose', and *Ledi Makbet Mtsenskovo uyezda* 'A Lady Macbeth of Mtsensk';

1933-34: <u>Laurens Hammond</u>, US engineer and inventor, designed first *electronic organ* as an electronic keyboard instrument, which derived from the harmonium, pipe organ and theatre organ;

1933-73: <u>Benjamin Britten</u>, English composer, created *A Boy was Born*; song cycles *Our Hunting Fathers*, and *On This Island*; *Violin Concerto, Sinfonia da Requiem, The Young Person's Guide to the Orchestra, String Quartets , Cello Symphony, Cello Sonata*; *'Spring' Symphony*; operas *Peter Grimes, Billy Budd*, and *Gloriana*; 'chamber operas' including *The Turn of the Screw*, 'children's operas'; and later operas *A Midsummer Night's Dream, Owen Wingrave*, and *Death in Venice*;

1943-88: <u>Leonard Bernstein</u>, US conductor, pianist and composers, was author of musical comedies *On the Town*, and *West Side Story*; ballets *The Dybbuk, Songfest*, and *Halil*; and operetta *Candide*;

1948-83: <u>Hans Werner Henze</u>, German composer, being committed to movements in Germany, Italy, and Cuba, was author of thirteen full-length and three one-act operas such as *Das Wundertheater* 'The Wonder Theatre'; *Boulevard Solitude*; *König Hirsch* 'King Stag'; *Elegy for Young Lovers*; *The Bassarids*; *We Come to the River*; and *The English Cat*; also of chamber ballets, nine symphonies, string quartets, concertos and other orchestral, chamber, vocal and piano music;

1949-60: <u>Theodor Adorno</u>, German social philosopher and musicologist, produced sociological writings on music, mass-culture and art, including *Philosophie der neuen Musik*; *Versuch über Wagner*; *Dissonanzen*; and *Mahler*; contributing to historical musicology;

1951-92: <u>Iain Ellis Hamilton</u>, Scottish composer, created *Clarinet Quintet*; *Clarinet Concerto*; two *Symphonies*; *Symphonic Variations*; several orchestral and chamber pieces; operas *The Royal Hunt of the Sun*; *The Cataline Conspiracy*; *Lancelot*; and *Raleigh's Dream*; as well as the wind octet *Antigone*;

1954-86: <u>Iannis Xenakis</u>, Romanian-Greek-born French composer, produced *Metastasis* for orchestra, as well as instrumental and other orchestral pieces, and developed *stochastic music*, a highly complex style incorporating mathematical concepts of chance and probability;

and <u>Mikis Theodorakis</u>, Greek composer, was author of the ballet *Antigone*, had a prolific musical output including oratorios, ballets, song cycles, as well as music for film scores, one of his songs being the acclamatory *Zorba the Greek*;

1962-69: <u>The Beatles</u>: <u>John Lennon</u>, <u>Paul McCartney</u>, <u>George Harrison</u> and <u>Ringo Starr</u>, English pop group, became well-known by songs including *Please, please me*; *She loves you*; *Yesterday*; *Eleanor Rigby*; *Yellow submarine*; and *Hey Jude*;

1965-92: <u>Richard Stoker</u>, English composer, pianist and writer, produced chamber and orchestral music in a serial style, and created cantata *Ecce Homo* 'Behold the Man'; opera *Johnson Preserv'd*; *Piano Concerto*; and *Chinese Canticle*; as well as published *Open Window – Open Door*, and novels *Tanglewood*, and *Diva*;

1968-2010: <u>Morton Subotnick</u>, US composer of electronic music, composed *Siver Apples of the Moon*, the first electronic work commissioned by a record company, Nonesuch; and released many multimedia CD-ROM (Compact Disc Read-Only Memory), including *All My Hummingbirds Have Alibis*, and *Echoes from the Silent Call of Girona*; and <u>Max Vernon Mathews</u>, US electrical engineer and pioneer in computer music, developed the 'Conductor' programme for *real-time tempo*, *dynamic* and *timbre control* of a *pre-input electronic score*; by which became the 'father of computer music';

21st century: <u>*Contemporary classical music composers*</u> (*compositions*) – US <u>Henry Brant</u> (*Ice Field*); Italian <u>Luciano Berio</u> (*Sequenza XIV for cello*); French <u>Pierre Boulez</u> (*Incises*); Germans <u>Hans Werner Henze</u> (*Phaedra*), and <u>Karlheinz Stockhausen</u> (*Sonntag aus Licht*, *Klang*, and *Fünf weitere Sternzeichen*); British <u>Peter Maxwell Davies</u> (*Naxos Quartets*, *Kommilitonen!*, *Symphonies 8* and *9*), <u>Graham Waterhouse</u> (*Der Handschuh*, *Rhapsodie Macabre*, and *Zeichenstaub*), and <u>Thomas Adès</u> (*The Tempest*); Estonian <u>Arvo Pärt</u> (*Symphony No. 4*); Slovakian <u>Peter Machajdik</u> (*Namah*); Danish <u>Frederik Magle</u> (symphonic suite *Cantabile*); British-US <u>Tarik O'Regan</u> (*Heart of Darkness*); Iranian <u>Mehdi Hosseini</u> (*Concerto for String Quartet and Chamber Orchestra*); Sri Lankan <u>Dinesh Subasinghe</u> (*Karuna Nadee*).

Musicologists proposed as *musical eras*:
Prehistoric;
Ancient (before AD500);
Early (500-1760);
Common practice (1600-1900); and

Modern and contemporary (1900-present).

In European art music, there were considered the *periods*:

<u>*Early*</u> (*Medieval* 500-1400, and *Renaissance* 1400-1600);

<u>*Common practice*</u> (*Baroque* 1600-1760, *Classical* 1730-1820, and *Romantic* 1815-1910);

<u>*Modern and contemporary*</u> (the *20th century* 1900-2000, *Contemporary* 1975-present; and the *21st century* 2000-present).

Despite the musicologists' expertise, these eras and periods are far from covering the complete scale of music that is extended from its origin to its fate. Data presented at the end of chapter 7 indicate that music emerged 73230 years after humans' origin, as the initial time $200000 - 73230 \approx 126770$ years ago, and will last until the eve of human fate at 204350 years from humans' origin, as the final time $t_\bullet = 204350 - 73230 = 131120$ years from the origin of music, or $131120 - 126770 = 4350$ years after the present.

The later and then better established sequences in development of music can be roughly delimited by transitional times as:

Historical music \ 1400...1250BC $\approx$ 1325BC \

Ancient classic music \ AD300...375 $\approx$ AD337 \

Medieval music \ AD1640...1650 $\approx$ AD1645 \ *Modern music*;

whereby the approximated transitional times would be 3339; 1677; 369 years ago respectively. As the origin of music was 126770 years ago, it follows that these changes in musical development happened at the times $126770 - 3339 = 123431$; $126770 - 1677 = 125093$; $126770 - 369 = 126401$ years from the origin of music, so that their ratios to the final time $t_\bullet \approx 131120$ years from the same origin are $123431/131120 \approx 0.941$; $125093/131120 \approx 0.954$; $126401/131120 \approx 0.964$. These ratios can be recognized in Background to time, Table *(k)* for $z/8 = k$ sequences, as follows

$$t_{1.75}/t_\bullet \approx 0.9413755;$$
$$t_{1.875}/t_\bullet \approx 0.9540453;$$
$$t_2/t_\bullet = 0.9640276.$$

Consequently, the selected transitional times, in years from the origin of music, are reconsidered as

$$t_{1.75} = t_\bullet \cdot tanh(1.75) \approx 123433;$$
$$t_{1.875} = t_\bullet \cdot tanh(1.875) \approx 125094;$$
$$t_2 = t_\bullet \cdot tanh(2) \approx 126403.$$

Applying the formula $t_k = t_\bullet \cdot tanh(k)$ for earlier values of argument k, the timeline, sequences, and intrinsic characteristics of music in its development are presented as:

z/8 = k	Time years)		Sequences of music
	from the origin $t_k =$ $t_\bullet \cdot tanh(k)$	from present t_k - 126770	
	131120	+4350	
...	...	...	
2.125	127432	+662 (AD2675)	–
2	126403	-367 (AD1646)	_Modern music
1.875	125094	-1676 (AD337)	_Medieval music
1.75	123433	-3337 (1324BC)	_Ancient classic music
1.625	121331	-5439 (3426BC)	_Historical music
1.5	118683	-8087 (6074BC)	_Social music
1.375	115363	-11407	_ Late band music
1.25	111227	-15543	_Intermediate band music
1.125	106116	-20654	_ Early band music
1	99860	-26910	_Later instrumental music
0.875	92296	-34474	_ Intermediate instrumental music
0.75	83281	-43489	_Earlier instrumental music
0.625	72719	-54051	_Choral kind of music
0.5	60593	-66177	_ Song full kind of music
0.375	46988	-79782	_ Song kind of music
0.25	32114	-94656	_ Song-like kind of music
0.125	16305	-110465	_ Early singing
0	0	-126770	_ Lulling and lamenting

z/8 = k	Time from origin $t_k = t_\bullet \cdot tanh(k)$ (years)	Period $\tau_k =$ $(t_\bullet^2 - t_k^2)/(2t_\bullet)$ (years)	Frequency $f_k =$ $(2t_\bullet)/(t_\bullet^2 - t_k^2)$ (years)$^{-1}$	Angular speed $\omega_k =$ $2\pi/\tau_k = 2\pi \cdot f_k$ (years)$^{-1}$
	131120	0	∞	∞
...	...	...	...	...
2.125	127432	3636.1	0.0002750	0.0017280
2	126403	4632.2	0.0002159	0.0013564
1.875	125094	5887.5	0.0001699	0.0010672
1.75	123433	7461.7	0.0001340	0.0008421
1.625	121331	9423.6	0.0001061	0.0006668
1.5	118683	11847.2	0.0000844	0.0005304
1.375	115363	14810.2	0.0000675	0.0004242
1.25	111227	18384.0	0.0000544	0.0003418
1.125	106116	22619.9	0.0000442	0.0002778
1	99860	27533.7	0.0000363	0.0002282
0.875	92296	33076.2	0.0000302	0.0001900
0.75	83281	39112.0	0.0000256	0.0001606
0.625	72719	45395.1	0.0000220	0.0001384
0.5	60593	51559.4	0.0000194	0.0001219
0.375	46988	57140.7	0.0000175	0.0001100
0.25	32114	61627.3	0.0000162	0.0001020
0.125	16305	64546.2	0.0000155	0.0000973
0	0	65560.0	0.0000153	0.0000958

12. Exploration

In general, an act or process of exploring or investigating, an examination, and the investigation of unknown regions is called *exploration* (from Latin *exploro* 'to search out, investigate, explore'). In the present chapter, exploration is referred as the act of searching or moving from a known to an unknown place (including to outer space) for the purpose of getting information or discovery of new resources.

In its course, exploration was closely related to human migration that usually followed it and widely extended in new territories. A very simple representation of exploration can be obtained by analyzing its reason R_j, enforced by a justly determined purpose $P\,^j_i$, to reach a geographical or extraterrestrial objective O_i, and therefore it would be expressed by an equation such as

$$O_i = P\,^j_i \cdot R_j.$$

Exploration and human migration emerged in Africa by travels up to its north-eastern limit, from where modern humans successively travelled to the Near East and Middle East, and further to South and East Asia, then to Australia and Europe, the Far East and North America, South America and Pacific Islands; later expeditions covered the entire globe, and were followed by spatial missions for Moon landing and investigation of other planets and their satellites in the Solar System.

Therefore, the exploration of new places, on the Earth and in outer space, followed a long lasting course, which was marked by discoveries of new lands and seas, descriptions and writings, elaborations of books, maps, charts and atlases, and later by aeronautical and spatial travels, which are briefly presented below in chronological order.

100000 years ago: Migration of modern humans from the *North-east Africa*, across the *Isthmus of Suez*, to the *Near East* in Western Asia;

100000-90000 years ago: Migrating humans were present in the *Near East*, as known by their remains found at *Qafzeh* in Israel and dated 92000 years ago;

90000-50000 years ago: Modern humans travelled to suitable areas in the *Middle East*, and *South Asia*;

60000-50000 years ago: Groups of humans were moving eastwards to *Southeast Asia*, and to the *Far East*;

50000 years ago: Migration of humans from South-east Asia, across the *East Indies* and *Papua Guinea*, to *Australia*, during the temporarily lowered sea level at Torres Strait, as evidenced by their oldest remains and graves such as the burials at Lake Mungo;

45000-40000 years ago: Human groups travelled from the Near East, across the straits of *Bosporus* and *Dardanelles*, to *Europe*, as evidenced by their remains found in some Romanian, Slovenian, German, French and Iberian caves;

40000-15000 years ago: Palaeolithic people moved from *North-east Asia*, across the *Bering Land Bridge*, into *Western Alaska* and then southwards into *North America*, by at least two consecutive migrations, following either the *coastal route* along the ice-free Pacific coast line, or the *continental route* along an open land corridor between two ice sheets to arrive directly into the region east of Rocky Mountains; the last southward migration was evidenced by a series of stone tools, a mastodon tusk scarred by a circular cut marks from a knife, and by later remains at Clovis site in New Mexico;

35000-30000 years ago: Humans populated the *Far East*, as attested by their remains found in China, Japan and Philippines, and meanwhile *central Asia* and *Siberia*;

13000-11000BC: Presence of human people in *Peru*, *Argentina*, and *South Chile* was evidenced by their remains and man-made objects;

10000-4000BC: *Colonization of Pacific islands* by peoples coming from the Far East and Indonesian islands;

5000-3000BC: In Northern Hemisphere, Old World's civilizations evolved according to their geographical position: at latitudes of less than 15° in *West Africa, East Africa, South India, South Indo-China*, and *Malaysia*; 15°...35° in *North Africa, Saharan-Arabian Sub-deserts, North India*, and *North Indo-China*; 23°...40° on *Nile, Tigris-Euphrates, Indus*, and *Yangtze-Hwang-ho* valleys; 45°...55° in *Europe, Eurasian Steppes, Mongolian Plateau*, and *Manchurian Plain*; and more than 55° in *North Europe* and *North Asia*. Development of peoples living at 23°...40° in wetter conditions, with pottery techniques and irrigation systems, was about simultaneous with evolvement of peoples living at 15°...35° in relatively improved conditions, with domesticated animals or cultivated plants, and at 45°...55° in relatively warmer conditions, with extensive grazing areas for increased number of herbivores; these conditions led to a significant increase of populations, which have been ready to start the following *major migrations* when these environmental conditions would be changed;

3000-2000BC: By mid-3rd millennium BC, climatic conditions became about simultaneously drier in Saharan-Arabian Sub-deserts, and cooler in Eurasian Steppes; consequently, _Semites_ from South and _Indo-Europeans_ from North began their invasions to attractive areas of first civilizations, which changed the course of history for ever;

2750BC: Further expansion of _Egyptians_ in south of their Kingdom during the 2nd Dynasty; _Cushites_ in North-east Africa; _Akkadians_, _Sumerians_ and _Elamites_ in Middle East; _Caucasians_ at South Caucasus; _Berbers_ (descending from Caucasoid horse-breeding nomad people) in North and North-west Africa; _Indo-Europeans_ in an area extended from Europe through North Pontic-Caspian Steppes to Central Asia; and _Finns_ in North Europe;

2275-2250BC: Migration of _Hittites_ and _Iranians_ from their Indo-European lands to Asia Minor and to Middle East respectively;

2000BC: Invasion of _early Greeks_ (Indo-European speakers) into the Peloponnese; settlement of _Hittites_ in Anatolia; migration of _Bantu_ people south from Central Africa; arrival of the first settlers in _New Guinea_; and the beginning of penetration into _Melanesia_ by group of _immigrants from Indonesia_;

1850BC: During the 12th Dynasty of Middle Kingdom in Egypt, Harappa on Indus Valley, and Elamites at North of Persian Gulf; _Iranians_ extended eastwards of the Caspian Sea; _Kassites_, _Guti_, and _Hurrians_ between South-east Black Sea and North-west Persian Gulf; _Amorites_ stretched westwards of Elamites to North-east Mediterranean; _Berbers_ and _Hamites_ moved westwards of Egypt into North Africa; _Semites_ extended eastwards of Egypt in Arabian Peninsula; _Hittites_ in central Anatolia, and _Luvians_ in West Anatolia; _Minoans_ and _Pelasgians_ in Crete and Peloponnese; _Greeks_ in Pindus, and _Illyrians_ settled northwards of Pindus along the Adriatic coast to the Danube; _Italics_ in West and South of Italian Peninsula; _Celto-Ligurians_ extended northwards from Pyrenees and Italian Peninsula to Atlantic, North Sea, Jutland, Baltic Sea, and central Europe; _Thraco-Cimmerians_ moved from Aegean Sea across Carpathians to central Europe, Black Sea, Caucasus, and Caspian Sea; _Slavs_ stretched northwards from Thraco-Cimmerians to Balts; _Balts_ extended northwards to the Baltic Sea; _Teutones_ in Jutland and South Scandinavia; and _Finns_ in North Europe; because of these movements and settlings of new comers, the original non-Indo-European inhabitants of West Europe were restricted only into a few areas of

North Scotland, South-west France, and North-east Spain, surviving as far ancestors of Picts and Basques;

After 1750BC: *Slavery* was recorded in Babylon, China, and India;

1600BC: *Iranian expansion* from Central Asia to North-east, to South-east as *Aryans*, and to South-west as *Mitanni*;

1595BC: Conquest of Babylon by *Hittites*; and rise of the first Dynasty of Babylon;

1550-500BC: *Phoenicians* traded throughout the Mediterranean Sea and Asia Minor, sailing to *North Africa* where they founded colonies, and even further to western coasts of Europe and then to *British Isles* from where they got tin for their artefacts;

1500BC: As an Indo-European branch, the *Aryans* were settled in *India*;

1450BC: Conquest of mainland Greece by so-called *Mycenaean people* was accomplished, and they took control on the Aegean Sea;

1350BC: Ascendancy of *Hittite Empire* in the Near East;

1300BC: Arrival of first settlers in *Fiji*, *Tonga*, and *Samoa*;

1300-700BC: *Cimmerians* extended from North-west of Caspian Sea and North of Black Sea to *Pannonia*;

1270BC: *Israelites* escaped from their captivity in Egypt and travelled to the *Promised Land* in present-day Israel;

1200BC: Collapse of Hittite Empire as a result not only of disputes between its leaders, but also because of invasion of the *Aegean peoples* 'Peoples of the sea', such as Thracian *Phrygians*, and Neo-Hittite *Luvians*, in the Near East;

c.1190BC: Sack and burning of *Troy* by Greeks, as the final act of their trade rivalry;

1116BC: *Assyrians* advanced with their military forces to *Babylon* and conquered it;

1100BC: Invasion of *Greece* by *Dorians*, and meanwhile rise of *Assyrian Empire*;

1100-1000BC: Under the pressure of Assyrians from East, Indo-Europeans from North, and Egyptians from South, groups of people from the East Mediterranean coast migrated on sea westwards to South Europe and North Africa, bearing their cultural heritage and alphabetic script that spread around the Mediterranean; the migrants were *Etruscans*, initially called by Greeks *Tyrrhenians*, because they came from somewhere North of the old cities Tyre and Sidon, settling on the 'Tyrrhenian' coast in Etruria, West of Apennines between the rivers Tiber and Arno; and *Phoenicians*, later called Puns and also Carthaginians, on the North African and Iberian coasts;

1050BC: Beginning of Rig-vedic *Aryan expansion*, with horses and light chariots, from Indus to Ganges in India;

850-750BC: Colonizing the West Mediterranean, Phoenician navigators founded *Carthage* (814 BC), according to legend by Queen Dido, *Utica* and *Gades* (Cádiz) on the North African and South Iberian coast;

750BC: Migrating *Kush warriors* advanced into *Egypt* and conquered it, founding their own dynasty;

750-550BC: Under pressure from Assyrian, and then Babylonian Empires, the Phoenicians continued their expansion, and subsequently founded colonies such as *Citium* and *Idalium* in Cyprus, *Thenae* and *Hadrumetum* on the North African coast, as well as *Lilybaeum* and *Panormus* on the North-west coast of Sicily;

750-500BC: Because of fast growing population, and political oppression at home, Greek migration developed by foundation of *Ionian colonies* on the North coast of the Aegean Sea, the South coast of Asia Minor, the North-east African coast, coasts of the Sea of Marmara and of the Black Sea, the North coast of Sicily, the West Italian coast, and on the North-west Mediterranean coast; *Achaean colonies* on the South Italian coast; *Aeolian colonies* on the North-east coast of Aegean Sea; *Corinthian colonies* on the North-west coast of Aegean Sea, the West coast of Adriatic Sea, and the East coast of Sicily; *Theraean* and *Rhodean colonies* on the North African coast, and the South coast of Sicily; *Megaraean* or *Spartan colonies* on the South coast of Black Sea, and in South Italy; by these colonies, the Mediterranean basin was divided between the Greeks and the Phoenicians;

700BC: Stony carved *world map* was discovered in Babylon, showing an ancient viewpoint of the known world (now in British Museum, London);

700-600BC: Under pressure from Scythians, *Cimmerians* migrated to Asia Minor, while *Scythians* replaced them and extended westwards up to the Carpathians;

689BC: Assyrian King Sennacherib advanced with his army, defended the Babylonians and sacked their capital *Babylon*;

234

609BC: _Babylonians_ and _Medes_ sacked _Nineveh_, and so marked the end of Assyrian Empire;

600BC: _Phoenician voyage_ by ships _round Africa_, sponsored by the Egyptian pharaoh; and _Imago Mundi_, a Babylonian world map, reconstructed by Eckhard Unger, which shows Babylon on the Euphrates, surrounded by a circular landmass and displaying Assyria, Urartu and several cities, in turn surrounded by a 'bitter river' (Oceanus), with seven islands arranged around it so as to form a seven-pointed star (National Maritime Museum in Greenwich, London);

600-334BC: Rise and expansion of _Median_, and then _Persian Empire_;

580-550BC: Anaximander, Ionian philosopher, first attempted to draw a _map of inhabited world_ grouping around theAegean at the centre, and all surrounded by the ocean, and showing the curvature of the Earth's surface, but visualizing it as a cylinder rather than a sphere;

525BC: Conquest of _Egypt_ by _Persians_;

c.500BC: Hecataeus of Miletus, pioneer Greek historian and geographer, drawn a _world map_, describing the Earth as a circular plate with an encircling Ocean and Greece in the centre of world;

500-200BC: _Sarmatians_ from North of the Caspian Sea started to push _Scythians_ westwards, and replaced them;

450BC: Nanno the Navigator, Carthaginian navigator and commander, explored the Atlantic coast of the North-west Africa, and founded the fortresses of _Thymiaterion_, _Acra_ (Agadir), and _Migdol_ (Mogador) 'Watchtower';

450-440BC: Herodotus, Greek historian, travelled in _Thrace_, on the _coasts of Black Sea_, in _Persia_, _Tyre_, _Egypt_, and _Cyrene_, then visited _Sicily_ and _Lower Italy_, collecting not only historical material but also precious geographic and ethnographic information as inserted in _Histories_;

400-270BC: From their homelands in central Europe, _Celts_ migrated to South into _Italy_; to East into _Pannonia_, _Carpathian Basin_, and even over Carpathians up to _River Don_; and to South-east along the Danube into _Illyria_, _Macedonia_, _North Greece_, _Thrace_, and _Asia Minor_;

335-323BC: _Macedonians_, under the command of their young King Alexander I, crossed the Hellespont, invaded _Anatolia_, subdued the _Eastern cost of Mediterranean_, were welcomed in _Egypt_, conquered the entire _Persian Empire_, advanced into _West India_, and marched through *Gedrosia* (Balochistan); as the result Alexander became the first great conqueror of Asia, and *Hellenism* spread widely in the ancient world;

330-320BC: Pytheas of Marseilles, Greek navigator and geographer, was commissioned to reconnoitre a new trade route to the tin and amber markets of northern Europe, sailed past Gibraltar along the coasts of Spain and Gaul, *circumnavigated Great Britain* via Shetland Islands, and even further to 'Thule' formerly identified as Iceland but more probably northern Norway;

264BC: Start of the *Punic Wars* between Rome and Carthage, followed by displacements of population;

220-200BC: Eratosthenes, Greek mathematician, astronomer and geographer, drew an improved *world map*, incorporating information from the campaigns of Alexander the Great and his successors, in which Asia became wider; he was also the first geographer to use *parallels* and *meridians* within cartographic depictions, attesting to his understanding of the spherical nature of the Earth;

219-210BC: Xu Fu, Chinese court sorcerer, led two voyages for exploring the *Eastern Seas*;

200BC: North-west America was inhabited by *Anasazi people*;

200-70BC: Pressed from the East, *Scythians* were completely replaced by *Sarmatians* who split into the branches of *Iazygians* and *Roxolani*;

138-125BC: Zhang Qian, Chinese statesman, was sent by the emperor Wudi to find allies in west against the marauding Xiongnu, travelled through Xiongnu's territory and, after twelve years as prisoner, headed west to *North of Bactria* and *Central Asia*, pioneering establishment of the *Silk Road* that became functional from 106BC;

118BC: Eudox of Cyzicus, Greek navigator, following the request of Egyptian King Ptolemy VIII, explored the *Arabian Sea* to obtain trade information;

102-101BC: After ravaging many areas of central Europe, migratory tribes of *Teutones* and *Cimbri* tried to penetrate into Italy, but each of them was defeated by the Roman general Gaius Marius;

30BC: *Romans* extended their controlled territories conquering *Egypt*;

29BC-AD20: Strabo, Greek geographer and Stoic, visited *Corinth*, explored the *Nile*, and then settled in Rome, where he wrote the 17-volume *Geographica*, a work of great value for extensive observations and copious references to predecessors;

27BC: Following a steadily expansion, *Roman Empire* was founded by <u>Octavianus Augustus</u>;

AD1-300: As a branch of the Niger-Congo group of people, <u>*Bantu*</u> started and developed one of the largest migrations in human history, extending firstly to *East Africa*, and eventually to *South Africa*;

10-250: Teutonic tribes from *Scandia* (South Scandinavia and Jutland) began their South and South-east migrations, and differentiated as <u>*Jutes*</u>, <u>*Angles*</u>, <u>*Frisians*</u>, <u>*Saxons*</u>, <u>*Franks*</u>, <u>*Alemanni*</u>, <u>*Marcomanni*</u>, <u>*Quadi*</u> and <u>*Vandals*</u>, neighbouring southwards Roman Empire; <u>*Lombards*</u> and <u>*Burgundians*</u>, neighbouring eastwards Balts and Slavs; <u>*Gepids*</u> and <u>*Goths*</u>, neighbouring southwards Roman Empire and northwards Slavs;

40-45: <u>Pomponius Mela</u>, earliest Roman geographer, depicted a *geographical view* of the world, including China, and of Europe, delimited the shores of *Codanus sinus* (South-western Baltic Sea), and described the countries adjoining the Southern coasts of the *Mediterranean Sea*;

89-93: <u>Gaius Cornelius Tacitus</u>, Roman historian, travelled widely through Germany searching for peoples and customs there, which were recorded in his *Germania*, a monograph of great ethnographical value;

94-160: Already penetrated into North China, the Altaic nomads of <u>*Hsiung-nu*</u> were broken by Han China, and started their movement westwards as a people later called <u>*Huns*</u>;

100: Arrival of first settlers in *Hawaiian Islands*;

100-120: <u>Marinus of Tyre</u>, Greek geographer, cartographer and mathematician in Roman Syria, improved the construction of maps and developed a *system of nautical charts*, and first assigned to each place a proper *latitude* and *longitude*, using the parallel of Rhodes for measurements of latitude, and a 'Meridian of the Isles of the Blessed (Canary or Cape Verde Islands)' as zero meridian; thus founding *mathematical geography*;

122-126: Construction of *Hadrian's Wall*, called after the Roman emperor <u>Publius Aelius Hadrian</u>, to defend Roman Britannia against attacks from North outsiders; this wall running from Wallsend on Tyne to Maryport in West Cumbria can be seen even today;

130-160: <u>Claudius Ptolemy</u>, Egyptian astronomer and geographer, produced the great work entitled *Geographia*, containing a catalogue of places, with latitude and longitude, general descriptions, details of noting position of places; and made a *map of world* and other maps;

142-200: *Antonine Wall*, called after the Roman emperor <u>Titus Aurelius Antoninus Pius</u>, was built as a line of fortification and defence against the North peoples (*Picts*) in northern Britannia, as the North-west frontier of Roman Empire, between Clyde and Forth rivers in Scotland;

220-300: In order to prevent incursions of Turkish and Mongol peoples, the *Qin Dynasty Great Wall*, was ordered by Emperor <u>Qin Shihuang</u>, the unifier of China, and built as a continuous defensive system of around 8-metre height and more than 2200-kilometre length, from West Gansu to the Gulf of Liaodong;

311-316: Renewed <u>*Hsiung-nu*</u> attacks, and sack of the Chinese capitals Loyang and Chang'an;

370: <u>*Black Huns*</u> first appeared in Europe, pushing the Goths to the Roman Empire;

410-455: <u>*Visigoths*</u> and <u>*Vandals*</u> invaded Italy, Spain, and North Africa;

411-585: On their move, the <u>*Sueves*</u> founded a kingdom in Galicia, which survived until it was absorbed into the Visigoth kingdom;

440: Coming from Manchuria, <u>*Hsien-pi*</u> people invaded North China;

442-533: Recognized by the imperial government of Constantinople, the *Vandal kingdom* was established in North Africa, where survived until being re-conquered by Justinian's general <u>Belisarius</u>, after the Battle of Ad Decimum;

443-534: <u>*Burgundians*</u> coming from North installed along the Rhône Valley, in Gaul;

447-453: <u>*Huns*</u> under their king <u>Attila</u>, the 'Scourge of God', devastated all territories between the Black Sea and Mediterranean, defeated the Roman army, overran Thrace, Macedonia and Greece, invaded Gaul where they were defeated by the Roman general <u>Flavius Aëtius</u> and the Visigoth king <u>Theodoric I</u> at Battle of Catalaunian Fields, retreated to Pannonia, but invaded again Italy where Rome was saved only by huge bribes from Pope St Leo I; finally Attila was murdered by his Burgundian bride, princess <u>Ildeco</u>;

449-663: South and East of Great Britain were gradually occupied by German tribes of <u>*Angles*</u>, <u>*Saxons*</u>, <u>*Frisians*</u> and <u>*Jutes*</u>; meanwhile <u>*Franks*</u> moved to South-east France;

460-550: Following the fall of the Huns' power, <u>*Slavs*</u> migrated southwards to the Danube where they were stopped from penetrating the Eastern Roman Empire;

480-84: *White Huns* (Ephthalites) destroyed the Gupta Empire of India; and then invaded Persia, killing the Sasanian emperor, but the Persian Empire survived;

489-552: *Ostrogoths* moved south into Balkans and into Italy, where they took control, but did not succeeded in reconciliation with Romans;

496-510: Migrating *Thuringians*, *Alemans* and *Visigoths* were defeated by the Merovingian ruler Clovis, who also extended his authority over Thuringians;

522-550: Cosmas Indicopleustes, merchant and monk from Alexandria, during the reign of Justinian I the Great, voyaged to *Ethiopia* and *India* for proving authenticity of the world biblical account, and completed *Topografia Christiana* 'The Christian Topography' with a *famous map of world*;

530-560: St Brendan (the Navigator), Irish abbot and traveller, made his legendary voyage to a land of saints far to the west and north, possibly to *Hebrides* and the *Northern Isles*, or even Iceland;

550-601: *Blue (Celestial) Turks* drove the Juan-juan, later called *Avars*, westwards out of Mongolia, and thus Avars migrated to Europe, invaded Frankish lands, formed a large khanate North of the Danube, but were defeated by the Byzantines; meanwhile, *Lombards* invaded North Italy;

568: Zemarchus, Byzantine general, travelled to *Samarkand* and *Western Turkish Khanate*, bringing back precious information for the Eastern Roman Empire;

620-650: During the decline of the Avar Khanate, *Slavs* took advantage to penetrate into the Balkans, occupying Macedonia, despite the opposition of the Eastern Roman Empire, and differentiating as *Slovens*, *Croats* and *Serbs*;

630-900: Under pressure of the Khazar Khanate, and then of Magyars extending North of Black Sea, the *Volga Bulgars* remained behind, while the *Danube Bulgars* moved south-westwards initially to the Danube, and then to the Balkans and East Pannonia;

632-712: Spread of Islam enabled the vast *Arab expansion* both eastwards to Persia and Afghanistan, and westwards to North Africa and Iberian Peninsula;

640-660: <u>Xuan Zang</u> (Hsüan Tsang), Chinese Buddhist traveller, made a 16-year pilgrimage through *China* and *India*, his travels forming the basis for the novel *Xiyou zhi* 'Monkey' written by <u>Wu Cheng'en</u> in 1593;

640-1450: In other conditions than in antiquity, *slavery* persisted in the Arab lands, and in central Europe, where many Slavs were captured and taken as slaves to Germany (hence the derivation of the word *slave*); slave-owning societies included Ottoman Empire, Crimean Khanate, Inca Empire in Peru, as well as Sokoto Caliphate and Hausa in Nigeria; then Mongols, Kazakhs, Turkic groups, and some native American peoples such as Comanche and Creek also kept slaves; after the Reconquista, Portuguese and Spanish enslaved the captured Muslims and imported Africans starting with prince <u>Henry the Navigator</u>;

721-759: The advance of <u>*Arabs*</u> into South-west Europe was stopped by the Frankish king <u>Charles Martel</u> in the Battle of Poitiers, and then they were expelled over the Pyrenees;

758-973: <u>*Saracen (Arab) raids*</u> from Islamic Spain and North Africa, were followed by attacks and occupations of large areas in South Europe, such as Sardinia, Balearic Islands, Corsica, Sicily, Taranto, Bari, South coast of France, and even monasteries and towns in Italy, including Rome itself, but forces of the Byzantine Empire drove them from Italy;

772-800: Movements of *migratory peoples* in *West central Europe* were stopped, checked, and controlled by Franks under their King <u>Charlemagne</u>, who subdued Saxons, Lombards, Avars, and many Slavs, following a Christianizing policy;

793-968: <u>*Viking raids*</u>, *attacks*, and *colonisations* in Britain, Ireland, on the North, West, and South coasts of France, as well as in the Iberian Peninsula;

876-1015: <u>*Scandinavian invaders*</u> occupied much of Britain, instituting their rule known as 'Danelaw' in the East and North England, despite the remarkable success of Wessex's King <u>Alfred the Great</u> to reassess Saxon status in country by a treaty formalizing partition of England;

899-955: Migrating from the North of Black Sea and through Pannonia, <u>*Magyars*</u> routed the Italian army, defeated and destroyed the Bavarian army, plundered areas of Bavaria, Swabia, Thuringia and Francia, finally being decisively defeated by the Holy Roman Emperor <u>Otto I</u> at Lechfeld;

970-1020: <u>Erik the Red</u> and his son <u>Leif the Lucky</u>, Norwegian Viking explorers, the father explored *Iceland* and *Greenland*, founding colonies there, and his son landed in 'Vinland' on *northern American coast*; both of them being subject of Icelandic sagas;

970-1223: Crushing the Khazars in South-east Russia, Grand Prince <u>Svyatoslav</u> involuntarily opened the way to fierce *Pechenegs* (Turkic nomads) to dominate the South Russian steppes, until they were displaced westwards by equally warlike *Cumans*, also called *Polovtsy*, who sacked Kiev and steadily colonized territories in South Russia and North of Black Sea upto the Carpathians, but their military force together with that of the Russians were defeated by *Mongols* in the Battle of Kalka;

1020-50: *Anglo-Saxon Mappa Mundi*, is a pigment-on-vellum *world map* containing the earliest known, relatively realistic depiction of the British Isles; being ultimately based on a model dating from Roman times, and showing the provinces of the Roman Empire of which 'Britannia' was one (British Library, London);

1059-94: After taking control of North France, *Viking invaders*, called *Normans*, constituted there a duchy and extended their power over Sicily and South Italy where they established the Kingdom of the Two Sicilies, as well as over England following Battle of Hastings (1066) under the command of <u>William the Conqueror</u> who crowned himself King of England;

1075: <u>Shen Gua</u>, Chinese administrator, engineer and scientist, produced *Dream Pool Essays*, which included a large *atlas of China and foreign regions*, and also made a *three-dimensional raised-relief map*;

1150-54: <u>Muhammad al-Idrisi</u>, Arab geographer, as requested by the Norman King <u>Roger II of Sicily</u>, he elaborated *Tabula Togeriana*, a map of the whole known world at his time, incorporating the knowledge of Africa, Indian Ocean and the Far East, gathered by Arab merchants and explorers, and showing most of Eurasia, but only the northern part of Africa; his map became most important for European exploration of Asia;

1160-73: <u>Benjamin of Tuleda</u>, Navarrese Jewish Rabbi, visited *Syria, Palestine, Baghdad, Persia,* and *Arabian Peninsula*, collecting there data for later use;

1206-23: *Mongols*, ruled by <u>Genghis Khan</u>, overran the empire of *North China*, conquered the *Kara-Khitai Khanate* from Lake

Balkhash to Tibet, took *Bokhara, Samarkand* and *Khwarezm*; two of Genghis' lieutenants penetrated northwards from southern shore of the Caspian through *Georgia* into *southern Russia* and *Crimea*; meanwhile one of his generals completed the conquest of all *northern China* except Honan; thus Genghis became the second great conqueror of Asia, his empire stretching from the Black Sea to the Pacific;

1235: *Ebstorf Mappa Mundi*, an old map printed by the German Gervase of Ebstorf, painted on 30 goatskins sewn together, and measuring 3.6 x 3.6 metres, centred on Jerusalem with east on the top (copy in British Museum, London);

1237-42: *Mongols* returned in strength and struck the middle and upper Volga, overcame the Volga Bulgars, annihilated the people of South-west Russia, sacked Kiev, and imposed tribute upon conquered lands; meanwhile Mongol campaigns were carried out over principalities of Volhynia and Galicia, as well as deep into Polish principalities and Hungarian Kingdom, culminating with their victory against a German-Polish army at Legnica;

1240-42: Invasions of *Swedes* and *Germans* to drive Russia from the Baltic Sea ended with decisive victories of the Russian Prince of Novgorod Alexander Nevski (Alexander of Novgorod) on the River Neva and Lake Peipus respectively; details on these victories were recorded in the *Second Pskovian Chronicle* (1260-80) where are inserted 'Tales of the Life and Courage of the Pious and Great Prince Alexander';

1271-95: Marco Polo, Venetian merchant and traveller, together with his father and uncle crossed *Central Asia* and the *Gobi Desert* to *China*, where they met Kublai Khan who sent Marco as an envoy to *Yunnan*, northern *Burma, Karakorum, Cochin-China*, and southern *India*; then he served as Governor of *Yang Chow* and helped to subdue the city of *Saianfu*; finally Polos left the imperial court, sailed to *Persia*, and from there back to Venice; eventually, as prisoner at Genoa, Marco wrote *Divisament dou Monde*, an important account of his travels;

1280: Richard de Bello of Haldingham, Prebendary of Lincoln, elaborated *Mappa Mundi*, a large map (65 by 53 inch), painted in Lincoln, England, and worked in vellum, embodying the medieval belief about the world and being centred on Jerusalem, with East at top;

1300-1484: Of Turkish origin, the *Ottomans* from Anatolia gradually expanded conquering territories from West Asia Minor to Dardanelles, then the East Balkans, Serbia, Bulgaria, Greece, Constantinople and Herzegovina, as well as making vassals Wallachia, Moldavia, Bosnia and Bujak;

1311: Petrus Vesconte, Genoese cartographer, was author of the oldest signed *Portolan chart*;

1325-54: Ibn Battutah, Moroccan traveller and geographer, visited *Mecca, Persia, Mesopotamia, Asia Minor, Bokhara, India, China, Sumatra*, southern *Spain* and *Timbuktu*, then dictating the entertaining history of his journeys, entitled *Rihlah* 'Travels';

1328-39: Wang Dayuan, Chinese explorer, made two major trips by ship: one along the South China Sea, visiting many places in *Southeast Asia*, reaching *South Asia* and landing in *Sri Lanka* and *India*; and another one to *North Africa* and *East Africa*;

1389: *Da Ming Hun Yi Tu*, or *Amalgamated Map of the Great Ming Empire*, was made for the first Ming emperor, as a world coloured map with China at the centre, and Europe, half-way round the globe, and painted on 17 square metres of silk;

1392-1405: *Turko-Mongols*, commanded by their Khan Timur Lenk (Tamerlane), subdued nearly all *Persia, Georgia*, and the *Tatar Empire*, conquered the *territories between Indus and lower Ganges*, won *Damascus* and *Syria* from the Mamluk sovereigns of Egypt, and a big part of the *Ottoman Empire*; Timor was the third great conqueror of Asia, and his Timurid dynasty ruled until 1507 over modern-day Iran, Afghanistan, much of central Asia, Pakistan, India, Mesopotamia, Anatolia and the Caucasus;

1409-1598: Military expeditions of the Chinese General Ch'iu Fu against the *Mongol invaders* from North; then *Japanese pirate raids* on the East coast of China, and invasion of Korea by forces under command of Hideyoshi Toyotomi;

1410-30: Zheng He, Chinese admiral, made seven voyages to explore *Arabia, East Africa, India, Indonesia* and *Thailand*;

1411-15: Albertinus de Virga, Venetian cartographer, produced a *circular world map*, drawn on a piece of parchment 69.6 x 44 centimetres and about 44 centimetres in diameter;

1418-56: Henry the Navigator, Portuguese prince, discovered the *Madeira Islands, Azores* and *Cape Verde Islands*, thus preparing the way for new explorations along the western coast of Africa, and eventually to India;

1440-70: <u>Afanasy Nikitin</u>, Russian traveller and merchant, was one of the first Europeans to travel to *India*, and to document his visit there;

1450-1870: *Slave trade* was practiced by the Portuguese, Spanish, Dutch, French and British, by capturing at least 11.5 million West and East Africans, transporting them overseas, and selling the survivors in Americas, India, and Muslim world; meanwhile the Arab slave trade flourished along the East coast of Africa; all slaves suffering cruelty and indignity of the inhuman and shameful trade; finally slavery was first abolished in British Empire and in USA;

1459: *Fra Mauro map*, one great medieval European map, was made by the Venetian monk <u>Fra Mauro,</u> as a circular planisphere drawn on parchment and set in a wooden frame about two metres in diameter, under a commission by King Alphonso V of Portugal, which did not survive, but its copy was completed by Andrea Bianco;

1480-98: <u>João Fernandes Lavrador</u>, Portuguese explorer, after reaching Labrador, that bears his name, he charted the coasts of South-western *Greenland* and of the adjacent North-eastern part of *North America*;

1486: <u>Bartolomeu Diaz</u>, Portuguese navigator and explorer, following western coast of Africa, reached the *Cape of Good Hope*, so finding a route to India;

1492: <u>Martin von Behaim</u>, German mariner, cosmographer, astronomer, geographer and explorer in service to the King of Portugal, made the *Erdapfel globe* (German *Erdapfel* 'earth apple, potato), the oldest surviving terrestrial globe, constructed of a laminated linen ball reinforced with wood and overlaid with a *map* painted by <u>Georg Glockendon</u> (Germanic Museum of Nuremberg, Germany);

1492-1504: <u>Christopher Columbus</u>, Genoese explorer, became famous for his four voyages: first, on the flagship *Santa Maria* attended by caravels *Pinta* and *Niña*, to *San Salvador*, *Cuba* and *Haiti*; second to *Guadeloupe*, *Montserrat*, *Antigua*, *Puerto Rico* and *Jamaica*; third, to *Trinidad* and the *mainland of South America*; and fourth, to *Honduras* and *Nicaragua*;

1497-1500: <u>John Cabot</u>, Genoese-born British navigator and explorer, under letters patent from King Henry VII, he sailed in search of a route to Asia, and was the first European to see *Newfoundland*, probably Cape Breton Island, Nova Scotia, and claimed North America for England;

1497-1503: <u>Vasco da Gama</u>, Portuguese navigator, in search for a route from Portugal to India, proceeded to an expedition from Lisbon, first sailing round the Cape of Good Hope, establishing the colony *Malindi* on eastern African coast, crossing Indian Ocean, and arriving at *Calicut*, where hostile population murdered 40 Portuguese, but da Gama escaped, travelled back to Lisbon, and then returned with a 20-ship squadron, founded colonies *Mozambique* and *Sofala*, and bombarded Calicut;

1497-1713: Conquest and colonization of *North America* by <u>*English*</u> with settlements in Newfoundland, Virginia, Maine, New Hampshire, New York, Massachusetts Bay, Maryland, Rhode Island, Connecticut, Carolina, New Jersey, Rupert's Land, and Pennsylvania; by <u>*Spanish*</u> with settlements in Mexico, Guatemala, Honduras, and Pensacola; by <u>*French*</u> with settlements in Nova Scotia, New France, and Louisiana; and by <u>*Swedish*</u> with settlement in Delaware;

1499-1505: <u>Amerigo Vespucci</u>, Florentine-born Spanish explorer, made an expedition to the New World, searching the *coast of Venezuela*; the name of *America* being given by the German cartographer M Waldseemüller after publication of an account of Amerigo's travels, based on his letters *Quattuor Americi navigations* 'Four Voyages', the author's name being Latinized as *Americus*;

1500: <u>Juan de la Cosa</u>, Spanish cartographer, explorer and conquistador, made several maps of which the only survivor is his *Mappa Mundi*, the first known European cartographic representation of the *Americas* (Museo Naval of Madrid, Spain);

1500-01: <u>Pedro Álvarez Cabral</u>, Portuguese navigator, commanding a fleet of 13 vessels and drifting into the South American current of the Atlantic Ocean, he was carried to the unknown coast of *Brazil*, claiming it on behalf of Portugal, and then travelled to India, diverting to *Mozambique* that he first described;

1500-20: <u>Johannes Werner</u>, German scientist, refined and promoted the *Werner map projection*;

1502-10: <u>Ludovico de Varthema</u>, Italian traveller and writer, led journeys through *Arabia*, *Persia*, *India*, and across the Pacific to *Spice Islands*, all accounted in *Itinerario de Lodovico de Varthema Bolognese* 'Travels of Ludovico de Varthema';

1502-15: <u>Afonso Albuquerque the Great</u>, Portuguese Viceroy of Indies, following a route around South Africa and landing on the Indian Malabar Coast, he conquered *Goa*, *Ceylon*, *Malacca* and *Ormuz Island*, which formed together *Portuguese East India*;

1506-13: <u>Martin Waldseemüller</u>, German cartographer, constructed *globes* and made *world maps*, including *Orbis Typus Universalis*, as the first maps using the name America;

1507-28: <u>Diogo Rodriguez</u>, Portuguese explorer, travelled on the Indian Ocean, where he discovered the island of *Mauritius*, and then, on the newly discovered route to Goa, sailed along the Mascarene Islands where *Rodrigues Island* was named after him;

1508: <u>Conrad Peutinger</u>, German scholar and antiquary, got a 4th or early 5th century copy of a Roman Road map, now known as *Tabula Peutingeriana* 'Peuntinger Table', which was printed by Jan Moretus at Antwerp, and is conserved at Österreichische Nationalbibliothek at Hofburg in Vienna;

1512: <u>Luis Ponce de León</u>, Spanish explorer, on a quest for the legendary 'Fountain of Youth', he discovered and explored *Florida*;

1512-13: <u>Vasco Núñez de Balboa</u>, Spanish explorer, led an expedition to *Darién* in Central America, where he first crossed the *Isthmus of Panama*, and was the first European to have a *view of Pacific Ocean*;

1518-21: <u>Hernán Cortés</u>, Spanish conquistador, conquered the *Aztec Empire* (Mexico) for Spain;

1519-22: <u>Ferdinand Magellan</u>, Portuguese navigator, and <u>Juan Sebastian del Cano</u>, Basque navigator, achieved the first circumnavigation, with five ships including flagship *Santa Maria de la Victoria*, coasting Patagonia, passing through *Magellan Strait*, reaching the *Pacific Ocean* and then *Philippines*, where Magellan was killed by the local people; only one of the ships completed the *first circumnavigation of the world*, being taken back to Spain by the captain JS del Cano;

1519-54: <u>Lopo Homem</u>, Portuguese cartographer and cosmographer, made *world maps*, depicting regions such as Libya, Ethiopia, Guinea, Americas - identified as 'Mundus Novus Brazil', and Asia (now in National Library of France); as well as his *Maritime Chart* (National Library of Portugal);

1524-26: <u>Giovanni da Verrazano</u>, Italian navigator and explorer, entered the service of King Francis I of France, and led an expedition to North America, exploring the coast from *Cape Fear*, South Carolina, northward to *Cape Breton*, Newfoundland, and becoming the first European to enter the *New York Bay*;

1527: <u>Diego Ribeiro</u>, Portuguese cartographer working for Spain, made the first scientific world map, the *Padrón real*, based on empiric latitude observations; this map also shows, for the first time, the real extension of the *Pacific Ocean*, and the *North American coast* as a continuous one;

1531-1814: *South America* was conquered and colonized initially by <u>*Spanish*</u> in Peru and Chile, then by <u>*Portuguese*</u> in eastern South America, and finally by <u>*French*</u>, <u>*Dutch*</u> and <u>*English*</u> in French Guiana, Surinam, and Guyana respectively;

1532-37: <u>Francisco Pizarro</u>, Spanish soldier, conquered the *Inca Empire* (Peru) also for Spain;

1534-41: <u>Jacques Cartier</u>, French navigator, made three voyages to North America searching for a westerly route to Asia, and discovered the *St Lawrence River*;

1536-94: <u>Gerardus Mercator</u>, Flemish geographer and map-maker, produced a terrestrial globe, a map of the Holy Land, maps of many parts of Europe, including Great Britain, introduced *Mercator's map projection*, in which path of a ship steering on constant bearing is represented by a straight line on map, used for nautical charts, and published the first *Atlas of Europe* covered by a drawing of Atlas holding a globe on his shoulders, so that the term 'atlas' became applied to any book of maps;

1539-41: <u>Hernando de Soto</u>, Spanish explorer, set out to explore *Florida*, and then travelled through much of present-day *Georgia*, *Carolina*, *Tennessee*, *Alabama*, and *Oklahoma*, being the first European to see the *Mississippi River*;

1560-80: <u>Yermak Timofeyevich</u>, Russian Cossack, under the rule of Tsar <u>Ivan the Terrible</u>, he led the Russian conquest of *Siberia*;

1563-70: <u>Juan Fernández</u>, Spanish navigator, discovered *Fernández Island* in Pacific, as well as *San Felix* and *San Ambrosio Islands*;

1564-73: <u>Abraham Ortelius</u>, Flemish cartographer and geographer, published the first modern atlas entitled *Theatrum Orbis Terrarum* (now in Basel University Library), then a two-sheet *map of Egypt*, an eight-sheet *map of Asia*, a six-sheet *map of Spain*, and contributed to the *Map of England and Wales*;

1567-95: <u>Alvaro de Mendaña de Neyra</u>, Spanish explorer, searching for 'Terra Australis', discovered the *Solomon Islands* and *Tuvalu*, and then the *Marquesas Islands*;

1570: <u>Jean Cossin</u>, French cartographer, elaborated the *Cosmographic*

Map of Universal Description of the World, Dieppe, first using the *sinusoidal 'Mercator equal-area'* projection (National Library of France);

1577-80: <u>Francis Drake</u>, English navigator, entered the Pacific Ocean with five ships, two of them being lost, but he sailed north to Vancouver, failed to find a Northwest Passage to Atlantic, turned south and moved across the Pacific, refitted in Java, headed to the Cape of Good Hope, and returned to England, becoming the *first Englishman to circumnavigate the world*;

1581-1800: <u>*Russian*</u> vast East *expansion* in *Asia* took place from European Russia, through West and central Siberia, to central Asia, and to the Far East including Kamchatka;

1583: <u>Humphrey Gilbert</u>, English explorer, landed in *Newfoundland* and claimed it for the Crown, as well as established a colony at *St John's*;

1585-95: <u>Walter Raleigh</u>, English courtier and navigator, travelled to America, despatching a settlement to Roanoke Island in North Carolina; then with five ships explored the *coasts of Trinidad* and sailed up *Orinoco*;

1597: <u>Willem Barents</u>, Dutch navigator, searching for a North-east Passage, died *off Novaya Zemlya*, where he had the winter quarters which were found undisturbed in 1871 together with part of his journal that was recovered in 1875 by another expedition; so that that Arctic sea was named after him *Barents Sea*;

1600-20: <u>William Adams</u>, English navigator, on the Dutch vessel *de Liefde*, reached *Japan*, and then became an agent of the Dutch East India Company;

1603-08: <u>Samuel de Champlain</u>, French explorer, made a voyage to Canada and founded *Quebec*, mapping many new areas, and becoming known as 'founder of Canada', where *Lake Champlain* was named after him;

1606: <u>Luis Vaez de Torres</u>, Breton-born Spanish navigator, discovered the islands called now *Vanuatu*, from where he first sailed westward through the *Straits of Torres*, between New Guinea and Australia, and sighted the tip of *Cape York* on the mainland of Australia, which he took to be another island;

1609-10: <u>Henry Hudson</u>, English navigator, discovered the *Hudson River*, following it for 150 miles to Albany; in another voyage he reached Greenland and arrived at the waters named after him *Hudson*

Strait and *Hudson Bay*;

1615-16: <u>William Baffin</u>, English navigator, discovered *Baffin Bay*, and then *Lancaster, Smith,* and *Jones Sounds* (passages of water) between the Arctic and Pacific Oceans;

1616: <u>Willem Corneliszoon Schouten</u>, Dutch mariner, was the first to cross the *Drake Passage*, discovering and rounding *Cape Horn*, so named after his birthplace;

1618-22: <u>Hessel Gerritsz</u>, Dutch engraver and cartographer, produced a chart of the *Indonesian islands* and the *North-west coast of Australia*, and then a *map of the Pacific Ocean*, which were extensively consulted by AJ Tasman on his voyage around Australia and to New Zealand;

1631-79: <u>Jean Baptiste Tavernier</u>, French traveller, made voyages to *Constantinople, Persia, Aleppo, Malta,* and *Italy*, then to *Syria, Ispaham, Agra,* and *Golconda,* and finally to *Hindustan, Batavia, Bantam* and *Holland by Cape*, bringing information for eastern trade; his published works being *Six Voyages*, and *Recueil*;

1642-44: <u>Abel Janszoon Tasman</u>, Dutch navigator, discovered *Tasmania, New Zealand, Tonga,* and *Fiji*, also explored the *Gulf of Carpentaria* and the *North-western coast of Australia*;

1644-1890: After North-west *Australia* was discovered by the Dutch navigator AJ Tasman, and its Eastern coast was surveyed by the English navigator J Cook, <u>*British*</u> gradually colonized the continent, while most of <u>*Aborigines*</u> moved inland;

1652: <u>Claes Jansz Visscher</u>, Dutch cartographer, completed a *World Map*, included in 1959 edition of Hendrick Doncker's *Sea Atlas* (now National Library of Australia);

1675-78: <u>Nicolae Milescu</u>, Moldavian writer, diplomat and traveller, made a long *journey to China*, which was recorded in his three-volume *Traveller's Notes from China*, an influential work for the Russian explorers of East Siberia;

1682: <u>René Robert Cavelier La Salle</u>, French explorer, pioneering the search of Canada, travelled along Ohio and Mississippi to the sea, claiming the lands for France as *Louisiana* after King Louis XIV;

1713-62: <u>John Harrison</u>, English inventor and horologist, after the British government offered prizes for discovery of a method to determine the longitude accurately, he developed a *marine chronometer* that, in a voyage to Jamaica, determined the *longitude* within 18 geographical miles;

1728-41: <u>Vitus Jonassen Bering</u>, Danish navigator, led the Russian 600-strong Great Northern expedition to determine whether Asia and America continents were joined, investigated *Siberian coast* and *Kuril Islands*; then in another expedition sailed from Okhotsk towards the American continent, and noticed land northwards, but was forced to return because of sickness and storms, wrecked on Avatcha island, later called after him *Bering Island*, the same as the *Bering Sea* and *Bering Strait*;

1730-45: <u>Pierre Gaultier Verendrye</u>, French explorer, travelling from Nipigon on Lake Superior over much of unknown Canada, discovered *Rainy Lake*, *Lake of the Woods*, *Lake Winnipeg*, and then reached the *Mandan country* south of Assiniboine River, *upper Missouri*, *Manitoba* and *Dakota*;

1750-1805: In *India*, after <u>*Portuguese*</u>, <u>*French*</u>, <u>*British*</u>, <u>*Dutch*</u> and <u>*Danish*</u> settlements were established, the British power extended in South, East, and North of the subcontinent;

1766-68: <u>Samuel Wallis</u>, English explorer, as naval officer during a circumnavigation of globe, he discovered *Tahiti* and *Wallis Islands*;

1768-78: <u>James Cook</u>, English navigator, became well-known for his three expeditions: first, on *Endeavour* of Royal Society, to Pacific, for observing transit of Venus across the Sun, and continuing by circumnavigation and chart of *New Zealand*, survey of *eastern coast of Australia*, sail through strait between Australia and New Guinea, Java and the Cape of Good Hope; second, on *Resolution* and *Adventure*, for searching stretch northwards of the Antarctic lands, followed by discovery of *New Caledonia*; and third and last, for identifying a passage round northern coast of America from Pacific, in which a number of *Pacific islands* and the *western coast of North America* were surveyed from 45°N to the Bering Strait;

1771-72: <u>Yves Joseph de Kerguélen-Trémarec</u>, French aristocrat and naval officer, on a voyage to search for Terra Australis, discovered a group of islands, named *Kerguélen's Islands*, in the South Indian Ocean;

1788-93: <u>Alexander Mackenzie</u>, Scottish explorer, established Fort Chipewayan on Lake Athabasca, in Canada, then discovered *Mackenzie River*, following it to the sea and becoming first European to cross the *Rocky Mountains* to the Pacific Ocean;

1794-95: <u>George Vancouver</u>, English navigator and explorer, surveyed in Australia and New Zealand, then charted in detail along the *western coast of North America,* and sailed round *Vancouver Island*;

1801-03: <u>Matthew Flinders</u>, English explorer, circumnavigated Australia and explored the *Australian coastline*; *Flinders River* in Queensland and the *Flinders Ranges* in South Australia being named after him;

1819-21: <u>Fabian Gottlieb Benjamin Bellingshausen</u>, Russian explorer, made an expedition around the world and also made discoveries in Pacific Ocean, sailing up to 70°S in Antarctic where *Bellingshausen Sea* was called after him;

1822-23: <u>James Weddell</u>, English navigator, sailed up to 74°15' South by 34°17' West in *Antarctica*, and discovered the *Weddell Sea* and *Weddell Quadrant*, as well as an unknown kind of seal;

1824-28: <u>Hamilton Hume</u>, Australian explorer, intending to find an overland passage from Lake George to southern coast, discovered part of the *Murray River*, and had the first sighting of Australia's highest mountain, called *Mount Kosciusko*; then participated in discovery of the *Darling River*;

1824-52: *European possessions in SE Asia* started with *Portuguese*, *Spanish*, *Dutch*, *English*, and *French* establishments, and culminated with *Dutch* driving all their rivals out of the spice trade, and *Spanish* taking over the Philippines;

1828-40: <u>George Simpson</u>, Canadian explorer, performed an overland journey round the world, and discovered *Simpson's Falls*, and *Cape George Simpson*;

1829-33: <u>John Ross</u>, accompanied by his nephew <u>James Clark Ross</u>, Scottish explorers and naval officers, made an expedition in search of the Arctic Northwest Passage, during which *Boothia Peninsula, King William Land*, and *Gulf of Boothia* were discovered and named by him;

1831-36: <u>Robert Fitzroy</u>, English naval officer, made a scientific survey of *South American waters*, and circumnavigated in HMS *Beagle*, accompanied by English naturalist CR Darwin, who visited Tenerife, Cape Verde Islands, Brazil, Montevideo, Tierra del Fuego, Buenos Aires, Valparaiso, Chile, Galapagos, Tahiti, New Zealand, Tasmania and Keeling Islands; during that voyage, Darwin started his studies of coral reefs, fauna, flora and geology, as basis of his great *The Origin of Species by Means of Natural Selection*;

1839-49: <u>James Clark Ross</u>, Scottish explorer, led an expedition to Antarctic, during which *Victoria Land* and volcano *Mt Erebus* were discovered; and another expedition to Baffin Bay, his name being given to *Ross Island, Ross Sea*, and *Ross's Gull*;

1844-54: <u>Roderick Impey Murchison</u>, Scottish geologist, did exploration work in Africa and Australia, and discovered *Murchison Falls* in Uganda, and *Murchison River* in Western Australia;

1852-71: <u>David Livingstone</u>, Scottish missionary and traveller, and <u>Henry Morton Stanley</u>, British-US explorer, discovered *Lake Ngami, Victoria Falls* of Zambezi, *Lakes Shirwa* and *Nyasa*, and also *Lakes Mweru* and *Bangweulu*, as well as published *Missionary Travels*, and *The Zambezi and its Tributaries*;

1860-64: <u>John Hanning Speke</u>, English explorer, and <u>James Augustus Grant</u>, Scottish explorer, looking for the sources of Nile, they discovered and explored the African *Lake Victoria Nyanza*, tracking the Nile flowing out of it;

1866-68: <u>Ernest-Marie-Louis Doudart de Lagrée</u> and <u>Francis Garnier</u>, French explorers, participated at the *Mekong River Expedition* charged with assessing the possibility of a navigable waterway to China; made the first survey of ruined temples of *Angkor*; and mapped 4,991 kilometres of unknown territory in *Cambodia*, and *Yunnan*;

1875-78: <u>Verney Lovett Cameron</u>, English explorer, was the first European to cross *Africa from coast to coast*, and then travelled overland to India in order to prove the feasibility of a Constantinople (Istanbul) – Baghdad railway;

1876-78: <u>Pierre Savorgnan de Brazza</u>, French explorer, searched the *Ogowe River*, and the *North of Congo*, and founded *Brazzaville* on the north shore of Stanley Pool;

1876-89: *Theory of Colonialism* showing that settlements abroad should be treated as 'property domains exploited for benefit of the mother countries';

1881-90: <u>Joseph Thomson</u>, Scottish explorer, was the first European to reach *Lake Nyasa (Malawi) from North*, travelling on to Lake Tanganyika, discovered *Lake Baringo*, and *Mount Elgon*; and later explored *Sokoto* in North-west Nigeria and *Upper Congo*;

1886-1909: <u>Robert Edwin Peary</u>, US naval commander and explorer, participated at eight Arctic expeditions, exploring Greenland and the region later called *Peary Land*, and attained to reach the *North Pole*;

1888-95: <u>Fridtjof Nansen</u>, Norwegian explorer, biologist and oceanographer, made adventurous journey across Greenland from east to west, then started with his specially built sealer *Fram*, travelling to New Siberian islands and across ice to highest latitude of 86°14' N, and overwintering in Franz Josef Land;

1891-1908: <u>Frederick Albert Cook</u> and <u>Robert Edwin Peary</u>, US explorers, led an Arctic expedition to *Greenland*, then Cook claimed to have made first ascent of the highest mountain in North America, namely *Mount McKinley*, as reported in his writing *To the Top of the Continent*, and in another expedition Cook apparently reached the *North Pole*;

1901-12: <u>Robert Falcon Scott</u>, English explorer, <u>Edward Adrian Wilson</u>, English physician, naturalist and explorer, and <u>Lawrence Edward Grace Oates</u>, English explorer, took part in the National Antarctic expedition led by Scott searching the Ross Sea, who discovered *King Edward VII Land*; then another expedition in Terra Nova, with a sledge party including EA Wilson and LEG Oates, when they reached the *South Pole*, only a month later than the Norwegian expedition led by REG Amundsen;

1901-16: <u>Ernest Henry Shackleton</u> and <u>Robert Falcon Scott</u>, British and English explorers respectively, were in an Antarctic expedition, on *Discovery*; then another expedition commanded by EH Shackleton, reaching a point at 97 miles from the *South Pole*, and during a further Shackleton's expedition, on his ship *Endurance*, that was crushed in ice, by means of sledges and boats, reaching *Elephant Island* from where a perilous voyage of 800 miles was made to South Georgia;

1902-11: <u>Roald Engelbreth Gravning Amundsen</u>, Norwegian explorer, sailed the Northwest Passage from east to west in the smack *Gjöa* and located the *Magnetic North Pole*, and first reached the *South Pole*, one month ahead of Captain RF Scott;

1907: <u>Ernest Henry Shackleton</u>, British explorer, and <u>Douglas Mawson</u>, Australian explorer and geologist, undertook an Antarctic expedition, reaching a point 97 miles from the *South Pole*, which was at that time a record;

1908-11: <u>Frederick Albert Cook</u> and <u>Robert Edwin Peary</u>, US explorers, ascended the highest mountain in North America, namely *Mount McKinley*, as reported in their work *To the Top of the Continent*, then reached the *North Pole*, and published *My Attainment of the Pole*;

1909-14: <u>Alistair Mackay</u>, Scottish doctor and polar explorer, <u>Douglas</u>

Mawson, Australian explorer and geologist, and Edgeworth David, Welsh Australian geologist and polar explorer, were first to reach the *Magnetic South Pole*; followed by an *Australasian expedition in Antarctic* led by Mawson for charting 3,220 kilometres of coast named after him *Mawson Coast*;

1914-45: Widespread *migrations caused by the First World War* and *the Second World War*, and also by *decolonization*;

1921-24: Knud Johan Victor Rasmussen, Danish explorer and ethnologist, in support of theory that Inuits and North American Indians are both descendants of migratory tribes from Asia, he did a cross-examination of Greenland and Bering Strait by dog sledge to visit Inuit groups along route, where *Knud Rasmussen Land* was called after him;

1926-39: Lincoln Ellsworth, US explorer, achieved flights over both the *North Pole* and the *South Pole*, and an Antarctic expedition, claimed thousands of square miles of territory for USA, now called *Ellsworth Land*;

1945-51: Jean Malaurie, French explorer and anthropogeographer, made many scientific expeditions to Arctic, several documentary films on Inuit people, and also reached the *Magnetic North Pole*;

1953-67: Edmund Percival Hillary, New Zealand mountaineer and explorer, achieved the first conquest of *Mt Everest* up to its summit, first overland trip to the *South Pole* using tracked vehicles, and the first ascent of *Mt Herschel* in Antarctica;

1955-89: Wally Walter William Herbert, British explorer, participated in Falkland Islands Dependencies Survey in *Antarctica*, followed by expeditions to *Lapland*, *Svalbard* and *Greenland*, also participated in the New Zealand Antarctic expedition, surveying large areas of *Queen Maud Range*; achieved the *first surface crossing of Arctic Ocean* from Alaska to Spitsbergen via *Pole* as the longest sustained sledge journey in history, and had several attempts to circumnavigate Greenland; as well as wrote a number of books, including *The Noose of Laurels*;

1961: Yuri Alekseyevich Gagarin, Soviet cosmonaut, was the first man to *travel in space*, and to complete a circuit of Earth in the *Vostok spaceship satellite*;

1968-69: Wally Herbert, British explorer, made the first surface crossing of the Arctic Ocean, a journey of 6115 kilometres from Alaska to Spitsbergen via the Pole, the first undisputed man to reach the *North Pole on foot*;

1969: <u>Neil Alden Armstrong</u>, <u>Buzz Aldrin</u>, and <u>Michael Collins</u>, US astronomers, set out in Apollo 11, a successful *Moon-landing expedition* and recorded the *first men steps on Moon*;

1983-98: <u>Richard Hobe</u>, US naval scientist, wrote and published *Marine Navigation Workbook: Piloting and Celestial and Electronic Navigation*;

1992: *The Prehistoric Exploration and Colonization of the Pacific*, was written by <u>Geoffrey Irwin</u> and published by Cambridge University;

2008: *The East Asian 'Mediterranean': Maritime Crossroads of Culture, Comerce and Human Migration* was edited by <u>Angela Schottenhammer</u> in Germany;

<u>*NASA current mission*</u>: *Advanced Composition Explorer* (Major mission of the Explorer programme).

According to specifications given in chapter 7, exploration emerged in Africa 105566 years ago, and will last up to 4350 years after the present, i.e. its final time will be $t_{\bullet}$ = 105566 + 4350 = 109916 years from the origin of exploration. In order to find out and delimit the sequences of exploration, there are selected only the better dated ones, which are presented with their transitional times as:

$$2200...1500BC \approx 1850BC \setminus Ancient\ conquests\ and\ colonisations$$
$$\setminus 300BC...AD140 \approx 80BC \setminus Major\ invasions$$
$$\setminus AD1160...1460 \approx AD1310 \setminus Great\ explorations;$$

corresponding to 2014 + 1850 ≈ 3864 years ago; 2014 + 80 ≈ 2094 years ago; 2014 - 1310 ≈ 704 years ago; or 105566 - 3864 ≈ 101702; 105566 - 2094 ≈ 103472; 105566 - 704 ≈ 104862 years from the origin of exploration respectively. The last transitional times divided by the final time $t_{\bullet}$ = 109916 years from the origin of exploration result in the approximate values 101702/109916 ≈ 0.925, 103472/109916 ≈ 0.941, 104862/109916 ≈ 0.954, which can be found in the same succession in Background to time, Table *(k)* for $z/8 = k$ sequences, as $t_{1.625}/t_{\bullet}$ = 0.9253462; $t_{1.75}/t_{\bullet}$ = 0.9413755; $t_{1.875}/t_{\bullet}$ = 0.9540453; whereby their accurate values, in years from the origin of exploration

$$t_{1.625} = t_{\bullet} \cdot tanh(1.625) \approx (109916) \cdot (0.9253462) \approx 101710;$$
$$t_{1.75} = t_{\bullet} \cdot tanh(1.75) \approx (109916) \cdot (0.9413755) \approx 103472;$$
$$t_{1.875} = t_{\bullet} \cdot tanh(1.875) \approx (109916) \cdot (0.9540453) \approx 104865,$$

corresponding to 105566 - 101710 = 3856, 105566 - 103472 = 2094, 105566 - 104865 = 701 years ago; or 1842BC, 80BC, and AD1312 respectively. Continuing the procedure, the exploration timeline, sequences, and intrinsic characteristics are reconstituted below.

$z/8 =$ k	Time (years)		Sequences of exploration
	from the origin $t_k = t_\bullet \cdot tanh(k)$	from present t_k - 105566	
	109916	+4350	
	...	...	
2	105962	+396 (AD2410)	–
1.875	104865	-701 (AD1313)	_Great explorations
1.75	103472	-2094 (80BC)	_Major invasions
1.625	101710	-3856 (1842BC)	_Ancient conquests + colonisations
1.5	99490	-6076 (4062BC)	_ Invasions of agricultural areas
1.375	96707	-8859 (6845BC)	_Pacific islands
1.25	93234	-12332	_Inland explorations
1.125	88955	-16611	_South America
1	83711	-21855	_Mesoamerica
0.875	77370	-28196	_North Eurasia
0.75	69813	-35753	_Far East and central Asia
0.625	60959	-44607	_East Asia and North America
0.5	50794	-54772	_Australia and Europe
0.375	39389	-66177	_Southeast Asia and Indonesia
0.25	26920	-78646	_South Asia
0.125	13668	-91898	_Middle East
0	0	-105566	_ North Africa and Near East

$z/8 =$ k	Time from origin $t_k = t_\bullet \cdot tanh(k)$ (years)	Period $\tau_k =$ $(t_\bullet^2 - t_k^2)/(2t_\bullet)$ (years)	Frequency $f_k =$ $(2t_\bullet)/(t_\bullet^2 - t_k^2)$ (years)$^{-1}$	Angular speed $\omega_k =$ $2\pi/\tau_k = 2\pi \cdot f_k$ (years)$^{-1}$
	109916	0	∞	∞
	...	...	...	...
2	105962	3882.9	0.0002575	0.0016182
1.875	104865	4934.9	0.0002026	0.0012732
1.75	103472	6255.1	0.0001599	0.0001018
1.625	101710	7899.7	0.0001266	0.0007954
1.5	99490	9931.5	0.0001007	0.0006327
1.375	96707	12415.3	0.0000805	0.0005061
1.25	93234	15416.1	0.0000649	0.0004076
1.125	88955	18962.4	0.0000527	0.0003314
1	83711	23081.2	0.0000433	0.0002722
0.875	77370	27727.6	0.0000361	0.0002266
0.75	69813	32787.2	0.0000305	0.0001916
0.625	60959	38054.2	0.0000263	0.0001651
0.5	50794	43221.6	0.0000231	0.0001454
0.375	39389	47900.4	0.0000209	0.0001312
0.25	26920	51661.5	0.0000194	0.0001216
0.125	13668	54108.2	0.0000185	0.0001161
0	0	54958.0	0.0000182	0.0001143

13. Inhabitation

The act of dwelling in or living permanently in a place, called *inhabitation* (from Latin *inhabito* 'to inhabit'), took place after the humans explored new areas, where they learned to recognize topographical features with their characteristics, to find hospitable places and to adapt in inhospitable ones, to look for sources of water and food, to locate places for hunting and fishing, or for sheltering and hiding, as well as for avoiding danger and feeling pleasure.

Early human settlements were dependent on the proximity to water and other natural resources, such as fertile land for naturally growing crops and grazing livestock, or prey for hunting. However, humans have a great capacity for altering their habitats by various methods, such as through cultivation of plants, irrigation or desiccation of lands, construction of houses, temples and sanctuaries, roads and bridges, transport and manufacturing of goods.

Exploration was usually followed by migration at various scales of expansion and duration from a region to another, as a result of growing population in relation to natural resources, enslavement, human trade, or worsening social and political conditions. By difference from the natural migrations of animals and plants, which take place in accordance with the general tendency of increasing habitable entropy

$$d\psi/d\zeta = k \cdot d(\ln \zeta)/d\zeta = k/\zeta \geq 0$$

to re-establish a local or areal equilibrium, where ψ, ζ, and k are the habitable entropy, habitable probability, and Boltzmann's constant respectively; human migration and inhabitation could produce great regional disequilibria.

Along with human inhabitancy, populations increased firmly in the attractive regions and slowly in the unattractive regions, and despite famines, diseases, smallpox, measles, influenza, and other plagues, the population increased in an about exponential manner. In first systematic studies, the increase of population was considered to follow the natural law of growth, according to which its rate of growth in time t could be proportional to its size N, i.e.

$$dN/dt = b \cdot N, \text{ or } dN/N = b \cdot dt,$$

where the coefficient b has been presumed to be constant. In this case, an integration from the time of first available information t_0 when

$N(t_o) = N_o$ to a certain time $t_i > t_o$ when $N(t_i) > N_o$ leads to the equation

$$ln(N/N_o) = b \cdot (t - t_o), \text{ or } N = N_o \cdot exp[b \cdot (t - t_o)].$$

Despite of disastrous events in human existence, the recorded data show that the growth of world population was faster than predicted by this equation, so that it must be reconsidered with a variable coefficient

$$b_i(t) = [ln(N_i/N_o)]/(t_i - t_o).$$

An attentive analysis of the available data indicates that indeed the coefficient $b_i(t)$ increased from 0.00034 to 0.00072 year $^{-1}$, becoming gradually more than double during the last 10000 years. Consequently, there are real perspectives of a supra-population for which the world is not entirely prepared.

Later development of inhabitation involved the introduction of *census* (meaning in Latin 'valuation of every Roman citizen's estate' and deriving from *censere* 'to determine, assess'). In modern understanding, a census is *the procedure of systematically acquiring and recording information about the members of a given population*, the term being used mostly in connection with national population and housing censuses.
Some data relating to inhabited areas and regions, demographical evolution, censuses, diseases, cities, as well as contributions to unveil, study, reconstitute, and simulate this evolution are altogether chronologically presented below.

90000-85000 years ago: *Stable inhabitation* in the Near East and the Middle East took place after human migration from Africa;

85000-55000 years ago: *Eastward inhabitation* of South and South-east Asia, and then of the Indonesian islands;

70000 years ago: *World population* was less than 15 thousand;

55000-40000 years ago: *Southward inhabitation* of Australia;

40000-30000 years ago: *North-westward inhabitation* of Europe, *Northward inhabitation* of East Asia to its North-east extremity, and from there *Eastward inhabitation* of Alaska;

16000-10000 years ago: *Southward inhabitation* of North America, Mesoamerica and South America;

11000-10000BC: *World population* was probably 1 million;

9000BC: *World population* rose up to around 3 million;

8000BC: At the dawn of agriculture, *World population* reached about 5 million; and the early city-like *Mureybet*, in Syria, had a population of around 500;

7500-3500BC: *Inhabitation of the earliest cultures' areas - Peru* (Norte Chico), *Mesoamerica* and *Balsas River*, in America; - *Yangtze River* (Pengtoushan) and *Yellow River* in China; - *Egypt's Western Desert* (Nabta Playa) on the way to the River Nile; - *Rivers Tigris* and *Euphrates* in South Mesopotamia; - *Indus Valley* in India; all of them with favourable environmental conditions for developing the world's first great civilizations;

7000BC: Increase of the *world population* to about 7 million; and *Çatalhöyük*, in Turkey, had around 1000 inhabitants;

6800BC: *Jericho*, near the Jordan River in Palestinian territories, became the earliest known walled city in the world, and had a population of 2000-3000;

6500-6000BC: *Lepinski Vir*, in Serbia, is the oldest known urban settlement in Europe, discovered on the right bank of the Danube during the construction of the Romanian-Serbian hydro-power plant at the Iron Gates, and relocated on a higher level to be protected after the dam was built; the town had a central square-like building and other 136 buildings disposed as a horseshoe around the central one;

6000BC: *World population* rose up to about 10 million;

5000BC: Development of the first civilizations was based on a *world population* of around 15 million; and the city *Byblos*, in Lebanon, was founded;

5000-4200BC: Transition from *Pre-agricultural* to *Agricultural inhabitation*;

4700-4200BC: *Solnitsata*, in Provadia, Bulgaria, Europe's oldest prehistoric town was unearthed, featuring two-storey houses defended by a wall, and inhabited by 350-500 people;

4200BC: Foundation of the city *Susa*, in Iran;

4000BC: *World population* reached around 20 million; *Faiyum*, in Lower Egypt, and *Sidon*, in Lebanon, were among the early cities in the world;

4000-3500BC: *Uruk*, in Iraq, was a Sumerian city-state with a population increasing from 5000 to 10000 inhabitants;

4000-3000BC: *Plovdiv*, in Bulgaria, one of the oldest cities in Europe was founded;

3800BC: *Dobrovody*, in Ukraine, was an old urban settlement with c.10000 citizens, covering a total area of about 2.5 square kilometres;

3700BC: *Eridu*, in Iraq, another Sumerian city-state had a population between 6000 and 10000;

3650BC: *Gaziantep*, in Turkey, was a city inhabited by Bronze Age people who first used chariots;

3500BC: *Mycenae*, in Peloponnese, Greece, was founded and subsequently developed as a city of Mycenaean power;

3200-3100BC: *Censuses in Egypt* during the early Pharaonic period; and *Luxor*, in Egypt, was a well populated city;

3100BC: *Memphis*, in Egypt, was a city inhabited by more than 30000 people;

3000BC: During the rise of ancient classical civilizations, the *world population* was around 25 million; and the early cities *Jerusalem*, in Israel, *Kirkuk*, in Iraq, and *Zurich*, in Switzerland, were founded;

2568BC: *Giza*, in Egypt, was founded and then became a central city of Egyptian civilization;

2500-2200BC: Transition from *Agricultural* to *Centralized inhabitation*;

2205BC: Foundation and rise of the city *Xi'an*, in China, preserving until today its famous Terracotta Army;

2030BC: *Ur*, in Iraq, was a city-state inhabited by 65000 people;

2000BC: *World population* rose up to 35 million; and the city of *Lisbon*, in Portugal, was founded;

1700-1600BC: *Censuses in the early Mesopotamian and Greek city states*;

1500BC-AD1300: *Inhabitation of islands* - Tonga, Fiji, Samoa, Hawaii, Madagascar, New Zeeland;

c.1300BC: After the exodus from Egypt, *early Israelite censuses* were mentioned in Bible (the books of Exodus and Numbers);

1000BC: Increase of the *world population* to around 50 million; *Thebes*, in Egypt, was a city with population of 50000-60000; and *Babylon*, in Iraq, was inhabited by about 100000 people;

612BC: *Babylon*, in Iraq, was the first city to have a population above 200000;

560-500BC: <u>Servius Tullius</u>, the 6th King of Rome, instituted the *Roman censuses*, usually carried out every five years, at which time the number of arm-bearing citizens was counted at around 80,000; during the early Roman Republic, the census was a list that kept track of *all adult males fit for military service*;

550-310BC: Transition from *Centralized inhabitation* to *Major cities*;

500BC: Rise of the *world population* up to about 100 million;

440-430BC: <u>Herodotus</u>, Greek historian, during his travels to collect data for his great narrative history, also got information related to *demography*;

430-429BC: *Bubonic plague* was recorded in Western Anatolia;

420-404BC: <u>Thucydides</u>, Greek historian, gave *demographical details* in his *De Bello Peloponnesiaco* 'History of the Peloponnesian Wars';

410-380BC: <u>Hippocrates</u>, Greek physician, wrote *Airs, Waters, Places*, containing shrewd observations about the geography of diseases and the role of environment in shaping the health of a community; *Epidemics III*, examining epidemics in a population and offering case histories of patients with acute diseases; and *The Sacred Disease*, elaborating a rigorous defence of naturalistic causes of diseases;

300BC: *Indian census* during the reign of Emperor <u>Chandragupta Maurya</u>;

100BC: *Alexandria*, in Egypt, and *Rome*, in Italy, had populations of about 400000 each;

AD1: *World population* increased to around 200 million; and the population of *Rome*, in Italy, was 1 million;

AD2: *Chinese census* during the Han Dynasty, resulting in 57.67 million people registered in 12.36 million households;

16: *Plague* (possibly malaria) in Southern central China;

160-180: *Antonine plague*, so-called after the Roman emperor <u>Antoninus Pius</u>, occurred in Anatolia;

162: *Plague* (possibly malaria) spread from Southern central to North-western China;

251-266: *Plague* (probably smallpox or measles) spread from North Africa to Italy, North-west Greece, and central Balkans;

312-322: *Plague* (possibly smallpox or measles) in Northern China;

500: The city *Constantinople* (Istanbul), in Turkey, had a population of 450000-500000;

542-543: *Bubonic plague of Justinian*, so-called after the Byzantine emperor Justinian I the Great, spread from Arabia to Egypt and the Byzantine Empire;

552: *Plague* (possibly smallpox or measles) in Korea and Western Japan;

c.640: Hazrat Umar, the 2nd Rashidun caliph, ordered the *first census in the caliphate*, which was followed by other regular censuses;

645-664: *Bubonic plague* spread from Byzantine Empire through Europe to England and Ireland;

700: *Chang'an*, in China, was a city with population of around 1 million;

775: *Baghdad*, in Iraq, had a population of about 1 million;

950-1100: Transition from *Major cities* to *Metropolitan areas*;

1000: *World population* was around 310 million;

1086: Completion of *Domesday Book*, an example of Norman efficiency, recording the great survey of much of England and Wales which was ordered by William I the Conqueror to determine the landholders and their taxations;

1120-50: William of Conches, Norman French scholastic philosopher, wrote *De philosophia mundi*, covering not only physics, astronomy and meteorology, but also geography and medicine, the last one dealing chiefly with procreation and childbirth;

1200-1396: *Population of China* decreased approximately from 123 to 65 million because of famine and plagues;

1338-50: *Black Death*, one of the most disastrous cases of Bubonic plague, spread from Asia through Asia Minor and North Mediterranean to Crimea and Europe;

1340-50: *World population* decreased approximately from 450 to 370 million because of famine and plagues, including the *Great Famine* and *Black Death* in Europe;

1500: Total *population of Americas* was 50-100 million;

1644: *China* was inhabited by about 150 million people;

1662: John Graunt, English haberdasher, wrote *Natural and Political Observations Made upon the Bills of Mortality*, using the analysis of mortality rolls in early modern London, attempting to create a system to warn of the onset and spread of bubonic plague in the city, and statistically estimating the population of London;

1665: *Black Death* extended in London and much of England;

1693: <u>Edmond Halley</u>, English astronomer and mathematician, published a *study on trade winds and monsoons*, and *Breslau Table of Mortality*, founding life insurance mathematics and annuities;

1700-1900: *Europe's population* increased from about 100 to 400 million;

1749: *First modern census* accomplished in Sweden, as a continuing complete count taken accurately at regular intervals;

1750: *World population* was of 791 million; and *Indian subcontinent's population* of 125 million;

1771-72: <u>Richard Price</u>, Welsh moral philosopher, published *Observations on Reversionary Payments*, helping to establish a scientific system for life insurance and pensions, and *An Appeal to the Public on the subject of the National Debt*;

1798: <u>Thomas Robert Malthus</u>, English economist and clergyman, published *Essay on the Principle of Population*, pointing out that, if unchecked, population would be subject to exponential growth, and this increasing tendency could overpass the growth in food production and in means of subsistence;

1800: *World population* was 978 million; and the British Parliament passed the *Census Act*, followed by the first official census in England and Wales;

1838-60: <u>Augustus De Morgan</u>, British mathematician and logician, was one of the founders of the London Mathematical Society, formulated *De Morgan's laws*, introducing mathematical induction, and wrote *On the Application of Probabilities to Life Contingencies*;

1850: Rise of the *world population* up to 1262 million, of *China's population* up to 430 million people, and of *Indian subcontinent's population* also up to 430 million people;

1872-1911: <u>Louis-Adolphe Bertillon</u>, French physician, statistician, anthropologist and inventor of anthropometry, published works such as *Les Mouvements de la population dans les divers États d'Europe et notamment en France*; *Démographie figurée de la France*; as well as *Alcoholism and Ways of Combating It judged from Experience*, and *The Depopulation of France* (both translated from French in English);

1877-91: Joseph Körösi, Austrian-Hungarian statistician, published his works entitled *Statistique internationale de grandes villes*; *Projet d'un Recensement du Mond*; and *Demologische Beiträge*, by which he developed the statistical approach of demography;

1880-1920: Further *development of demography* and use of the *methods and techniques of demographic analysis* due to Anders Nicolas Kaier, Richard Böckh, Émile Durkheim, Wilhelm Lexis and Luigi Bodio;

1892-96: *Black Death* spread from South China to West India;

1900: *World population* increased to 1650 million; and London, in the UK, had a population of 6.48 million;

1945: After the Second World War, *Russian population* was about 90 million;

1950: *World population* was 2519 million; and *New York*, in the USA, had a population of 12.46 million;

1953: *China's population* was of 580 million; and *Indian subcontinent's population* also of 580 million;

1955: Rise of the *world population* up to 2756 million;

1955-61: Louis Henry, French historian, founded *historical demography* and *one-place study fields*, and wrote *Some data on natural fertility*, Eugenics Quarterly, 8;

1956: Louis Henry and Michel Fleury, French historians and demographers, published their work entitled *Des registres paroissiaux à l'histoire de la population. Manuel de dépouillement et l'exploitation de l'etat civil ancien*;

1958: Ansley Johnson Coale and Edgar Hoover, US demographer analysts, published their work *Population Growth and Economic Development in Low-Income Countries*, Princeton University Press;

1960: *World population* reached 2982 million;

1965: Increase of the *world population* to 3335 million; and David Glass, English sociologist, and David Eversley, German-born British social researcher, published their book *Population in History: Essays in Historical Demography*, London: Edward E. Arnold;

1966-67: Ansley Johnson Coale, US demographer, and Paul Demeny, US editor on demography, published works including *Regional Model Life Tables and Stable Populations*; and *Methods of Estimating Basic*

Demographic Measures From Incomplete Data; New York: Academic Press;

1967-72: <u>Ansley Johnson Coale</u>, US demographer, was author of demographical works *Factors associated with the development of low fertility: An historic summary*, Proceedings of the World Population Conference, Belgrade, Vol. 2, New York: United Nations; and *The Growth and Structure of Human Populations: A Mathematical Investigation*, Princeton University Press;

1970: *World population* was 3692 million;

1975: Rise of the *world population* up to 4068 million; and *Tokyo*, in Japan, had a population of 23 million;

1980: *World population* was 4435 million;

1982: <u>Dennis Willigan</u> and <u>Katherine Lynch</u>, US demographers, produced the book *Sources and Methods of Historical Demography*, New York: Academic Press;

1985: Increase of the *world population* to 4831 million;

1990: *World population* was 5263 million;

1991: By disintegration of the USSR, the *population of Russia* remained about 148 million; <u>Ansley Johnson Coale</u>, US demographer and expert on Population Trends, wrote *Excess Female Mortality and the Balance of the Sexes in the Population: An Estimate of Number of 'Missing Females'*, Population and Development Review, 17(3); and <u>Samuel Hulse Preston</u>, US demographer and sociologist, wrote together with <u>Patrick Heuveline</u> and <u>Michel Guillot</u>, demographic researchers, the book *Demography: Measuring and Modeling Population Processes*, New York: Blackwell; and also published with <u>Michael Haines</u>, US researcher, another book entitled *Fatal Years: Child Mortality in Late Nineteenth Century America*, Princeton University Press;

1995: Rise of *world population* up to 5674 million;

1996: <u>George Armelagos</u>, <u>Kathleen Barnes</u> and <u>James Lin</u>, US anthropologists, published their work *Disease in Human Evolution: The Re-Emergence of Infectious Disease in the Third Epidemiological Transition*, National Museum of Natural History Bulletin for Teachers, 18(3);

2000: *World population* increased to 6070 million; and <u>Samuel Preston</u>, <u>Heuveline Patrick</u> and <u>Michel Guillot</u>, US demographers, were authors of *Demography: Measuring and Modelling Population Processes*, Blackwell Publishing;

2002: <u>Dennis Stanford</u> and <u>Bruce Bradley</u>, US demographical researchers, suggested that the American *Clovis people* could have inherited technology from the *Solutrean people* who lived in southern Europe 21000-15000 years ago, and wrote *Ocean Trails and Prairie Path? Thoughts about Clovis Origins*, Memoirs of the California Academy of Sciences, 27;

2004: <u>Robert William Fogel</u>, US economic historian and scientist, published his work *The Escape from Hunger and Premature Death, 1700-2100: Europe, America, and the Third World* (Cambridge Studies in Population, Economy and Society in Past Time), Cambridge University Press;

2005: Rise of the world population up to 6454 million;

2010: *World population* was 6972 million, and the UN and US Census Bureau estimated that world population will increase until 2020 to 7657, until 2030 to 8321, until 2040 to 8874, and until 2050 to 9306 million;

2011: <u>Sven Kunisch</u>, <u>Stephan Boehm</u> and <u>Michael Boppel</u>, Swiss researchers, edited *From Grey to Silver: Managing the Demographic Change Successfully*, Springer-Verlag, Berlin-Heidelberg; and <u>Josef Ehmer</u>, <u>Jens Ehrhardt</u> and <u>Martin Kohli</u>, German researchers, edited *Fertility in the History of the 20th Century: Trends, Theories, Policies, Discourses*, Historical Social Research, 36(2);

2012: Top ten *largest cities*, with population in million -
Tokyo (Japan), 37.22; *Delhi* (India), 22.65;
Mexico City (Mexico), 20.45; *New York-Newark* (the USA), 20.35;
Shanghai (China), 20.21; *São Paulo* (Brazil), 19.92;
Mumbai (India), 19.74; *Beijing* (China), 15.59;
Dhaka (Bangladesh), 15.39; *Kolkata* (India), 14.40;

2013-14: *World population* rose up to 7178 million; the *population by continents* was given as -
Asia 4283,
Africa 1100,
Europe 738,
North America 557,
South America 399,
Oceania 38 million, *Antarctica* 5 thousand;
and the twenty *most populated countries* were
China 1356, *India* 1236, *US* 319,

Indonesia 254, *Brazil* 203, *Pakistan* 196,
Nigeria 177, *Bangladesh* 166, *Russia* 142,
Japan 127, *Mexico* 120, *Philippines* 108,
Ethiopia 97, *Vietnam* 93, *Egypt* 87,
Germany 81, *Iran* 81, *Turkey* 82,
Congo-Kinshasa 77, and *Thailand* 68 million.

In accordance with the table presented at the end of chapter 7, inhabitation outside Africa emerged 86668 years ago, as its initial time, and will last until the eve of human fate 4350 years after the present, so that its final time is estimated as $t_\bullet$ = 86668 + 4350 = 91018 years from the origin of inhabitation.

In the development of inhabitation and demography, the later sequences, as more undoubtedly established, are delimited by times of transition, as follows:

Pre-agricultural inhabitation \ 5000...4200BC ≈ 4600BC \
Agricultural inhabitation \ 2500...2000BC ≈ 2250BC \
Centralized inhabitation \ 550...310 ≈ 430BC \
Major cities \ AD950...1100 ≈ AD1025 / *Metropolitan areas*;

or 6614; 4264; 2444; 989 years ago; which represent approximately 86668 - 6614 = 80054; 86668 - 4264 = 82404; 86668 - 2444 = 84224; 86668 - 989 = 85679 years from the origin of inhabitation respectively. The last times divided by the final time of inhabitation $t_\bullet$ = 91018 years from the same origin result in the ratios 80054/91018 ≈ 0.880; 82404/91018 ≈ 0.905; 84224/91018 ≈ 0.925; 85679/91018 ≈ 0.941. As these values of ratios are similar to those given in Background to time, Table *(k)* for $z/8 = k$ sequences, where they are given as

$$t_{1.375}/t_\bullet = tanh(1.375) = 0.8798267;$$
$$t_{1.5}/t_\bullet = tanh(1.5) = 0.9051483;$$
$$t_{1.625}/t_\bullet = tanh(1.625) = 0.9253462;$$
$$t_{1.75}/t_\bullet = tanh(1.75) = 0.9413755;$$

it follows that the transitional times from one sequence to another can be calculated, in years from the origin of inhabitation, as:

$$t_{1.375} = t_\bullet \cdot tanh(1.375) ≈ 80080;$$
$$t_{1.5} = t_\bullet \cdot tanh(1.5) ≈ 82385;$$
$$t_{1.625} = t_\bullet \cdot tanh(1.625) ≈ 84223;$$
$$t_{1.75} = t_\bullet \cdot tanh(1.75) ≈ 85682.$$

Continuing the calculation of transitional times backward, the timeline, sequences, and intrinsic characteristics in course of inhabitation and its associated demography are established and presented in the two tables below.

$z/8 = k$	Time (years)		Sequences of inhabitation
	from the origin $t_k = t_\bullet \cdot tanh(k)$	from present $t_k - 86668$	
	91018	+4350	
...	...	...	
1.875	86835	+167 (AD2181)	_
1.75	85682	-986 (AD1028)	_Metropolitan areas
1.625	84223	-2445 (431BC)	_Major cities
1.5	82385	-4283 (2269BC)	_Centralized inhabitation
1.375	80080	-6588 (4574BC)	_Agricultural inhabitation
1.25	77209	-9459 (7445BC)	_Pre-agricultural inhabitation
1.125	73661	-13007	_World population 1 to 6 million
1	69319	-17349	_Inhabit. of food-rich areas
0.875	64068	-22600	_Inhabit. in North America
0.75	57810	-28858	_Inhabit. of North Asia
0.625	50479	-36189	_Inhabit. of Far East, Central Asia
0.5	42061	-44607	_Inhabitation of Europe
0.375	32617	-54051	_Inhabitation of Indonesia, Australia
0.25	22292	-64376	_Inhabit. of South + Southeast Asia
0.125	11318	-75350	_World population < 15000
0	0	-86668	_Rare population and inhabitation

$z/8 = k$	Time from origin $t_k = t_\bullet \cdot tanh(k)$ (years)	Period $\tau_k = (t_\bullet^2 - t_k^2)/(2t_\bullet)$ (years)	Frequency $f_k = (2t_\bullet)/(t_\bullet^2 - t_k^2)$ (years)$^{-1}$	Angular speed $\omega_k = 2\pi/\tau_k = 2\pi \cdot f_k$ (years)$^{-1}$
	91018	0	∞	∞
...	...	...	...	...
1.875	86835	4086.9	0.0002447	0.0015374
1.75	85682	5179.6	0.0001931	0.0012131
1.625	84223	6541.4	0.0001529	0.0009605
1.5	82385	8223.6	0.0001216	0.0007640
1.375	80080	10280.8	0.0000973	0.0006112
1.25	77209	12761.5	0.0000784	0.0004924
1.125	73661	15702.0	0.0000637	0.0004002
1	69319	19112.4	0.0000523	0.0003287
0.875	64068	22960.1	0.0000436	0.0002737
0.75	57810	27150.0	0.0000368	0.0002314
0.625	50479	31511.1	0.0000317	0.0001994
0.5	42061	35790.4	0.0000279	0.0001756
0.375	32617	39664.7	0.0000252	0.0001584
0.25	22292	42779.1	0.0000234	0.0001469
0.125	11318	44805.3	0.0000223	0.0001402
0	0	45509.0	0.0000220	0.0001381

14. Knowledge

The facts, information, and skills acquired through experience or/and education, or the theoretical or/and practical understanding of a subject define *knowledge* (from Old English *cnāwan*, Latin *(g)nosco, (g)noscere* 'to know', Greek *γιγνωσκω* 'to understand'). In philosophy, the study of knowledge is called *epistemology* (Greek *ἐπιστήμη* 'knowledge, understanding' and *λόγος* 'study of') as a philosophical branch that investigates the origin, nature, methods, and limits of human knowledge, and also refers to the theory of knowledge, as well as questioning what knowledge is and how it can be acquired, and the extent to which any given subject or entity can be known. In 1750s, the German philosopher AG Baumgarten introduced the term *gnosiology* (from Greek *γνῶσις* 'knowledge' and *λόγος* in sense of 'word, discourse'), meaning the study of knowledge, that is currently used mainly in regard to Eastern Christianity. Knowledge is also considered the basis of general culture, such as the French statesman ÉM Herriot expressed as *La culture, c'est ce qui reste quand on a tout oublié* 'Culture is what is left when we have forgotten everything else' (c.1950).

In course of human evolution, knowledge acquisition involved complex cognitive processes, such as perception, communication, association, and reasoning, deriving from the human capacity of acknowledgment.

General knowledge K results from the summation Σ of all kinds of information i, and can be expressed in the simplified mathematical form

$$K = \Sigma_i K_i,$$

where each K_i represents the product of its quality Q^l_i multiplied by the exponential function of argument m^i_l and its quantity Q^n_i multiplied by the exponential function of argument m^i_n, i.e.

$$K_i = [(Q^l_i \cdot exp \, m^i_l) \cdot (Q^n_i \cdot exp \, m^i_n)]_i.$$

Interpreting the ratio of qualitative to quantitative information $(Q^l_i \cdot exp \, m^i_l)/(Q^n_i \cdot exp \, m^i_n) = \rho_i$ as a 'knowledge efficiency', the relationship between Q^l_i and Q^n_i can be rewritten as

$$(Q^l_i \cdot exp \, m^i_l) = \rho_i \cdot (Q^n_i \cdot exp \, m^i_n).$$

Therefore, knowledge of kind i is expressed in the form

$$K_i = \rho_i \cdot (Q^n_i \cdot exp \, m^i_n)^2,$$

and general knowledge can be evaluated by the expression

$$K = \Sigma_i\,K_i = \Sigma_i\,\rho_i\cdot(Q^n{}_i\cdot exp\;m^i{}_n)^2.$$

Developed on cerebrum (white matter), the wrinkled cerebral cortex (grey matter) covers a quite large surface storing various kinds of knowledge organized into topological areas $j = 1, 2,...$ of mental kinds of energy E_j, each of them resulting from its intrinsic intensity I_j multiplied by its extension $A^j{}_j$ within the cerebral cortex. By this simplified interpretation, the energy afferent to each area is

$$E_j = A^j{}_j\cdot I_j,$$

while its variation

$$dE_j = I_j\cdot dA^j{}_j + A^j{}_j\cdot dI_j$$

represents the changes in distribution of mental energy afferent to any topical area, where E_j, $A^j{}_j$, and I_j are expressed in joules, square metres, and joules per square metre respectively. The development of these topological areas depends on both general and particular understanding, preference, perception, and judgement, in rapport with cultural standards of a given community.

Knowledge is based on experience and good judgment, altogether constituting *wisdom* (Old English *wīs* coming from *wit* 'ingenuity, intelligence, sense') embodied by ancient Greeks as the goddess *Athena*. Using the notations experience x, knowledge y, and good judgment z, wisdom w can be expressed, for example, as the geometric mean

$$w = (x^a\cdot y^b\cdot z^c)^{1/3}.$$

Synonymous with sagacity, discernment, or insight, wisdom is usually defined as: deep understanding and realization of people, things, events, or situations, resulting in the ability to choose, act, or inspire to consistently produce optimum outcome with minimum time, energy, or thought; ability for optimal application of perceptions and knowledge to obtain the appropriate resolutions; as well as understanding what is true or right for optimum judgement and action. According to recent research in positive psychology, wisdom is defined as the coordination of 'knowledge and experience', and 'its deliberate use to improve well being', on the basis of the criteria: a wise person has self-knowledge; a wise person seems sincere and direct with others; others ask wise people for advice; and a wise person's actions are consistent with his/her ethical beliefs.

Knowledge represents not only an increased amount of information in certain fields of interest or activity, but also a wide comprehension and interconnection of various kinds of information which lead to an improved view of the world as a whole. Thus, among various kinds of knowledge can be identified, sometimes unexpectedly, fundamental links relating laws, principles, or concepts of thermodynamics, ecology, geology, biology, theory of mind, morality, sociology, and legislation, which can by roughly described as follows:

Zeroth - 'Could be systems in equilibrium with each other' (thermodynamics), 'Same causes produce same effects in past as in present (geology), 'Common origin of all forms of life' (biology), '*Ex nihilo nihil*: Out of nothing - nothing comes' (theory of mind), 'Could be societies with same state of equality' (sociology);

First - 'Conservation of internal energy in isolated systems' (thermodynamics), 'Everything is connected to everything' (ecology), 'Earth's energy upheld by conversion of its different forms' (geology), 'Coexistence of organisms in their variety' (biology), 'Relative stability of cerebrocortical area' (theory of mind), 'Conservation of good and bad behaviours as a whole' (morality), 'Preservation of equality in separate societies' (sociology), 'Justice and injustice are parts of same system' (legislation);

Second - 'Tendency of increasing entropy' (thermodynamics), 'Everything must go somewhere' (ecology), 'Rock cycling by magmatic, sedimentary, metamorphic processes' (geology), 'Life evolution by natural selection' (biology), 'Creativity results from mental energy per unit area of cerebral cortex' (theory of mind), 'General tendency of diminishing bad behaviour' (morality), 'In the best society there is still place for better' (sociology), 'Justice develops by reducing injustice' (legislation);

Third - 'Inaccessibility of absolute zero temperature' (thermodynamics), 'Nature knows best' (ecology), 'Untouchable static state of the Earth' (geology), 'Unsustainable over-population' (biology), 'Convertibility of mental energy by death' (theory of mind), 'Bad behaviour never entirely disappears' (morality), 'Complete equality can never be reached' (sociology), 'Perfect justice could never exist' (legislation).

The meaning or significance and stages in development of knowledge and wisdom followed a long and interesting historical course marked by steps in human evolution and by valuable contributions from great religions, mythologies, philosophies, and thoughts, as well as from thinkers, wise persons, scientists, encyclopaedists, etc., from which some are mentioned below.

80000-60000 years ago: *Ancestral knowledge* of humans who had come from Africa and experienced life in Asia;

18000-17000 years ago: *Shamanic knowledge* was characteristic for people mainly feeding from hunting;

15000-11000BC: *Prehistoric knowledge* emerged at groups of people domesticating animals and cultivating plants;

5000-4900BC: *Basic ancient knowledge* which preceded the first civilizations in the world;

3500-2900BC: *Sumerian* and *Egyptian knowledge* were associated with *Enki* 'God of wisdom and intelligence', and *Saa* 'God of wisdom' respectively;

3000-2200BC: Transition from *Sumerian* and *Egyptian knowledge* to *Initial ancient knowledge*;

2000-1300BC: Abraham and Moses, Hebrew prophets of the Old Testament; based on their legacy, the word wisdom was mentioned 222 times in the Old and New Testaments of the Bible;

1370BC: Akhenaten, Egyptian king of the 18th dynasty, said '*To be satisfied with a little, is the greatest wisdom; and he that increaseth his riches, increaseth his cares; but a contented mind is a hidden treasure, and trouble findeth it not*';

1000-400BC: Transition from *Initial ancient knowledge* to *Classic ancient knowledge*;

930BC: Solomon, King of Israel, asked God for *wisdom* (*2 Chronicles* 1), finding wisdom with 'fear of the Lord' (*Proverbs* 1:7 and 9:10), such as '*Wisdom* calls aloud in the street, she rises her voice in the public squares' (*Proverbs* 1:20), and urged readers to obtain and to increase in *wisdom* (*Proverbs* and *Psalms*);

800-600BC: *Ascendancy of knowledge in the classic ancient world*;

550BC: *Taoism* (*Daoism*), China, considered *wisdom* as adherence to the *Three Treasures: charity, simplicity*, and *humility*;

530-480BC: *Buddhism*, Nepal and India, considered that a *wise person* is endowed with *good bodily conduct, good verbal conduct*, and *good mental conduct*;

485-480BC: Confucius, Chinese philosopher, thought 'By three methods we may learn *wisdom*: first, by *reflection*, which is the noblest; second, by *imitation*, which is the easiest; and third, by *experience*, which is the bitterest; *Wisdom*, compassion, and courage are the three universally recognized moral qualities of men'; and 'Love of learning is akin to *wisdom*. To practice with vigour is akin to humanity. To know to be shameful is akin to courage';

468-458BC: <u>Aeschylus</u>, Greek tragic dramatist, said *'Memory is the mother of all wisdom'*;

420-410BC: <u>Euripides</u>, Greek tragic dramatist, pointed that *'Cleverness is not wisdom'*;

410BC: <u>Sophocles</u>, Athenian tragedian, thought *'Wisdom* is the supreme part of happiness', and *'Wisdom* outweighs any wealth';

400-370BC: *<u>Socratic method</u>* was to ask for definitions of familiar concepts such as justice, courage and piety, to elicit contradictions in the responses of interlocutors, and thus to demonstrate their ignorance , which was claimed to share; <u>Socrates</u> said that *'Wisdom* begins in wonder', and 'The only *true wisdom* is in *knowing* you know nothing'; and *<u>Platonism</u>* taught that 'Most excellent man, are you who are a citizen of Athens, the greatest of cities and the most famous for *wisdom* and power, not ashamed to care for the acquisition of wealth and for reputation and honour, when you neither care nor take thought for *wisdom* and truth and the perfection of your soul?'; <u>Plato</u> wrote *Theaetetus*, in which he defined knowledge as 'justified true belief', asserting '*knowledge* as nothing but perception', '*knowledge* as *true judgment*', '*knowledge* as a *true judgment with an account*', and specifying that a statement must meet three criteria in order to be considered *knowledge*, namely it must be *justified*, *true*, and *believed*; he produced *Phaedo* as a basis of knowledge theory, in terms of 'But if we are guided by me we shall believe that the soul is immortal and capable of enduring all extremes of good and evil, and so we shall hold ever to the upward way and pursue *righteousness with wisdom* always and ever, that we may be dear to ourselves and to the gods both during our sojourn here and when we receive our reward';

335-330BC: <u>Diogenes of Sinope</u>, Greek philosopher and moralist, stated *'Wise kings* generally have *wise counsellors*; and he must be a *wise man* himself who is capable of distinguishing one';

330-322BC: <u>Aristotle</u>, Greek philosopher, metaphysicist and biologist, achieved encyclopaedic knowledge including *logic, ethics, politics, rhetoric, physics, zoology* and *classification of life forms*, and founded *Natural philosophy*, defining *wisdom* as the *understanding of causes*, i.e. knowing why things are a certain way, which is deeper than merely knowing that things are a certain way;

86-50BC: <u>Posidonius the Athlete</u>, Greek philosopher and scientist, became known by his *encyclopaedic work* consisting of poly-mathematics, geography, astronomy, and meteorology;

60BC: <u>Titus Lucretius</u>, Roman poet, wrote 'So vital strength of his spirit won through, and he made his way far outside the flaming walls of the world and ranged over the measureless whole, both in *mind* and *spirit*';

50BC: <u>Marcus Tullius Cicero</u>, Roman orator and statesman, set 'Nobody can give you *wiser advice* than yourself';

35-20BC: <u>Quitus Horace</u>, Roman poet and satirist, wrote '*Wisdom* is not wisdom when it is derived from books alone';

AD50: <u>Epictetus</u>, Greek philosopher and moralist, thought 'It is the nature of the *wise* to resist pleasures, but the foolish to be a slave to them';

50-60: <u>Lucius Annaeus Seneca</u>, Roman philosopher and statesman, declared '*No man was ever wise* by chance';

60: <u>St Paul the Apostle</u>, Christian missionary, preached 'For to one is given by the Spirit the word of *wisdom*; to another the word of *knowledge* by the same Spirit' (1st Epistle to the Corinthians);

77: <u>Pliny the Elder</u>, Roman scholar, wrote 160 volumes of manuscript, used for his universal 37-volume encyclopaedia *Historia Naturalis*, covering natural history, architecture, medicine, geography, geology, and all aspects of the world around;

100-128: <u>Decimus Junius Juvenal</u>, Roman lawyer and satirist, became known for his saying 'Never does nature say one thing and *wisdom* another';

597-600: <u>St Augustine of Canterbury</u>, Italian prelate and first English Archbishop, thought 'Patience is the companion of *wisdom*';

600-1000: Transition from *Classic ancient knowledge* to *Intermediate knowledge*;

618-633: <u>St Isidore of Seville</u>, Spanish prelate and Doctor of Church, the last of Latin Fathers of Church, wrote a 20-volume encyclopaedia of knowledge, entitled *Etymologiae*, a standard work for scholars throughout the Middle Ages;

622-632: <u>Muhammad</u>, Prophet, founder of *Islam*, produced Qur'an (Koran), stating 'He gives wisdom to whose He wills, and whoever has been *given wisdom has certainly been given much good*. And none will remember except those of understanding';

830: <u>Rabanus Maurus Magnentius</u>, Carolingian philosopher, wrote the popular encyclopaedia *De universo or De rerum naturis*, a work of avant-garde for following encyclopaedists;

900-1100: *Norse mythology*, in which the god *Odin* was especially known for his *wisdom*, often acquired through various hardships and ordeals involving pain and self-sacrifice;

960: <u>Johannes Thurmayr Aventinus</u>, German humanist scholar and historian, produced the remarkable *Encyclopedia orbisque doctrinarum, hoc est omnium atrium, scientiarum, ipsius philosophiae index ac divisio*;

1000-1200: *Inuit tradition* contains the idea according to which the aim of teaching is to develop *wisdom*, so that a person becomes *wise*, when he can see what needs to be done and to do it successfully without being told what to do;

1260: <u>St Thomas Aquinas</u>, Italian scholastic philosopher and theologian, regarded *wisdom* to be the *father*, i.e. the 'cause, measure, and form', of all virtues;

1520-1600: *Catholicism, Lutheranism* and *Puritanism* included preaching that *wisdom* is one of the seven gifts of the Holy Spirit;

1597-1625: <u>Francis Bacon</u>, English philosopher and statesman, wrote *Meditationes Sacrae*, including the famous aphorism *Nam et ipsa scientia potestas est* 'For also knowledge itself is power'; and also 'If you dissemble sometimes your *knowledge* of that you are thought to know, you shell be thought, another time, to know that you know not', 'The *knowledge* of man is as the waters, some descending from above, and some springing from beneath; the one informed by the light of nature, the other inspired by divine revelation', 'A prudent question is one-half of *wisdom*', and '*Wise men* make more opportunities than they find';

1599: <u>Samuel Daniel</u>, English poet, wrote 'Soul of the world, *knowledge*, without thee / What hath the earth that truly glorious is? / Why should our pride make such a stir to be, / To be forgot?';

1610: <u>John Donne</u>, English poet, thought 'Oh wrangling schools, that search what fire / Shall burn this world, had none the *wit* / Unto this *knowledge* to aspire, / That this her fever might be it?';

1610-11: <u>William Shakespeare</u>, English dramatist and poet, expressed his ideas such as 'A spider steeped, and one may drink, depart, / And yet partake no venom, for his *knowledge* / Is not infected; but if one present / Th'abhorred ingredient to his eye, make known / How he hath drunk, he cracks his gorge, his sides, / With violent hefts. I have drunk, and seen the spider';

1625-28: <u>Fulke Greville</u>, English poet and courtier, versified 'The mind of man is this world's true dimension, / And *knowledge* is the measure of the mind; / And as the mind in her vast comprehension / Contains more worlds than all the world can find, / So *knowledge* doth itself far more extend / Than all the minds of men can comprehend';

1644: <u>John Milton</u>, English poet, said that 'Where there is much desire to learn, there of necessity will be much arguing, much writing, many opinions; for opinion in good men is but *knowledge* in the making';

1659-68: <u>Jean Baptiste Poquelin Molière</u>, French playwright, thought that 'A *wise* man is superior to any insult which can be put upon him, and the best reply to unseemly behaviour is patience and moderation';

1665: <u>Comte de Bussy-Rabutin</u>, French soldier and poet, wrote that *L'amour vient de l'aveuglement, / L'amitié de la connaissance* 'Love comes from blindness, / Friendship from *knowledge*';

1687-1714: <u>Gottfried Wilhelm Leibniz</u>, German philosopher and mathematician, stated that *C'est Dieu qui est la dernière raison des choses, et la connaissance de Dieu n'est pas moins la principe des sciences, que son essence et sa volonté sont les principes des êtres* 'It is God who is the ultimate reason of things, and the *knowledge* of God is no less the beginning of science than his essence and will are the beginning of beings'; and *Mais la connaissance des vérités nécessaries et éternelles est ce qui nous distingue des simples animaux et nous fait avoir la Raison et les sciences, en nous élevant à la connaissance de nous-mêmes et de Dieu. Et c'est ce qu'on appelle en nous Âme Raisonnable, ou Esprit* 'It is the *knowledge* of necessary and eternal truths which distinguishes us from mere animals, and gives us Reason and the sciences, raising us to *knowledge* of ourselves and of God. It is this in us which we call the rational soul or Mind';

1697: <u>Daniel Defoe</u>, English writer and adventurer, wrote 'Why then should women be denied the benefits of instruction? If *knowledge* and *understanding* had been useless additions to the sex, God almighty would never have given them capacities';

1728: <u>Ephraim Chambers</u>, English encyclopaedist, published 2 folio volumes of *Cyclopaedia, or Universal Dictionary of Arts and Sciences*, inspiring D Diderot's great French Encyclopédie;

1739-58: <u>Alexander Gottlieb Baumgarten</u>, German philosopher, published two main books, namely *Metaphysica* where he defined taste, in its wider meaning, as the ability to judge according to the senses, instead of according to the intellect; and two-volume *Aesthetica* in which he introduced the term *gnosiology*;

1746: <u>Philip Dormer Stanhope Chesterfield</u>, English statesman and man of letters, thought that 'The *knowledge* of the world is only to be acquired in the world, and not in a closet';

1746-54: <u>Étienne Bonnot de Condillac</u>, French philosopher and psychologist, published *Essai sur l'origine des connaissances humaines* 'Essay on the Origin of *Human Knowledge*', and *Traité des sensations* 'Treatise on Sensations';

1751-55: <u>Charles-Louis de Secondat Montesquieu</u>, French philosopher and jurist, wrote an essay on taste (Goût) included in the French *Encyclopédie*, revealing the author's wide knowledge;

1751-76: <u>Denis Diderot</u>, French writer, and <u>Jean le Rond d'Alembert</u>, French philosopher and mathematician, produced the major work of Enlightenment in 35 volumes, entitled *Encyclopédie, ou Dictionnaire raisonné des sciences, des arts et des métiers* 'Encyclopaedia, or Critical Dictionary of Sciences, Arts and Trades';

1755: <u>Benjamin Franklin</u>, US statesman and diplomat, wrote that 'The doors of *wisdom* are never shut';

1755-70: <u>Samuel Johnson</u> (Dr Johnson), English writer, critic and lexicographer, published his great *Dictionary*, which contains a series of memorable definitions and quotations, disdaining an offer of patronage from Lord Chesterfield, and thinking that 'So it is in travelling; a man must carry *knowledge* with him, if he would bring home *knowledge*';

1780-90: <u>Johann Wolfgang von Goethe</u>, German poet and dramatist, wrote that 'Ignorant men raise questions that *wise men* answered a thousand years ago', and '*Wisdom* is found only in truth';

1790-93: <u>William Blake</u>, English poet, said that 'I was in a printing house in Hell, and saw the method in which *knowledge* is transmitted from generation to generation';

1805-10: <u>Napoleon I Bonaparte</u>, Emperor of France, considered that 'The *truest wisdom* is a *resolute determination*';

1805-11: <u>Friedrich Arnold Brockhaus</u>, German publisher, founded *Brockhaus*, initially called 'Konversations-Lexikon', which became a prestigious German encyclopaedia;

1817: <u>George Gordon Noel Byron</u>, English poet, wrote that 'Sorrow is *knowledge*: they who know the most / Must mourn the deepest o'er the fatal truth, / The Tree of *Knowledge* is not that of Life';

1817-54: <u>Adam Black</u>, Scottish publisher, was owner of the

Encyclopaedia Britannica, after A Constable's failure, and had rights to publish W Scott's novels;

1818: <u>John Keats</u>, English poet, versified that 'O fret not after *knowledge* – I have none, / And yet my song comes native with the warmth. / O fret not after *knowledge* – I have none, / And yet the Evening listens';

1819-41: <u>Arthur Schopenhauer</u>, German philosopher, produced the major work *The World as Will and Idea*, including reflections on *theory of knowledge* and implications for philosophy of nature, aesthetics and ethics;

1820-24: <u>Thomas Jefferson</u>, 3rd President of USA, pointed that 'Honesty is the first chapter in the book of *wisdom*';

1828: <u>Thomas Babington Macaulay</u>, English writer and politician, said that '*Knowledge* advances by steps, and not by leaps';

1829-33: <u>Francis Lieber</u>, German-born US philosopher and jurist, created the 13-volume *Encyclopaedia Americana*, which was later worldwide used as a source of information about personalities and definitions;

1834: <u>William Whewell</u>, English philosopher and scientist, wrote that 'Nature, so far as it is the object of scientific research, is a collection of facts governed by laws: our *knowledge of nature* is our *knowledge of laws*';

1834-40: <u>Victor Marie Hugo</u>, French poet and author, was quoted with '*Wisdom* is a sacred communion';

1842: <u>Alfred Tennyson</u>, English poet, versified that 'This grey spirit yearning in desire / To follow *knowledge* like a sinking star, / Beyond the utmost bound of human thought';

1847: <u>Thomas De Quincey</u>, English critic and essayist, was quoted with 'There is first the *literature of knowledge*, and secondly, the *literature of power*';

1849: <u>Matthew Arnold</u>, English poet and critic, versified that 'Others abide our question. Thou art free. / We ask and ask: Thou smilest and art still, / Out-topping *knowledge*';

1850-60: <u>Charles John Huffam Dickens</u>, English novelist, wrote that 'There is a *wisdom of the head*, and a *wisdom of the heart*';

1854: <u>James Frederick Ferrier</u>, Scottish philosopher, published his book *The Institutes of Metaphysic*, and introduced the term *epistemology* into English;

1861: <u>Herbert Spencer</u>, English revolutionary philosopher, thought that '*Science* is *organized knowledge*';

1866-76: <u>Pierre Athanase Larousse</u>, French publisher, lexicographer and encyclopaedist, produced a major work in 17 volumes entitled *Grand dictionnaire universel du XIXe siècle*, a combined dictionary and encyclopaedia;

1872: <u>Oliver Wendell Holmes</u>, US physician, poet and essayist, set that 'It is the *province of knowledge* to speak and it is the *privilege of wisdom* to listen';

1873-1892: <u>Friedrich Wilhelm Nietzsche</u>, German philosopher, produced works such as *Unzeitgemässe Betrachtungen* 'Thoughts Out of Season', *Die Fröhliche Wissenschaft* 'The Joyful Wisdom', *Also sprach Zarathustra* 'Thus Spake Zarathustra'; and was quoted with 'Does *wisdom* perhaps appear on the earth as a raven which is inspired by the smell of carrion?';

1875-78: <u>Benjamin Disraeli</u>, English statesman, thought that 'The *wisdom of the wise* and the *experience of the ages* are perpetuated by quotations';

1877: <u>Thomas Henry Huxley</u>, English biologist, mentioned in writing that 'If *a little knowledge is dangerous*, where is the man who has so much as to be out of danger?';

1891-1902: <u>George Bernard Shaw</u>, Irish dramatist and critic, published extensively books including *The Quintessence of Ibseanism*; *The Perfect Wagnerite*; and *Man and Superman*; as well as wrote that 'We are made *wise* not by recollection of our past, but by the *responsibility for our future*';

1901-09: <u>Theodore Roosevelt</u>, 26th President of the USA, said that '*Nine-tenth of wisdom* is being *wise in time*';

1928-62: <u>Walter Lippmann</u>, US journalist, is quoted with the phrase 'It *requires wisdom* to understand *wisdom*: the music is nothing if the audience is deaf';

1930: <u>Holbrook Jackson</u>, English bibliophile and literary historian, thought that 'Pedantry is the dotage of *knowledge*';

1930-40: <u>William Faulkner</u>, US novelist, wrote that 'The *end of wisdom* is to dream high enough to lose the dream in the seeking of it';

1947-53: <u>Édouard Marie Herriot</u>, French statesman and man of letters, mentioned *La culture, c'est ce qui reste quand on a tout oublié*

'Culture is what is left when we have forgotten everything else', which was inserted in his posthumous *Notes et Maximes*, Hachet, Paris;

1948: <u>Walter Hutchinson</u>, English editor, recognizing that a new encyclopaedia was widely needed to reflect the immense changes in knowledge from the late 19th-early 20th centuries, he founded *The Hutchinson Encyclopedia*;

1948-1956: <u>Bertrand Arthur William Russell</u>, English philosopher, mathematician and writer, published *Human Knowledge: Its Scope and Limits*; developed the *Theory of Knowledge*; defined *knowledge* as belief which is in agreement with facts, showing that no one knows what a belief is, no one knows what a fact is, and no one knows what sort of agreement between them would make a belief true; also published *Logic and Knowledge: Essays*;

1928-62: <u>Robert Lee Frost</u>, US lyric poet, was recognized as a major poet in the country on the basis of his volumes of poetry, including *West-Running Brook*; *A Witness Tree*; *Steeple Bush*; and *In the Clearing*; stating that 'A poem begins in delight and ends in *wisdom*';

1970-2013: *Knowledge base* was founded as a store of information or data that is available to draw on, and also as the understanding of a set of facts, assumptions, and rules which a computer system has available to solve a problem;

1972-89: <u>Stephen Jay Gould</u>, US palaeontologist, with palaeontologist <u>Niles Eldgredge</u>, posited the theory of 'punctuated equilibrium', and has also championed the idea of 'hierarchical evolution', criticizing the 'adaptationist program' and emphasizing that many characters of organisms are not 'adaptive' in the strict sense; he popularized his ideas in a monthly column in Natural History magazine and a series of collected essays, including *Ever Since Darwin*, and then in *Wonderful Life*, concluding that '*Human thought* is suited to the *environment of Evolutionary Adaptation*';

1979: <u>Richard McKay Rorty</u>, US philosopher, published *Philosophy and the Mirror of Nature*, a forceful and dramatic attack on foundationalist, metaphysical aspirations of traditional philosophy; and showed that 'The eventual demarcation of philosophy from science was made possible by the notion that philosophy's core was the '*theory of knowledge*', a theory distinct from the sciences because it was their foundation...Without this idea of a theory of knowledge, it is hard to imagine what philosophy could have been in the age of modern science';

1986-2000: End of *Intermediate knowledge*, extending from medieval through Renaissance and Post-Renaissance to modern times; and the beginning of *Computer-knowledge base*, which enables an easy access to information and knowledge in the following times;

1993: <u>Daniel Povinelli</u>, US cognitive scientist, published *Reconstructing the evolution of mind*, intending to replace conventional models of the evolution of human mind by a new one based on the *evolution of metacognition*;

2001: <u>Jimmy Donal Wales</u>, US Internet entrepreneur, <u>Larry Sanger</u>, US Internet project developer, and <u>Ben Kovitz</u>, Jewish computer programmer, founded *Wikipedia*, until now including over 21 million freely usable articles in 285 languages, written by about 40 million registered users and numerous anonymous worldwide contributors.

Basic knowledge emerged after the humans spread out from Africa to Asia, by accumulation of more experience and capacity of judgment 70207 years ago (see the table of chapter 7) as the initial time of knowledge, and will last until 4350 years from present into the future, as the final time $t_\bullet$ = 70207 + 4350 = 74557 years from the origin of knowledge.

According to the above data, the latest sequences in development of knowledge can be delimited as follows:

Sumerian and Egyptian knowledge \ 3000...2200BC ≈ 2600BC \
Initial ancient knowledge \ 1000...400BC ≈ 700BC \
Classic ancient knowledge \ AD600...1000 ≈ AD800 \
Intermediate knowledge \ AD1986...2000 ≈ AD1993 \
Computer-knowledge base;

corresponding to 4614; 2714; 1214; 21 years ago; or 70207 - 4614 ≈ 65593; 70207 - 2714 ≈ 67493; 70207 - 1214 ≈ 68993; 70207 - 21 ≈ 70186 years from the origin of knowledge respectively. These sequential times from the origin of knowledge divided by the final time $t_\bullet$ = 74557 years from the same origin result in the approximate values 65593/74557 ≈ 0.880; 67493/74557 ≈ 0.905; 68993/74557 ≈ 0.925; and 70186/74557 ≈ 0.941; which can be found in the same succession in Background to time, Table *(k)* for $z/8 = k$ sequences, as
$t_{1.375}/t_\bullet = tanh(1.375) = 0.8798267$; $t_{1.5}/t_\bullet = tanh(1.5) = 0.9051483$; $t_{1.625}/t_\bullet = tanh(1.625) = 0.9253462$; $t_{1.75}/t_\bullet = tanh(1.75) = 0.9413755$
respectively, which lead to the accurate times in years from the origin of knowledge:
$t_{1.375} = t_\bullet \cdot tanh(1.375) = 65597$; $t_{1.5} = t_\bullet \cdot tanh(1.5) = 67485$;
$t_{1.625} = t_\bullet \cdot tanh(1.625) = 68991$; $t_{1.75} = t_\bullet \cdot tanh(1.75) = 70186$.

Using the formula $t_k = t_\bullet \cdot tanh(k)$ for earlier values of the argument k, the other transitional times can be calculated, so that finally the knowledge timeline, sequences, and intrinsic characteristics are finally presented in the next tables.

$z/8 =$	Time (years)		Sequences of knowledge advance
k	from origin $t_k = t_\bullet \cdot tanh(k)$	from present $t_k - 70207$	
	74557	+4350	
...	...	...	
1.875	71131	+924 (AD2938)	–
1.75	70186	-21 (AD1993)	_Computer-knowledge base
1.625	68991	-1216 (AD798)	_ Intermediate knowledge
1.5	67485	-2722 (708BC)	_Classic ancient knowledge
1.375	65597	-4610 (2596BC)	_Initial ancient knowledge
1.25	63245	-6962 (4948BC)	_Sumerian + Egyptian knowledge
1.125	60339	-9868 (7854BC)	_Basic ancient knowledge
1	56782	-13425	_Transitional knowledge
0.875	52481	-17726	_Prehistoric knowledge
0.75	47355	-22852	_Shamanic knowledge
0.625	41349	-28858	_Worldwide knowledge amounting
0.5	34454	-35753	_Knowledge in Far East
0.375	26718	-43489	_Knowledge acquirement in Europe
0.25	18260	-51947	_Knowledge in Australia
0.125	9271	-60936	_Knowledge acquirement in Asia
0	0	-70207	_Ancestral knowledge

$z/8 =$ k	Time from origin $t_k = t_\bullet \cdot tanh(k)$ (years)	Period $\tau_k =$ $(t_\bullet^2 - t_k^2)/(2t_\bullet)$ (years)	Frequency $f_k =$ $(2t_\bullet)/(t_\bullet^2 - t_k^2)$ (years)$^{-1}$	Angular speed $\omega_k =$ $2\pi/\tau_k = 2\pi \cdot f_k$ (years)$^{-1}$
	74557	0	∞	∞
...	...	...	...	...
1.875	71131	3347.3	0.0002987	0.0018771
1.75	70186	4242.9	0.0002357	0.0014809
1.625	68991	5358.2	0.0001866	0.0011726
1.5	67485	6736.6	0.0001484	0.0009327
1.375	65597	8421.6	0.0001187	0.0007461
1.25	63245	10453.9	0.0000957	0.0006010
1.125	60339	12862.3	0.0000777	0.0004885
1	56782	15656.1	0.0000639	0.0004013
0.875	52481	18807.7	0.0000532	0.0003341
0.75	47355	22239.7	0.0000450	0.0002825
0.625	41349	25812.5	0.0000387	0.0002434
0.5	34454	29317.6	0.0000341	0.0002143
0.375	26718	32491.2	0.0000308	0.0001934
0.25	18260	35042.4	0.0000285	0.0001793
0.125	9271	36702.1	0.0000272	0.0001712
0	0	37278.5	0.0000268	0.0001685

15. Navigation

The science relating to sailings, ships, sailors, or the sea is called *nautics* (from Latin *nauticus* 'of a sailor, nautical', Greek ναυτικος 'of a ship'), which includes sailing, shipbuilding, nautical research and design, as well as construction and equipment of a wide range of water-crafts such as boats, sailboats, barges, ships, steamships, screw-steamers, submarines, container ships, cruise liners, carriers, tankers, nuclear submarines, and drill ships. Along with nautics, *navigation* (from Latin *navigare* 'to sail, go by sea, steer a ship' deriving from *navis* 'ship' and *agere* 'to drive') is the act, science, or art of conducting ships, aircrafts, and spacecrafts, especially the finding of position and determination of course by astronomical observations and mathematical computations.

As many achievements in nautics and navigation are due to the exploration, science and industry, which are separately approached in chapters 12, 20 and 21, the present chapter mainly refers to the emergence and development of various types of ships and other water-crafts, as well as docks and ports; their designers, constructors, architects, and innovators; and nautical historians, scientists and writers, such as displayed below.

60000-50000 years ago: *First known construction of ships or floating vessels* was practiced by South-eastern Asians who arrived on Borneo and other islands of Indonesia travelling by sea during an ice age when the sea level was lower and distances between islands shorter;

50000 years ago: *First Aborigines arrived in Australia by boats*, coming from New Guinea and other Indonesian islands, and crossing the Lombok Strait to Sahul;

35000-33000 years ago: *Japanese were plying* the Pacific waters *in boats*;

10000-4000BC: *Travels by sea from island to island* led to colonization of the *Pacific islands* by people coming from Indonesia and the Far East;

5000-4000BC: Earliest representation of a *ship under sail* on a painted disc was found in Kuwait; and invention of the *first sailboat* that was long and narrow but manoeuvrable by several rowers at one time;

3200BC: Earliest known *depictions of sails*, in ancient Egypt, where *reed boats* were sailing upstream against the Nile's current;

3000BC: *Ships built by early Egyptians* were attested by discovery of a *ship hull* about 25-metre long;

3000-2000BC: *Arabic navigation* in the Persian Gulf and Indian Ocean;

3000-1450BC: *Minoan maritime power* was based on *strong wooden ships* powered by sails and oars, establishing long-distance sea routes for trade with Egypt, Anatolia, Mesopotamia, Greece, and Spain; part of those ships were also used to save many Minoans after the devastating eruption of the Thera volcano (1400 or 1375BC);

2500BC: *Khufu ship*, a 43.6-metre vessel, was found sealing into a pit at the foot of the Great Pyramid of Giza; and the oldest known *tidal dock built during the Harappan civilization* at Lothal, Gujarat in India, from where ships had established trade with Mesopotamia;

1300-1200BC: *Trading by ship* was practiced by Greeks and Phoenicians in the Mediterranean Sea;

1200-900BC: *Phoenician maritime power* was based on *galleys*, man-powered sailing vessels, and on the newly created *biremes*; that maritime power, with centres such as Tyre and Sarepta, extended over the Mediterranean to North Africa, Sicily, and Iberian peninsula;

850-750BC: Homer's referred to *seafaring* and around 1000 *ships* used during the Trojan War;

700-500BC: *Rectangular barges*, known as 'castle ships', with multiple decks and guarded ramparts, were constructed during the ancient Zhou Dynasty, as recorded in the naval history of China;

490-479BC: *Greek maritime power* during the Persian Wars was mainly due to *triremes*, including 380 Greek ships armed with sharp brass battering rams at their prows, using the manoeuvre called *diekplous*, namely a line abreast with one wing occasionally leading, which decisively defeated the mighty Persian vessels at the strait of Salamis (480BC), saving not only Greece but also all Europe from an Asian conquest;

335-325BC: Crates of Chalkis, Greek engineer, designed and directed large-scale construction of the *port of Alexandria*, dedicated to Alexander the Great;

264-146BC: *Roman maritime power* during the Punic Wars developed after capturing a Punic warship, which was dragged ashore and precisely copied section by section, particularly intriguing was the powerful bronze beak or ram that was attached to the ship's bows at water level; in order to increase the efficiency of their fleet, the Romans invented the so-called *corvus* (Latin 'crow, raven'), a military boarding device used in naval warfare for direct access of soldiers to an enemy's ship; then constructed a fleet of 100 *quinqueremes* and 20 *triremens*; and later assembled 160 vessels to support Scipio Africanus' army which defeated Hannibal's army at the Battle of Zama (202BC); subsequently controlling the naval operations in the Mediterranean until the end of Punic Wars;

240-220BC: Archimedes, Greek mathematician and inventor, studied the equilibrium positions of floating bodies, stated the *principle of Archimedes*, and thus founded *hydrostatics*;

AD115-117: Apollodorus of Damascus, Roman architect, constructed in Rome *Trajan's Column*, displaying a *Roman trireme* used to cross the Danube to Dacia during the 101-106 wars;

650-850: *Kentoushi delegation* of the Tan Dynasty of China recorded *Japanese navigation routes*, including ocean navigation;

900-1200: *Viking long ships* of which remains were found in many places from Scandinavia to Britain, Iceland and Greenland, as well as on western and southern coasts of continental Europe;

1220-30: *Construction of Portuguese caravels* began, the caravels being later summoned to perform the duties of exploration;

1436: Michael of Rhodes, Venetian navigator, was author of a remarkable manuscript, as the world's first known *treatise on shipbuilding*, including descriptions and illustrations of three kinds of *galleys* and of two kinds of *round ships*;

1552: *Lloyd's Register* in London first used the term *nautical*, whence the name of science *nautics*;

1581: Crostóbal de Barros, Spanish ship designer, wrote *Discución de prototipos de galleon*;

1611: Tomé Cano, Spanish expert in ship building, published *Arte para fabricar, fortificar y aparejar naos de guerra y merchant*, Sevilla: La Laguna;

1620-24: Cornelis Jacobszoon Drebbel, Dutch-born British inventor, constructed a rudimentary *submarine* propelled by means of oars, which was improved and successfully tested in Thames;

1640-55: Blaise Pascal, French mathematician and physicist, studying the fall of mercury columns with increased altitude, postulated the principle that pressure increases proportionally with depth in a liquid, called *Pascal's principle*;

1640-1858: *Sail ships of the line* - adapted for the line-of-battle tactic, as ships supporting each other with broadsides rather than meléeing;

1759-1922: *HMS Victory* is a 104-gun first-rate ship of the line of the Royal Navy, which was commissioned by Chatham Dockyard, and completed with 3560 tonne displacement, 57 metre gundeck, 15.8 metre beam, 8.8 metre draught, 6.55 metre depth of hold, 5440 square metre sails, and able to reach 15-17 kilometre per hour speed; she was launched in 1765, and became most famous as Lord <u>Nelson</u>'s flagship at the 1805 Battle of Trafalgar; in 1922 she was moved to a dry dock at Portsmouth, England, and preserved as a museum ship;

1775: <u>David Bushnell</u>, North American naval engineer, designed the *first military submarine* 'Turtle', accommodating a single person, which was the first verified submarine capable of independent underwater operation and movement and the first using *screws for propulsion*;

1775-1882: Period of the *US Old Navy*;

1794-1814: <u>Robert Fulton</u>, US engineer, developed a *submarine torpedo boat*, launched the steam vessel *Clermont* on Hudson River, and constructed the *world's first steam warship* 'Fulton the First';

1822-45: <u>Claude Louis Marie Henri Navier</u>, French physicist and civil engineer, and independently <u>George Gabriel Stokes</u>, Irish mathematician and physicist, studied the force opposing bodies in their passage through a viscous fluid, and discovered the *Navier-Stokes equations* of hydrodynamics;

1838-45: <u>Isambard Kingdom Brunel</u>, English engineer and inventor, designed *Great Western*, the first steamship built to cross the Atlantic Ocean, and *Great Britain*, the first ocean screw-steamer made entirely of wrought iron;

1839-43: <u>Francis Pettit Smith</u>, English inventor, built the *first successful screw-propelled steamer* 'Archimedes', and then the *first screw warship* 'Rattler' for Royal Navy;

1859-1905: *Steam battleships* - equipped with iron instead of wood, steam replacing sail power, rotatable turrets instead of fixed cannons, and explosive shells replacing cannonballs;

1882-1941: Period of the *US New Navy*;

1884-97: <u>Charles Algernon Parsons</u>, Irish engineer, developed the *high-speed steam turbine*, which was used to build the *first turbine-*

driven steamship called 'Turbinia';

1900-06: <u>George Charles Vincent Holmes</u>, English naval designer and historian, wrote and published *Ancient and Modern Ships*;

1902-14: <u>Alexander Carlisle</u>, Irish shipbuilder and engineer, designed *Olympic class ocean liners*; and <u>Thomas Andrews Jr</u>, Irish businessman shipbuilder, was chief naval architect of the super liner 'RMS Olympic' launched in 1910, of her sister ship oc*ean liner* 'RMS Titanic' launched in 1011, and then of *Britannic* completed in 1914 and serving as a hospital ship in the First World War;

1906-44: *Dreadnoughts* - battleships of Dreadnought-type 'all-big-gun' replaced the previous types of battleships;

1910-14: British transatlantic ocean liners *RMS Olympic, RMS Titanic*, and *HMHS Britannic* were successively launched;

1936-41: *Bismarck* was the first of two Bismarck-class battleships built for the German Kriegsmarine, which was laid down at the Blohm & Voss shipyard in Hamburg and launched two and a half years later, being completed in August 1940; with displacement 41700 tonnes, length 241.6 metres, beam 36 metres, draft 9.3 metres, 3 geared turbines, 3 three-blade screws, power 112 megawatts, and able to reach a speed of 55.6 kilometres per hour; she and her sister ship *Tirpitz* were the largest battleships ever built by Germany; Bismarck conducted only one offensive operation, in May 1941, then she was to break into the Atlantic Ocean and raid Allied shipping from North America to Great Britain, was neutralized by a sustained bombardment from a British fleet, and sunk with heavy loss of life;

1937: Dutch ocean liner *Nieuw Amsterdam* was built in Rotterdam for the Holland America Line, this liner being considered to have been Holland America's finest ship;

1950-54: <u>Hyman George Rickover</u>, US naval engineering officer, led the team who successfully adapted nuclear reactors as a means of ship propulsion, the first vessel so equipped being the USS *Nautilus*, the world's *first nuclear submarine*;

1960-70: <u>Jeorme Goldman</u>, Friede & Goldman, Inc.'s naval architect and marine engineer, pioneered the development of offshore technology; designed a substantial portion of offshore-drilling units, such as one of the first *jack-up drilling units*, the first *catamaran drilling ship, semi-submersibles*, and introduced high standards of strength, stability and seaworthiness in *offshore engineering*;

1979-2010: *Seawise Giant* was delivered by Sumitomo Heavy

Industries, Ltd. at their Oppama shipyard in Yokosuka, Kanagawa, Japan, as a 418000 ton supertanker, with displacement of 657019 tonnes and a draft of 24.6 metres; which was sunk during the 1980-88 Iran-Iraq War, but was later salvaged and restored into service, used as a floating storage and offloading (FSO) unit, moored off the coast of Qatar in the Persian Gulf at the Al Shaheen Oil Field; she had initially a Greek owner, then was brokered with Hong Kong Orient Overseas Container Line, and finally was sent for scrapping in India;

1990: Publication of *European Naval and Maritime History, 300-1500*, written by <u>Archibald Lewis</u> and <u>Timothy Runyan</u>, with data relating to Roman and Byzantine naval power and shipping; Muslim naval and maritime power; Latin Western naval power and shipping, notably Irish, Frisian, Viking, English, Hanseatic and Iberian naval tradition;

1998-99: <u>Iijima Yukito</u>, Japanese maritime technologist, collected early evidences on ship building, and wrote the paper *History of Navigational Technique and its Development in Japan*, Toba Shosen Koto Senmon Gakko Kiyo;

1999: *Vaka: Saga of a Polynesian Canoe* was written by <u>Thomas Davis</u> and refers to a great Polynesian voyaging canoe, as well as 'Takitumu' and the people who sailed across the Pacific Ocean;

2004: <u>Thomas Cutler</u>, US surface warfare officer, edited the book *Dutton's Nautical Navigation*, referring to <u>Benjamin Dutton</u>'s expertise, which unveils how navigational charts are replaced by vector images on computer screens, magnetic compasses enhanced by digital flux gate technology, chronometers joined by atomic clocks;

2005: <u>Samuel EuGene Mark</u>, US nautical archaeologist, published *Homeric Seafaring*, a comprehensive history of Homer's references to ships and seafaring, revealing the way that Greeks built ships and approached the sea between 850 and 750BC, Texas A&M University Press;

2005-13: <u>Brian Lavery</u>, British naval historian, wrote a series of books entitled *The Island Nation: A History of Britain and the Sea*, Conway Maritime Press; *Empire of the Seas*, Conway Publishing; *Ship: 5,000 Years of Maritime Adventure*, and *Conquest of the Ocean*, Dorling Kindersley;

2006: <u>Paul D'Arcy</u>, US historian, published *The people of the sea: environment, identity and History of Oceania*, University of Hawaii Press;

2011: *World's largest shipbuilding country* became South Korea with a global market share of 51.2%, producing large vessels such as cruise liners, super tankers, wide range of shipping (LNG) carriers, drill ships and large container ships; being followed in the *world's shipbuilding market share* by China, Japan and Philippines;

2012: <u>Thomas Cutler</u>, US surface warfare officer, published *A Sailor's History of the US Navy*, US Black Engineer and Information Technology;

2013-14: *World's largest 'ship'*, a floating liquefied natural gas platform, named *Prelude*, and owned by Shell; she is 1601 feet (488 metres) long, i.e. 150 feet longer than the Empire State Building, 243 feet wide, weighs approximately 600000 tonnes, and expected to produce 3.6 million tonnes of liquefied natural gas per year; her storage tanks have a capacity equivalent to about 175 Olympic swimming pools, and the ship also has three 6700-horsepower engines; it was built up within a dock in South Korea and will be operational in less than three years.

The specifications presented at the end of chapter 7 and the above data lead to the settings: 1st, navigation originated 56157 years ago, as its initial time, and will last until the final time $t_\bullet$ = 56157 + 4350 = 60507 years from its origin; and 2nd, the later and then more accurate sequences of navigation can be delimited by the approximate times as

*Pacific inter-insular sailing \ 3500...2100BC ≈ 2800BC *
*Red Sea - Indian Ocean sailing \ 1300...500BC ≈ 900BC *
*Mediterranean sailing \ AD450...800 ≈ AD625 *
*Overseas seafaring \ AD1790...1900 ≈ AD1845 *
Widespread overseas trade routes.

Expressed in years ago, these transitional times become 4814; 2914; 1389; 169; or in years from the origin of navigation they are 56157 - 4814 = 51343; 56157 - 2914 = 53243; 56157 - 1389 = 54768; 56157 - 169 = 55988 respectively. The transitional times from the origin of navigation divided by the final time $t_\bullet$ = 60507 years from the same origin result in the following approximate values: 51343/60507 ≈ 0.849; 53243/60507 ≈ 0.880; 54768/60507 ≈ 0.905; 55988/60507 ≈ 0.925. Such successive values of ratios can be found in Background to time, Table *(k)* for $z/8 = k$ sequences, where they are given as: $t_{1.25}/t_\bullet$ = $tanh(1.25)$ = 0.8482836; $t_{1.375}/t_\bullet$ = $tanh(1.375)$ = 0.8798267; $t_{1.5}/t_\bullet$ = $tanh(1.5)$ = 0.9051483; $t_{1.625}/t_\bullet$ = $tanh(1.625)$ = 0.9253462. Therefore, the selected transitional times are calculated with the accurate values in years from the origin of navigation:

$$t_{1.25} = t_\bullet \cdot tanh(1.25) = 51327;\ t_{1.375} = t_\bullet \cdot tanh(1.375) = 53236;$$
$$t_{1.5} = t_\bullet \cdot tanh(1.5) = 54768;\ t_{1.625} = t_\bullet \cdot tanh(1.625) = 55990.$$

At full scale, the timeline, sequences, and intrinsic characteristics of navigation are displayed in the tables below.

$z/8 =$ k	Time (years)		Navigation sequences
	from origin $t_k =$ $t_\bullet \cdot tanh(k)$	from present $t_k - 56157$	
	60507	+4350	
...	...	...	
1.75	56960	+803 (AD2817)	_Widespread trade routes
1.625	55990	-167 (AD1847)	_Overseas seafaring
1.5	54768	-1389 (AD625)	_Mediterranean sailing
1.375	53236	-2921 (907BC)	_Red Sea - Indian Ocean sailing
1.25	51327	-4830 (2816BC)	_Pacific inter-insular sailing
1.125	48968	-7189 (5175BC)	_Use of log boats for navigation
1	46082	-10075 (8061BC)	_Later canoes
0.875	42591	-13566	_Sea-going dugout canoes
0.75	38431	-17726	_One sheet canoes
0.625	33557	-22600	_Slightly larger boats
0.5	27961	-28196	_Small boats
0.375	21683	-34474	_River and sea dugouts
0.25	14819	-41338	_Navigation by reed boats
0.125	7524	-48633	_Aborigines crossed the sea
0	0	-56157	

$z/8 =$ k	Time from origin $t_k =$ $t_\bullet \cdot tanh(k)$ (years)	Period $\tau_k =$ $(t_\bullet^2 - t_k^2)/(2t_\bullet)$ (years)	Frequency $f_k =$ $(2t_\bullet)/(t_\bullet^2 - t_k^2)$ (years)$^{-1}$	Angular speed $\omega_k =$ $2\pi/\tau_k = 2\pi \cdot f_k$ (years)$^{-1}$
	60507	0	∞	∞
...	...	...	...	...
1.75	56960	3443.0	0.0002904	0.0018249
1.625	55990	4348.4	0.0002300	0.0014449
1.5	54768	5466.8	0.0001829	0.0011493
1.375	53236	6834.1	0.0001463	0.0009194
1.25	51327	8483.6	0.0001179	0.0007406
1.125	48968	10438.7	0.0000958	0.0006019
1	46082	12705.5	0.0000787	0.0004945
0.875	42591	15263.6	0.0000655	0.0004116
0.75	38431	18048.8	0.0000554	0.0003481
0.625	33557	20948.2	0.0000477	0.0002999
0.5	27961	23792.9	0.0000420	0.0002641
0.375	21683	26368.4	0.0000379	0.0002383
0.25	14819	28438.8	0.0000352	0.0002209
0.125	7524	29785.7	0.0000336	0.0002109
0	0	30253.5	0.0000331	0.0002077

16. Meteorology

The interdisciplinary scientific study of weather, climate, and atmosphere in general is called *meteorology* (Greek μετέωρος 'lofty, high in the sky' from μετα 'above', ἑωρ 'to lift up' and λόγος 'study of'). In areas inhabited from many millennia ago, humans noticed atmospheric phenomena such as the successive seasons, winds, tides, floods, rain, showers, clouds, lightning, thunder, and other atmospheric processes which directly or indirectly affected their life. By difference from climate, that means how the atmosphere varies over relatively long periods of time; the weather means what conditions of the atmosphere are over a short period of time. Awareness of weather fluctuations probably appeared during the climate change from an interglacial episode to a glacial one around 50000-40000 years ago, when humans needed to know how the weather will be, in order to protect themselves and to provide their food. The interest in weather conditions increased about 22000-18000 years ago when the last glaciation reached its extremity, and became significant approximately 14000-13000 years ago when the temperature was increasing.

Much later, *weather forecasting* was developed as the application of science and technology to predict the state of the atmosphere. Weather forecasts consist of collecting quantitative data about the current state of the atmosphere on a given place, and using scientific understanding of atmospheric processes to project how the atmosphere will evolve on that place. Such a forecast was first based upon barometric pressure, weather conditions, and sky condition, and later on computer-based models that take many atmospheric factors into account.

Weather forecasts are required for a variety of public uses, such as private sector, agriculture, forestry, utility companies, marine, air traffic, and military applications. Following users' requests, *numerical weather prediction* was developed by sampling the state of the fluid at a given time, and applying the equations of fluid dynamics and thermodynamics to estimate the state of the fluid at some time in the future. The main inputs from country-based weather services are represented by surface observations from automated *weather stations* at ground level over land and from *weather buoys* at sea; the stations transmit hourly in METAR reports or every six hours in SINOP reports. Sites launch *radiosondes*, which rise through the troposphere and well into the stratosphere. In areas where traditional data sources are not available, the data are obtained from *weather satellites*.

Information on precipitation location and intensity are provided by *meteorological radar*, and on wind speed and direction are provided by *Pulse Doppler weather radar*. For weather forecasting, there are in use global climate models and analogue models.

The development of weather warning and meteorology was marked by conceptions, treatises, studies, and other works due a series of peoples and personalities, some of them being mentioned below.

950-850BC: *Meteorological knowledge in India* was amounted to levels unveiled by treatises such as the *Upanishads* containing discussion about the processes of cloud formation and rain, as well as the causes of seasonal cycles by the movement of the Earth around the Sun;

650BC: *Babylonians* interpreted and predicted the *weather* from cloud patterns and from astrology;

350BC: Aristotle, Greek philosopher and scientist, wrote *Meteorologica* 'Meteorology', explaining the *hydrologic cycles* as being caused by the movement of Earth around the Sun, and describing weather patterns, whence he was considered the *founder of meteorology*;

320-290BC: Theophrastus, Greek philosopher, compiled a work, concerning *weather forecasting*, entitled *Book of Signs*;

AD40-45: Pomponius Mela, early Latin geographer, formalized the *climatic zone system*, and divided the known Earth into five zones, of which two only were habitable; his main work was an unsystematic compendium in three volumes entitled *De Situ Orbis* 'A Description of the World';

560-585: Varāhamihira, Indian Shrigaud Bahmin astronomer and mathematician, produced the encyclopaedic *Brihat-Samhita*, covering a wide range of subjects, including rainfall and clouds, and providing evidence about the knowledge of atmospheric processes; he also wrote the treatise *Pancha Siddhantica*, first mentioning *ayanamsa* (shifting of the equinox) calculated as 50.32 seconds;

870-890: Abu Hanifah Dinawari, Muslim polymath, geographer and historian, wrote *Kitâb al-anwâ'* 'Book of Weather', and *Kitâb al-nabât* 'Book of Plants', in the second one applying meteorology to agriculture; he referred to the seasons and rain, the *anwa* (heavenly bodies of rain) and atmospheric phenomena such as winds, thunder, lightning, snow, floods, valleys, rivers, lakes, wells and other sources of water;

1020-40: <u>Alhazen</u> (Abu 'Ali al-Hasan ibn al-Haytham), Muslim scientist, polymath, astronomer and philosopher, produced works such as *Horizontal Sundials*, and *On the Rainbow and Halo*; made detailed scientific study of the *annual inundation of the Nile River*; showed that atmospheric refraction is also responsible for *twilight*, estimating that twilight begins when the Sun is 19 degrees below the horizon; and estimated the maximum possible height of the *Earth's atmosphere* as 52000 *passuum* (about 79 kilometres);

1120-50: <u>William of Conches</u>, Norman French scholastic philosopher, produced *De philosophia mundi*, containing discussions on *meteorology* such as a description of air becoming less dense and colder as the altitude increases, and attempted to explain the circulation of air in connection with the circulation of oceans;

1295-1310: <u>Theodoric of Freiburg</u>, theologian and physicist, made first correct geometrical analyses providing an accurate explanation for the *rainbow*, and afterwards wrote *De iride et radialibus impressionibus* 'On the Rainbow and the impressions created by irradiance';

1300-11: <u>Kamal al-Din al-Farisi</u>, Persian physicist, mathematician and scientist, became known for giving the first mathematically satisfactory explanation of the *rainbow*, and proposed a model where the ray of light from the Sun was refracted twice by a water droplet, one and more reflections occurring between the two refractions;

1683-86: <u>Edmond Halley</u>, English astronomer and mathematician, apart of his great astronomical works, established the mathematical law connecting barometric pressure with heights above sea level, and published studies on magnetic variations, *trade winds* and *monsoons*;

1739: *Poor Richard's Almanack*, a yearly almanac published by the North American inventor and scientist <u>Benjamin Franklin</u>, who adopted the pseudonym of 'Poor Richard' for this purpose, contained the calendar, *weather*, poems, sayings, astronomical and astrological information;

1792: *Old Farmer's Almanac*, edited by the US publisher <u>Robert Thomas</u>, comprised *weather forecasts*, *tide tables*, *planting charts*, astronomical data, recipes, and articles on a number of topics including gardening, sports, astronomy and farming;

1830-50: <u>Francis Beaufort</u>, British naval officer and hydrographer, working in the navy, devised the *Beaufort scale* of wind force and a *tabulated system of weather registration*;

1855-63: <u>Robert Fitzroy</u>, English naval officer and meteorologist, set up a network of telegraph stations for rapid collection of *meteorological observations*; pioneered the making *weather charts*; introduced a set of *symbols* for wind speed and direction, pressure and temperature; and the term *synoptic chart*; analyzed the *Royal Charter storm* of 1859; initiated a system of *gale warnings* for shipping; wrote *The Weather Book*, and invented the *Fitzroy barometer*;

1880-83: <u>Ezekiel Stone Wiggins</u>, Canadian weather and earthquake predictor, wrote *Architecture of the Heavens*, and produced *Wiggins' storm herald, with almanac*, both of them being based on his astronomical calculations and theories that storms, unusual tides, earthquakes and cyclones were all caused by 'planetary attraction';

1898: <u>Vilhelm Friman Koren Bjerknes</u>, Norwegian mathematician and meteorologist, formulated the *circulation theorem* and applied it to a study of atmospheric and ocean processes; as well as deduced the equations enabling *calculation for a developing cyclone*, based on both thermal energy and baroclinicity;

1922-50: <u>Lewis Fry Richardson</u>, English physicist and meteorologist, in the first attempt at weather forecasting, he observed different points in the atmosphere and used fundamental equations to calculate the conditions at these points six hours ahead; his ideas formed the basis on which more powerful computers now perform numerical weather prediction; his results were presented as *Weather Prediction by Numerical Process*; he also researched the turbulence, devising the *Richardson number* of atmospheric turbulence;

1925: *First public radio forecasts* in US were made by <u>Edward Rideout</u> on *WEEI*, and the *Edison Electric Illuminating Station*;

1935-50: <u>John Von Neumann</u>, Hungarian-born US mathematician, was known by development of the theory of linear operators, set theory and design of earliest computers, but also by making the *first computerized weather forecast*, and writing the paper *Numerical Integration of the Barotropic Vorticity Equation*, opening the way for advanced weather forecasting;

1940-47: *Television forecasts* were beginning with <u>James Fidler</u> in Cincinnati, and then on *DeMont Television Network*;

1955: Practical use of *numerical weather prediction* began being spurred by the development of *programmable electronic computers*;

1963: <u>Edward Norton Lorenz</u>, US mathematician and meteorologist, showed that long range forecasts, those made at a range of two weeks or more, are impossible to definitively predict the state of atmosphere, owing to the chaotic nature of the involved fluid dynamics equations, and proved the *limit of predictability* to be about 10-14 days;

1970-2009: Some of the <u>*Later published books*</u> concerning the weather and meteorology: <u>Barbara Tufty</u> - *1001 Questions Answered about Storms and Other Natural Disasters*, New York, Dodd Mead; <u>Miles Harris</u> - *Oportunities in Meteorology*, New York, Educational Books; <u>Nigel Calder</u> - *The Weather Machine*, New York, Viking Press; <u>Antony Smith</u> - *The Weather*, Hutchinson; <u>Alan Watts</u> - *The Weather Handbook*, Thomas Reed Publications; <u>Jack Williams</u> - *The AMS Weather Book: The Ultimate Guide to America's Weather*, University of Chicago Press;

<u>*NASA current mission*</u>: *Calipso* (Mission to provide climate observations for improving the prediction of climate change).

A timeline of meteorology can be established taking into account both the specifications at the end of chapter 7, and the above data, as: 1st, the origin of meteorology was 44368 years ago, as its initial time, and its end will be 4350 years from present into the future, i.e. the final time will be $t_\bullet$ = 44368 + 4350 = 48718 years from the origin of meteorology; and 2nd, the later and then more accurate times delimiting the sequences of meteorology are

Contemplative meteorology \ 1250...800BC ≈ 1025BC \
Observational meteorology \ 400...30BC ≈ 215BC \
Earlier weather study \ AD400...620 ≈ AD510 \
Later weather study \ AD1100...1220 ≈ AD1160 \
Weather prediction \ AD1690...1800 ≈ AD1745 \
Meteorological forecasting.

The selected transitional times represent 3039; 2229; 1504; 854; 269 years ago; and then 44368 - 3039 ≈ 41329; 44368 - 2229 ≈ 42139; 44368 - 1504 ≈ 42864; 44368 - 854 ≈ 43514; 44368 - 269 ≈ 44099 years from the origin of weather scrutiny respectively. Divided by the final time $t_\bullet$ = 48718 years from the same origin, the last times lead to the following approximate values: 41329/48718 ≈ 0.848; 42139/48718 ≈ 0.865; 42864/48718 ≈ 0.880; 43514/48718 ≈ 0.893; 44099/48718 ≈ 0.905. Comparing these values to those given in

Background to time, Table *(l)* for *z/16 = l* sequences, where they are succeeding as

$$t_{1.25}/t_{\bullet} = tanh(1.25) = 0.8482836; \quad t_{1.3125}/t_{\bullet} = tanh(1.3125) = 0.8649066; \quad t_{1.375}/t_{\bullet} = tanh(1.375) = 0.8798267; \quad t_{1.4375}/t_{\bullet} = tanh(1.4375) = 0.8931933; \quad t_{1.5}/t_{\bullet} = tanh(1.5) = 0.9051483;$$

and noticing the similarity of ones to others; the selected transitional times are evaluated, in years from the origin of weather scrutiny, as follows:

$$t_{1.25} = t_{\bullet}\cdot tanh(1.25) = 41327; \quad t_{1.3125} = t_{\bullet}\cdot tanh(1.3125) = 42137; \quad t_{1.375} = t_{\bullet}\cdot tanh(1.375) = 42863; \quad t_{1.4375} = t_{\bullet}\cdot tanh(1.4375) = 43515; \quad t_{1.5} = t_{\bullet}\cdot tanh(1.5) = 44097.$$

Continuing this procedure backward, the timeline, sequences, and intrinsic characteristics of advance in weather scrutiny and meteorology are given in the next tables.

z/16 = l	Time (years)		Sequences
	from origin t_l = $t_{\bullet}\cdot tanh(l)$	from present t_l - 44368	
	48718	+4350	
...	...	...	
1.5625	44617	+249 (AD2263)	—
1.5	44097	-271 (AD1743)	_Meteorological forecasting
1.4375	43515	-853 (AD1161)	_Weather prediction
1.375	42863	-1505 (AD509)	_Later weather study
1.3125	42137	-2231 (217BC)	_Earlier weather study
1.25	41327	-3041 (1027BC)	_Observational meteorology
1.1875	40426	-3942 (1928BC)	_Contemplative meteorology
1.125	39428	-4940 (2926BC)	_Indian-Chinese weather matter
1.0625	38322	-6046 (4032BC)	_Sumerian + Egyptian weather matter
1	37103	-7265 (5251BC)	_Expecting floodings
0.9375	35762	-8606 (6592BC)	_Waiting for rainy seasons
0.875	34293	-10075	_Agriculture scheduled on weather
0.8125	32688	-11680	_Nomadic seasonal moving
0.75	30943	-13425	_Weather change in animal behaviour
0.6875	29054	-15314	_Weather turned to warmer
0.625	27019	-17349	_Preparing for winter or summer
0.5625	24838	-19530	_Early weather awarenes
0.5	22513	-21855	_Warning for cold (glacial) weather
0.4375	20051	24317	_Times for gathering
0.375	17458	-26910	_Times for animal feeding
0.3125	14747	-29621	_Warning for drier weather
0.25	11932	-32436	_Sensory meteorology
0.1875	9029	-35339	_Times for animal hunting
0.125	6058	-38310	_Noticing the migrators' behaviour
0.0625	3041	-41327	_Adapting to weather changes
0	0	-44368	_Early obsevations of weather

z/16 = l	Time from origin $t_l = t_\bullet \cdot tanh(l)$ (years)	Period $\tau_l = (t_\bullet^2 - t_l^2)/(2t_\bullet)$ (years)	Frequency $f_l = (2t_\bullet)/(t_\bullet^2 - t_l^2)$ (years)$^{-1}$	Angular speed $\omega_l = 2\pi/\tau_l = 2\pi \cdot f_l$ (years)$^{-1}$
	48718	0	∞	∞
...	...	...	...	...
1.5625	44617	3928.4	0.0002546	0.0015994
1.5	44097	4401.8	0.0002272	0.0014274
1.4375	43515	4925.2	0.0002030	0.0012757
1.375	42863	5503.2	0.0001817	0.0011417
1.3125	42137	6136.5	0.0001630	0.0010239
1.25	41327	6830.4	0.0001464	0.0009199
1.1875	40426	7586.3	0.0001318	0.0008282
1.125	39428	8404.2	0.0001190	0.0007476
1.0625	38322	9286.8	0.0001077	0.0006766
1	37103	10230.4	0.0000977	0.0006142
0.9375	35762	11233.2	0.0000890	0.0005593
0.875	34293	12289.4	0.0000814	0.0005113
0.8125	32688	13392.8	0.0000747	0.0004691
0.75	30943	14532.4	0.0000688	0.0004324
0.6875	29054	15695.5	0.0000637	0.0004003
0.625	27019	16866.6	0.0000593	0.0003725
0.5625	24838	18027.4	0.0000555	0.0003485
0.5	22513	19157.3	0.0000522	0.0003280
0.4375	20051	20150.1	0.0000496	0.0003118
0.375	17458	21231.0	0.0000471	0.0002959
0.3125	14747	22127.0	0.0000452	0.0002840
0.25	11932	22897.8	0.0000437	0.0002744
0.1875	9029	23522.3	0.0000425	0.0002671
0.125	6058	23982.3	0.0000417	0.0002620
0.0625	3041	24264.1	0.0000412	0.0002589
0	0	24359.0	0.0000411	0.0002579

17. Medicine

The study or/and practice of preventing, diagnosing, and treating diseases or disorders of the body or mind, for restoring or preserving health, comprehensively including surgery and obstetrics, is called *medicine* (from Latin *medicina* 'the art of healing').

Medicine emerged and developed by observations, descriptions, studies, explanations, experiments, researches, discoveries, and techniques, either orally transmitted or written, due to peoples, teams, and personalities, some of them being mentioned in chronological order below.

40000-30000 years ago: *Cave paintings* are showing primary knowledge of the anatomy of animals, suggesting that the dwellers applied it to their own bodies, and indicating knowledge of *anatomy*, as one of the oldest branches of medicine;

30000-20000 years ago: *Surgery* emerged by early *trepanation*, as a surgical medicine, being evidenced by cave paintings and Gravettian remains from the Upper Palaeolithic;

18000-15000 years ago: Early known *treatment of injuries and traumas* was dealing with *internal organs systems* such as heart, lungs, liver, gastro-intestinal tract, kidneys, urinary tract, brain, spinal column, muscles, joints and nerves;

8500-7200BC: *Prehistoric medicine* incorporated plants (herbalism), animal parts, and minerals for *treating diseases*;

4000BC: Urlugaledin, first named Sumerian surgeon, possessed a personal *seal depicting two knives encircled by medicinal plants* (now in Louvre, Paris);

3100-2700BC: *Transition from prehistoric to ancient medicine* took place in Mesopotamia, Egypt, India, and China;

2650-2600BC: Imhotep 'He who comes in peace', Egyptian polymath and physician, was the first physician in history known by name from Egyptian hieroglyphs, and author of a *medical treatise*, containing anatomical observations, aliments and cures;

1800BC: *Kahun Gynaecological Papyrus*, dealing with *women's health* and referring to gynaecological diseases, fertility, pregnancy and contraception, was found at El-Lahun, Egypt, by Flinders Petrie in 1893;

1792-1750BC: <u>Hammurabi</u>, Amorite King of Babylon, became known for his *Code*, written on a tablet (now in Louvre, Paris), which contains specific legislation regulating *surgeons* and *medical compensation*, as well as malpractice and victim's compensation;

1600-1300BC: *Egiptian medical papyri* - *Edwin Smith Surgical Papyrus*, as the oldest known *surgical treatise*, showing the heart, vessels, liver, spleen, kidneys, hypothalamus, uterus and bladder (New York Academy of Medicine); *Hearst Papyrus*, dealing with *urinary system, blood, hair* and *bites* (Bancroft Library, University of California, Berkeley); *Ebers Papyrus*, relating to the heart as the centre of blood supply with vessels attached for every member of the body, and containing 700 magical formulas and remedies (Library of University of Leipzig, Germany); *London Medical Papyrus*, containing *recipes* and *magical spells* for various aliments (British Museum, London); *Brugsch Papyrus*, dealing with *contraception* and *fertility* (Berlin Museum);

650-630BC: <u>Ashurbanipal</u>, Assyrian scholar-king, had an epic poem, in which is mentioned that the legendary king Gilgamesh attributed his recurring dreams to his goddess-mother Ninsun, as *first known recorded dream interpretation*;

c.600BC: <u>Sushruta</u>, Indian surgeon, was known as 'the father of surgery', *teaching* and *practicing* it *on the banks of the Ganges*, near today Benares in Northern India;

550BC: <u>Daniel</u> (Belteshazzar), Judean exile and prophet, performed *first nutritional experiment*, mentioned in Bible's 'Book of Daniel', which he made at Nebuchadnezzar's court;

475BC: <u>Anaxagoras</u>, Ionian philosopher, stated that food is absorbed by human body as containing *homeomerics* which involve existence of nutrients;

420-380BC: <u>Hippocrates</u>, Greek physician, founded *scientific medicine*, making observations about geography of disease, as inserted in works *Airs, Waters, Places*; the role of environment in shaping community's health and epidemics in population, contained in *Epidemics III*; and naturalistic causes of diseases, presented in *The Sacred Disease*;

330-322BC: <u>Aristotle</u>, Greek philosopher and scientist, believed that *sense perception* is only means of human knowledge;

300-290BC: <u>Herophilus</u>, Greek anatomist, was the first known to *dissect the human body*, to compare it with that of other animals, also to describe the brain, liver, spleen, sexual organs and nervous system, dividing the latter into sensory and motor, and also was the first to *measure the pulse*, for which he used a water clock;

100BC-AD100: *Charaka Samhitā*, the oldest known text of *Ayurveda*, was related to the *ayurvedic medicine*, as a system of traditional medicine native to the Indian subcontinent and a form of alternative medicine;

50BC: <u>Asclepiades</u>, Greek physician, is associated with the doctrine that disease results from discord in corpuscles of body, and the belief in *healing by dreams*, which was applied for curing patients;

c.AD100: <u>Aretaeus</u>, Greek physician, produced a great work represented by eight books, the first four discussing the *causes and symptoms of diseases*, and the others referring to the *cure*;

175: <u>Claudius Galenus</u>, Greek physician, elaborated a physiological system, whereby body's three principal organs – *heart, liver*, and *brain* – were central to living processes; performed *audacious operations*, including *brain* and *eye surgeries*; informed about medical *diagnostic aid by pulse*; and wrote *De usu partium* 'The Uses of the Parts' as a hymn to Creator, showing that the body's organs are perfectly adapted to functions which they serve;

250-350: *Suśruta Samhitā*, a later text of *Ayurveda*, which together with the earlier *Charaka Samhitā* covered general medicine, paediatrics, surgery, ophthalmology, mental diseases, toxicology, elixirs and aphrodisiacs;

300-600: *Transition from ancient to medieval medicine* in Europe, the Near East, Indian subcontinent, the Far East, and South-eastern Asia;

610-620: <u>Chao Yuanfang</u>, physician during the Sui Dynasty, was in charge of compiling the *Zhubing Yuanhou Zonglun* 'General Treatise on the Aetiology an Symptoms of Diseases', greatly valued for centuries as a means of categorizing and describing diseases;

984: <u>Tamba Yasuyori</u>, Japanese physician, wrote the oldest surviving Japanese medical text, the 30-volume *Ishinpō*, based on the Chinese *Zhubing Yuanhou Zonglun*, which referred to acupuncture, internal medicine, dermatology, otolaryngology, surgery, pharmacology, gynaecology, obstetrics, paediatrics and health;

c.1000: <u>Avicenna</u> ('Abd Allah ibn Sina), Persian philosopher and physician, pioneered the experimental medicine, interpreting the Aristotle's settings, and wrote *al-Qānūn fī at'-t'ibb* 'Canon of Medicine', a medical textbook that remained a standard work for the next few centuries;

1543: <u>Andreas Vesalius</u>, Belgian anatomist, elaborated the great work *De Humani Corporis Fabrica* 'On the Structure of the Human Body', with excellent descriptions and drawings of bones, nervous system, and thalamus;

1548-78: <u>Li Shizen</u>, Chinese pharmaceutical naturalist and biologist, completed *Ben Cao Gang Mu* 'Great Pharmacopoeia', including more than 11,000 prescriptions, and recording many instances of sophisticated Chinese medicine, for example the use of *mercury-silver amalgam for tooth fillings*, not introduced to Europe until the 19th century;

1611-14: <u>Sanctorius</u> (Santorio Santorio), Italian physiologist and physician, wrote *Ars de statica medicina*, first systematic study of basal metabolism, consisting of an elaborate series of measurements of his own weight, food intake, and excreta;

1646-51: <u>William Harvey</u>, English physician, discovered *blood circulation*, showing that heart is a muscle functioning as a pump and effecting blood movement through body via lungs by means of arteries, the blood then returning through veins to heart;

1765: <u>Lazaro Spallanzani</u>, Italian biologist and naturalist, disproved the long-established theory of spontaneous generation, showing that *broth, boiled thoroughly and hermetically sealed, remained sterile*;

c.1770: <u>Antoine Laurent Lavoisier</u>, French chemist, discovered details of metabolism, and demonstrated that *oxidation of food is source of body heat*;

1778-79: <u>William Cullen</u>, Scottish chemist and physician, wrote *First Lines of the Practice of Physic*, emphasizing the importance of the nervous system in causation of disease, and coining the word *neurosis* to describe a group of nervous diseases;

1780-85: <u>John Brown</u>, Scottish physician, introduced the *Brunonian system* of medicine, according to which all diseases are divided into 'sthenic' and 'asthenic', depending or not depending respectively on an excess of excitement;

1780-98: <u>Samuel Hahnemann</u>, German physician, experimented with the curative power of bark, as source of quinine, leading to the conclusion that drugs produce a very similar condition in healthy persons as in sick ones; formulated the famous principle *similia similibus curantur* 'like cures like', contrasting to belief of allopathic 'ordinary' practitioners, and thus founded *homeopathy*;

1812-17: <u>James Parkinson</u>, English physician and amateur palaeontologist, described *appendicitis* and *perforation*, recognizing latter condition as a cause of death; and was the first to describe paralysis agitans, known as *Parkinson's disease*;

1816: <u>François Magendie</u>, French physiologist, observed that dogs fed only carbohydrates and fat are losing their body protein dying in a few weeks, while dogs fed protein are surviving, and thus identified protein as an *essential dietary component*;

1816-19: <u>René Théophile Hyacinthe Laënnec</u>, French physician, invented the *stethoscope*, founded *chest medicine*, and published *Traité de l'auscultation médiate* 'On Mediate Auscultation';

1820: <u>Friedrich Accum</u>, German chemist, published his *Treatise on Adulteration of Food and Culinary Poisons*, arousing public opinion against unclean food and dishonest trading;

1840: <u>Justus von Liebig</u>, German chemist, had many contributions in medicine, identifying the chemical makeup of *carbohydrates* (sugars), *fats* (fatty acids), and *proteins* (amino acids);

1846-88: <u>William Benjamin Carpenter</u>, English biologist, had the idea of 'unconscious cerebration', and published *Principles of Human Physiology*; *The Microscope and its Revelations*; *Principles of Mental Physiology*; as well as *Nature and Man*;

1855-65: <u>Claude Bernard</u>, French physiologist, discovered that the body fat can be synthesized from carbohydrate and protein, and thus showed that energy in blood *glucose* can be stored as fat, or *glycogen*;

1856-63: <u>Gregor Johann Mendel</u>, German-Czech Austrian monk and botanist, observed that organisms inherit traits via *discrete units of inheritance*, later called genes; as well as formulated *Mendel's first law* 'law of segregation', and *Mendel's second law* 'law of independent assortment', as basic concepts of modern genetics;

1858: <u>Henry Gray</u>, English anatomist and surgeon, published the first edition of *Gray's Anatomy*, which was updated and completed in following editions, becoming one of the best books on anatomy in the world;

1859-81: <u>Charles Robert Darwin</u>, English naturalist, published treatises including *The Descent of Man and Selection in Relation to Sex*; *The Expression of the Emotions in Man and Animals*; and *The formation of Vegetable Mould through the action of Worms*;

1861-70: <u>Paul Broca</u>, French surgeon and anthropologist, located the motor speech centre in the brain, since known as the *convolution of Broca* 'Broca's gyrus', his anthropological investigations giving strong support to CR Darwin's theory of evolutionary descent of man;

1862-85: <u>Louis Pasteur</u>, French chemist, disproved the theory of spontaneous generation of bacteria from inorganic matter, founding *modern bacteriology*; discovered that *fermentation* is essentially due to organisms and not spontaneously generated; as well as set the 'germ theory of disease', showing that *disease is communicable through spread of micro-organisms*;

1863-1920: <u>Wilhelm Max Wundt</u>, German physiologist and psychologist, studied the nervous system and senses, the relations of physiology and psychology, logic and other subjects; and published *Vorlesungen über die Menschen und Thierseele* 'Lectures on the Mind of Humans and Animals', *Grundriss der Psychologie* 'Outlines of Psychology', and ten-volume *Völkerpsychologie* 'Ethnic Psychology'; founding *experimental psychology*;

1867: <u>Joseph Lister</u>, English surgeon, introduced *antiseptic surgery*, revolutionizing modern surgery by soaking instruments and surgical gauzes in carbonic acid;

1873-1918: <u>Camillo Golgi</u>, Italian cytologist, discovered how to stain *nerve tissue using silver nitrate*, and the *Golgi bodies* in animal cells which, through their affinity for metallic salts, are readily visible under the microscope, and thus opened up a new field of research into the *central nervous system*, *sense organs*, *muscles* and *glands*;

1874-85: <u>Carl Wernicke</u>, German neurologist and psychiatrist, published *Der Aphasische Symtomencomplex* 'The Aphasic Syndrome', revealing *aphasia* 'loss of speech' as marked by a severe defect in understanding of speech, known as *sensory aphasia*, and showing that this kind of aphasia is typically localized in the so-called *Wernicke's area* of left temporal lobe;

1877-1908: <u>Victor Babes</u>, Romanian physician, biologist and bacteriologist, discovered a parasitic sporozoan of ticks, named *Babesia*, causing a rare and severe disease called *babesiosis*; published the treatise *Les bactéries et leur rôle dans l'anatomie et l'histologie pathologiques des maladies infectieuses* 'Bacteria and their role in the histopathology of infectious diseases'; and identified cellular inclusions in rabies-infected nerve cells, later called *Babes-Negri bodies*;

1880-85: <u>Herman Ebbinghaus</u>, German experimental psychologist, made researches concerning memory to investigate higher mental processes, and discovered so-called *forgetting curve* that relates memory failure to time; he also published *Über das Gedächtnis* 'Memory';

1880-96: <u>Charles Louis Alphonse Laveran</u>, French physician and parasitologist, studied *malaria*, discovering that parasite was spread through mosquito bites, and also other tropical diseases including *sleeping-sickness, leishmaniasis,* and *kala-azar,* showing the role played by *protozoa* in causing diseases;

1882-95: <u>Robert Koch</u>, German physician and pioneer bacteriologist, discovered the *tubercle bacillus* that causes tuberculosis, and the *cholera bacillus,* and also formulated the essential scientific principles, known as *Koch's postulates,* establishing *clinical bacteriology* as a medical science;

1883-84: <u>Georg Theodor August Gaffky</u>, German bacteriologist, was the first to isolate a pure culture of *typhoid bacillus,* and, together with R Koch, discovered the *Vibrio* responsible for cholera;

1884: <u>Kanehiro Takaki</u>, Japanese physiologist, observed that Japanese sailors were suffering of *beriberi,* while British sailors and Japanese officers were not affected, and introduced *vegetables* and *meats* in Japanese sailors' diet;

1890-1905: <u>Albrecht Kossel</u>, German physiological chemist, investigated chemistry of cells and proteins, and explained that, in blood leukaemia, *guanide* derives from decomposed young nucleated erythrocytes;

1890-1907: <u>Eduard Buchner</u>, German chemist, attested that the absence of oxygen is not necessary for fermentation, and discovered *enzymes*;

1891-1931: <u>George Frederick Stout</u>, English philosopher and psychologist, made important contributions to psychology and philosophy of mind by publications such as *Analytic Psychology*; *Manual of Psychology*; and *Mind and Matter*;

1892: <u>William Osler</u>, Canadian-British physician, published his work *The Principles and Practice of Medicine,* a textbook codifying scientific clinical practice;

1892-1914: <u>Santiago Ramón y Cajal</u>, Spanish physician and histologist, worked on the microstructure of nervous system, revealing *how nerve impulses are transmitted to brain,* and wrote articles and books including two-volume *Estudios sobre la degeneración y regeneración del sistema nervioso* 'The Degeneration and

Regeneration of the Nervous System';

1894-1901: <u>Ioan Cantacuzino</u>, Romanian physician and bacteriologist, made relevant discoveries in treatment of *cholera*, *epidemic typhus*, *tuberculosis*, and *scarlet fever*; worked on immunity and invertebrates, and coined the term *contact immunity*;

1895: <u>Albert Calmette</u>, French bacteriologist, discovered the *BCG vaccine* 'Bacille Calmette-Guérin', then used in *inoculation against tuberculosis*;

1897: <u>Christiaan Eijkman</u>, Dutch physician and pathologist, observed that chickens feeding in native Java diet of white rice had symptoms of *beriberi*, but remained healthy when feeding *unprocessed brown rice* with outer bran intact; only two decades later it was proved that outer rice bran contains vitamin B_1, also known as thiamine;

1899-1910: <u>Joseph Goldberger</u>, US physician and epidemiologist, demonstrated that *pellagra is a nutritional disorder* caused by an unbalanced diet and cured by addition of fresh milk, meat or yeast, but not an infectious disease;

1900-23: <u>Sigmund Freud</u>, Austrian neurologist, founded *psychoanalysis*; replaced unconscious with *long-term structured memory*; and made the theory of *division of unconscious mind* into *Id, Ego*, and *Super-Ego*;

1901-40: <u>Karl Landsteiner</u>, Austrian-born US pathologist, discovered the four *major human blood groups* 'A, O, B, AB', then the *'M' and 'N' groups*, and finally the *rhesus 'Rh' factor*;

1905: <u>William Bateson</u>, English geneticist, studied the process of *linkage*, by which some genes are inherited together; and coined the word *genetics*;

1906: <u>August Paul von Wassermann</u>, German bacteriologist, worked on bacteriology and chemotherapy, and established the *Wassermann's blood-serum test* for producing syphilis antibodies in blood and reacting with known antigens to form a chemical complex;

1906-15: <u>Harvey Washington Wiley</u>, US food chemist, worked on the analysis of foods, promoted the *Pure Food and Drug Act* of 1906, and published a textbook entitled *Not by Bread Alone*;

1907: <u>Alois Alzheimer</u>, German psychiatrist and neuropathologist, achieved a full clinical and pathological description of pre-senile dementia known as *Alzheimer's disease*;

1907-21: <u>Frederick Gowland Hopkins</u>, English biochemist, recognized accessory food factors, now called vitamins, in association with lactate production in muscle contraction, and discovered *glutathione*;

1910-11: <u>Thomas Hunt Morgan</u>, US geneticist and biologist, demonstrated that certain traits are kinked, although they are not always inherited together; suggesting that *genes are carried on X chromosome*, and they can cross over to other chromosomes at a *rate of crossing-over* useful as measure of distance along chromosome;

1910-20: <u>Otto Fritz Meyerhof</u>, US biochemist, and <u>Archibald Vivian Hill</u>, English physiologist, discovered the fixed relationship between *consumption of oxygen* and *metabolism of lactic acid* in muscle; and explained the *production of heat in muscle* respectively;

1911: <u>Eugen Bleuler</u>, Swiss psychiatrist, published a study on *schizophrenia* 'splitting of the mind', coining the word 'schizophrenia'; and <u>John Hughlings Jackson</u>, English neurologist, postulated that the *evolution of nervous system proceeds from simplest centres to most complex structures*;

1912: <u>Casimir Funk</u>, Polish-born US biochemist, coined the term *vitamin*, a vital factor in diet, coming from words 'vital' and 'amine', because vitamins are preventing diseases such as scurvy, beriberi, and pellagra;

1913: <u>Elmer Verner McCollum</u>, US biochemist, was the first to discover the fat soluble *vitamin A*, and the water soluble *vitamin B*; and used the term *vitamin C* as a then-unknown substance preventing scurvy;

1914-47: <u>Edward Calvin Kendall</u>, US chemist, <u>Tadeus Reichstein</u>, Swiss chemist, and <u>Philip Showalter Hench</u>, Polish-born US physician, isolated *thyroxine, cortisone*, and 29 related *steroids* from adrenal cortex; synthesized new steroids, especially those of *adrenal gland*; and investigated the role of *cortisone*, respectively; altogether showing biology and therapeutic uses of suprarenal hormones;

1915: <u>Alfred Henry Sturtevant</u>, <u>Thomas Hunt Morgan</u> and <u>Hermann Joseph Müller</u>, US geneticists, established the basis for chromosomal theory of heredity, as published in *The Mechanism of Mendelian Inheritance*;

1917: <u>Julius Wagner-Jauregg</u>, Austrian neurologist and psychiatrist, discovered the therapeutic value of *malaria inoculation* in treatment of *dementia paralytica*;

1919-30: <u>Charles Scott Sherrington</u> and <u>Edgar Douglas Adrian</u>,

English physiologists, examined the *sensory systems* in animals and humans, investigated the *response mechanisms of receptors and sense organs*, and worked on recording and analysing the *information in the central nervous system*;

1920-22: <u>Dame Harriette Chick</u>, English nutritionist, established that *sunlight* and *dietary cod-liver oil*, rich in vitamin D, *could eliminate childhood rickets*; and <u>Adolf Otto Reinhold Windaus,</u> German chemist, researched the *structure of cholesterol*, and synthesized *vitamin D*, which is structurally related to cholesterol;

1920-23: <u>Frederick Grant Banting</u>, Canadian physiologist, and <u>John James Rickard Macleod</u>, Scottish physiologist, purifying a pancreatic extract which lowered blood sugar levels, they discovered the *hormone insulin* as main treatment for *diabetes*;

1921-30: <u>Joseph Erlanger</u> and <u>Herbert Spencer Gasser</u>, US physiologists, using a new powerful electrical equipment to dissect and analyse the nature and function of nerve fibres, they discovered the *fundamental properties of neural conduction of impulses*, and deduced that the *speed of impulse is proportional to diameter of nerve fibre*;

1925-42: <u>Paul Hermann Müller</u>, Swiss chemist, was interested and worked for promotion of new insecticides, and then discovered and developed the production of *dichlorodiphenyltrichloroethane* 'DDT', an extremely toxic substance usable in a variety of disease carriers and plant pests;

1928-35: <u>Henry Hallett Dale</u>, English physiologist, and <u>Otto Loewi</u>, German pharmacologist, discovered *acetylcholine*, an organic molecule acting as a neurotransmitter; and identified several possible *transmitter substances*, distinguishably acetylcholine, respectively, relating to the chemical transmission of nerve impulses;

1928-43: <u>Alexander Fleming</u>, Scottish bacteriologist, and <u>Howard Walter Florey</u>, Australian pathologist, discovered *penicillin*, as a curious mould on a culture of staphylococci, having unsurpassed antibiotic powers; and developed an antibiotic penicillin, by *clinical testing in USA* where it was put into mass production, so that by end of the Second World War had already saved many lives;

1928-57: <u>Carl Gustav Jung</u>, Swiss psychiatrist, stated that the content of dreams is related to the dreamer's unconscious desires, and *dreams are messages to the dreamer*, arguing that dreamers should pay attention for their own good;

1929-35: <u>Edward Adelbert Doisy</u>, US biochemist, researched reproduction and hormones, and studied female sex hormones *estrone*, *estriol*, and *estradiol*, delineating *four stages of endocrinology* as recognition of gland, detection of *hormone*, its extraction and purification, and finally its structure and synthesis;

1929-40: <u>Ernst Boris Chain</u>, British biochemist, encountering an A Fleming's paper on penicillin, he discovered that *penicillin was not an enzyme but a new small molecule*, and then worked to improve its purification;

1930-39: <u>John Howard Northrop</u>, US biochemist, isolated the *first bacterial virus*, and found the first *relation between biological function of an enzyme* and *its chemical properties*;

1930-49: <u>Albert Claude</u>, Belgian biologist, developed *cell fractionalization* using a high-powered centrifuge, isolating a *tumour agent from cancerous cells*, studying normal cells, and separating various 'organelles' such as *nucleus*, *mitochondria* and *microsomes*, later known as ribosomes;

1932-43: <u>Hans Adolf Krebs</u>, German-born British biochemist, described the *urea cycle* whereby carbon dioxide and ammonia from urea in presence of liver slices; then elucidated the *citric acid cycle* 'Krebs's cycle' of energy production;

1937-38: <u>Paul Karrer</u>, Swiss chemist, elucidated the structures of *vitamins E, K,* and *B_2* 'riboflavin'; and studied the chemistry of vitamin C and *biotin*;

1939-44: <u>Selman Abraham Waksman</u>, US biochemist, discovered the anti-cancer drug *actinomycin*, first anti-tuberculosis drug *streptomycin*, and several other anti-bacterial agents;

1940-75: <u>Baruj Benacerraf</u>, Venezuelan-born US immunologist, <u>Jean Dausset</u>, French immunologist, and <u>George Davis Snell</u>, US geneticist, searched on cells involved in body's defence against foreign substances as *antigens* showing that the response to them is genetically determined; investigated transfusion responses and antibody production leading to *tissue typing*, greatly reducing rejection risks in organ transplant surgery; and studied the genes responsible for rejection *tissue transplants in mice*, respectively;

1942-54: <u>Vincent du Vigneaud</u>, US biochemist, synthesized *thiamine*, *penicillin*, and two neurohypophysial peptide hormones, *oxytocin* and *vasopressin*;

1942-56: <u>Dorothy Mary Hodgkin</u>, British crystallographer, determined the *structure of penicillin*, and the *structure of vitamin B_{12e}* used to fight pernicious anaemia;

1943: <u>Konrad Emil Bloch</u>, German-born US biochemist, researched glucose, showing that in animal fatty acids cannot be converted into sugars, and discovered the direct *metabolic relationship between cholesterol* and *bile acids*;

1944: <u>Oswald Theodore Avery</u> and <u>Maclyn McCarty</u>, US bacteriologists and geneticists, showed that the non-virulent rough-coated strain could be transformed into a virulent smooth strain in presence of some dead bacteria, leading to *identification of molecule responsible for transformation*, later called DNA;

1945-49: <u>Walter Rudolf Hess</u>, Swiss physiologist, and <u>António Egas Moniz</u>, Portuguese neurologist, established methods of stimulating localized areas of brain by means of fine needle electrodes to study *brain function*; and used dyes in X-ray localization of brain tumours developing *prefrontal lobotomy* for control of schizophrenia and other mental diseases;

1945-70: <u>Bernard Katz</u>, British biophysicist, worked on mechanisms of neural transmission, showing that *chemical neurotransmitters are stored in nerve terminals* and *released in specific portions*, called 'quanta', *when stimulated by arrival of neural impulse*;

1946-52: <u>John Franklin Enders</u>, US bacteriologist, <u>Thomas Huckle Weller</u>, US virologist, and <u>Frederick Chapman Robbins</u>, US physiologist and paediatrician, discovered the ability of *poliomyelitis viruses* to grow in cultures of various types of tissue cells;

1948-54: <u>George Herbert Hitchings</u>, US biochemist, preparing and testing RNA and DNA bases and amino acids as growth factors, he evidenced a folic acid antagonist, leading to discovery of drugs to alleviate *gout*, combat *cancer* and *malaria*, and synthesis of successful anti-leukaemia drug *6-mercaptopurine*, followed by *azathioprine* which suppresses body's immune system to enable organ transplantation from an unrelated donor;

1948-67: <u>Severo Ochoa</u>, US geneticist, studied the *energetics of carbon dioxide fixation in photosynthesis*; achieved first *synthesis of artificial RNA*, solved *amino acid genetic code*, and identified a number of base triplets, as well as studied the *direction of protein synthesis along DNA*, and also the *first amino acid in a peptide sequence*;

1949: <u>Elizabeth Lee Hazen</u>, US microbiologist, and <u>Rachel Fuller</u>

Brown, US biochemist, developed the first useful antifungal antibiotic, called *nystatin*;

1950-65: François Jacob and Jacques Lucien Monod, French biochemists, with André Michel Lwoff, French microbiologist, showed that *genes are controlled by a system of other genes which regulate certain enzymes*, and developed the *theory of operon system*, whereby a regulator gene controls other genes by binding to specific section of DNA strand;

1950-75: Edward Lawrie Tatum, US biochemist, and Joshua Lederberg, US biologist and geneticist, described *transduction in bacteria*, whereby bacterial virus transfers part of its DNA into a host bacterium, leading to development of techniques for *manipulation of genes* and evidence of *sexual process of conjunction* in bacteria reproduction;

1950-77: Rosalyn Yalow, US biophysicist, researched on diabetes, developing *radioimmunoassay* 'RIA', an ultrasensitive method of measuring concentrations of body's substances, and suggesting that in adult diabetics antibodies, which inactivate injected insulin, are formed;

1951-53: Maurice Hugh Frederick Wilkins, British physicist, and Rosalind Elsie Franklin, English X-ray crystallographer, produced an *X-ray diffraction picture of DNA*, thus contributing to the double helix model of DNA by their *X-ray data of DNA fibres*;

1951-63: John Carew Eccles, Australian neurophysiologist, recorded the *depolarization of a post-synaptic muscle fibre* in response to a neural stimulus, identified *inhibitory neurons*, and demonstrated how *inhibitory synapses control flow of information* within the nervous system;

1953: Francis Harry Compton Crick, English molecular biologist, and James Dewey Watson, US biologist, worked on *structure of DNA* (deoxyribonucleic acid), finding that biological molecule contained in cells carries genetic information; elaborated the famous model of a *double-helical molecule*, consisting of two strands of nucleotide bases wound around a common axis in opposite directions, and suggesting a simple method for duplication, i.e. if strands are separated, new partner strands are reconstructed for each based on sequence of the old strand;

1953-63: Wilbur Sutherland, US biochemist, evidenced that a molecule known as *cyclic-AMP* 'cyclic adenosine monophosphate' activates glycogen-glucose transformation, and proposed that two hormones initiate entire process by inducing production of c-AMP, thus discovering a new principle, namely the *second messenger theory* of hormonal action;

1953-90: <u>Hans Jürgen Eysenck</u>, British psychologist, published *Uses and Abuses of Psychology*, *Know Your Own IQ*; *Race, Intelligence and Education* as a controversial view on racial differences in intelligence; and autobiographical *Rebel with a Cause*;

1954: <u>Jonas Edward Salk</u>, US virologist, produced the *Salk vaccine* against poliomyelitis, initially widely and recently restrictively used to prevent polio cases;

1954-58: <u>William Howard Stein</u> and <u>Stanford Moore</u>, US biochemists, developed a column chromatographic method for identification and quantification of *amino acid mixtures* in proteins and physiological tissues, analysed the *base sequence of RNA*, and studied a *novel protease from streptococcus*, showing that its molecular structure differed from that of plant protease papain;

1955-65: <u>Roger Wolcott Sperry</u>, US neuroscientist, pioneering the behavioural investigation of 'split-brain' animals and humans, he established that each hemisphere possesses specific higher functions: the *left side* controlling verbal activity and processes such as writing and reasoning; whereas the *right side* is more responsive to music, face and voice recognition;

1955-90: <u>Edmond Henri Fischer</u> and <u>Edwin Gerhard Krebs</u>, US biochemists, studied the *enzyme phosphorylase*, showing that conversions to and from compounds of phosphorus are involved in activating *glycogen phosphorylase*, as a biological regulatory mechanism;

1956: <u>George Emil Palade</u>, Romanian-born US cell biologist, developed a method of separating cell components, known as *cell fractionation*, and identified these components as *mitochondria*, *endoplasmatic reticulum*, *Golgi apparatus*, and *ribosomes*, showing that *protein synthesis occurs on strands of RNA in ribosomes*; thus contributing to the development of genetics;

1957: <u>Ana Aslan</u>, Romanian biologist and physiologist, studied and experimented on the aging process, leading to the foundation of *gerontology* and *geriatrics*, as well as to the production of stimulant substances such as *gerovital* and *aslavital*, widely applied in a number of countries;

1959-67: <u>Arthur Kornberg</u>, US biochemist, discovered the *DNA polymerase*, an enzyme synthesizing new DNA; and made a *synthesis of viral DNA*;

1962-70: Christian René de Duve, Belgian biochemist, discovered *lysosomes*, small organelles within cells which contain enzymes, whose malfunction causes some metabolic diseases, such as cystinosis;

1962-75: Daniel Nathans, US microbiologist, pioneered the use of restriction enzymes to fragment DNA molecules, and produced *first genetic map*, identifying location of *specific genes on DNA*;

1963-75: Donnall Thomas, US physician and haematologist, used *tissue-typing techniques* and drugs for suppressing immune system, in order to operate *bone marrow transplants* in treatment of leukaemia, aplastic anaemia, and certain genetic diseases;

1964: Baruch Samuel Blumberg, US biochemist, discovered *Australia antigen*, and its association with hepatitis B, known as HBV virus, which was rapidly applied for screening blood donors;

1966: Marshall Warren Nirenberg and Robert William Holley, US biochemists; with Har Gobind Khorana, US molecular chemist, approached the problem of 'code dictionary' by synthesizing a nucleic acid with a known base sequence, found which amino acid it converted to protein; and completed the *deciphering full code*;

1968-78: Patrick Christopher Steptoe, English gynaecologist and reproduction biologist, and Robert Geoffrey Edwards, British physiologist, worked on the problem of *in vitro fertilization* of human embryos, which latter resulted in the birth of a baby after this kind of fertilization and implantation in her mother's uterus;

1969-72: Gerald Maurice Edelman, US biochemist, and Rodney Robert Porter, English biochemist, investigated a number of *antibody forms in different vertebrates*, proposed the *bilaterally symmetrical four-chain structure* as basis of all immunoglobulins, and furthermore evidenced a typical Y-shaped human immunoglobulin *IgG* antibody molecule;

1970: David Baltimore, US microbiologist, discovered the *reverse transcriptase enzyme* which can transcribe DNA into RNA, allowing scientists to manipulate genetic code;

1970-77: Roger Guillemin, US physiologist, isolated and identified the chemical structures of *hypothalamic hormones*, principally hormone stimulating thyroid gland, and hormones releasing and inhibiting growth hormone;

1973-78: <u>Hamilton Othanel Smith</u>, US molecular biologist, obtained enzymes from bacteria which would split genes to give genetically active fragments as *restriction enzymes* allowing possibility of genetic engineering of a new kind; and isolated 'type II' enzymes which would split a DNA strand at a specific and predictable site, allowing to establish *nucleotide sequence of DNA*;

1973-2009: <u>Ralph Marvin Steinman</u>, Canadian immunologist and cell biologist, discovered the *dendritic cell and its role in adaptive immunity*, coining the term 'dendritic cells';

1974-90: <u>Hugh John Forster Cairns</u>, English molecular biologist, demonstrated that *cancer develops from a single abnormal cell* probably initiated by mutation of DNA sequence, but its progression depends on multiple factors such as smoking, diet, and hormones, and *does not require further alteration to the cell's DNA*;

1975-82: <u>Niels Kai Jerne</u>, English immunologist, <u>Georges Jean Franz Köhler</u>, German immunologist, and <u>Cesar Milstein</u>, Argentinean-born British molecular biologist and immunologist, searched into immune system, examining creation of antibodies, explaining development of *T-lymphocytes* and formulating *network theory* of interacting lymphocytes and antibodies; studied the pattern of inheritance of *hybridoma cells* showing that structural mutants of immunoglobulins could be formed by hybridomas; and developed the technique of *monoclonal antibodies* by fusing together different cells to maintain antibody production, respectively;

1976-77: <u>John Allan Hobson</u> and <u>Robert McCarley</u>, US psychiatrists, formulated the *activation-synthesis hypothesis* as neurobiological theory of dreams, proposing that dreams result from *brain activation during 'rapid eye movement' sleep*;

1977: <u>Frederich Sanger</u>, English biochemist, developed the technology of chain-termination DNA, for reading *full nucleotide sequence of DNA molecule*; <u>Thomas Cech</u>, US biochemist, discovered that *protein-free precursor RNA performs its own cleavage and splicing, acting in manner of an enzyme*, but modifying molecule in process; and <u>Phillip Allen Sharp</u>, US molecular biologist, invented the *mapping technique* used extensively in analysis of RNA molecules, leading to discovery that *genes are split into several sections, separated by stretches of DNA* known as 'introns' which appear to carry no genetic information;

1978: <u>Michael Smith</u>, Canadian biochemist, discovered 'site-specific muta-genesis', a technique allowing to alter the genetic code through *mutations induced at specific locations*, whereas all previous methods of mutation had produced only random mutations;

1979-2004: <u>John Robin Warren</u>, Australian pathologist, and <u>Barry James Marshall</u>, Australian physician and microbiologist, showed that the bacterium *Helicobacter pylori* is the cause of most *peptic ulcers* and its relation with *gastritis* and *stomach cancer* ; thus reversing decades of medical doctrine holding that ulcers were caused by stress, spicy food, and too much acid;

1980-86: <u>Paul Delos Boyer</u>, US biochemist, <u>John Ernest Walker</u>, English biochemist, and <u>Jens Christian Skou</u>, Danish biophysicist, elucidated the *enzymatic mechanism* underlying synthesis of *adenosine triphosphate* (ATP); and also ion-transporting enzymes, which enable the transfer of energy in cells;

1981-92: <u>Alfred Goodman Gilman</u>, US pharmacologist, and <u>Martin Rodbell</u>, US biochemist, discovered the *G-proteins* that enable cells to respond to signals from other cells or from outside body;

1981-99: <u>Roy John Britten</u> and <u>Eric Harris Davidson</u>, US molecular biologists, studied genomes of higher organisms, elucidated genome organization, and showed that they contain DNA strands organized into *unique, single-copy DNA sequences* (coding for single genes), *moderately repetitive DNA* (coding for gene families), and highly repetitive sequences which are repeated hundreds of thousands of times in genome;

1983: <u>Kary Banks Mullis</u>, US biochemist, discovered *polymerase chain reaction* 'PCR', allowing tiny quantities of DNA to be copied millions of times for practical analysis; and <u>Luc Montagnier</u>, French molecular biologist, discovered *HIV* 'Human immunodeficiency virus', showing that mycoplasms (bacteria-like organisms) might play a crucial part in *progression from HIV infection to symptomatic AIDS* 'Acquired Immune Deficiency Syndrome';

1983-95: <u>John Maynard Smith</u>, English geneticist and evolutionary biologist, developed a new phase of mathematical understanding of evolutionary processes, in particular the application of game theory to behavioural ecology, published *Evolution and the Theory of Games*, searched mutations and recombination in *human mitochondrial DNA*; and wrote the influential book *Theory of Evolution*, as well as *Evolutionary Genetics*, and *The Major Transitions in Evolution*;

1983-2007: <u>Herald zur Hausen</u>, German virusologist, researched and discovered the *human papilloma viruses* (HPV) causing *cervical cancer*;

1984: <u>Michael Stuart Brown</u> and <u>Joseph Leonard Goldstein</u>, US molecular geneticists, elucidated the *gene sequence with several mutations*, which codes for low-density lipoproteins as receptor, opening up *possibilities of synthesizing drugs to control cholesterol metabolism*;

1989-2000: <u>Paul Greengard</u>, US biochemist, <u>Arvid Carlsson</u>, Swedish pharmacologist, and <u>Eric Kandel</u>, US biologist, searched into *signal transduction in the nervous system*, enabling the development of a range of new drugs;

1990-98: <u>Louis Joseph Ignarro</u>, <u>Robert Francis Furchgott</u>, and <u>Ferid Murad</u>, US pharmacologists, made important research of the role of nitric oxide in communicating signals within cardiovascular system, contributing significantly to *treatment of heart disease*; and and then contributed to the development of a drug called *sildenafil citrate* with commercial name 'Viagra';

1995-2000: <u>Sydney Brenner</u>, British molecular biologist, <u>Howard Robert Horvitz</u>, US biologist, and <u>John Edward Sulston</u>, British biologist, discovered the *genetic regulation of organ development and programmed cell death*;

1995-2005: <u>Andrew Zachary Fire</u> and <u>Craig Cameron Mello</u>, US biologists, discovered *RNA interference-gene silencing by double-stranded RNA*;

1995-2008: <u>Elizabeth Helen Blackburn</u>, Australian-born US biologist, <u>Carolyn Widney Greider</u>, US molecular biologist, and <u>Jack William Szostak</u>, Canadian-US biologist, discovered how *chromosomes* are protected by *telomeres* and *enzyme telomerase*;

1997-2009: <u>Robert Geoffrey Edwards</u>, English physiologist, did pioneering work in *reproductive medicine*, and developed *in vitro fertilization*;

2000-10: <u>Jules Alphonse Hoffman</u>, Luxembourg-born French biologist, was dedicated to the field of biology using *insects as model organisms*; and <u>Bruce Alan Beutler</u>, US immunologist and geneticist, discovered key sensors of microbial infection in mammals, demonstrating that one of the mammalian *Toll-like receptors* (TLR) act as the membrane-spanning component of the *lipopolysaccharides* (LPS) receptor complex, and also worked on pathogenesis of sterile inflammatory and autoimmune diseases such as *systemic lupus erythematosus*; both of them contributing to *activation of innate immunity*;

315

2001-11: <u>John Bertrand Gurdon</u>, English developmental biologist, and <u>Shinya Yamanaka</u>, Japanese physician and researcher of adult stem cells, discovered that *mature cells can be reprogrammed to become pluripotent*;

2003: <u>Eugen Tarnow</u>, US neuro-psychoanalyst, suggested that dreams are ever-present *excitations of long-term memory*, thus reworking Freud's theory;

2004: <u>Jie Zhang</u>, Chinese psychiatrist, proposed that dreaming is a result of brain activation and synthesis, while the *function of sleep is to process, encode and transfer data from short-term memory to long-term memory*;

2005-12: <u>James Edward Rothman</u> and <u>Randy Wayne Schekman</u>, US cell biologists, and <u>Thomas Christian Südhof</u>, German-US biochemist, discovered the *machinery regulating vesicle traffic*, a major transport system in cells.

According to the table of chapter 7, and to data above, medicine emerged and evolved as:

1st, the origin of medicine was 34619 years ago, as the initial time, and its end will be 4350 years from present into the future, so that the final time is

$$t_\bullet = 34619 + 4350 = 38969 \text{ years from its origin; and}$$

2nd, In the development of medicine, the later and then better estimated transitional times from one sequence to another can be considered the following:

Egyptian-Indian medicine \ 500...40BC ≈ 270BC \
Greco-Roman medicine \ AD300...600 ≈ AD450 \
Medieval medicine \ AD1000...1200 ≈ AD1100 \
Chinese-European medicine \ AD1600...1760 ≈ AD1680 \
Modern medicine.

Thus, the approximate transitional times represent 2284; 1564; 914; 334 years ago; and 34619 - 2284 ≈ 32335; 34619 - 1564 ≈ 33055; 34619 - 914 ≈ 33705; 34619 - 334 ≈ 34285 years from the origin of medicine respectively. The last transitional times divided by the final time $t_\bullet$ = 38969 years from the same origin result in ratios 32335/38969 ≈ 0.830; 33055/38969 ≈ 0.848; 33705/38969 ≈ 0,865; 34285/38969 ≈ 0.880; which are roughly similar to the accurate values given in Background to time, Table *(l)* for *z/16* sequences, in the same succession, as

$$t_{1.1875}/t_\bullet = tanh(1.1875) = 0.8298019;$$

$$t_{1.25}/t_\bullet = tanh(1.25) = 0.8482836;$$
$$t_{1.3125}/t_\bullet = tanh(1.3125) = 0.8649066;$$
$$t_{1.375}/t_\bullet = tanh(1.375) = 0.8798267;$$

so that the accurate values of selected transitional times, in years from the origin of medicine, are:

$$t_{1.1875} = t_\bullet \cdot tanh(1.1875) = 32337;$$
$$t_{1.25} = t_\bullet \cdot tanh(1.25) = 33057;$$
$$t_{1.3125} = t_\bullet \cdot tanh(1.3125) = 33705;$$
$$t_{1.375} = t_\bullet \cdot tanh(1.375) = 34286.$$

Calculating the transitional times for earlier sequences, the timeline and intrinsic characteristics in development of medicine are presented as in the two tables below.

$z/16 = l$	Time (years)		Sequences of medicine
	from origin $t_l = t_\bullet \cdot tanh(l)$	from present $t_l - 34619$	
	38969	+4350	
...	...	...	
1.4375	34807	+188 (AD2202)	_Modern medicine
1.375	34286	-333 (AD1681)	_Chinese and European medicine
1.3125	33705	-914 (AD1100)	_Medieval medicine
1.25	33057	-1562 (AD452)	_Greco-Roman medicine
1.1875	32337	-2282 (268BC)	_Egyptian and Indian medicine
1.125	31538	-3081 (1067BC)	_Egyptian medical papyri
1.0625	30654	-3965 (1951BC)	_Egyptian-Babylonian remedies
1	29679	-4940 (2926BC)	_Sumerian medicine
0.9375	28606	-6013 (3999BC)	_Sumerian surgery
0.875	27430	-7189 (5175BC)	_Prehistoric medicine
0.8125	26147	-8472 (6458BC)	_Medical practices
0.75	24751	-9868 (7854BC)	_Treatments with plants, minerals
0.6875	23240	-11379	_Dietary knowledge
0.625	21612	-13007	_ Illness treatments
0.5625	19868	-14751	_Observations on diseases
0.5	18008	-16611	_ Beginning of internal medicine
0.4375	16038	-18581	_Treatment of injuries and traumas
0.375	13965	-20654	_ Treatment of traumas
0.3125	11796	-22823	_ First treatment of injuries
0.25	9544	-25075	_Continuing surgical medicine
0.1875	7222	-27397	_Beginning of surgical medicine
0.125	4846	-29773	_ Practice of trepanation
0.0625	2432	-32187	_ Oldest evidence of anatomy
0	0	-34619	_

$z/16 = l$	Time from origin $t_l = t_\bullet \cdot tanh(l)$ (years)	Period $\tau_l = (t_\bullet^2 - t_l^2)/(2t_\bullet)$ (years)	Frequency $f_l = (2t_\bullet)/(t_\bullet^2 - t_l^2)$ (years)$^{-1}$	Angular speed $\omega_l = 2\pi/\tau_l = 2\pi \cdot f_l$ (years)$^{-1}$
	38969	0	∞	∞
...	...	...	...	...
1.4375	34807	3939.7	0.0002538	0.0015948
1.375	34286	4401.6	0.0002272	0.0014275
1.3125	33705	4908.5	0.0002037	0.0012801
1.25	33057	5463.5	0.0001830	0.0011500
1.1875	32337	6067.7	0.0001648	0.0010355
1.125	31538	6722.5	0.0001488	0.0009347
1.0625	30654	7427.9	0.0001346	0.0008459
1	29679	8182.7	0.0001222	0.0007679
0.9375	28606	8985.1	0.0001113	0.0006993
0.875	27430	9830.6	0.0001017	0.0006391
0.8125	26147	10712.6	0.0000933	0.0005865
0.75	24751	11624.3	0.0000860	0.0005405
0.6875	23240	12554.7	0.0000797	0.0005005
0.625	21612	13491.5	0.0000741	0.0004657
0.5625	19868	14419.7	0.0000693	0.0004357
0.5	18008	15323.7	0.0000653	0.0004100
0.4375	16038	16184.2	0.0000618	0.0003882
0.375	13965	16982.2	0.0000589	0.0003700
0.3125	11796	17699.2	0.0000565	0.0003550
0.25	9544	18315.8	0.0000546	0.0003430
0.1875	7222	18815.3	0.0000531	0.0003339
0.125	4846	19183.2	0.0000521	0.0003275
0.0625	2432	19408.6	0.0000515	0.0003237
0	0	19484.5	0.0000513	0.0003225

18. Sociology

The study of structure and functioning of human society is named *sociology* (Latin *socius* 'partner, comrade, associate, ally' and Greek *λογία* 'study of' from *λόγος* 'word, speech, story'), this term being first coined by French essayist EJ Sieyès in 1780. Its subject is generally focused on a group of people involved with each other through persistent relations, or a large grouping sharing the same geographical or social territory, subject to the same political authority and dominant cultural expectations, and altogether is called *society* (Latin *societas* 'partnership, fellowship, association, alliance' deriving from *socius*). Human societies are characterized by patterns of relationships, as social relations, between individuals who share a distinctive culture and institutions; a given society being described as the sum total of such relationships among its constituent members. In the social sciences, a larger society involves stratification and dominance patterns in subgroups.

At a certain stage of social development, the integration of individuals into a group, such as community or society, can be described taking into account that any individual I is identifiable by distinctive attributes a_i, each of them amplified or diminished by a specific index b^i, so that $I = a^{b1}{}_1 \cdot a^{b2}{}_2 \cdot ... \cdot a^{bi}{}_i \cdot ... \cdot a^{bm}{}_m = \Pi^m{}_1\ a^{bi}{}_i$. According to the equality $ln(a^{b1}{}_1 \cdot a^{b2}{}_2 \cdot ... \cdot a^{bi}{}_i \cdot ... \cdot a^{bm}{}_m) = b^1 \cdot ln\ a_1 + b^2 \cdot ln\ a_2 + ... + b^i \cdot ln\ a_i + ... + b^m \cdot ln\ a_m$, or equivalently $ln(\Pi^m{}_1\ a^{bi}{}_i) = \Sigma^m{}_1\ (b^i \cdot ln\ a_i)$, any individual can be described in the form $ln\ I = ln(\Pi^m{}_1\ a^{bi}{}_i) = \Sigma^m{}_1\ (b^i \cdot ln\ a_i)$, or $I = \Pi^m{}_1\ a^{bi}{}_i = exp[\Sigma^m{}_1\ (b^i \cdot ln\ a_i)]$. Individuals I_1, I_2,... I_j,... I_n with their efficiencies Ie_j multiplied by coefficients of composition c^j are constituents $C_j = c^j \cdot Ie_j$ of a group G of efficiency $Ge = \Sigma^n{}_1\ c^j \cdot Ie_j$.

Sociology is closely related to social science, called *economics* (Greek *οίκονομία* 'management of a household, administration', from *οίκος* 'house' and *νόμος* 'custom, law') that studies how rational individuals, groups, and organizations manage scarce resources, which have alternative uses, to achieve desirable ends. Today, there are in use *macroeconomic models*, as analytical tools designed to describe the operation of the economy in a country or a region, such as: simple theoretical, empirical forecasting, dynamic stochastic general equilibrium, or agent-based computational macroeconomic models.

The origin of society is still disputed, maybe in sub-Saharan Africa, in Asia, or in Australia, around 27000-25000 years ago, but its development is largely agreed on the basis of archaeological discoveries, and marked by remains, thoughts, works, theories, writings, historical and sociological studies, social activities, and recently by econometric analyses and computer-modelling in economics, from which some are chronologically mentioned below.

26000 years ago: *The mammoth hunter* of Dolni Vestonice, in Southern Moravia, was associated with a recently discovered *ivory figurine*, indicating some social features of humans at that time;

26000-22000 years ago: *Tribal* and *Patriarchal societies*;

22000 years ago: *The Venus* of Willendorf, in Austria, was discovered as a *carving of a woman*, without facial features, fat, with pendulous breasts and a huge belly; other similar figures have been found *across Europe* from France to Russia, which indicate a kind of commonality;

22000-17000 years ago: *Clan-based societies*, *Hunting-gathering societies* and *Fishing societies*;

15000-10000BC: *Herding societies*, *Animal-domesticating societies* and *Pastoral societies* in Africa, Asia, Europe, Australia and America;

10000-7500BC: *Horticultural societies* and *Farming societies* in Africa and Asia;

7500-5000BC: *Agrarian societies* and *Irrigation-based agricultural societies* in North-east Africa, the Near East, South and South-east Asia, the Far East, Europe, and Americas;

5000-4000BC: *Trading societies* on the Eastern coasts of the Mediterranean, the Near East and Middle East, coasts of Red Sea and Indian Ocean, coasts of Western Pacific, South and East Asia;

4000-3000BC: *First hierarchically organized societies* emerged in Mesopotamia, Egypt, India and China;

3000-2000BC: *Societies based on city-states* occurred in Mesopotamia, Egypt, Aegean area, Indus Valley, China, and Mesoamerica;

2000-1000BC: *Ancient imperial societies* developed in the Near East and Middle East, India and China, as well as in Mesoamerica;

1000-250BC: *Classic ancient societies* extending from North Africa and South Europe, through the Near East and Middle East, to South and East Asia;

485-480BC: <u>Confucius</u>, Chinese philosopher, compiled *Analects*, including thoughts about the importance of *social roles*;

443-425BC: <u>Herodotus</u>, Greek historian, according to his views on society and social development, described *democracy* in contrast with *oligarchy* and *monarchy*;

424-404BC: <u>Thucydides</u>, Greek historian, analysed social organization, and considered that *democracy* was that delivered by the Athenian statesman Pericles;

c.400BC: <u>Plato</u>, Greek philosopher, coined the term *governance*, deriving from Greek κυβερνάω 'to steer', and concluded that the only hope for Greek cities was to trust in philosopher-kings, who have knowledge of goodness and are able to lead others to goodness;

250BC-AD450: *Hellenistic-Roman societies* were extended in West Asia, South and West Europe, as well as in North Africa;

AD450-1100: *Feudal societies* occurred in Europe, Egypt, the Near East and Middle East, India, China, South-east Asia, and probably in Mexico and Peru;

1100-1660: *Revived societies* during pre-Renaissance and Renaissance extended their authority from Americas in West to the Far East and Pacific islands in East, and from Siberia in North to Africa and Australia in South;

1380-1400: <u>Ibn Khaldun</u>, North African Arab philosopher, historian and politician, produced a monumental history of the Arabs, with its introduction *Muqaddima* as the first work to advance social-scientific reasoning on *social cohesion* and *social conflict*, explaining the rise and fall of states by waxing and waning of *asabiya* 'solidarity'; thus he is considered the first *sociologist*;

1576-87: <u>Jean Bodin</u>, French political philosopher, wrote the major work *Les Six Livres de la République* 'The Six Books of a Commonweal', expounding that property and the family form the basis of society, and showing that a limited monarchy is the best possible form of government; he also wrote *Colloquium Heptaplomeres*, presenting a plea for religious tolerance through the device of a conversation between a Jew, a Muslim, a Lutheran, a Zwinglian, a Roman Catholic, an Epicurean, and a Theist;

1600-1720: *Early origins of industrial societies* in Western Europe, following a series of discoveries and innovations which led to development of mining, textiles, machinery, canals, etc.;

1662: <u>John Graunt</u>, English haberdasher and demographer, published *Natural and Political Observations Made upon the Bills of Mortality*, comprising *elements of statistics*;

1759-76: <u>Adam Smith</u>, Scottish economist and philosopher, wrote *Theory of Moral Sentiments*, and *Inquiry into the Nature and Causes of the Wealth of Nations*, as basis of the theory of society in tradition of Scottish moral philosophy;

1766: <u>Anne Robert Jacques Turgot</u>, French economist and politician, produced *Reflexions sur la formation et la distribution des richesses* 'Reflections on the Formation and Distribution of Wealth', as the best outcome of the Physiocratic school, a topic that was largely anticipated by A Smith;

1769-90: <u>Edmund Burke</u>, Irish statesman and philosopher, published political thoughts in his works *Observations on the Present State of the Nation*, and *On the Causes of the Present Discontents*; speeches on *American Taxation* and *On Conciliation with America*; he also wrote *Reflections on the Revolution in France*;

1780-89: <u>Emmanuel Joseph Sieyès</u>, French cleric and political theorist, coined the word *sociology* in an unpublished manuscript (later published as *Des Manuscrits de Sieyès*), and wrote the pamphlets *Essai sur les privileges* 'Essay on the Privileged' and *Qu'est-ce que le tiers-état* 'What is the Third Estate';

1798: <u>Thomas Robert Malthus</u>, English economist and clergyman, published *Essay on the Principle of Population*, pointing natural tendency of population to increase faster than means of subsistence, with significance in *sociology*;

1809-19: <u>David Ricardo</u>, English political economist, wrote the pamphlet *The High Price of Bullion, a Proof of the Depreciation of Banknotes*, and published *Principles of Political Economy and Taxation*;

1813-49: <u>Robert Owen</u>, Welsh social and educational reformer, was author of *A New View of Society*, showing that character was formed by social environment, and of *Revolution in Mind and Practice*; he also organized the *Grand National Consolidated Trades Union*;

1824-25: <u>William Thompson</u>, Irish economic theorist, published *An Enquiry into the Principles of the Distribution of Wealth Most Conducive to Human Happiness*, as well as *Appeal of One Half of the Human Race, Women, against the Pretentions of the Other Half, Men, to Retain them in Political, and thence in Civil and Domestic, Slavery*;

1830-54: <u>Auguste Comte</u>, French philosopher and social theorist, produced *Cours de Philosophie positive*, and *Système de politique positive* 'System of Positive Philosophy', expounding the laws of social evolution, describing the organization and hierarchy of all branches of human knowledge, and establishing the *science of society* as the basis for social planning and regeneration;

1835-56: Alexis Charles de Tocqueville, French historian and political scientist, wrote the penetrating political study *De la démocratie en Amérique* 'Democracy of America', extensive diary *Journeys to England and Ireland*, and the volume *L'Ancien Régime et la Révolution* 'The Old Regime and the Revolution';

1835-71: Adolphe Jacques Quételet, Belgian statistician and astronomer, published *Sur l'homme* 'A Treatise on Man and the Development of His Faculties', and *L'Anthropométrie*, advocating the use of statistics to formulate *social laws*;

1837-59: Henry Charles Carey, US political economist, wrote three-volume *Principles of Political Economy*, and also three-volume *Principles of Social Science*, regarding free trade as an ideal, but impossible in the existing state of US industry;

1840-58: Pierre Joseph Proudhon, French journalist and political theorist, produced *Qu'est-ce que la propriété* 'What is Property', the great work *Système des contradictions économiques* 'System of Economic Contradictions', and three-volume *De la justice dans la Révolution et dans l'église* 'On Justice in the Revolution and the Church';

1859-67: Karl Marx, German social, political and economic theorist, rejected A Comte's positivism, and attempted to develop a *science of society*, as expressed in *Zur Kritik der politischen Ökonomie*, and in *Das Kapital*;

1860-1900: Jean Henry Dunant, Swiss philanthropist, and Frédéric Passy, French economist, contributed in the foundation of *International Committee for the Red Cross*, along with *Inter-Parliamentary Union*; and organized the first *Universal Peace Congress* respectively;

1862-93: Herbert Spencer, English revolutionary philosopher, as an advocate of *Social Darwinism*, he produced the major nine-volume *System of Synthetic Philosophy*, bringing together metaphysics, ethics, biology, psychology and sociology, and also coining the phrase 'survival of the fittest'; his other works include *Social Statistics*; *Education*; *The Man Versus the State*; and *Autobiography*;

1872-78: Friedrich Wilhelm Nietzsche, German philosopher and writer, developed characteristic themes of vehement repudiation of

Christian and liberal ethics, detestation of democratic ideals, celebration of *Übermensch* 'superman' who can create and impose his own law, death of God and life-affirming *will to power*;

1882-1911: Wilhelm Dilthey, German philosopher and historian of ideas, made distinction between the natural sciences *Naturwissenschaften*, offering explanations of physical events through causal laws, and the human sciences *Geisteswissenschaften*, offering understanding 'Verstehen' of events in terms of human intentions and meanings;

1894-97: Émile Durkheim, French sociologist, produced methodological writings such as *Les Règles de la méthode sociologique* 'The Rules of Sociological Method', based on 'social facts' which should be treated as 'things' for explaining solely by reference to other social facts, not in terms of any individual person's actions;

1899-1905: Thorstein Veblen, US economist and social critic, made an unorthodox analysis of US social and economic institutions in his major work *The Theory of the Leisure Class*, and coined the term *conspicuous consumption* to describe pointless acquisitiveness through which leisure class declares its privileges;

1900-14: Georg Simmel, German sociologist and philosopher, was the principal representative of German *sociological formalism*, which emphasized the form of a phenomenon rather than its nature or content, encouraged the growth of an independent sociology, defining its boundaries with other disciplines; his works included *Philosophy of Money*, and a collection of essays published as *Georg Simmel: On Women, Sexuality and Love*;

1904: Max Weber, German sociologist, regarded as one of the founders of sociology, he became best known for his work *Die protestantische Ethik und der Geist des Kapitalismus* 'The Protestant Ethic and the Spirit of Capitalism';

1919-46: John Maynard Keynes, English economist, produced *The Economic Consequences of the Peace*, setting out views against harsh economic terms imposed on Germany in the Versailles Treaty; *Treatise of Probability*, exploring logical relationships between calling something 'highly probable' and a 'justifiable induction'; the great work *A Treatise on Money*; and the revolutionary *General Theory of Employment, Interest and Money*, pioneering theory of full employment; as well as he had an important role in formulation of the *Bretton Woods agreements*, and in establishment of the *International*

Monetary Fund;

1924-25: <u>Joseph Austen Chamberlain</u>, English politician, and <u>Charles Gates Dawes</u>, US diplomat, played important roles in negotiations for the *Locarno Pact*, regarding German reparation payments;

1929-45: <u>Karl Mannheim</u>, German sociologist, founded the *sociology of knowledge*, and published *Ideology and Utopia*; *Man and Society in an Age of Reconstruction*; *Freedom, Power and Democratic Planning*; and *Diagnosis of Our Time*;

1930-44: <u>Pitirim Sorokin</u>, US sociologist, founded the department of sociology at Harvard, developed a theory dividing socio-cultural systems into 'sensate' and 'ideational' types; and wrote *Sociology of Revolution*; *Crisis of our Age*; *Russia and the United States*; and *Fads and Foibles of Modern Sociology*;

1932-70: <u>Lionel Charles Robbins</u>, English economist, became well-known by his works *An Essay on the Nature and Significance of Economic Science*; *The Economic Problem in Peace and War*; *Classical Political Economy*; and *The Evolution of Modern Economic Theory*;

1935-48: <u>George Horace Gallup</u>, US public opinion pollster, founded the American Institute of Public Opinion, evolving in *Gallup Polls* for testing state of public opinion, and published *Public Opinion in a Democracy*, and *Guide to Public Opinion Polls*;

1937-69: <u>Talcott Parsons</u>, US sociologist, promoted *action theory*, and published *The Structure of Social Action*; *The Social System*; *Sociological Theory and Modern Society*; and *Politics and Social Structure*;

1939-62: <u>Jan Tinbergen</u>, Dutch economist, analysed the USA Depression in his *Business Cycles in the USA 1919-32*, as an economic advice to the League of Nations, and published *Econometrics*; *Economic Policy: Principles and Design*; and *Shaping the World Economy*;

1941-83: <u>George Joseph Stigler</u>, US economist, was author of books such as *Production and Distribution Theories*; *The Theory of Price*; *The Citizen and the State*; and *The Economist as Preacher*;

1944-63: <u>Gunnar Myrdal</u>, Swedish economist and politician, produced the classic study of race relations in USA, entitled *An American Dilemma*, and also wrote *Beyond the Welfare State*, and *The Challenge of Affluence*;

1945-75: <u>Paul Felix Lazarsfeld</u>, Austrian-born US sociologist, established the Bureau of Applied Social Research at Columbia, and promoted *quantitative methodology*, popular culture in mass communications, *political sociology*, as well as *applied sociology*;

1947-50: <u>George Catlett Marshall</u>, US soldier and politician, as Secretary of State, originated the *Marshall Aid plan* for post-war reconstruction of Europe 'ERP';

1947-91: <u>Herbert Alexander Simon</u>, US economist, pioneered research into *decision-making process* in economic organization, and published *Administrative Behaviour*; *Models of Man*; *Human Problem Solving*; *Reason in Human Affairs*; and *Models of My Life*;

1949-77: <u>Robert King Merton</u>, US sociologist, founded the *sociology of science* in its modern form, and wrote *Social Theory and Social Structure*; *On Theoretical Sociology*; *Science, Technology and Society in Seventeenth-Century England*; *The Sociology of Science*; and *The Sociology of Science in Europe*;

1950-75: <u>Mother Teresa of Calcutta</u>, Roman Catholic nun and missionary, founded the *Order of the Missionaries of Charity*, which became a pontifical congregation, running over 650 charity houses in 124 countries; she also worked with lepers and established a leper colony called *Shanti Nagar* 'Town of Peace' near Asansol, Bengal;

1950-80: <u>Lawrence Klein</u>, US economist, became known for his *computer-based models* that help governments forecast the future and act accordingly, and for *econometric models* and their application to the analysis of *economic fluctuations* and *economic policies*;

1951-75: <u>James Edward Meade</u>, English economist, contributed to international trade, by his two-volume *The Theory of International Economic Policy*, four-volume *Principles of Political Economy*, as well as *The Intelligent Radical's Guide to Economic Policy*;

1957-65: <u>Martin Luther King</u>, US clergyman and civil rights leader, founded the *Southern Christian Leadership Conference*, with the *great march on Washington* and the memorable 'I have a dream' speech, and espoused the *philosophy of non-violence and passive resistance*, which led to securing passage of the Civil Rights Act and Voting Rights Act;

1982-90: <u>John Charles Harsanyi</u>, Hungarian-born US economist, <u>John Forbes Nash</u>, US mathematician and economist, and <u>Reinhard Selten</u>, German-born US economist, pioneered *game theory with applications in economy*, as presented by JC Harsanyi and R Selten in *A General*

Theory of Equilibrium Selection in Games, and by JF Nash in his *mathematical development of this theory*;

1993-2011: Jean-Louis Brillet, researcher at the French National Institute of Statistics and Economic Studies, formalized his experience in *structural modelling* into a coherent strategy, both global and detailed, *econometric modelling, international macroeconomics*, and production of operational models using *EViews software*;

1995-2010: Lucian Liviu Albu, Romanian economist, published national and international studies on economy, including books such as *The Transition of Economy or Transition of Economics?*, and *Non-Linear Macroeconomics and Forecast –Theory and Application*, concerning integration of Eastern European countries into the European Union; and also developed a series of models in economics and econometrics, such as *A dynamic model to estimate the 'pure' productivity*, MPRA 13425, University Library of Munich;

2004: Peter Charles Bonest Phillips and Viv Hall, New Zealand economists, published their paper *Early development of econometric software at the University of Auckland*, Cowles Foundation for Research in Economics at Yale University, New Haven;

Current types of models in economics: *Stochastic models* - autoregressive models in which the *stochastic process* satisfies some relation between present and past values, usually based on *statistics* to formulate and test hypotheses or estimate various parameters; *Non-stochastic models* - either purely qualitative (e.g. involving in some aspect economics, undoubtedly of social science theory), or quantitative (e.g. with hyperbolic co-ordinates and/or specific forms of *functional relationships* between the variables); and *Qualitative models* - such as *scenario planning* in which possible future numbered events are played out, or *non-numerical decision tree analysis*).

Among the human abilities presented at the end of chapter 7, society originated 26653 years ago, as its initial time, and will last until about 4350 years from the present into the future, so that its final time would be $t_\bullet = 26653 + 4350 = 31003$ years. Meanwhile, according to the social development described in this chapter, the sequences in development of society can be delimited less accurately for earlier and more accurately for later ones. Later times of transition from one sequence to another can be approximated as follows:

Classic ancient \ 250BC \ *Hellenistic-Roman* \ AD450 \
Feudal \ AD1100 \ *Revived* \ AD1660 \ *Industrial societies.*

327

These transitional times represent 2264; 1564; 914; 354 years ago; and 26653 - 2264 $\approx$ 24389; 26653 - 1564 $\approx$ 25089; 26653 - 914 $\approx$ 25739; 26653 - 354 $\approx$ 26299 years from the origin of society. The last of these times divided by the final time $t_\bullet$ = 31003 years from the same origin result in ratios 24389/31003 $\approx$ 0.787; 25089/31003 $\approx$ 0.809; 25739/31003 $\approx$ 0.830; 26299/31003 $\approx$ 0.848; which approximate quite well the ratios given in Background to time, Table (l) for $z/16 = 1$ sequences, as $t_{1.0625}/t_\bullet$ = $tanh(1.0625)$ = 0.7866188; $t_{1.125}/t_\bullet$ = $tanh(1.125)$ = 0.8093011; $t_{1.1875}/t_\bullet$ = $tanh(1.1875)$ = 0.8298019; $t_{1.25}/t_\bullet = tanh(1.25) = 0.8482836$; therefore the transitional times are calculated, in years from the origin of society, as follows:

$$t_{1.0625} = t_\bullet \cdot tanh(1.0625) = (31003) \cdot (0.7866188) = 24388;$$
$$t_{1.125} = t_\bullet \cdot tanh(1.125) = (31003) \cdot (0.8093011) = 25091;$$
$$t_{1.1875} = t_\bullet \cdot tanh(1.1875) = (31003) \cdot (0.8298019) = 25726;$$
$$t_{1.25} = t_\bullet \cdot tanh(1.25) = (31003) \cdot (0.8482836) = 26299.$$

Continuing this procedure for earlier values of argument k, the society timeline, sequences, and intrinsic characteristics are found such as:

$z/16 = l$	Time (years)		Sequences in social development
	from the origin $t_l = t_\bullet \cdot tanh(l)$	from present $t_l - 26653$	
	31003	+4350	
...	...	...	
1.3125	26815	+162 (AD2176)	$-$
1.25	26299	-354 (AD1660)	_Industrial societies
1.1875	25726	-927 (AD1087)	_Revived societies
1.125	25091	-1562 (AD452)	_Feudal societies
1.0625	24388	-2265 (251BC)	_Hellenistic-Roman societies
1	23612	-3041 (1027BC)	_Classic ancient societies
0.9375	22758	-3895 (1881BC)	_Ancient imperial societies
0.875	21823	-4830 (2816BC)	_Societies based on city-states
0.8125	20802	-5851 (3837BC)	_Hierarchic societies
0.75	19692	-6961 (4947BC)	_Trading societies
0.6875	18489	-8164 (6150BC)	_Irrigation-based societies
0.625	17194	-9459 (7445BC)	_Agrarian societies
0.5625	15806	-10847	_Farming societies
0.5	14327	-12326	_Horticultural societies
0.4375	12760	-13893	_Pastoral societies
0.375	11110	-15543	_Animal-domesticating societies
0.3125	9385	-17268	_Herding societies
0.25	7593	-19060	_Fishing societies
0.1875	5746	-20907	_Hunting-gathering societies
0.125	3855	-22798	_Clan-based societies
0.0625	1935	-24718	_Patriarchal societies
0	0	-26653	_Tribal societies

$z/16 =$ l	Time from origin $t_l = t_\bullet \cdot tanh(l)$ (years)	Period $\tau_l =$ $(t_\bullet^2 - t_l^2)/(2t_\bullet)$ (years)	Frequency $f_l =$ $(2t_\bullet)/(t_\bullet^2 - t_l^2)$ (years)$^{-1}$	Angular speed $\omega_l =$ $2\pi/\tau_l = 2\pi \cdot f_l$ (years)$^{-1}$
	31003	0	∞	∞
...	...	...	...	...
1.3125	26815	3905.1	0.0002561	0.0016090
1.25	26299	4347.1	0.0002300	0.0014454
1.1875	25726	4827.9	0.0002071	0.0013014
1.125	25091	5348.3	0.0001870	0.0011748
1.0625	24388	5909.3	0.0001692	0.0010633
1	23612	6510.0	0.0001536	0.0009652
0.9375	22758	7148.7	0.0001399	0.0008789
0.875	21823	7820.9	0.0001279	0.0008034
0.8125	20802	8522.8	0.0001173	0.0007372
0.75	19692	9247.7	0.0001081	0.0006794
0.6875	18489	9988.4	0.0001001	0.0006290
0.625	17194	10733.7	0.0000932	0.0005854
0.5625	15806	11472.4	0.0000872	0.0005477
0.5	14327	12191.1	0.0000820	0.0005154
0.4375	12760	12875.7	0.0000777	0.0004880
0.375	11110	13510.9	0.0000740	0.0004650
0.3125	9385	14081.0	0.0000710	0.0004462
0.25	7593	14571.7	0.0000686	0.0004312
0.1875	5746	14969.0	0.0000668	0.0004197
0.125	3855	15261.8	0.0000655	0.0004117
0.0625	1935	15441.1	0.0000648	0.0004069
0	0	15501.5	0.0000645	0.0004053

19. Construction

The process, manner or result of building and assembling infrastructure is called *construction* (Latin *constructio* 'putting together, building, construction' from *construo, construere* 'to heap up together, construct, build up, arrange'). The activity of construction was based on identification and examination of various kinds of wood, stone, clay, other natural materials, then metals, alloys, cements, mortars, concretes and synthetic materials with their properties, in order to be used for marking places of shelter, burial, ritual, ceremony, celebration, and other significant events, as well as for protecting themselves, or for preserving, preparing and storing goods. This activity was developed by building structures such as: huts, houses, tombs, tumuli, forts, henges, monuments and villages; towns with fortifications, towers, bridges and harbours; cities with citadels, palaces, temples, and markets; vehicle roads and trade routes; pyramids, ziggurats, mausoleums, gardens, amphitheatres, dams, lighthouses, canals, aqueducts, drains and sewers; churches, mosques and cathedrals, and later silos, railways, tunnels, metros, highways, skyscrapers, airports, cosmodromes and so on. In general, there are residential, industrial, commercial, and civil constructions. As towns and then cities grew, the constructors became professional craftsmen, like bricklayers and carpenters, and later they were organized into guilds. Today, the activity of construction is managed by a project manager, and supervised by a constructor manager, design engineer, construction engineer or project architect.

In the course of constructional development there was a series of successive prominent features and representative works, as well as statesmen, builders, masons, engineers, designers, architects, millwrights, and so on; some of them only being mentioned below.

20000 years ago: *Theopetra cave*, in Greece, preserves a stone wall, as a remnant of the oldest known *stone structure* in the world;

20000-14000 years ago: *Shelters and houses, Earthen-constructions, Wood type constructions, Wood and stone constructions*;

14000-12000 years ago: *Stone structures*;

9500-8400BC: *Göbekli Tepe*, in South-eastern Anatolia, Turkey, still exhibits a *ceremonial structure* with pillars and circles, considered the *oldest standing monument* in the world;

8000BC: *Zeolots of Jerico*, in Israel, known by remains of a *defensive-agricultural structure* possibly belonging to an old village;

7400-6200BC: *Çatalhöyük*, in Anatolia, Turkey, contains *village ruins* with dwelling levels, unveiling the constructional techniques of the time, and spaces for manufacturing pottery and wool textiles;

5800-3000BC: *Khirokitia*, in Cyprus, was a *collective settlement* with surrounding *fortifications* for communal protection;

4850BC: *Barnenez*, in France, was used as a *passage grave*, being the oldest known building in Europe;

4700-4500BC: *Tumulus of Bougon*, near La-Mothe-Saint-Héray, in Poitou-Charentes, and *Tumulus Saint-Michel*, at Carnac, in Bretagne-Morbihan; both in France, were *Megalithic necropolises*;

4500BC: *First temple* built by Sumerians;

3700BC: *Neolithic Temples of Ġgantija*, in Gozo, Malta, were built in Malta's Copper Age and considered the second oldest free-standing monuments in the world, with walls of coralline limestone blocks, some weighing as much as 20 tonnes, quarried from a hill on the other side of a nearby valley, carried on smaller spherical stone balls, structured in typical clover-leaf shape with inner facing blocks, which contains five semi-circular apses, showing traces of plaster, connected with a central passage;

3650BC: *West Kennet Long Barrow*, in Avebury, England, is a *Neolithic chamberd tomb*, one of the oldest such structures in Britain;

3500BC: *Sechin Bajo*, in Peru, was used as a *Plaza*, and is the oldest known building in Americas;

2800-1400BC: *Stonehenge* on the Sainsbury Plain, in Wiltshire, England, is a *megalithic monument* which existed before the Bronze Age, and was developed by a double ring of bluestones transported on an impressive distance from South Wales, and then by a ring of 'sarsen' sandstones carried on a shorter distance from its location;

2700BC: *Temple of Ba'alat Gebal*, in Lebanon, was dedicated to the Byblos' Goddess of the City named *Astarte*;

2687-2668BC: *Step Pyramid of Djoser*, at Saqqara, North-west of Memphis, in Egypt, was designed by the earliest known architect Imhotep, and constructed of dressed masonry with six mastabas (of decreasing size) as the world's *oldest monumental structure*, 62-metre tall and base of 109x125 metres, and with underground tunnels about 5.5-kilometre long;

2600BC: *Mehrgarh*, in Pakistan, are mud brick storage structures, as the oldest known constructions in Asia;

2589-2566BC: *Great Pyramid of Giza*, also known as 'Pyramid of Khufu', in Egypt, 146.5 metre tall, was built with about 2.3 million limestone blocks transported from nearby quarries, and originally covered with casing stones to form a smooth outer surface, which was considered the 1st Wonder of Ancient World;

2500BC: *Mesoamerican pyramid*, at Chiapa de Corzo, in Mexico, is the oldest built pyramid in Central America when the Zoque civilization arose, as a *three-story-tall pyramid* with a top tomb coated head-to-toe in sacred red pigment;

2200-2100BC: *Sialk ziggurat*, in Kashan, Iran, erected as the oldest known *ziggurat* (Akkadian *ziqqurat*, coming from *zaqāru* 'to build on a risen area'), a massive structure having the form of a terraced step pyramid of successively receding stores or levels;

2100BC: *Ziggurat of Ur*, in Sumeria (now in Tell el-Mukayyar, Iraq), was a *mud brick and baked brick temple* constructed with three levels of terraces, originally 25-30 metres high, built by King Ur-Nammu for the moon goddess *Nanna*;

2000BC: *Dolmen de Viera*, at Antequera, Málaga, Spain, is a *Megalithic tomb* covered by a tumulus (mound) 50 metres in diameter;

2000-1300BC: *Knossos*, in Crete, Greece, was the Minoan capital with a vast *Palace* and surrounding villas, dependent buildings and cemeteries, as the most important construction of *Minoan civilization*;

1250BC: *Chogha Zanbil ziggurat*, in Khuzestan province of Iran, was erected by King Untash-Napirisha as an *Elamite complex* that is the best preserved structure of this kind;

1850-1580BC: Transition from *Pyramids and ziggurats* to *Palaces and megarons*;

1000-820BC: Transition from *Palaces and megarons* to *Temples and lighthouses*;

800-600BC: *Cuicuilco Circular Pyramid*, in Mexico, was a *ceremonial centre* as the oldest known building in North America;

c.600BC: *Hanging Gardens of Babylon* were built in Mesopotamia, and considered the 2nd Wonder of Ancient World;

590-570BC: *Temple of Artemis at Ephesus* was constructed in Asia Minor, and counted as the 3rd Wonder of Ancient World;

447-406BC: *Parthenon* and *Erechtheion*, on Acropolis in Athens, Greece, are two *great temples*, built under the supervision of the mason and sculptor <u>Phidias</u>, as Pericles' employee; the first temple being erected by the architects <u>Ictinos</u> and <u>Callicrates</u>, and dedicated to the maiden goddess *Athena*; and the second dedicated to both *Athena* and *Poseidon*;

c.350BC: *Mausoleum of Halicarnassus* was built in Asia Minor, and counted as the 5th Wonder of Ancient World;

335-325BC: <u>Crates of Chalkis</u>, Greek engineer, supervised construction of the new *city* and *port of Alexandria*, for Alexander the Great, as well as of *drainage, irrigation* and *water supply* works;

270-250BC: *Pharos of Alexandria* was erected in North Egypt, and considered the 7th Wonder of Ancient World;

230-100BC: Transition from *Temples and lighthouses* to *Aqueducts, amphitheatres and bridges*;

30-12BC: <u>Marcus Vipsanius Agrippa</u>, Roman general, statesman, designer and constructor, built a large aqueduct named *Aqua Virgo* 'The Virgin's Aqueduct' by which the water of a spring previously discovered, in keeping with the legend by a young maiden, was brought to the city; he also supervised the construction of several *canals* and *aqueducts*, about 700 *wells* for fresh water, more than 100 drinking *fountains*, and 130 *tanks of water*;

AD25: <u>Marcus Vitruvius Pollio</u>, Roman architect and military engineer, wrote *De Architectura* 'On Architecture' as the only Roman treatise on architecture still extant;

70-80: *Colosseum* (*Coliseum*), in Rome, Italy, is an elliptical amphitheatre built of concrete and stone, as the largest amphitheatre in the world; its construction began under the emperor <u>Titus Flavius Vespasian</u>, and was completed under his successor and heir <u>Flavius Sabinus Titus</u>;

101-102: <u>Apollodorus of Damascus</u>, Roman architect, designed and constructed *Trajan's Bridge* across Danube, at Drobetae (now in South-west Romania), for access to Dacia during the Roman-Dacian Wars, as one of the most impressive constructions of that time;

126: *Pantheon*, in Rome, Italy, is an imposing 43.3 metre tall *temple* rebuilt by the emperor <u>Publius Aelius Hadrian</u>, and completed with granite columns brought from Egypt and shaped in Corinthian style, and with a large inside *rotunda* under a confined unenforced concrete dome (the largest in the world) open to the sky by a central *oculus* (opening); this temple being erected on the site of a previous MV Agrippa's temple dedicated to all the gods of Rome and destroyed except for its façade;

470-550: Transition from *Aqueducts, amphitheatres and bridges* to *Churches and mosques*;

537-1931: *Hagia Sophia* (from Greek *Ἁγία Σοφία* 'Holy Wisdom', Latin *Sancta Sophia*, Turkish *Ayasofya*), in Istanbul, Turkey, was an Eastern Orthodox cathedral, the largest in the world, designed by Isidore of Miletus and Anthemius of Tralles; then transformed by Ottoman Turks into the principal mosque of Istanbul from 1453; and finally opened as a museum in 1935; it is sized as 82-metre length, 73-metre width, and 55-metre height;

622-1000: *Great mosques* such as *Al-Masjid al-Nabawi* 'Mosque of the Prophet' (400500 m^2) in Medina, Saudi Arabia; *Masjid al-Haram* 'The Sacred Mosque' (356800 m^2) in Mecca, Saudi Arabia; *Al-Aqsa Mosque* 'The Farthest Mosque' (144000 m^2) in Jerusalem, Israel; *Imam Reza Shrine* 'Shrine of Imām Reza' (598657 m^2) in Mashhad, Iran, the largest in the world; and *Al-Azhar Mosque* 'Mosque of the Most Resplendent' (7800 m^2) in Cairo, Egypt;

792: *Palatine Chapel*, in Aachen, Germany, is a centrally planned, *domed chapel*, consecrated in honour of *Virgin Mary*, which is the only standing component of the once great Charlemagne's *Palace of Aachen*;

1100-50: Transition from *Churches and mosques* to *Cathedrals and great palaces*;

1163-1345: *Notre Dame de Paris* is a Gothic style 90-metre high cathedral, located on the Île de la Cité in Paris, which was accomplished by builders including Jean de Chelles, Pierre de Montreuil, Pierre de Chelles, Jean Ravy, and Jean le Bouteiller;

1248-1880: *Hohe Domkirche St. Petrus* (Cologne Cathedral) is a Roman Catholic church in Köln 'Cologne', which is the biggest cathedral in Germany; it was built under the direction of Master Gerhard and Master Michael in Gothic style and sized as 144.6-metre length, 86.2-metre width and 157.4-metre antenna spire;

1386-1965: *Duomo di Milano* 'Milan Cathedral', is a Gothic-style cathedral, the largest in Italy, 158.5-metre length, 92-metre width, and 108-metre height, which was constructed under the supervisions of architects and engineers such as Simone da Orsenigo, Zeno da

Campione, Giovannio de Grassi, Fabio Mangone and Francesco Maria Ricchino;

1401-1528: *Catedral de Santa Maria de la Sede* 'Seville Cathedral', in Andalusia, Spain, is a Roman Catholic cathedral, the largest in Gothic style, of 135-metre length, 100-metre width, 42-metre height, and 105-metre spire height, which was built up with contribution of architects including Alonso Martínez, Pedro Dancart, Carles Galtés de Ruan, and Alonso Rodríguez;

1420-45: Filippo Brunelleschi, Italian architect and sculptor, designed and supervised construction of the *Dome of Santa Maria del Fiore*; *S Spirito*; *S Lorenzo*; and the *Spedale degli Innocenti* 'Foundling Hospital'; all of them in Florence, Italy;

1485: Leon Battista Alberti, Italian architect, wrote the ten-volume *De re aedificatoria*, a treatise explicitly showing *Roman wall construction*;

1564-1860s: *Palais des Tuileries*, royal and imperial palace in Paris, on the right bank of the River Seine, was initially erected by the architect Philibert de l'Orme, a great master of the French Renaissance, exposing an immense façade of 266 metres; further achievements were carried out until its demolition started in 1871 during suppression of the Paris Commune and completed in 1883;

1608-40: Jan Adrianszoon Leeghwater, Dutch hydraulic engineer and millwright, supervised *drainage of the largest lake* in northern Netherlands, of 17,000 acres in extent and up to 3 metres in depth, using *multi-stage scoop-wheel water-lifting systems*; he and also carried out many *drainage projects* in Holland, France, Germany, Denmark and Poland;

1621-52: Cornelius Vermuyden, English drainage engineer, was commissioned to repair a breach of the Thames at Dagenham, and also to construct a *draining system* for 122,000 hectares of *Bedford Level*;

1629: *Weibbe Hayes Stone Fort* is a *defensive fort*, considered as the oldest building in Australia;

1660-1700: Transition from *Cathedrals and great palaces* to *Great structures and skyscrapers*;

1663-1710: Christopher Wren, English architect, designed the *Chapel at Pembroke college* in Cambridge, the *Sheldonian Theatre* in Oxford; supervised the construction of the new *St Paul's Cathedral* in London; also designed the *Royal Exchange Greenwich Observatory*, the

Ashmolean Museum in Oxford; and used iron hangers to suspend floor beams at *Hampton Court Palace*, South-west of London;

1675-1720: *St Paul's Cathedral* is an English Baroque two-towered cathedral in the City of London, 158-metre length, 37-metre nave width, and 111-metre height, which was designed by architect Christopher Wren; and later decorated by painter James Thornhill in the 18th century; illustrated by German painter Julius Schnorr von Carolsfeld in the 19th century; provided with sculptures by William Reid Dick and John Skelton, and stained glass windows by designer Brian Thomas in the 20th century;

1678-88: Sébastien le Prestre de Vauban, French military engineer, constructed a *cordon of fortresses* and the magnificent *Aqueduct of Maintenon*, in France;

1759-72: James Brindley, English engineer and canal builder, designed and supervised the construction of *Worsley-Manchester*, *Grand Trunk*, *Birmingham*, *Chesterfield* and other *canals*, altogether 365 miles in length;

1790-1828: Thomas Telford, Scottish civil engineer, constructed the *Masonry Arch Bridge* over Severn at Montford, *Pont-Cysyllte Aqueduct* on Ellesmere Canal, *Caledonian Canal* in Scotland, *Wrought-Iron Menai Suspension Bridge* in Wales, and *St Katherine's Docks* in London;

1816-20: John Loudon McAdam, Scottish inventor and engineer, remade roads with crushed stone bound with gravel, so raising them to improve drainage, as a system named after him *macadam*;

1830-59: Isambard Kingdom Brunel, English engineer and inventor, designed and constructed the *Thames Tunnel*, *Hungerford Suspension Bridge* over Thames at Charing Cross, and *Clifton Suspension Bridge*; tunnels, bridges and viaducts for the *Great Western Railway*; and also the *docks* of Bristol, Monkwearmouth, Cardiff, and Milford Haven;

1850-59: Robert Stephenson, English mechanical and structural engineer, constructed the *Britannia Tubular Bridge*, bridges at Conway and Montreal, *High Level Bridge* at Newcastle upon Tyne, and *Royal Border Bridge* at Berwick;

1880-92: Gustave Eiffel, French engineer, designed notable bridges and viaducts, and also the *Eiffel Tower* in Paris, the world's highest building until 1930;

1882-92: Anghel Saligny, Romanian engineer, designed and supervised construction of the *Cernavoda Bridge* over Danube, the longest in Europe when it was finished; and the *Harbour of Constanta*, Romania's chief-port;

1899: *Cape Adare Huts*, were constructed as *explorers' huts*, the oldest buildings in Antarctica;

1931-65: <u>Othmar Hermann Ammann</u>, Swiss-born US structural engineer, designed the greatest suspension bridges, namely *G Washington Bridge* (1060 m) in New York, *Golden Gate Bridge* (1260 m) in San Francisco, and *Verrazano Narrows Bridge* (1280 m) in New York;

1932: <u>Ralph Freeman</u>, English civil engineer, designed *Sydney Harbour Bridge*, a 1670 ft span construction in Australia, as well as *long-span suspension bridges* over estuaries of Forth, Severn, and Humber rivers in the UK;

1943: *Pentagon* in Arlington, Virginia, USA, is a large office building with surface of about 600,000 square metres, from which 340,000 square metres used as offices, designed by architect <u>George Bergstrom</u>;

1959: *Great Buildings* of Beijing, China, were completed and became operational as *Great Hall of the People*; *National Museum of China*; *Cultural Palace of Nationalities*; *Beijing Railway Station*; *Workers Stadium*; *Diaoyutai State Guesthouse*; *Minzu Hotel*; and *Chinese People's Revolutionary Military Museum*;

1966: The 553-metre high *CN Tower* in Toronto City was built as the first true skyscraper in Canada;

1973: *Sydney Opera House* in Sydney, Australia, a big building of 183 by 120 metres sized and 65 metres high, on a surface of 18,000 square metres;

1997: *Palace of the Parliament* in Bucharest, Romania, is considered the world's largest civilian building, that measures 270 by 240 metres, 86 metres height and 92 metres underground, with floor space of 340,000 square metres, which was designed and supervised by a group of 700 architects led by <u>Anca Petrescu</u>;

2005-09: Construction of the 322-metre high *Q1-Gold Coast* in Australia; and of the 270-metre high *Cullinan I 'North Tower'* in Hong Kong;

2010: Building of the 302-metre high *Capital City Moscow Tower*; 265-metre high *8 Spruce Street* in New York City; and 261-metre high *Sapphire* in Istanbul;

2012: Construction of the 414-metre high *Princess Tower* in Dubai; 310-metre high *The Shard* in London; and 302-metre high *Gramercy Residences* at Makati in Philippines;

2013: *The New Century Global Centre in Chengdu*, China, is a megastructure of 1,700,000 square metres, considered the largest building in the world.

According to the table of chapter 7, the activity of construction emerged 20207 years ago, as its initial time, and will last until 4350 years forward from present, i.e. its final time would be

$t_\bullet$ = 20207 + 4350 = 24557 years from the origin of this activity.

It developed in sequences of predominant/representative structures, from which the later ones are delimited by the transitional times:

Great pyramids and ziggurats \ 1850...1580BC $\approx$ 1715BC \
Palaces and megarons \ 1000...820BC $\approx$ 910BC \
Temples and lighthouses \ 230...100BC $\approx$ 165BC \
Aqueducts, amphitheatres and bridges \ AD470...550 $\approx$ AD510 \
Churches and mosques \ AD1100...1150 $\approx$ AD1125 \
Cathedrals and great palaces \ AD1660...1700 $\approx$ AD1680 \
Great structures and skyscrapers.

The above transitional times represent 3729; 2924; 2179; 1504; 889; 334 years ago; and 20207 - 3729 $\approx$ 16478; 20207 - 2924 $\approx$ 17283; 20207 - 2179 $\approx$ 18028; 20207 - 1504 $\approx$ 18703; 20207 - 889 $\approx$ 19318; 20207 - 334 $\approx$ 19873 years from the origin of construction; the last of these times divided by the final time $t_\bullet$ = 24557 from the same origin resulting in 16478/24557 $\approx$ 0.671; 17283/24557 $\approx$ 0.704; 18028/24557 $\approx$ 0.734; 18703/24557 $\approx$ 0.762; 19318/24557 $\approx$ 0.787; 19873/24557 $\approx$ 0.809. Similar values of these ratios are given in Background to time, Table *(1)* for $z/16 = 1$ sequences, in the same succession, as

$$t_{0.8125}/t_\bullet = tanh(0.8125) = 0.6709671;$$
$$t_{0.875}/t_\bullet = tanh(0.875) = 0.7039056;$$
$$t_{0.9375}/t_\bullet = tanh(0.9375) = 0.7340715;$$
$$t_1/t_\bullet = tanh(1) = 0.7615942;$$
$$t_{1.0625}/t_\bullet = tanh(1.0625) = 0.7866188;$$
$$t_{1.125}/t_\bullet = tanh(1.125) = 0.8093011;$$

whereas the accurate values of the considered transitional times, in years from the origin of construction, are more accurately calculated:

$$t_{0.8125} = t_\bullet \cdot tanh(0.8125) = 16477;$$
$$t_{0.875} = t_\bullet \cdot tanh(0.875) = 17286;$$
$$t_{0.9375} = t_\bullet \cdot tanh(0.9375) = 18027;$$
$$t_1 = t_\bullet \cdot tanh(1) = 18702;$$

$$t_{1.0625} = t_\bullet \cdot tanh \ (1.0625) = 19317;$$
$$t_{1.125} = t_\bullet \cdot tanh \ (1.125) = 19874.$$

Such calibrated, the timeline, sequences, and intrinsic characteristics in development of construction are completed as:

$z/16 = l$	Time (years)		Sequences of construction
	from origin $t_l = t_\bullet \cdot tanh(l)$	from present t_l - 20207	
	24557	+4350	
...	...	...	
1.1875	20377	+170 (AD2184)	_
1.125	19874	-333 (AD1681)	_Great structures and skyscrapers
1.0625	19317	-890 (AD1124)	_Cathedrals and great palaces
1	18702	-1505 (AD509)	_Churches and mosques
0.9375	18027	-2180 (166BC)	_Aqueducts, amphitheatres, bridges
0.875	17286	-2921 (907BC)	_Temples and lighthouses
0.8125	16477	-3730 (1716BC)	_Palaces and megarons
0.75	15597	-4610 (2596BC)	_Great pyramids and ziggurats
0.6875	14645	-5562 (3548BC)	_Megalithic monuments and temples
0.625	13619	-6588 (4574BC)	_Neolithic temples and tombs
0.5625	12520	-7687 (5673BC)	_Tumuli and megalithic necropolises
0.5	11348	-8859 (6845BC)	_Fortified settlements
0.4375	10107	-10100	_Villages and tumuli
0.375	8800	-11407	_Defensive + agricultural structures
0.3125	7434	-12773	_Ceremonial stone structures
0.25	6014	-14193	_Stone structures
0.1875	4551	-15656	_Wood and stone constructions
0.125	3054	-17153	_Wooden type constructions
0.0625	1533	-18674	_Earthen-constructions
0	0	-20207	_ Shelters and houses

$z/16 = l$	Time from origin $t_l = t_\bullet \cdot tanh(l)$ (years)	Period $\tau_l = (t_\bullet^2 - t_l^2)/(2t_\bullet)$ (years)	Frequency $f_l = (2t_\bullet)/(t_\bullet^2 - t_l^2)$ (years)$^{-1}$	Angular speed $\omega_l = 2\pi/\tau_l = 2\pi \cdot f_l$ (years)$^{-1}$
	24557	0	∞	∞
...	...	...	...	...
1.1875	20377	3824.2	0.0002615	0.0016430
1.125	19874	4236.5	0.0002360	0.0014831
1.0625	19317	4680.9	0.0002136	0.0013423
1	18702	5157.0	0.0001939	0.0012184
0.9375	18027	5661.8	0.0001766	0.0011098
0.875	17286	6194.6	0.0001614	0.0010143
0.8125	16477	6750.7	0.0001481	0.0009307
0.75	15597	7325.4	0.0001365	0.0008577
0.6875	14645	7911.6	0.0001264	0.0007942
0.625	13619	8502.0	0.0001176	0.0007390
0.5625	12520	9086.9	0.0001100	0.0006915

0.5	11348	9656.5	0.0001036	0.0006507
0.4375	10107	10198.6	0.0000981	0.0006161
0.375	8800	10701.8	0.0000934	0.0005871
0.3125	7434	11153.3	0.0000897	0.0005633
0.25	6014	11542.1	0.0000866	0.0005444
0.1875	4551	11856.8	0.0000843	0.0005299
0.125	3054	12088.6	0.0000827	0.0005198
0.0625	1533	12230.7	0.0000818	0.0005137
0	0	12278.5	0.0000814	0.0005117

20. Science

The systematic study of structure and behaviour of physical and natural world through observation and experiment is called *science* (from Latin *scientia* 'knowledge, acquaintance, skill'). In modern use, science often refers not only to the knowledge itself, but also to the way of pursuing knowledge.

Scientific interests and studies emerged since environmental knowledge has been sufficiently accumulated to make a detailed analysis of a subject or situation, so that its origin can be dated back to around 15000 years ago. Initially transmitted from generation to generation in an oral tradition (e.g. cultivation of maize in Southern Mexico, millet in Transcaucasia and China, and barley in Near East) from 9000-7000BC, scientific studies were diversified and developed in ancient Mesopotamia, Egypt, India, China, Greece, and Rome from about 3500BC to AD400, then sciences advanced in Renaissance Europe between 1400 and 1640, and furthermore across the world until present.   .

In a general representation, the branches of science are grouped into *formal sciences* (logic and mathematics), *physical sciences* (physics and chemistry), *life sciences* (cellular biology and functional biology), *social sciences* (psychology and sociology), and *Earth-space sciences* (geosciences and astronomy).

Apart from the branches approached in other chapters of the present work, the scientific studies and sciences in this chapter mainly refer to mathematics, mechanics, thermodynamics, chemistry, and electricity, which were developed by people and personalities, discoveries, principles, laws and theories, such as those displayed below in chronological order.

4000-2110BC: *Sumerians* formulated the earliest concepts in *algebra* and *geometry*, and developed a system based on *units of ten*; divided a *circle into units of sixty*, which is the basis of time up to the modern world; and developed a system of *weights* and *measures* serving in the ancient world until Roman times;

2750-1350BC: *Egyptian science* was represented by the physician Merit-Ptah, who worked during the 2nd Dynasty; the astronomer Nakht, serving during the reign of Tuthmose IV; and the physician Penthu, who was chief of physicians during the reign of Pharaoh Akhenaten;

c.2000BC: *Rhind Mathematical Papyrus* was discovered in Egypt, as probably the first written mathematical work;

1900BC: *Plimpton 322*, a Babylonian mathematical tablet, recorded a number of Pythagorean triplets such as (3, 4, 5) and (5, 12, 13);

580-560BC: Thales of Miletus, Ionian astronomer and geometer, identified *water as original substance and basis of universe*, worked on static electricity, leading to the observation that amber becomes *magnetic by friction*, in contrast to minerals such as magnetite which need no rubbing, and formulated *Thales' theorem*, and other important mathematical settings, justifying author's nomination as 'father of science', and his inclusion in the traditional canon of 'Seven Wise Men';

530-500BC: Anaximenes, Ionian thinker, proposed *air as the first principle and basic form of matter*, which could be transformed into other substances by condensation and expansion; and Pythagoras, Greek philosopher and mathematician, discovered the *relation of numbers*, stated *Pythagoras' theorem* of right-angled triangles, and postulated that the *Earth is spherical in shape*;

c.450BC: Empedocles, Greek philosopher and poet, wrote *On Nature*, describing a cosmic cycle where basic elements *Earth*, *Air*, *Fire*, and *Water* periodically combine and separate under the influence of dynamic forces, as well as first demonstrated that *air has weight*;

450-430BC: Anaxagoras, Ionian philosopher, stated that *matter is infinitely divisible*, and that *order is produced from chaos* by intelligent principles;

c.400BC: Kidinnu (Cidenas), Babylonian astronomer and mathematician, as mentioned on scribes' clay tablets, developed arithmetical methods to compute the changing length of daylight in course of a year, and to predict appearances and disappearances of the Moon, planets and eclipses of the Sun and Moon, calculating a value for the *solar year* that is in use for today's calendars;

305-270BC: Epicurus, Greek philosopher, developing the theory of Democritus, assumed that the *world operates on mechanical principles*, and that the human soul and body are combinations of atoms that dissolve and perish together;

c.300BC: Euclid, Greek mathematician, wrote *Elements* of geometry, in 13 books, the earliest substantial Greek mathematical treatise to have survived, representing a model of rigorous mathematical exposition for centuries, covering geometry of lines in the plane, circles, ratios and the geometry of three dimensions; later called *Euclidean geometry*;

260-212BC: <u>Archimedes</u>, Greek mathematician and scientist, formulated the *lever and pulley functions*, and founded the science of *hydrostatics*, studying equilibrium positions of floating bodies and discovering that a body weighed when immersed in a fluid shows a loss of weight equal to weight of fluid it displaces, called *principle of Archimedes*; discovered formulae for areas and volumes of *spheres*, *cylinders*, *parabolas*, and other figures, by methods *anticipating theories of integration*, and *involving infinitesimals* as a heuristic tool to obtain results prior a rigorous proof;

230-200BC: <u>Eratosthenes</u>, Greek mathematician and geographer, invented the 'sieve of Eratosthenes' method for listing the *prime numbers*, and a mechanical method of *duplicating the cube*, as well as measured the obliquity of the ecliptic and the *circumference of the Earth* with considerable accuracy;

180-145BC: <u>Aristarchus of Samothrace</u>, Alexandrian grammarian and scientist, became famous for his theory of the Earth's motion, maintaining not only that the Earth revolves on its axis but that it travels in a circle around the Sun, anticipating the theory of N Copernicus; he also developed a method for determining the relative distances of the Sun and Moon; as well as wrote many commentaries and treatises, and elaborated a *heliocentric model* for the Solar System;

AD75-77: <u>Pliny the Elder</u>, Roman scholar, and <u>Scribonius Largus</u>, Roman physician and writer, attested that numbing effect of *electric shocks* delivered by catfish and torpedo rays could travel along conducting objects;

c.250: <u>Diophantus of Alexandria</u>, Greek mathematician, wrote *Arithmetica*, dealing with solution of algebraic equations, useful in many problems with no uniquely determined solution, later known as *Diophantine problems*;

449: <u>Aryabhata</u>, Indian mathematician, was known by his *Aryabhatiya*, introducing *trigonometric functions* such as sine, 'versine', cosine and inverse sine, *trigonometric tables*, as well as techniques and *algorithms of algebra*;

1100: <u>Jia Xian</u>, Chinese mathematician, described Pascal's triangle and applied it for *binomial coefficients*;

1220-60: <u>Jordanus de Nemore</u>, French-German physicist, studied *component forces*, and *inclined planes*, and formulated the *principle of mechanical work*, and the *concept of static moment*;

1245: <u>Ch'in Chiu-shao</u>, Chinese mathematician, solved *cubic equations*, and deduced solutions of *equations of order higher than 3*;

1269: <u>Petrus Peregrinus</u>, French scientist and soldier, evidenced the *magnetic poles* by marking the ends of a round natural magnet, invented a *compass with graduated scale*, and described *magnetism* while pivoting needles in glass boxes;

1340-50: <u>Jean Buridan</u>, French scholastic philosopher, wrote on mechanics, optics, and particularly on logic, giving his name to the famous problem of decision-making called *Buridan's Ass*, where an ass faced with two equidistant and equally desirable bales of hay starves to death because there are no grounds for preferring to go to one bale rather than the other;

1530-40: <u>Paracelsus</u>, German alchemist and physician, discovered many techniques which became standard laboratory practice, such as *concentrating alcohol* by freezing it out of its solution;

1550-70: <u>Girolamo Cardano</u>, Italian mathematician, naturalist, physician and philosopher, wrote *Ars Magna* 'The Great Skill', giving for the first time formulae to solve *cubic and quartic equations*;

1582-1638: <u>Galilei Galileo</u>, Italian astronomer, mathematician and natural philosopher, discovered that *all bodies fall at the same rate* when air resistance is not present, i.e. acceleration of gravity is constant (at ground level), and a body moving along an inclined plane has a constant acceleration; demonstrated the parabolic trajectories of projectiles; as well as defended the Copernican system in *Dialogue on the Two Principal Systems of the World*, and completed *Discourses on the Two New Sciences*, discussing at length the principles of mechanics;

c.1600: <u>Michael Sendivogius</u>, Polish alchemist, philosopher and medical doctor, pioneered chemistry, and distilled *oxygen* 'the elixir of life' obtained by warming saltpetre, but existing in free-state as a gas, to an extent of about 21% volume in atmosphere; and <u>William Gilbert</u>, English physician, first used the terms *electricity, electric force*, and *electric attraction*, and stated that amber is not only substance which when rubbed attracts light objects;

1614: <u>John Napier</u>, Scottish mathematician, wrote *Mirifici Logarithmorum Canonis Descriptio* 'Description of the Marvellous Canon of Logarithms', introducing *natural logarithms*;

1617: <u>Henry Briggs</u>, English mathematician, introduced *logarithms in base 10*, as an important simplification for practical use of logarithms in calculation;

1619-26: <u>Edmund Gunter</u>, English mathematician and astronomer, invented instruments bearing his name, such as the 22 yard long 100-link *Gunter's chain*, which was a forerunner of the modern slide rule *Gunter's line*, the 2 foot rule with scales of chords *Gunter' scale*, and the portable *Gunter's quadrant*, as well as introduced the words *cosine* and *cotangent* into the language of trigonometry;

1635: <u>Bonaventura Cavalieri</u>, Italian mathematician, introduced method of *indivisibles*, developing idea of figures made up of lines in order to determine their areas, thus pioneering integral calculus;

1640-69: <u>Blaise Pascal</u>, French mathematician and physicist, elaborated the famous *theorem on a hexagon inscribed in a conic*; obtained patent of a *calculating machine*; wrote on area of cycloid heralding invention of *integral calculus*; and, after noticing the fall of mercury columns with increased altitude, postulated *the principle of transmission of fluid-pressure* 'Pascal's law, or principle' stating that the pressure exerted anywhere in a confined incompressible fluid is transmitted equally through the fluid such that the pressure variations remain the same;

1642-44: <u>Evangelista Torricelli</u>, Italian physicist and mathematician, discovered that, because of atmospheric pressure, water will not rise above 33 feet in a suction pump, and stated the *fundamental principles of hydromechanics*;

1648-60: <u>Johann Rudolph Glauber</u>, German physician, discovered *hydrochloric acid*, produced *nitric acid*, therapeutic *Glauber's salt* (sodium sulphate), as well as *acetone, benzene,* and *alkaloids*;

1654: <u>Pierre de Fermat</u>, French mathematician, and <u>Blaise Pascal</u>, French mathematician and physicist, founded *probability theory*, applicable for a random variable which can only take specific discrete values, possibly infinite in number;

1660-78: <u>Christiaan Huygens</u>, Dutch physicist, developed the doctrine of accelerated motion under gravity, discovered the *laws of collision of elastic bodies*, stated the undulatory theory by the *principle of Huygens*, first proposed the *undulatory theory of light*, discovered *polarization*, promoted the *wave theory*, and the *wave nature of light*;

1661-70: <u>Robert Boyle</u>, Irish physicist and chemist, published *Sceptical Chymist*, criticizing the theories of matter and defining chemical element as practical limit of chemical analysis; formulated *Boyle's law*, stating that pressure and volume of gas are inversely proportional; studied calcinations of metals, properties of acids, alkalis, and specific gravity; and first prepared 'phosphorus';

1669: Hennig Brand, German alchemist, discovered *phosphorus* 'light bearer', as a white waxy substance present in urine, which glowed in dark;

1670-80: Robert Hooke, English experimental philosopher and architect, discovered the relationship between stress and strain in elastic bodies, known as *Hooke law*, and formulated the simplest theory of the arch, balance-spring of watches, and anchor escapement of clocks;

1673-79: Edmé Mariotte, French physicist and physiologist, formulated the *laws of elastic and inelastic collisions*, and restated the Boyle-Mariotte law, using it to estimate the *height of the atmosphere*;

1680-1716: Gottfried Wilhelm Leibniz, German philosopher and mathematician, and Isaac Newton, English scientist and mathematician, invented *infinitesimal calculus*, and provided original contributions in statistics, probability theory, and calculating machines;

1687: Isaac Newton, English scientist and mathematician, produced the great *Philosophiae Naturalis Principia Mathematica* 'The Mathematical Principles of Natural Philosophy' where *three laws of motion* are unveiled: 1st, a body in a state of rest or uniform motion will remain in that state until a force acts on it; 2nd, an applied force is directly proportional to the acceleration it induces; and 3rd, for every 'action' force which one body exerts on another, there is an equal and opposite 'reaction' force exerted by second body on first;

1690-1713: Jacques Bernoulli, Swiss mathematician, introduced the term *integral*; did important contributions to *probability theory*, including *law of large numbers*; as well as elaborated *permutation theory*, and introduced *Bernoulli numbers* as coefficients of exponential series;

1697-1718: Abraham de Moivre, French mathematician, wrote *The Doctrine of Chances* on probability theory, and set the fundamental formula for complex numbers, known as *de Moivre's theorem*, relating exponential and trigonometric functions;

1715: Brook Taylor, English mathematician, published *Methodus incrementorum*, containing *Taylor's theorem* on power series expansions;

1730: <u>James Stirling</u>, Scottish mathematician, wrote *Methodus differentialis*, enabling important advances in the theory of infinite series and finite differences, and gave *Stirling's formula of approximation* for factorial function;

1733-48: <u>Leonhard Euler</u>, Swiss mathematician, studied infinite series and differential equations, introduced or established new functions, including *gamma function* and *elliptic integrals*, created the *calculus of variations*, and wrote the treatise *Introductio in analysis infinitorum* and other treatises on differential and integral calculus and algebra;

1740-50: <u>Pierre Louis Moreau de Maupertuis</u>, French mathematician, formulated the *principle of least action*, showing that a mechanical system evolves such that its action is as small as possible;

1742: <u>Colin Maclaurin</u>, Scottish mathematician, wrote *Treatise on fluxions*, giving a systematic account of I Newton's approach to calculus; and introduced the *Maclaurin series*, as a B Taylor's series expansion of a function about zero;

1743-58: <u>Jean le Rond d'Alembert</u>, French philosopher and mathematician, formulated the *d'Alembert principle*, worked on partial differential equations applicable in studies on motion of vibrating strings, and celestial mechanics, and published *Traité de dynamique* 'Treatise of the Dynamics' developing the mathematical theory of Newtonian dynamics;

1746-48: <u>Benjamin Franklin</u>, US statesman, inventor and scientist, made distinction between *positive* and *negative electricity*, proved that *lightning* and *electricity* are identical, and suggested that buildings could be protected by *lightning-conductors*;

1751-58: <u>Axel Fredrik Cronstedt</u>, Swedish metallurgist and mineralogist, isolated *nickel*, as an element with magnetic properties, and published *Essay towards a System of Mineralogy*, in which minerals and stones were distinguished for first time and *chemical composition* was advocated as primary method of classification of minerals;

1754-61: <u>Joseph Black</u>, Scottish chemist, evidenced that *causticity of lime* and *alkalis* is due to absence of 'fixed air' (carbon dioxide) present in limestone and carbonates of alkalis;

1766: <u>Henry Cavendish</u>, English natural philosopher and chemist, analysing distinct 'factitious airs' of which normal atmospheric air is composed, he discovered that among these were 'fixed air' as carbon dioxide, and 'inflammable air' as *hydrogen*, the second one being first isolated by him;

1772: Carl Wilhelm Scheele, Swedish chemist, and Antoine Laurent Lavoisier, French chemist, investigated air and fire, and identified the 'inflammable gas' that was named *oxygen* and its importance in respiration, combustion and as a compound with metals; and Daniel Rutherford, Scottish physician and chemist, isolated and discovered 'noxious air', called *nitrogen*, by passing air through a carbon dioxide absorbing solution, and the remaining air did not support combustion;

1774-81: Carl Wilhelm Scheele, Swedish chemist, identified *barium*, *chlorine*, *manganese*, and *molybdenum*, as well as *citric acid, lactic acid, glycerol, hydrogen cyanide, hydrogen fluoride*, and *hydrogen sulphide*;

1775-81: Claude Louis de Berthollet, French chemist, discovered the connection between the manner in which a chemical reaction proceeds and the mass of reagents, and demonstrated that *chemical affinities are affected by temperature and concentration* of reagents;

1775-87: Alessandro Giuseppe Anastasio Volta, Italian physicist and inventor, discovered *electrophorus* as precursor of the induction machine, *condenser*, and *candle flame collector of atmospheric electricity*, before the invention of electrochemical battery;

1776-89: Antoine Laurent Lavoisier, French chemist, discovered *oxygen* and its importance, and published the masterpiece *Traité élémentaire de chemie* 'Treaty of Elementary Chemistry';

1780-88: Joseph Louis de Lagrange, French mathematician, worked on *calculus of variations*, number theory, differential equations, and algebraic equations, the last ones representing major steps in the development of group theory by *permutations of roots of an equation*, and completed *Traité de mécanique analytique* 'Treatise of Analytical Mechanics' in which mechanics is based entirely on variational principles, at a high degree of elegance;

1783: Don Fausto d'Elhuyar y de Suvisa and Juan José d'Elhuyar y de Suvisa, Spanish chemists and metallurgists, made experiments to isolate, and discovered *tungsten*, now known as wolfram, a hard, rare metal obtained from wolframite;

1783-95: Gaspard Monge, French mathematician and physicist, discovered that water resulted from an *electrical explosion of oxygen and hydrogen*, and wrote the treatise *Leçons de géométrie descriptive*, stating principles for general application of geometry to arts of construction, now called *descriptive geometry*;

1785: <u>Charles Augustin Coulomb</u>, French physicist, discovered the law of electrical attraction, called *Coulomb's law*, stating that the force between two small charged spheres is related to charges and distance between them;

1790-94: <u>Johan Gadolin</u>, Finnish chemist, investigated rare earth elements, analysing a new black mineral from Ytterby, Sweden, and isolated from it a rare earth mineral, called *yttria*, as an important step towards the identification of other undiscovered earth elements;

1790-95: <u>Luigi Galvani</u>, Italian physiologist, discovered so-called *animal electricity*, followed by use of discoverer's name in words such as 'galvanized' and 'galvanometer';

1790-1810: <u>Johan Gottlieb Gahn</u>, Swedish chemist and mineralogist, improving smelting methods and use of by-products, he isolated the metallic *manganese* and developed its preparation on a large scale; he also discovered *selenium*;

1794-1820: <u>Thomas Young</u>, English physicist, physician and Egyptologist, established the *wave theory of light*, and combined classical wave theory with the theory of colours to explain the *interference phenomenon* produced by ruled gratings, thin plates, and colours of the rainbow ;

1794-1825: <u>Adrien-Marie Legendre</u>, French mathematician, published *Éléments de géométrie*; *Essai sur la théorie des nombres* 'Essay on the Theory of Numbers', including discovery of the *law of quadratic reciprocity*; and *Traité des fonctions elliptiques* 'Treatise on Elliptical Functions';

1796-1825: <u>Pierre Simon Laplace</u>, French mathematician and astronomer, formulated the fundamental differential *Laplace equation* for gravitational attraction of spheroids, and completed *Mécanique céleste*, a great treatise on celestial mechanics;

1800: <u>William Nicholson</u>, English physicist and inventor, constructed the first *voltaic pile in England*; and observed that when the ends of leads from battery were immersed in water, bubbles of gas were produced, thus discovering *electrolysis*;

1801-25: <u>Carl Friedrich Gauss</u>, German mathematician, astronomer and physicist, wrote *Disquisitiones arithmeticae*, containing wholly new advances in number theory; first used the *method of least squares* in statistics; studied the *theory of errors* of observation; also worked on pure mathematics, including *differential equations, hypergeometric function*, *curvature of surfaces*, four different *proofs of fundamental theorem of algebra*, and six *quadratic reciprocity*; as well as studied celestial mechanics, resulting in the treatise *Theoria motus corporum coelestium* 'The Theory of the Motion of Celestial Bodies';

1803: <u>John Dalton</u>, English chemist and natural philosopher, formulated the law showing that in a mixture of gases each gas exerts same pressure as it would be only gas present in a given volume, named *Dalton's law*; which led to interpretation of chemical analyses in terms of relative weights of atoms of elements involved, and to the laws of chemical combination;

1803-18: <u>Jöns Jacob Berzelius</u>, Swedish chemist, discovered *cerium*, *selenium* and *thorium*; isolated silicon, zirconium and titanium; drew a table of atomic weights using oxygen as a base, and devised the *modern system of chemical symbols*;

1806-22: <u>Thomas Johann Seebeck</u>, Estonian-born German physicist, made studies on heating and chemical effects of colours of solar spectrum, investigated optical polarization in stressed glass, and also discovered *thermoelectricity*, initially called 'thermomagnetism', which occurs when an electric current is generated through application of heat to a junction of two metals;

1810-22: <u>Joseph de Fourier</u>, French mathematician, introduced the expansion of functions in trigonometric series, now known as *Fourier series*, by which almost any function of real variable can be expressed as a sum of sines and cosines of integral multiples of variable; stated the *Fourier transform* used in operational calculus; wrote *Théorie analytique de la chaleur* 'Analytical Theory of Heat' applying it to solve partial differential equations for heat conduction in solid bodies; and also discovered an important theorem on the roots of algebraic equations;

1810-40: <u>Friedrich Wilhelm Bessel</u>, German mathematician and astronomer, introduced *Bessel's equations* and *functions* for solving differential equations of cylindrical processes;

1811: <u>Amedeo Carlo Avogadro</u>, Italian physicist and chemist, formulated *Avogadro's law*, according to which equal volumes of all gases contain equal numbers of molecules when at the same temperature and pressure;

1818-21: <u>Augustin Jean Fresnel</u>, French physicist, invented multi-faceted lighthouse *Fresnel lens*, and the special prism called *Fesnel's rhomb* producing circularly polarized light; established the *undulatory theory of light*; as well as published brilliant papers relating the *polarization phenomena* to the *hypothesis of transverse waves*, and the three-volume *Œuvres Complètes* 'Complete Works';

1820-25: <u>Hans Christian Oersted</u>, Danish physicist, discovered the *magnetic effect produced by an electric current*, paving the way for electromagnetic discoveries of AM Ampère and M Faraday; made extremely accurate measurement of compressibility of water, and first isolated *aluminium*;

1820-40: <u>Siméon Denis Poisson</u>, French mathematical physicist, made important contributions to potential theory and transformation of equations in mechanics by means of *Poisson brackets*, and discovered the *Poisson distribution* as a special case of binomial distribution in statistics;

1822-30: <u>André Marie Ampère</u>, French mathematician and physicist, founded the *science of electrodynamics*, and published *Observations électro-dynamiques* 'Electrodynamic Observations', and *Théories des phénomènes électro-dynamiques* 'Theory of Electrodynamic Phenomena';

1822-45: <u>Claude Louis Marie Henri Navier</u>, French physicist and civil engineer, and independently <u>George Gabriel Stokes</u>, Irish mathematician and physicist, studied the force opposing small bodies in their passage through a viscous fluid, and discovered the *Navier-Stokes equations* of fluid mechanics (hydrodynamics);

1824: <u>Sadi Carnot</u>, French physicist, published his masterpiece *Réflexions sur la puissance motrice du feu et sur les machines propres à developer cette puissance* 'Reflections on the Motive Power of Fire, and on the Machines Appropriate to develop this Power', concerning on scientific principles to an analysis of *working cycle* and *efficiency of steam engine*;

1825-26: <u>Antoine Jérôme Balard</u>, French chemist, isolated and discovered *bromine*, which was recognized as an element, with similarities to iodine and chlorine, as one of the 'families' of elements;

1825-27: <u>Niels Henrik Abel</u>, Norwegian mathematician, introduced *Abel's integrals* and *functions*, which became a central theme of later analysis, emphasizing the analogy of elliptic functions with familiar trigonometric functions, and influencing the development of complex functions;

1826-51: <u>Charles François Sturm</u>, French mathematician, and <u>Joseph Liouville</u>, French mathematician, the first discovered *Sturm's theorem* concerning location of the roots of a polynomial equation and worked on *linear differential equations*; both of them established the *Sturm-Liouville theory* as a standard procedure to solve certain types of integral equations by developing *eigenfunctions*; and the second studied the *theory of differential equations, mathematical physics* and *celestial mechanics*, introduced new methods of investigating *transcendental numbers*, leading to *Liouville numbers*, set the *Liouville function* also important in number theory, and formulated *Liouville's theorem* used in complex analysis;

1827: <u>Georg Simon Ohm</u>, German physicist, formulated and published the law relating voltage, current and resistance in an electrical circuit, called *Ohm's law* of charge transport; and <u>Robert Brown</u>, Scottish botanist, first observed the movement of fine particles in a liquid, which was called *Brownian movement*;

1828-30: <u>Jean Baptiste Biot</u>, French physicist and astronomer, and <u>Félix Savart</u>, French physician and physicist, discovered the *law defining the intensity of magnetic field* produced at a given point *near a long straight current-carrying conductor*;

1828-33: <u>George Green</u>, English mathematician and physicist, published *An essay on the application of mathematical analysis to the theories of electricity and magnetism*; deduced *Green's theorem* relating integrals taken over a volume with those taken over the surface enclosing that volume; and created *Green's functions*, for solving partial differential equations;

1828-40: <u>Friedrich Wöhler</u>, German chemist, attempting to prepare ammonium cyanate from silver cyanate and ammonium chloride, obtained *urea*, the first natural product synthesized;

1829-35: <u>Carl Gustav Jacob Jacobi</u>, German mathematician, produced the first definitive book on *elliptic functions*, entitled *Fundamenta nova*, discovered many remarkable *infinite series*, as well as achieved important advances in study of differential equations, theory of numbers, and determinants;

1830-55: <u>János Bolyai</u>, Hungarian mathematician, and independently <u>Nikolai Ivanovich Lobachevski</u>, Russian mathematician, developed theories in which Euclid's parallel postulate did not hold, and thus founded hyperbolic geometry, or *Bolyai-Lobachevskian geometry*, as a *non-Euclidean geometry*;

1831-32: <u>Évariste Galois</u>, French mathematician, founded the *Theory of groups*, published by J Liouville in 1846, presenting solubility of equations by radicals, and the theory of algebraic equations and

Abelian integrals;

1831-34: <u>Michael Faraday</u>, English chemist and physicist, stated the *law of induction*, showing that the electromotive force produced around a closed path is proportional to the rate of charge of magnetic flux through any surface bounded by that path; founded *electric motor technology*; as well as defined the *laws of electrolysis*: 1st, mass of substance altered at an electrode is directly proportional to quantity of electricity transferred at that electrode; and 2nd, for a given quantity of electricity, mass of an elemental material altered at an electrode is directly proportional to element's equivalent weight;

1835-37: <u>Antoine César Becquerel</u>, French physicist, investigated the electrical properties of minerals, and first used *electrolysis to isolate metals from their ores*;

1839-45: <u>Christian Friedrich Schönbein</u>, German chemist, discovered *ozone*, and synthesized *nitro-cellulose* 'gun cotton', a highly inflammable fluffy white substance useful as an explosive;

1840-45: <u>James Prescott Joule</u>, English natural philosopher, experimenting on heat, discovered the *Joule effect* asserting that heat produced in a wire by an electric current was proportional to the resistance and to the square of the current; showed experimentally that heat is a form of energy; determined quantitatively the amount of mechanical (and later electrical) energy to be expended in the propagation of heat energy, establishing the *mechanical equivalent of heat*; also made a mathematical study of the *current through a resistance causing localized heating*, and first described the phenomenon of *magnetostriction*;

1840-50: <u>Augustin Louis Cauchy</u>, French mathematician, founded the theory of *functions of complex variable*, gave a definitive account of the theory of *determinants*, and developed the ideas of *permutation groups*;

1843-60: <u>William Rowan Hamilton</u>, Irish mathematician, introduced *quaternions*, showing that algebra of four dimensions was possible, and also *Hamilton function*, and *Hamilton-Jacobi equation*;

1849-90: <u>George Gabriel Stokes</u>, Irish mathematician and physicist, used spectroscopy to determine the chemical compositions of the Sun and stars, formulated *Stokes's law* relating to the force opposing a small sphere in its passage through a viscous fluid, and established *Stokes's theorem*, reducing certain surface integrals to line integrals, which is largely applied in vector calculus;

1850-58: <u>Lejeune Dirichlet</u>, German mathematician, did important contributions to *number theory* and the *Fourier series*, as well as excelled in *boundary value problems* for mathematical physics;

1850-60: <u>August Ferdinand Möbius</u>, German mathematician, investigated surfaces which can exist, and discovered the *Möbius strip*, one-sided surface formed by giving a rectangular strip a half-twist and then joining its ends together;

1851: <u>Jean Bernard Léon Foucault</u>, French physicist, demonstrated the *rotation of the Earth* by means of a freely suspended pendulum, and contributed to the development of mechanics;

1854-57: <u>Georg Friedrich Bernhard Riemann</u>, German mathematician, wrote *On the hypotheses that underlie geometry*, presenting a *n-dimensional curved space*; and introduced the *Riemann surface* to deal with multi-valued algebraic functions, as a key concept in analysis; his ideas were essential in the formulation of A Einstein's theory of general relativity, and led to the modern theory of differentiable manifolds, which now plays a vital role in theoretical physics;

1855-80: <u>Adolph Eugen Fick</u>, German physiologist, stated the *laws of diffusion*: 1st, mass of solute diffusing through unit area per second is proportional to concentration gradient; and 2nd, partial derivative of concentration with respect to time is proportional to second derivative of concentration with respect to distance;

1856: <u>Henry Darcy</u>, French hydrogeologist, discovered the proportionality of groundwater flow to hydraulic head gradient, called *Darcy's law* of hydraulic flow;

1857-61: <u>Aleksandr Mikhailovich Butlerov</u>, Russian chemist, contributed to the *theory of chemical structure*, formulating 'double bonds', and discovered *hexamine* and *formose reaction*;

1857-73: <u>James Clerk Maxwell</u>, Scottish physicist, published papers on *kinetic theory of gases*, and theoretically established the *nature of Saturn's rings*; and most importantly founded the *theory of electromagnetism*, extended in his *Treatise on Electricity and Magnetism*, where M Faraday's theory of electrical and magnetic forces was mathematically stated;

1858-90: <u>August Kekule von Stradonitz</u>, German chemist, found the solution of apparent irreconcilability of views on *tetravalency of carbon* with formula of benzene C_6H_6 revealing *cyclic nature of benzene molecule*, subsequently proposed delocalized rather than fixed *double bonds*; and elucidated the *structure of pyridine* as an important landmark in structural organic chemistry;

1859: Gustav Robert Kirchhoff, German physicist, and Robert Wilhelm Bunsen, German chemist and physicist, discovered *spectrum analysis*, which facilitated the identification of new elements, including *caesium* and *rubidium*;

1860-90: Karl Theodor Wilhelm Weierstrass, German mathematician, achieved advances in the theory of elliptic and Abelian functions, constructed the *first accepted example of a continuous but nowhere-differentiable function*, and demonstrated that *every continuous function could be uniformly approximated by polynomials*;

1861-71: William Crookes, English chemist and physicist, discovered the metal *thallium* and *sodium amalgamation* process, as well as wrote *Select Methods of Chemical Analysis*;

1863: Joules Dupuit, French civil engineer, published his *Études théoriques et pratiques sur le mouvement des eaux dans les canaux découverts et à travers les terrains perméables* 'Theoretical and practical studies on the water movement in open canals and through permeable terrains';

1863-69: Ludwig Valentin Lorenz, Danish physicist, gave a *mathematical description for light waves*, showing that double refraction can occur in certain conditions, and worked on the relationship between refraction and specific densities of media, leading to *Lorentz-Lorenz formula*;

1867-68: James Clerk Maxwell, Scottish physicist; and Ludwig Boltzmann, Austrian physicist, discovered the *Maxwell-Boltzmann distribution* and founded *Maxwell-Boltzmann statistics*, describing statistical distribution of particles over various energy states in thermal equilibrium, when temperature is high enough and density is low enough to render quantum effects negligible;

1868: Pierre Jules César Janssen, French astronomer, and Norman Lockyer, English astronomer, observing the solar prominences outside of a total eclipse, they postulated the existence of an unknown element named *helium* 'the Sun element';

1869: Dmitri Ivanovich Mendeleyev, Russian chemist, developed the *periodic table of elements*, arranged in ascending order of their atomic weight and showing that chemically similar elements tend to fall in same columns;

1869-70: John Tyndall, Irish physicist, discovered the *Tyndall effect* -

scattering of light by very small particles in the path of light, the scattered light being mainly blue; and suggested that the *blue colour of sky* is due to greater scattering of blue light by particles of atmospheric dust and water vapour;

1870-80: <u>Wilhelm Eduard Weber</u> and <u>Friedrich Wilhelm Georg Kohlrausch</u>, German physicists, defined the *ratio for charge of a capacitor* in electric and magnetic units, and investigated the *conductivity of electrolytic solutions*, leading to *Kohlrausch's law* on independent ion migration; and <u>Osborne Reynolds</u>, English physician and engineer, contributed to the field of hydrodynamics, improved *centrifugal pumps*, and developed the mathematical framework for *turbulence*, introducing *Reynolds number*, a dimensionless ratio characterizing dynamic state of a fluid;

1870-85: <u>Ernst Werner von Siemens</u>, German electrical engineer, devised numerous forms of galvanometer and other electrical instruments, and discovered the *self-acting dynamo*; as well as determined the *electrical resistance of different substances*;

1870-95: <u>Camille Jordan</u>, French mathematician, published *Traité de substitutions* 'Treatise on Substitutions', a standard work for many years; and *Cours d'analyse*, applying group theory to geometry and linear differential equations;

1870-1901: <u>Charles Hermite</u>, French mathematician, demonstrated that the base *e* of natural logarithms *is transcendental* (no solution of a polynomial equation with rational coefficients), and published works on *theory of numbers, elliptic functions, Abelian functions*, and *invariant theory*;

1872-1925: <u>Felix Klein</u>, German mathematician, worked for the *Erlanger Programm*, showing how different geometries could be classified *in terms of group theory*, and studied non-Euclidean geometry, function theory, and elliptic modular and automorphic functions;

1873: <u>Josiah Willard Gibbs</u>, US engineer, chemist and physicist, discovered the *chemical potential* that a substance has to produce in order to alter a system, originally described as the increase of mass energy divided by quantity of substance added in a given mass;

1873-1910: <u>Johannes Diderik van der Waals</u>, Dutch physicist, postulated the existence of intermolecular forces and a finite molecular volume; derived the *van der Waals equation* of state agreeing closely with experimental data; and discovered *van der Waals forces* for weak attractions between molecules, relating the pressure, volume and temperature of gases and liquids;

1877: <u>Ludwig Boltzmann</u>, Austrian physicist, formulated *Boltzmann's equation*, a fundamental diffusion equation based on particle conservation, showing that the rate of losses, including leakage out of region of interest and rate of disappearance by reactions of all kind, is equal to the rate of production from sources within region and rate of scattering into region, i.e. *increasing entropy corresponds to molecular randomness*;

1877-78: <u>Louis Paul Cailletet</u>, French physicist liquefied *oxygen, nitrogen, carbon monoxide, hydrogen, nitrogen dioxide,* and *acetylene*;

1877-86: <u>Jacobus Henricus van't Hoff</u>, Dutch physical chemist, developed the principles of *chemical kinetics*, applied thermodynamics to *chemical equilibria*, and set the equation for the effect of temperature on equilibria, known as *Van't Hoff isochore*; he also wrote about *osmotic pressure*;

1880-1884: <u>Georg Ferdinand Ludwig Philipp Cantor</u>, German mathematician, elaborated an *arithmetic of infinite*, extending the concept of cardinal and ordinal numbers to infinite sets, as required in modern analysis; and worked on classical analysis, particularly in trigonometric series;

1880-1890: <u>Sofya Vasilevna Kovalevskaya</u>, Russian mathematician, developed the theory of partial differential equation and Abelian integrals, and demonstrated the *fundamental theorem of algebra*;

1881-1888: <u>Jules Henri Poincaré</u>, French mathematician, produced the *Theory of automorphic functions*, non-Euclidean geometry, and complex functions, relating to importance of topological considerations in differential equations; as well as introduced basic ideas in modern topology, such as *triangulation, homology, Euler-Poincaré formula,* and *fundamental group*;

1881-1889: <u>Svante August Arrhenius</u>, Swedish physical chemist, determined and interpreted the electric conductivities of dilute solutions of electrolytes, initiating the *theory of ions*, and formulated the dependence of chemical reaction on temperature, called *Arrhenius equation*;

1882-1889: <u>Jules Henri Poincaré</u>, French, and <u>Leonhard Euler</u>, Swiss, mathematicians, studied the *three-body problem*, and motion of rigid bodies in three dimensions respectively, as well as elaborated the fundamental group, called the *Euler-Poincaré formula*, opening new directions in celestial mechanics;

1883-1894: <u>James Joseph Sylvester</u>, English mathematician, participated in the foundation of *algebraic theory of invariants*, which became a powerful tool in resolving physical problems, and also contributed to *number theory*;

1884-1899: <u>John Henry Poynting</u>, English physicist, introduced the *Poynting vector*, giving a simple expression for the rate of flow of electromagnetic energy; demonstrated that energy flow of electromagnetic waves could be calculated by an equation, known as *Poynting's vector*; as well as published *On the Mean Density of the Earth*, and two-volume *Textbook of Physics*;

1885: <u>William Stanley</u>, US electrical engineer, invented the *electric transformer*, and worked on a *long-range transmission system for alternating current*;

1886-1892: <u>Henry Moissan</u>, French chemist, first isolated *fluorine*, and founded *high-temperature chemistry* which has many industrial applications;

1886-1897: <u>Émile Picard</u>, French mathematician, worked on complex analysis, integral and differential equations, generated the *theory of complex surfaces and integrals*, and introduced the *method of successive approximations*, a powerful technique for determining whether solutions to differential equations exist;

1887-1908: <u>Wilhelm Ostwald</u>, German physical chemist, regarded as one of the founders of physical chemistry, studied electrolytic conductivity resulting in *Ostwald's dilution law*; also founded the journals *Zeitschrift für physikalische Chemie* and *Annalen der Naturphilosophie*, and published the two-volume *Lehrbuch der allgemeinen Chemie*;

1890-1930: <u>Vito Volterra</u>, Italian mathematician, worked on integral equations, studying *spaces of functions*, as well as on mathematical physics and mathematics of population, and then set the differential *Lotka-Volterra equations*, describing a simple 'predator-prey' population model;

1892: <u>Hendrik Antoon Lorentz</u>, Dutch physicist, deduced the formula for electromagnetic force, called the *Lorentz force*, which assembles forces of both electric and magnetic fields, according to the right-hand rule, i.e. if the thumb of the right hand points along the velocity and the index finger along the magnetic field, then the middle finger points along the total force; and <u>Aleksandr Mikhailovich Lyapunov</u>,

Russian mathematician, published *The General Problem of the Stability of Motion*, relating to methods in the theory of stability of dynamical systems;

1892-1930: <u>Karl Pearson</u>, English mathematician and scientist, published *The Grammar of Science*, worked on eugenics, mathematics, and biometrics, founded *modern statistical theory*; and also published *Life of Galton*, applying methods to study F Galton's work;

1893-1920: <u>Charles Proteus Steinmetz</u>, US electrical engineer, using complex numbers, developed the *mathematical theory of alternating currents*, and discovered *magnetic hysteresis* for calculating alternating current circuits and *lightning arresters* on high-power transmission lines;

1894: <u>John William Strutt Rayleigh</u>, English physicist, and <u>William Ramsay</u>, Scottish chemist, discovered *argon*, in conjunction with each other;

1895: <u>William Ramsay</u>, Scottish chemist, isolated an inert gas resembling argon by boiling a mineral called 'cleveite' and using spectroscopic analysis, thus discovering *helium* on the Earth;

1895-1897: <u>Pierre Curie</u>, French physicist, evidenced that ferromagnetic materials lose their property at a certain temperature, called *Curie point*; as well as established the relation of magnetic susceptibility of a paramagnetic material to absolute temperature, known as *Curie's law*;

1895-1922: <u>Georges Urbain</u>, French chemist, used fractional crystallizations to separate and isolate *samarium*, *europium*, *gadolinium*, *terbium*, *dysprosium*, *holmium*, *lutetium*, *ytterbium* and *hafnium*;

1895-1951: <u>Jan Łukasiewicz</u>, Polish mathematical logician, was author of *Principle of Contraction*, *multi-valued logic*, and *Aristotle's Syllogistic from the Standpoint of Modern Formal Logic*;

1896: <u>Antoine Henry Becquerel</u>, French physicist, during his study of fluorescent uranium salt, pitchblende, accidentally identified 'Becquerel rays' as a property of atoms, thus being the first to discover *radioactivity* and to prompt the beginning of the nuclear age;

1897: <u>Joseph John Thomson</u>, English physicist, discovered the *electron*, with its mass nearly 2000 times smaller than the mass of a hydrogen atom;

1898: <u>Marie Curie</u> and <u>Pierre Curie</u>, French physicists, worked on magnetism and *radioactivity*, coining this term, and isolated *radium*, showing that it emits electrically positive, negative, and neutral particles; and also *polonium*, named after Marie's native Poland;

1898-1913: <u>Jean Baptiste Perrin</u>, French physicist, demonstrated that suspended particles showing Brownian motion essentially obey the gas laws, determined a fairly accurate value for Avogadro number, and published *Les Atomes* which became a classic work;

1898-1920: <u>Morris William Travers</u>, English chemist, discovered *krypton*, *neon* and *xenon*, and worked on high-temperature furnaces and fuel technology;

1899: <u>David Hilbert</u>, German mathematician, established the abstract *axiomatic foundations of geometry*, with no attempt to define the basic terms but only to prescribe how they could be used;

1899-1900: <u>Max Karl Ernst Planck</u>, German theoretical physicist, analyzing the laws of thermodynamics and black body radiation, formulated *quantum theory*, according to which the emission and absorption of energy take place in small discrete instalments or *quanta*, and thus predicted phenomena inexplicable in classical Newtonian theory;

1900: <u>Georg Ferdinand Ludwig Philipp Cantor</u> and <u>Julius Wilhelm Richard Dedekind</u>, German mathematicians, founded the *theory of sets* as a basis in mathematics, which became essential for topology and modern analysis; and <u>Friedrich Ernst Dorn</u>, German chemist, noticing that radium apparently becomes less radioactive if swept with a current of gas, discovered *radon*;

1900-12: <u>Max Theodor Felix von Laue</u>, German physicist, applied the concept of *entropy to optics*, deduced the relationship between speed of light in flowing water and theory of special relativity, and also discovered *how X-rays are diffracted by atoms in crystals*;

1900-18: <u>Tullio Levi-Civita</u>, Italian mathematician, founded *tensor calculus*, by generalization of vectorial calculus; and used it for giving a tensorial expression of the general theory of relativity;

1902: <u>Oliver Heaviside</u>, English physicist, and independently, <u>Arthur Edwin Kennelly</u>, US engineer, predicted the existence of an ionized gaseous layer capable of reflecting radio waves, called 'Heaviside layer', now known as *ionosphere*;

1902-07: <u>Herman Minkowski</u>, German mathematician, discovered a new branch of number theory, geometry of numbers, and most importantly gave a precise *mathematical description of space-time* as it appears in the theory of relativity;

1902-15: <u>Henri Léon Lebesgue</u>, French mathematician, became famous for his *theory of integration*, which was first publicly presented by his dissertation *Intégrale, longueur, aire* 'Integral, length, area'; then wrote *Sur les séries trigonométrique*, containing three major theorems - a trigonometrical series representing a bounded function is a Fourier series; the nth Fourier coefficient tends to zero '*Riemann-Lebesgue lemma*'; and a Fourier series is integrable term by term; he also developed the theory of measure and integration, bearing his name and including the *Lebesgue-Stieltjes integral*;

1903-29: <u>Konstantin Eduardovich Tsiolkovsky</u>, Russian astrophysicist and rocket pioneer, published *Exploration of Cosmic Space by Means of Reaction Devices*, which established his reputation as 'the father of space flight theory', and provided the foundation of *multi-stage rocket technology*;

1903-34: <u>William Morris Davis</u>, US geomorphologist, after an expedition in Turkestan, made wide-ranging studies of the *role of rain in erosion, development of rivers, glacial erosion*, formation of *coral reefs, arid landscapes, elevation and subsidence* of land masses, and introduced the concept of *cycles of erosion*, thus opening the way for mechanical and mathematical approach in studies of erosion and sedimentation;

1904: <u>Philipp Eduard Anton Lenard</u>, German physicist, studied the magnetic deflection of cathode rays and their electrostatic properties, resulting in evidence that *atoms contain units of both positive and negative charge*; and <u>Joseph Valentin Boussinesq</u>, French mathematician and physicist, published *Recherches théoriques sur l'écoulement des nappes d'eau infiltrés dans le sol* 'Theoretical research on the flow of layers of water infiltrated in soil';

1904-08: <u>Ernst Friedrich Ferdinand Zermelo</u>, German mathematician, showed that the choice could be applied to *well-ordering of any set*, and gave the first obvious description of *set theory*, with many mathematical applications;

1904-18: <u>Paul Langevin</u>, French physicist, established the *relation of paramagnetic molecular movement to absolute temperature*, predicted *paramagnetic saturation*, and applied *sonar techniques to detect submarines* during the First World War;

1905-19: <u>Albert Einstein</u>, German-Swiss-US mathematical physicist, discovered *photoelectric effect*, a phenomenon resulting from absorption of photon energy by electrons, leading to their release from a surface, when photon energy exceeds the work function, or otherwise allowing conduction when incident energy exceeds an atomic binding energy; studied *Brownian motion*, giving empirical evidence for atomic theory and supporting the application of statistical physic; established the *special theory of relativity*, showing that the speed of light is independent of the motion of an observer, and postulating the relative motion and constant velocity of light or zero acceleration; discovered *mass-energy equivalence* (energy is equal to mass multiplied by square speed of light), which has been dubbed 'the world's most famous formula'; developed the *general theory of relativity*, extending relativity from constant to varying velocities or non-zero acceleration; and initiated *relativistic cosmology*, applying general relativity to entire universe;

1905-20: <u>Andrei Andreyevich Markov</u>, Russian mathematician, introduced the concept of *Markov chain*, as a series of events in which probability of a given event occurring depends only on immediately previous event, with many applications in physics and biology;

1908-11: <u>Heike Kamerlingh Onnes</u>, Dutch physicist, achieved the first *liquefaction* and then *solidification of helium*; and discovered that the *electrical resistance of metals* cooled to *near absolute zero disappears*, a phenomenon later called 'superconductivity';

1908-45: <u>Percy Williams Bridgman</u>, US physicist, made experiments on properties of solids and liquids under high pressure, obtained a *new form of phosphorus*, and demonstrated that at high pressures the *viscosity increases with the pressure* for most liquids;

1910-21: <u>Robert Andrews Millikan</u>, US physicist, refining oil drop technique, he evidenced that the charge on each droplet is always a multiple of the same basic unit as *charge of electron*; confirmed A Einstein's theoretical equations; and calculated an *accurate value for Planck's constant*;

1910-25: <u>Owen Willans Richardson</u>, English physicist, carried out famous work on *thermionics*, a term coined to describe the phenomenon of emission of electricity from hot bodies;

1911: <u>Ernest Rutherford</u>, New Zealand physicist, introduced the revolutionary concept of the *nuclear atom*, in which mass is concentrated in nucleus surrounded by planetary electrons;

1911-25: <u>Traian Lalescu</u>, Romanian mathematician, published *Algebraic Calculus*, and *Treatise of Analytical Geometry*; and also established the *Lalescu Sequence*;

1912-33: <u>Peter Joseph Wilhelm Debye</u>, US physicist and physical chemist, worked on dielectric constants, on molecular dipole moments, known as *Debyes*, and on theory of *strong electrolytes*; developed the *Debye-Scherrer X-ray diffraction powder method*, and the theory of *X-ray scattering by gaseous molecules*; as well as achieved theoretical treatments for *electro-optical Kerr effect*, and *adiabatic demagnetization*;

1912-66: <u>Victor Valcovici</u>, Romanian physicist and mathematician, wrote *Sur les principles de Hamilton et la moindre action*; *Propriétés minimales du mouvement des systèmes*; *Une extension des liaisons non holonomes et des principes variationnels*; *Nouveau fondement fonctionnel de la Mécanique*, and *The Variational Principles of Mechanics*;

1913: <u>Frederick Soddy</u>, English radio chemist, demonstrated that uranium decays into radium, and discovered *isotopes*, which have fundamental importance to all physics and chemistry; and <u>Harry Moseley</u>, English physicist, established that *properties of elements* are determined *by atomic number* rather than atomic weight;

1913-50: <u>Niels Henrik David Bohr</u>, Danish physicist, developed *Bohr's model* of atomic structure, explaining the spectrum of hydrogen by means of the atomic model and quantum theory, developed the *liquid drop model of the nucleus*, and introduced the *quantized energy level theory of the atom*;

1914-18: <u>Srinivasa Ramanujan</u>, Indian mathematician, gave an exact formula for the number of ways *an integer* can be written *as a sum of positive integers*;

1914-25: <u>James Franck</u>, US physicist, and <u>Gustav Ludwig Hertz</u>, German physicist, provided evidence that atoms would only absorb a fixed amount of energy, and thus demonstrated the *quantized nature of the atom's electron energy levels*;

1915-22: <u>George Neville Watson</u>, English mathematician, published his great *Treatise on the Theory of Bessel Function*, displaying in particular asymptotic expansions of Bessel functions, and introduced *Watson's quintuple product identity*;

1918: <u>Gheorghe Constantinescu</u>, Romanian inventor and scientist, founded the *theory of sonics*, a branch of continuum mechanics which describes the transmission of mechanical energy through vibrations, first presented in his work *A treatise on transmission of power by vibrations*;

1919-20: Ernest Rutherford, New Zealand physicist, provided evidence that *protons* are constituents of atomic nuclei, and proposed the existence of neutrons;

1920-40: Henri Léon Lebesgue, French mathematician, developed the *theory of measure* and *Lebesgue's integration*, as well as made important applications in complex analysis;

1920-72: Octav Onicescu, Romanian mathematician and physicist, published *Sopra gli spazi einsteinieni a gruppi di transformazioni*; *The Principles of Probability Theory*; and *Principles d'une Statistique informationnelle*; as well as *The Invariantive Mechanics*;

1921-27: Émile Félix Édouard Justin Borel, French mathematician, worked on *measure theory and probability*, and formulated the *theorem* and *product of convolution*;

1923-28: Louis-Victor Broglie, French physicist, became famous for his *wave-particle duality*, showing that particles can behave as waves, and thus opening the way to wave mechanics;

1924: Satyendra Nath Bose, Indian physicist, and Albert Einstein, German-Swiss-US mathematical physicist, founded *Bose-Einstein statistics*, corresponding to the condensate phenomenon that some particles should appear at very low temperature;

1924-31: Wolfgang Pauli, Austrian-Swiss theoretical physicist, formulated *Pauli's exclusion principle*, according to which there are no two electrons in an atom with exactly same state, i.e. with same quantum numbers, and predicted the existence of a *low-mass neutral particle*, called 'neutrino';

1924-43: Karl Terzaghi, Austrian-born US civil engineer, published *Magnum Opus, Erdbaumechanik* and *Theoretical Soil Mechanics*, formulating *Terzaghi's principle*, which states that when a rock is subjected to a stress, it is opposed by the fluid pressure of pores in the rock;

1925-27: Werner Karl Heisenberg, German theoretical physicist, re-interpreted classical mechanics, introducing a *matrix-based quantum mechanics* where phenomena must be describable both in terms of wave theory and quanta; as well as formulated the revolutionary *principle of indeterminacy* or *uncertainty principle*, showing that there is a fundamental limit to the accuracy to which certain pairs of variables, such as position and momentum, can be determined; and Charles Thomson Rees Wilson, Scottish pioneer of atomic and nuclear physics, devised the *cloud chamber* method of marking the track of alpha-particles and electrons, by which the movement and interaction of atoms could be observed and photographed;

1925-47: <u>Patrick Maynard Stuart Blackett</u>, English physicist, made the first photography of *nuclear collisions involving transmutation*, and developed CTR Wilson's cloud chamber, using it to confirm the existence of the *positron*, as antiparticle of the electron;

1926: <u>Enrico Fermi</u>, Italian-born US nuclear physicist, and independently <u>Paul Adrien Maurice Dirac</u>, English mathematical physicist, founded *Fermi-Dirac statistics*, describing energies of single particles in a system of many identical particles that obey W Pauli's exclusion principle, and applying it to particles with half-integer spin, called *fermions*;

1926-30: <u>Erwin Schrödinger</u>, Austrian physicist, inspired by L-V Broglie's wave-particle duality, originated *wave mechanics* as part of quantum theory with cerebrated *Schrödinger's wave equation*;

1927-33: <u>Vladimir Aleksandrovich Fock</u>, Soviet theoretical physicist, worked in quantum mechanics, generalized E Schrödinger's wave equation to relativistic case, and formulated the *Klein-Fock equation*; then developed quantum mechanics to multi-particle systems to solve wave equation for multi-electron atoms, called *Hartree-Fock technique*;

1927-35: <u>Clinton Joseph Davisson</u>, and <u>Lester Halbert Germer</u>, US physicists, studying electron scattering from a block of nickel when their vacuum system broke down, and finding familiar peaks of a diffraction pattern; they discovered *diffraction of electrons*, confirming L-V Broglie's theory of wave nature of particles;

1927-36: <u>George Paget Thomson</u>, English physicist, noticing that a beam of electrons could produce circular interference fringes, discovered *electron diffraction by crystals*;

1927-38: <u>Pascual Jordan</u>, German theoretical physicist, worked on the theory of *quantum mechanics in matrix representation*, showing how light could be interpreted as being composed of discrete quanta of energy, and contributed to the foundation of *quantum electrodynamics*;

1927-60: <u>Eugene Paul Wigner</u>, US theoretical physicist, had the idea of *parity conservation in nuclear interactions*, demonstrated that *strong nuclear force has very short range*, co-operated for *Breit-Wigner formula* which describes resonant nuclear reactions, and formulated the *Wigner theorem* concerning the conservation of angular momentum of electron spin;

1928: <u>Otto Diels</u>, German chemist, and <u>Kurt Alder</u>, German organic chemist, discovered the *reaction of an activated olefin with a diene to give a cyclic structure* 'Diels-Alder reaction' with predictable stereochemistry, a reaction of enormous synthetic value;

1928-30: <u>Paul Adrien Maurice Dirac</u>, English mathematician and physicist, formulated the *relativistic wave equation*, and from its negative energy solutions deduced and predicted the existence of *anti-matter*; he also published the classic work *The Principles of Quantum Mechanics*;

1930-35: <u>Max Born</u>, German physicist, using E Schrödinger's wave equation, deduced that the state of a particle, e.g. its energy or position, could only be predicted in terms of probabilities, and from this the existence of *quantum jumps between discrete states*;

1930-75: <u>John Hasbrouck Van Vleck</u>, US physicist, researched dielectric and magnetic susceptibilities, published *The Theory of Electric and Magnetic Susceptibilities*, and elucidated the *chemical bonding in crystals*;

1931-48: <u>Grigore Moisil</u>, Romanian mathematician and computer pioneer, published *On a class of systems of equations with partial derivatives from mathematical physics*, and *Logic and Theory of Proof*, as well as introduced multi-valued algebras, now known as *Łukasiewicz-Moisil algebras*;

1932: <u>Stefan Banach</u>, Polish mathematician, published *Théorie des opérations linéaires* 'Theory of Linear Operations', relating to a class of infinite dimensional linear spaces; <u>John Von Neumann</u>, US mathematician, worked on the theory of linear operators, gave a new axiomatization of set theory, and formulated a precise description of *quantum theory*; <u>Carl David Anderson</u>, US physicist, discovered the *positron*, positively charged electron-type particle, thus confirming the existence of antimatter; <u>Harold Clayton Urey</u>, US physical chemist, discovered *deuterium* 'heavy hydrogen', containing one proton and one neutron, by repeatedly distilling a sample of liquid hydrogen; and <u>James Chadwick</u>, English physicist, bombarding beryllium with alpha particles, discovered the *neutron*, a neutral particle whose mass is closed to that of a proton;

1932-48: <u>Norbert Wiener</u>, US mathematician, worked on stochastic processes and harmonic analysis, introduced the concepts later called *Wiener integral* and *Wiener measure*, and founded *cybernetics* as published in his *Cybernetics, or control and communication in the animal and the machine*;

1932-58: <u>Douglas Rayner Hartree</u>, English mathematician and physicist, researched computational methods applied to a wide variety of problems in atomic physics; invented the *method of self-consistent field in quantum mechanics*, and the automated control of chemical plants; also developed a *differential analyser*, and an analogue computer, as an early electronic digital computer;

1932-64: <u>Sin-Itiro Tomonaga</u>, Japanese physicist, wrote papers on *positron creation* and *annihilation*, and on high-energy *neutrino-neutron scattering*, as well as worked on relativistic quantum description of interaction between a photon and an electron, developing the theory of *quantum electrodynamics*;

1933-68: <u>Alfred Tarski</u>, US logician and mathematician, worked on mathematical logic, formulated the *Banach-Tarski paradox*, allowing any set to be broken up and reassembled into a set of twice size, as well as defined the *truth* in formal logical languages, as presented in his monograph *Der Wahrheitsbegriff in den Formalisierten Sprachen* 'The Concept of Truth in Formalized Languages';

1934: <u>Henri Coanda</u>, Romanian aeronautical engineer, obtained patent for *Coanda effect* in fluid dynamic, as entrainment of a *free jet alongside a curve surface*;

1934-52: <u>Walther Wilhelm Georg Bothe</u>, German physicist, developed an electric circuit to replace the laborious process of counting scintillations by eye, and the *coincidence technique* which allowed two particles to be associated with each other, then used to study cosmic rays and nuclear physics;

1934-78: <u>Peter Kapitza</u>, Soviet physicist, constructed a helium liquefier to investigate extraordinary *superfluidity of helium-2*, and liquefied both *hydrogen* and *helium*;

1935: <u>Charles Vernon Theis</u>, US hydrogeologist, published *The relation between the lowering of the piezometric surface and the rate and duration of discharge of a well using ground-water storage*, widely used in hydrogelogical calculations;

1935-47: <u>Hideki Yukawa</u>, Japanese physicist, and <u>Cecil Frank Powel</u>, English physicist, suggested that a strong short-range attractive interaction between neutrons and protons would overcome electrical repulsion between protons, and then discovered intermediate particles propagating interaction as charged *p-mesons* or *pions*;

1935-80: <u>Caius Iacob</u>, Romanian mathematician, published *Sur la détermination des fonctions harmoniques conjuguée par certains conditions aux limites. Applications à l'hydrodynamique; Une introduction mathématique à la mécaniques des fluides*; and *Theoretical Mechanics*;

1937-50: Isidor Isaac Rabi, US physicist, developed the *resonance method* for accurately determining *magnetic moments of fundamental particles*, and also contributed to development of radar, nuclear bomb, laser and atomic clock;

1938: Lise Meitner, Austrian physicist, and Otto Robert Frisch, British physicist, following their work on nuclear transmutation, proposed that the production of barium is a result of nuclear fission, and operated *neutron bombardment experiments*, showing that *mass converted into energy*;

1940: Edwin Mattison McMillan, US atomic scientist, and Philip Hauge Abelson, US physical chemist, synthesized *neptunium* by bombarding uranium with slow moving neutrons, at Berkeley Radiation Laboratory of University of California;

1940-41: Glenn Theodore Seaborg, and Edwin Mattison McMillan, US atomic scientists, synthesized *plutonium*, as a secondary product of radioactive decay of neutron-bombarded uranium, which led to the development of the atomic bomb;

1940-50: William Feller, Croatian-born US mathematician, worked on probability theory, and wrote *Introduction to probability theory and its applications*, starting from first principles, yet containing original research often leading to surprising results, and packing practical examples;

1944: John Von Neumann, US mathematician, published *The Theory of Games and Economic Behavior*, referring both to games of chance and to games of pure skill, such as chess;

1944-50: Glenn Theodore Seaborg, Leon Owen Morgan, Ralph James and Albert Ghiorso, US atomic scientists, synthesized *americium* and *curium*, which were produced by bombarding uranium or plutonium with neutrons in nuclear reactors;

1945-60: Lev Landau, Soviet physicist, elaborated the theory of *superfluidity*, or *zero viscosity*, of condensed matter, particularly helium, as a collective behaviour of atoms in a liquid;

1949-51: Edward Mills Purcell, US physicist, developed *nuclear magnetic resonance*, allowing resonances when nuclei were placed in a magnetic field;

1949-55: <u>Albert Ghiorso</u>, and <u>Glenn Theodore Seaborg</u>, US atomic scientists, identified *berkelium, californium, einsteinium, fermium, mendelevium, nobelium, lawrencium, rutherfordium*, and *dubnium*, at Berkeley Radiation Laboratory of University of California;

1949-66: <u>Mahdi Hantush</u>, US groundwater hydrologist, published works on *groundwater steady and unsteady flow in aquifers*, and gave an important solution of flow in leaky (recharged) aquifers;

1950-54: <u>Polykarp Kusch</u>, German-born US physicist, studying the large discrepancy between observed and predicted values when measuring the hyperfine splitting in atomic hydrogen, he achieved precise determination of the *electron's magnetic moment*;

1950-60: <u>Jacques Salomon Hadamard</u>, French mathematician, worked on and contributed to the *theory of complex functions*, differential geometry, and partial differential equations; he is also known by his *product, code, dynamical system, inequality, finite part integral, manifold, matrix, space*, and *three-circle theorem*; and <u>Robert Hofstadter</u>, US physicist, using a linear accelerator to probe nuclear structure and investigating the nuclear charge distribution, discovered that *protons and neutrons also contain inner structure*, now known to be due to quarks;

1953: <u>Christopher Kelk Ingold</u>, English chemist, used physical techniques, such as isotope effects and molecular spectroscopy, in organic chemistry, the results being presented in his monumental work *Structure and Mechanism in Organic Chemistry*;

1955: <u>Emilio Segrè</u> and <u>Owen Chamberlain</u>, US physicists, discovered the *anti-proton*, as an anti-particle of the proton, with identical mass but negative electric charge;

1955-64: <u>Richard Phillips Feynman</u>, US physicist, initiated the *path integral approach* for describing quantum processes, contributing to the development of quantum electrodynamics, and introduced *Feynman diagrams* providing a pictorial representation of particle interactions; and <u>Julian Schwinger</u>, US physicist, did fundamental work in *quantum electrodynamics*, with important consequences for the physics of elementary particles;

1955-80: <u>Kai Mann Börje Siegbahn</u>, Swedish physicist, studied the energies of electrons emitted from solids exposed to X-rays, which revealed sharp peaks at energies characteristic of materials, leading to the technique of *electron spectroscopy for chemical analyses*, and its application for liquids and gases; he also contributed to the development of *high-resolution electron spectroscopy*;

1956: <u>Frederick Reines</u> and <u>Clyde Lorrain Cowan</u>, US physicists, provided a definitive experimental evidence of *neutrino*, as nature's most elusive particle;

1956-79: <u>Georg Wittig</u>, German chemist, discovered that some *ylides* - organometallic compounds containing both positive and negative charges - react smoothly with aldehydes and ketones with creation of an *olefinic double bond*, leading to synthesis of numerous important compounds, including vitamin A, vitamin D, steroids, and prostaglandin precursors;

1960-62: <u>Maria Goeppert-Mayer</u>, German-born US physicist, and independently <u>Johannes Hans Daniel Jensen</u>, German physicist, based on the fact that certain nuclei are very stable, having 'magic numbers' of protons and neutrons, they developed the *nuclear shell model*;

1960-75: <u>Nevill Francis Mott</u>, English physicist, and <u>Philip Warren Anderson</u>, US physicist, worked on *electronic properties* of disordered matter, and *electronic structure* of disordered systems;

1960-86: <u>Richard Edward Taylor</u>, Canadian physicist, investigated the structure of nucleons by scattering high-energy electrons from nuclear targets, and established the *constituents of nucleons*, now known as 'quarks', as real entities, by determining some of their properties;

1960-90: <u>Georges Charpak</u>, French physicist, developed *gaseous particle detectors*, resulting in a crucial advance in detector technology by device of a *multi-wire proportional chamber*, which allows large-area detectors capable of operating at high rates to be built relatively cheaply, this development revolutionized high-energy physics experiments;

1962-70: <u>Leon Max Lederman</u> and <u>Melvin Schwartz</u>, US physicists, made experiments resulting in 20 muon events, and confirmed the existence of two distinct *neutrino types*, as a basis for idea that fundamental particles come in generations, with electron, muon and tau-lepton all having associated neutrinos, and therefore demonstrating the doublet structure of leptons through discovery of *muon neutrino*;

1963-64: <u>Murray Gell-Man</u>, US physicist, and independently <u>George Zweig</u>, US theoretical physicist, introduced the concept of *quarks* which have one-third integral charge and baryon number, being a fundamental constituent of matter which combine to form composite particles called 'hadrons' such as protons and neutrons;

1963-70: <u>Niels Aage Bohr</u> and <u>Benjamin Roy Mottelson</u>, Danish physicists, developed the *collective model of the nucleus*, which combined quantum-mechanical shell model of nucleus and the classical liquid drop model;

1964: British theoretical physicists <u>Peter Ware Higgs</u> and <u>Tomas Walter Bannerman Kibble</u>; Belgian theoretical physicist <u>François Englert</u> and US-Belgian theoretical physicist <u>Robert Brout</u>; US particle physicists <u>Gerald Stanford Guralnik</u> and <u>Carl Richard Hagen</u>; proposed the mechanism suggesting the existence of a particle called *Higgs boson* and its associated *Higgs field* which is pivotal to the Standard Model and other theories within particle physics;

1964-75: <u>Kenichi Fukui</u>, Japanese chemist, and <u>Roald Hoffmann</u>, US chemist, elaborated the *frontier orbital theory* for atomic structure affecting course of chemical reactions, which was widely used in rationalizing organic reactivity and preceded sophisticated computer calculations, as well as formulated rules of *conservation of orbital symmetry*;

1965-85: <u>Jerome Isaac Friedman</u>, US physicist, <u>Henry Way Kendal</u>, US physicist, and <u>Richard Edward Taylor</u>, Canadian physicist, working at Stanford linear accelerator, they investigated electron-scattering from protons and neutrons, and studied the *properties of quarks*, establishing that quarks have spin and fractional charges of $+\frac{2}{3}$ and $\frac{1}{3}$ times electronic charge, and thus providing first incontrovertible *evidence for quarks as real entities* rather than abstract mathematical concepts;

1969-86: <u>Jack Steinberger</u>, German-born US physicist, proved the existence of a *neutral pion* by observations of its decay at Berkeley synchrotron, measured the spin and parity of charged pion, and established the existence of *two distinct neutrino types*;

1970-78: <u>Herbert Charles Brown</u>, US chemist, elucidated the chemistry of *boron*, and used boron compounds in organic synthesis, by researches at University of Chicago; and <u>Sheldon Lee Glashow</u>, US physicist, <u>Abdus Salam</u>, Pakistani theoretical physicist, and <u>Steven Weinberg</u>, US physicist, contributed to the theory of *unified weak and electromagnetic interaction* between elementary particles, including prediction of the *weak neutral current*;

1971: <u>Robert Coleman Richardson</u>, <u>David Morris Lee</u>, and <u>Douglas Dean Osheroff</u>, US physicists, searched on *superfluidity of matter near freezing point*, and discovered the isotope *helium-3*;

1971-83: <u>Carlo Rubbia</u>, Italian-born US physicist, and <u>Simon Van der</u>

Meer, Dutch physicist, discovered the *W and Z bosons*, as field particles which transfer *weak nuclear interactions*;

1974: Albert Ghiorso and Kenneth Hulet, US scientists, discovered *unnilhexium*, later named *seaborgium*, as element 106 created in Super HILAC accelerator at Lawrence Berkeley Laboratory;

1975-80: Martin Lewis Perl, US physicist, working in experimental particle physics, established the existence of the elementary particle *tau lepton*;

1976-87: Pierre-Gilles De Gennes, French theoretical physicist, worked on molecules in *substances undergoing phase transitions*, as well as on *polymers* and *liquid crystals*;

1977: Klaus Von Klitzing, German physicist, discovered the two-dimensional electronic behaviour in which *quantum Hall effect* was clearly implied, causing a major revision of the *theory of electric conduction in strong magnetic fields*;

1980: James Watson Cronin, and Val Logsdon Fitch, US physicists, discovered violations of fundamental symmetry principles in the *decay of neutral K-mesons*;

1980-86: Johannes Georg Bednorz, German physicist, and Karl Alexander Müller, Swiss physicist, investigated a new range of material types based on oxides, detected superconductivity at temperature of 35 kelvin, then achieved superconductivity up to 90 kelvin with superconductors now able to operate using inexpensive and plentiful coolants, thus opening the way for *superconductivity in ceramic materials*;

1980-88: Hans Georg Dehmelt, German-born US physicist, and Wolfgang Paul, German physicist, developed a device known as *Penning trap*, or *Paul trap* respectively, to constrain electrons and ions within a small space for study, as a technique to advance accuracy in determining atomic properties such as the *magnetic moment of an electron*;

1980-96: Steven Chu, US physicist, Claude Nessim Cohen-Tannoudji, French physicist, and William Daniel Phillips, US physicist, developed *methods for cooling matter to very low temperatures and trapping atoms* by use of laser light;

1985-97: Robert Betts Laughlin, US physicist, Horst Ludwig Störmer, German-born US physicist, and Daniel Chee Tsui, Chinese-born US physicist, discovered a *new form of quantum fluid* with functionally charged excitations;

1985-98: <u>Gerardus t'Hooft</u> and <u>Martinus Veltman</u>, Dutch physicists, elucidated the quantum structure of *electroweak interactions* in physics;

1986-2000: <u>Göran Stenman</u>, Swedish geneticist, leading a team at the University of Gothenburg, discovered that *MYB-NFIB fusion gene* was found in all examined *adenoid cystic carcinomas*, and opened the way to develop a drug that can turn off this gene;

1989-2000: <u>Fabian Pascal</u>, Romanian-born US scientist, became known by his research on *Structured Query Language* 'SQL', and by published works *SQL and relational basics*, M&T Books; *Understanding relational databases with examples in SQL-92*, New York, Wiley; and *Practical issues in database management: a reference for the thinking practitioner*, Boston, Addison-Wesley;

1990-2000: <u>Eric Allin Cornell</u>, US physicist, <u>Wolfgang Ketterle</u>, German physicist, and <u>Carl Edwin Wieman</u>, US physicist, achieved the *Bose-Einstein condensation in dilute gases of alkali atoms* at very low temperatures, and developed early fundamental studies of *properties of condensates*;

1990-2008: <u>Andre Konstantin Geim</u>, Russian-born British-Dutch physicist, and <u>Konstantin Sergeyevich Novoselov</u>, Russo-British physicist, researching graphene - a substance composed of pure carbon, with atoms arranged in a regular hexagonal pattern similar to graphite, but in one-atom thick sheet -, they made groundbreaking experiments regarding *two-dimensional material graphene*;

1995-2002: <u>Alexei Alexeyevich Abrikosov</u>, Russian theoretical physicist, <u>Vitaly Lazarevich Ginzburg</u>, Russian physicist, and <u>Anthony James Leggett</u>, English physicist, did more pioneering work on the *theory of superfluids and superconductors*;

1995-2003: <u>David Jonathan Gross</u>, US physicist, <u>Hugh David Politzer</u>, US theoretical physicist, <u>Frank Anthony Wilczek</u>, US theoretical physicist and mathematician, discovered *asymptotic freedom in the theory of strong interaction*;

1995-2006: <u>Albert Fert</u>, French physicist, and <u>Peter Andreas Grünberg</u>, German physicist, studying magnetoresistance as dependence of electrical resistance of a sample on strength of external magnetic field, they discovered *giant magnetoresistance*, which is defined as a quantum mechanical magnetoresistance effect observed in thin-film structures composed of alternating ferromagnetic and non-magnetic conductive layers;

1997: <u>Marshal Anderson</u>, US geneticist, leading a group of researchers at Genetic Epidemiology of Lung Cancer, discovered how the RGS17 gene impacting cancer development could change clinical diagnosis and treatment of *breast cancer genes*, as published in the article *A proven genetic test could help us identify people at risk before the disease progresses*;

2000-06: <u>Yoichiro Nambu</u>, Japanese-born US physicist, discovered the mechanism of *spontaneous broken symmetry in subatomic physics*;

2001-07: <u>Makoto Kobayashi</u>, Japanese physicist, and <u>Toshihide Maskawa</u>, Japanese theoretical physicist, discovered the *origin of broken symmetry*, predicting the *existence of at least three families of quarks* in nature;

2002-09: <u>Preda Mihailescu</u>, Romanian mathematician, proved *Catalan's conjecture*, a number-theoretical conjecture, formulated by French mathematician Eugène Charles Catalan in 1844, that has stood unresolved for over a century; and then he proved *Leopoldt's conjecture*, introduced by German mathematician Heinrich-Wolfgang Leopoldt in 1962-57, for all number fields;

2005-11: <u>Serge Haroche</u>, French physicist, and <u>David Jeffrey Wineland</u>, US physicist, achieved groundbreaking experimental *methods that enable to measure and manipulate individual quantum systems*;

2012: <u>George Booth</u>, English chemist, published his work *Communication: Excited states, dynamic correlation functions and spectral properties from full configuration interaction quantum Monte Carlo*, Journal of Chemical Physics; <u>David Wales</u>, English physical chemist, wrote *Decoding the energy landscape: extracting structure, dynamics and thermodynamics*, Philosophical Transactions of the Royal Society; and <u>David Lonie</u>, US crystallographer, and <u>Eva Zurek</u>, US biochemist, published their work *Identifying duplicate crystal structures: XtalComp, an open-source solution*, Computer Physics Communications;

2013: <u>ATLAS and CMS teams at the Large Hadron Collider</u>, CERN 'The European Organization for Nuclear Research', discovered the *Higgs boson* 'God particle', a particle with mass of c.125 giga-electron-volts per square speed of light (GeV/c^2), mean lifetime $1.56 \cdot 10^{-22}$ second, and no electric charge, that decays into W and Z bosons, two photons, and possibly other unidentified particles; and <u>Marius Albu</u>, Romanian-born British geologist and physicist, completed the book *Conventional and Non-conventional Forms of Energy*, United p.c. publisher, European Union, related to cosmic, terrestrial, life, human, cerebral, and socioeconomic forms of energy.

On the basis of human abilities presented in the table of chapter 7, science originated 15033 years ago, as its initial time, and will last until
4350 years from present forward, as its final time $t_\bullet = 15033 + 4350 = 19383$ years from the origin of science. Data above indicate that the sciences were developed in successive sequences, from which the later and then better determined ones are delimited as follows:

Mechanical principles \ 800...600BC ≈ 700BC \
Geometry and hydrostatics \ 100BC...AD70 ≈ 15BC \
Arithmetic and trigonometry \ AD550...700 ≈ AD625 \
Equations and mechanics \ AD1150...1270 ≈ AD1210 \
Calculus, chemistry and electricity \ AD1700...1790 ≈ AD1745 \
Quanta, relativity and informatics.

Therefore, the approximate time limits are 2714; 2029; 1389; 804; 269 years ago; or 15033 - 2714 ≈ 12319; 15033 - 2029 ≈ 13004; 15033 - 1389 ≈ 13644; 15033 - 804 ≈ 14229; 15033 - 269 ≈ 14764 years from the origin of science. The last values of approximate transitional times divided by the final time $t_\bullet = 19383$ years from the same origin result in values 12319/19383 ≈ 0.636; 13004/19383 ≈ 0.671; 13644/19383 ≈ 0.704; 14229/19383 ≈ 0.734; 14764/19383 ≈ 0.762; which can be found in Background to time, Table *(l)* for $z/16 = l$ sequences, in the same succession, as

$$t_{0.75}/t_\bullet = tanh(0.75) = 0.6351490;$$
$$t_{0.8125}/t_\bullet = tanh(0.8125) = 0.6709671;$$
$$t_{0.875}/t_\bullet = tanh(0.875) = 0.7039056;$$
$$t_{0.9375}/t_\bullet = tanh(0.9375) = 0.7340715;$$
$$t_1/t_\bullet = tanh(1) = 0.7615942;$$

leading to the accurate values:

$$t_{0.75} = t_\bullet \cdot tanh(0.75) = 12311;$$
$$t_{0.8125} = t_\bullet \cdot tanh(0.8125) = 13005;$$
$$t_{0.875} = t_\bullet \cdot tanh(0.875) = 13644;$$
$$t_{0.9375} = t_\bullet \cdot tanh(0.9375) = 14229;$$
$$t_1 = t_\bullet \cdot tanh(1) = 14762 \text{ years from the origin of science.}$$

Continuing the same procedure for calculating the transitional times which separate the earlier sequences, the timeline, sequences, and intrinsic characteristics in development of science are reconstituted, as follows:

$z/16 = l$	Time (years)		Sequences of sciences
	from origin $t_l = t_\bullet \cdot \tanh(l)$	from present $t_l - 15033$	
	19383	+4350	
...	...	...	
1.0625	15247	+214 (AD2228)	_
1	14762	-271 (AD1743)	_Quanta, relativity and informatics
0.9375	14229	-804 (AD1210)	_Calculus, chemistry and electricity
0.875	13644	-1389 (AD625)	_Equations and mechanics
0.8125	13005	-2028 (14BC)	_Arithmetic and trigonometry
0.75	12311	-2722 (708BC)	_Geometry and hydrostatics
0.6875	11560	-3473 (1459BC)	_Mechanical principles
0.625	10750	-4283 (2269BC)	_Numbers and elementary geometry
0.5625	9882	-5151 (3137BC)	_Quantities and timing
0.5	8957	-6076 (4062BC)	_Astronomic observations
0.4375	7977	-7056 (5042BC)	_Lines, plane figures and shapes
0.375	6946	-8087 (6073BC)	_Formulation of simple rules
0.3125	5867	-9166 (7152BC)	_Physical-mathematical notions
0.25	4747	-10286	_Classifications
0.1875	3592	-11441	_Experiments
0.125	2410	-12623	_Observations on useful materials
0.0625	1210	-13823	_Simple arithmetic
0	0	-15033	_Counting and comparing

$z/16 = l$	Time from origin $t_l = t_\bullet \cdot \tanh(l)$ (years)	Period $\tau_l = (t_\bullet^2 - t_l^2)/(2t_\bullet)$ (years)	Frequency $f_l = (2t_\bullet)/(t_\bullet^2 - t_l^2)$ (years)$^{-1}$	Angular speed $\omega_l = 2\pi/\tau_l = 2\pi \cdot f_l$ (years)$^{-1}$
	19383	0	∞	∞
...	...	...	...	...
1.0625	15247	3694.7	0.0002707	0.0017006
1	14762	4070.2	0.0002457	0.0015437
0.9375	14229	4468.8	0.0002238	0.0014060
0.875	13644	4889.4	0.0002045	0.0012851
0.8125	13005	5328.7	0.0001877	0.0011791
0.75	12311	5781.9	0.0001730	0.0010867
0.6875	11560	6244.3	0.0001601	0.0010062
0.625	10750	6710.5	0.0001490	0.0009363
0.5625	9882	7172.4	0.0001394	0.0008760
0.5	8957	7622.0	0.0001312	0.0008244
0.4375	7977	8.050.0	0.0001242	0.0007805
0.375	6946	8446.9	0.0001184	0.0007438
0.3125	5867	8803.6	0.0001136	0.0007137
0.25	4747	9110.2	0.0001098	0.0006897
0.1875	3592	9358.7	0.0001069	0.0006714
0.125	2410	9541.7	0.0001048	0.0006585
0.0625	1210	9653.7	0.0001036	0.0006509
0	0	9691.5	0.0001032	0.0006483

21. Industry

The production of economic goods or services within the economy is called *industry* (from Latin *industria* 'diligence, activity, zeal'). It is usually defined as commercial production and sale goods, a specific branch of manufacture and trade, the sector of an economy made up of manufacturing enterprises, industrial management, an energetic devotion to a task or an endeavour, and sometime as ongoing work or study associated with a specified subject or figure.

In a large sense, the industry emerged by discovery of copper, a naturally occurring, relatively pure metal, native to many places, and the earliest evidences of wrought native copper and gold are dated back to around 9000BC. Subsequent experiments with ores of native metals led to their extraction and frequent use beginning with gold c.6000BC, copper c.4200BC, and silver c. 4000BC, then continuing with lead c.3500BC, tin c.1750BC, iron (smelted) c.1500BC, and mercury c.750BC. Industry successively developed by copper metallurgy and agro-technique, earthenware-pottery and trade routing, weaving-sail technique and bronze metallurgy, irrigation technique and iron metallurgy, hydrotechnique and glassware, early machinery and gunpowder-cannon, and later engines, equipments and high technology. Manufacturing industry became a key sector of production and labour in European and North American countries during the Industrial Revolution, by replacing previous mercantile and feudal economies, due to the rapid advance in technology, such as production of steel and coal. The development of industry proceeded from steam-powered machines and electrical equipments through mechanized assembly lines and automatic devices to computers and robots, replacing human operators, and recording notable social, cultural, and environmental impacts, some of them with unpredictable consequences.

At the top level, industry is classified into sectors such as *primary or extractive*, *secondary or manufacturing*, *tertiary or services*, occasionally *quaternary or intellectual services*, and *quinquenary or culture and research* sectors. In market and finance research, the main systems of industry classification are *Market-based classification systems* including 'Global Industry Classification Standard', and *Industry Classification Benchmark*. Industries can also be classified by product, such as: Aerospace, Agriculture, Automotive, Chemical, Computer, Construction, Cultural, Defence, Electronic, Energy,

Entertainment, Extractive, Metallurgic, Financial Services, Fish, Food, Health Care, Heavy, Hospitality, Information, Insurance, Manufacturing, Mass Media, Meatpacking, Paper, Petrochemical, Petroleum, Poverty, Shipbuilding, Semiconductor, Software, Telecommunications, Trade, Water, and other industries.

A model of industrial promotion and development can be conceptualized in terms of *Acceptable* discoveries, inventions, innovations or ideas A_i, deriving from *Social* demands, needs, requirements or desires S_j and *Managerial-marketable* valuations, appreciations, decisions or approvals $M^j{}_i$, which can be related so that $A_i = M^j{}_i \cdot S_j$. Subsequently, the A_i become operational by their *Outputs* O_k and *Efficiencies* E^i, and then their specific production P_k can be expressed in the form

$$P_k = O_k \cdot E^i \cdot A_i = O_k \cdot E^i \cdot M^j{}_i \cdot S_j,$$

whereby the cumulative production can be represented as

$$P = \Sigma^n{}_{k=1} \, O_k \cdot E^i \cdot A_i = \Sigma^n{}_{k=1} \, O_k \cdot E^i \cdot M^j{}_i \cdot S_j.$$

The successive stages and developers of industry in its complexity can be briefly presented, in chronological order, as follows:

9000-8000BC: *Copper ores* were first discovered in the area Caucasus-Anatolia;

8000-7000BC: *Experiments with copper ores* were carried out by some inhabitants of South Caucasus and Anatolia;

7000-6000BC: *Copper metallurgy* first started in Anatolia and the Near East;

6000-5000BC: *Early bronze use* was recorded in the Near East; *Irrigation techniques* emerged in Mesopotamia;

5000-4300BC: *Bronze casting* and *plough use* started to be practiced in South-east Asia, especially in Anatolia and Mesopotamia, and then in Egypt, India and China;

4300-3500BC: *Bronze Age* began in the Near East; *Sail* was invented in Egypt; *Copper use* was evidenced in Thailand;

3700BC: *Wheel* was invented by Sumerians, as an essential part indispensable for transport, ceramics, and further industrial development;

3500-2600BC: *Bronze metallurgy* was practiced in South Asia and Thailand; *Glassmaking* started in Mesopotamia, South Asia and Europe; *Wheels* and *Wheeled vehicles* appeared in Mesopotamia, Indus Valley, Northern Caucasus, Balkans and Central Europe;

2600-1800BC: *Bronze working* became a widespread activity, being evidenced in Peru and Eastern Mediterranean area; and *Bronze metallurgy* spread from the Near East to Europe;

1800-1100BC: *Bronze technology* was promoted and developed in China and also in South-east Asia;

1100-430BC: *Hallstatt technology* emerged in Western central Europe; and *Ironmaking* appeared at Meroe on the Nile Valley, and at Taruga in Nigeria;

430BC-AD200: *La Tène technology* began in Western central Europe; and *Steelmaking* was practiced in Europe, the Near East, the Middle East and Far East;

c.250BC: <u>Archimedes</u>, Greek mathematician and inventor, devised the *Archimedean screw* used for rising water and many machines for defending Syracuse; and demonstrated the *powers of levers* in moving large weights, emphasized as 'Give me a firm spot on which to stand, and I shall move the Earth';

c.AD60: <u>Hero of Alexandria</u>, Greek mathematician, invented a primitive *steam turbine* 'aeolipile', a *fire engine pump*, a *coin-operated device*, and other machines and mechanisms, described in his *Pneumatical* 'Pneumatics';

200-800: *Steel weaponry* was developing in Europe, the Near East, Middle East and Far East;

724-727: <u>Yi Xing</u>, Chinese inventor, experimented *new devices and instruments*, constructed a *water-driven celestial sphere*, for which he first used *concentric gears* and shafts as part of mechanism, and made the first *Chinese clock* to strike hours and half-hours;

800-1345: *Gunpowder* was invented in China; and *Firearms* entered in use both in China and Europe;

1094: <u>Su Song</u>, Chinese astronomer and inventor, produced a *clock*, from which is known that it was housed in a tower of 33 feet in height, driven by a water wheel of 10 feet in diameter, and probably accurate to within 100 seconds a day;

1206: <u>al-Razaz al-Jazari</u>, Muslim polymath, inventor and mechanical engineer in Mesopotamia, after constructing *mechanical and hydraulic devices*, including a *reciprocating water pump* and its crankshaft, he produced the book *al-Jami' bain al-'ilm wa al-'amal al-nafi' fi sina'at al-hiyal* 'The Book of Knowledge of Ingenious Mechanical Devices', describing one hundred of such devices;

1345-1845: *Industrial revolution* was initiated in Europe, and transmitted through North and South America to the South, East, South-east Asia and Australia;

1540: <u>Vannoccio Vincenzio Agustino Luca Biringuccio</u>, Italian metallurgical engineer, wrote *De la Pirotechnia* 'Concerning Pyrotechnics' related to mining, metallurgy and other industrial processes, thus pointing out the level of industrial development in his time;

1555: <u>Georgius Agricola</u>, German mineralogist and metallurgist, practised as a physician and followed his interest in minerals to study mining, compiling the standard work *De Re Metallica*, a detailed record of 16th century mining, ore-smelting and metal working;

1588: <u>Agostino Ramelli</u>, Italian military engineer, published *Le diverse et artificiose machine del Capitano Agostino Ramelli* 'The Various and Ingenious Machines of Captain Agostino Ramelli', describing and illustrating almost 200 devices such as water pumps, cranes, grain mills, military bridges and ballistic engines;

1590: <u>Zacharias Janssen</u>, Dutch spectacle-maker, working on spectacles and lenses, invented the first *optical microscope* and *compound microscope*;

1604-20: <u>Cornelius Jacobszoon Drebbel</u>, Dutch-born British inventor, devised a *clock driven by changes in atmospheric pressure*, a new *method for manufacturing sulphuric acid*, a *thermostatto* regulating supply of air to furnace; he also devised and improved a rudimentary *submarine* that was successfully tested in Thames;

1608-10: <u>Hans Lippershey</u>, Dutch optician, working in optical devices, discovered that the combination of two separated long-focus and short-focus convex lenses can make distant objects appear nearer, and obtained patent for this type of *telescope*; also he evidenced that if this combination is reversed it becomes microscope;

1642: <u>Blaise Pascal</u>, French mathematician, physicist, theologian and man-of letters, studied the fall of mercury columns with increased altitude, and then invented the *barometer*, *hydraulic press*, and *syringe*, and most importantly the oldest *calculating machine*;

1644: <u>Evangelista Torricelli</u>, Italian physicist and mathematician, first described the *mercury barometer* 'torricellian tube', and also improved both telescopes and microscopes;

1650-54: <u>Otto von Guericke</u>, German engineer, invented a primitive *vacuum pump*, enabling new physical studies, and demonstrated the *effect of atmospheric pressure on a near vacuum*, consisting of two large metal hemispheres placed together and air within pumped out, which could not then be separated by two teams of eight horses, but fell apart when air was allowed to re-enter;

1654-63: Edward Somerset Worcester, English inventor, first devised a *steam water-pump*, published *Century of Inventions*, giving a brief account of a hundred inventions, and exposing his interest in machine and steam engine 'water pump' calculation;

1668: Isaac Newton, English scientist and mathematician, constructed *reflecting telescopes*, of a type further developed by later astronomers for studying the universe;

1675-90: Denis Papin, French scientist, created a *steam digester* as a prototype pressure cooker, investigated hydraulics and pneumatics, and constructed a *working model of an atmospheric condensing steam engine* on principles later developed by T Newcomen and J Watt;

1696-98: Thomas Savery, English inventor and military engineer, invented *rowing vessels* by means of paddle-wheels, and developed the first *practical high-pressure steam engine* for pumping water from mines;

1709-14: Gabriel Daniel Fahrenheit, German instrument-maker and physicist, produced high-quality meteorological instruments, devised an accurate *alcohol thermometer*, introduced the commercially successful *mercury thermometer*, and eventually devised a temperature scale calibrated at 32 and 96 degrees, and zero fixed at freezing point of ice and salt;

1712: Thomas Newcomen, English inventor, based on T Savery's atmospheric steam engine for pumping water from mines, he constructed the first *practical working engine*, that was widely used in collieries;

1712-1942: *Industrial Revolution* began in Great Britain, where the traditional agrarian economy was replaced with an economy dominated by machinery and manufacturing; this process transferred the balance of political power from landowners and peasants to industrial capitalists and urban workers, and subsequently spread through Europe, North America, Japan, and colonial empires; this revolution was based on innovations in textiles, engine manufacturing and iron making, and opened up an unprecedented development of mining, metallurgy, steam power, chemicals, machine tools, gas lighting, glass making, paper machine, transport, sail, canals, roads, railways, bridges, telecommunication, combustion engines, steamships, and aircrafts;

1713-62: <u>John Harrison</u>, English inventor and horologist, after the British government offered prizes for discovery of a method to determine longitude accurately, he developed a *marine chronometer* which, in a voyage to Jamaica, determined the *longitude* within 18 geographical miles, as well as invented the *gridiron pendulum*, *going fusee*, and *remontoir escapement*;

1723-30: <u>Henri Pitot</u>, French hydraulic and civil engineer, as superintendent of the *Canal du Midi*, constructed an *aqueduct for water supply of Montpellier*, and invented a device now known as *Pitot tube* by which relative speed of a fluid past orifice of tube can be measured;

1729-53: <u>Bernard Forest de Belidor</u>, French engineer, wrote the influential handbooks *Science des Ingénieurs* 'Science of Engineers', covering military engineering, and *Architecture Hydraulique* 'Hydraulic Architecture', covering civil engineering;

1730-33: <u>John Kay</u>, English inventor, patented an 'engine' for *twisting and cording* mohair and worsted, and also his *flying shuttle*, one of the most important inventions in the history of textile machinery, which was adopted by the Yorkshire woollen manufacturers;

1731-57: <u>René Antoine Ferchault de Réaumur</u>, French natural philosopher, invented a *thermometer* using a mixture of alcohol and water instead of mercury and calibrating it in the *Réaumur scale* of 80 degrees between freezing and boiling points of water; he also wrote the monumental *Description des arts et métiers* 'Description of Arts and Crafts', showing acquisition of wide knowledge and developing methods for producing *iron* and *steel*;

1741-45: <u>Jacques de Vaucanson</u>, French engineer and inventor, improved the machines for weaving and dressing silk, and invented the *first fully automatic loom*, controlled through a system of perforated cards;

1742: <u>Anders Celsius</u>, Swedish astronomer, invented the *centigrade scale of temperature*, using a mercury thermometer with 0° at melting point of ice and 100° at boiling point of water;

1764-68: <u>James Hargreaves</u>, English inventor, designed the *spinning-jenny*, an early type of spinning-machine with several spindles, and built a *spinning mill* at Nottingham;

1765: <u>James Watt</u>, Scottish engineer and inventor, improved the reciprocating steam engine and devised the *condensing steam engine*, with efficiency about three times more than the old atmospheric engines;

1774-79: <u>John Wilkinson</u>, English ironmaster and inventor, created the *cannon-boring machine*, which afterward was used to bore more accurate *cylinders for steam engines*;

1775-1800: <u>Alessandro Giuseppe Anastasio Volta</u>, Italian physicist and inventor, devised the *electrophorus*, as precursor of induction machine, *condenser, candle flame collector of atmospheric electricity*, and *electrochemical battery* 'voltaic pile';

1785: <u>Edmund Cartwright</u>, English inventor and clergyman, invented the *power loom* used for wool-combing machines, and then built a *weaving mill* at Doncaster;

1794-1814: <u>Robert Fulton</u>, US engineer, obtained patent for a double-inclined plane to supersede locks, invented a mill for sawing and polishing marble, developed a *submarine torpedo boat*, launched the steam vessel *Clermont* on the Hudson River, and constructed the *world's first steam warship* named *Fulton the First*;

1796-1815: <u>Richard Trevithick</u>, English inventor and engineer, devoted his life to improvement of *steam carriage* and invented such a *carriage*, constructed the *first steam railway locomotives* and a large number of *stationary steam engines*;

1801-08: <u>Joseph Marie Jacquard</u>, French silk-weaver, invented the *Jacquard loom*, which used *perforated cards* for controlling movement of warp threads, enabling ordinary workmen to produce most beautiful patterns in a style previously accomplished only with patience, skill and labour;

1805-34: <u>Johann Georg Bodmer</u>, Swiss inventor, produced the *percussion shell, cotton carding* and *spinning machine* in England, as well as an *opposed-piston steam engine*;

1815-29: <u>George Stephenson</u>, English railway engineer, devised the *colliery safety lamp* called 'Geordie', constructed the *first locomotive* 'Blucher', improved the locomotives and rails, and designed the triumphant 'Rocket' locomotive performing 30 mph speed;

1825-36: <u>William Sturgeon</u>, English scientist, constructed the first *practical electromagnet*, first *moving-coil galvanometer*, and various electromagnetic machines;

1826: <u>Joseph Nicéphore Niepce</u>, French chemist, introduced the first *permanent photographs*, making a stable image by a pewter plate coated with bitumen of Judea, a kind of asphalt which hardens on exposure to light;

1829-35: <u>Joseph Henry</u>, US physicist, built electromagnets and discovered the electromagnetic phenomenon of *self-inductance*, invented a precursor to the *electric doorbell* and *electric relay*, constructed the first *electromagnetic motor*, and noted the effects of *resistance on electric current*;

1832-44: <u>Samuel Finley Breese Morse</u>, US artist and inventor, had the idea and patented the first *magnetic telegraph* by which was sent the historic message 'What hath God wrought?', and also introduced *Morse code*;

1836-43: <u>Francis Pettit Smith</u>, English inventor, used his patent for *screw propeller*, to build the first *screw-propelled steamer*, called 'Archimedes', and then the first *screw warship*, called 'Rattler', for Royal Navy;

1840-55: <u>Henry Bessemer</u>, English metallurgist and inventor, developed a method for the production of *bronze powder* and *'gold' paint*, at a fraction of the cost of long-established German process; obtained a patent for the economical process to turn molten pig-iron directly into steel by blowing air through it in a *Bessemer converter*;

1843-45: <u>Carl Friedrich Gauss</u> and <u>Wilhelm Eduard Weber</u>, German scientists, studied the Earth's magnetism, developed the magnetometer, and invented the *electrodynamometer*;

1844: <u>Charles Goodyear</u>, US inventor, first produced the *vulcanized rubber*, which led to the development of rubber-manufacturing industry and well-known tyres;

1845-50: <u>Wilhelm Eduard Weber</u>, German scientist, after inventing the *electrodynamometer*, was the first to *apply the mirror and scale method* of reading deflections;

1845-present: *Modern industry* advanced and spread around the world;

1847: <u>Alfred Krupp</u>, German arms manufacturer, established the first *Bessemer steel plant*, becoming a foremost arms supplier to Germany and also to other countries in the world;

1847-67: <u>Ernst Werner von Siemens</u>, German electrical engineer, established factories for making *telegraphy equipment*, a business known as 'Siemens Brothers';

1848-79: <u>Joseph Wilson Swan</u>, English chemist, inventor and industrialist, made experiments on *carbonized paper filaments for electric lamps*, successfully demonstrated the *electric bulb*, and obtained patent for *carbon process for printing photographs* in permanent pigment;

1850-55: <u>Alexander Parks</u>, English chemist and inventor, constructed an *electroplating spider's web*, and obtained patent for *xylonite*, a form of celluloid;

1850-65: <u>Lines Yale</u>, US inventor and manufacturer, as a locksmith, invented various types of locks, including small cylinder *Yale locks*, later widely used in other countries;

1852-58: <u>Jean Bernard Leon Foucault</u>, French physicist, constructed the first *gyroscope* and the *Foucault prism*, and improved the *mirrors of reflecting telescopes*;

1852-60: <u>Isaac Merritt Singer</u>, US inventor and manufacturer, and <u>Elias Howe</u>, US inventor, improved the *single-thread, chain-stitch machine*, and desgined the *Howe needle*;

1859: <u>Jean Joseph Etienne Lenoir</u>, French inventor and engineer, created the first *practical internal-combustion gas engine* fuelled by coal gas and air, and constructed the *first car* to use it; and <u>Joseph Whitworth</u>, English engineer and inventor, produced a *gun of compressed steel*, with spiral polygonal bore, and the *standard screw-thread* called after him;

1864: <u>William Siemens</u>, British electrical engineer, and <u>Pierre Emilee Martin</u>, French metallurgist, improved the bulk steelmaking method, first using the *Process of open-hearth regenerative steel furnace* 'Siemens-Martin processes' derived from the fact that molten metal laid in a comparatively shallow pool on the furnace hearth;

1865-90: <u>Ernst Abbe</u>, German physicist, developed instruments for *measuring refractive indices of glass* and a *focometer to control performance of optical workshop*, invented the arrangement known as *Abbe's homogeneous immersion*, worked to improve the microscope, and founded the *diffraction theory of optical imaging*;

1866-75: <u>Alfred Bernhard Nobel</u>, Swedish chemist and manufacturer, invented a safe form of nitro-glycerine called *dynamite, smokeless gunpowder* and *gelignite*; created an industrial empire for manufacturing new substances, including *artificial gutta-percha*; and endowed the annual *Nobel prizes*;

1869-71: <u>Zénobe Théophile Gramme</u>, Belgian electrical engineer, constructed the *first successful direct-current dynamo*, incorporating the ring-wound armature called *Gramme ring*, and improved it as the *first electric generator* to be used commercially for electroplating and electric lighting;

1873: Hermann Wilhelm Vogel, German chemist, invented the *orthochromatic photographic plate* and studied spectroscopic photography, designing a *photometer*;

1876: Nikolaus August Otto, German engineer, invented the *four-stroke internal-combustion engine*, with the sequence of operation called 'Otto cycle';

1876-87: Alexander Graham Bell, US inventor, after experimenting with various acoustical devices and producing first intelligible telephonic transmission, he obtained patent for *telephone*, founded the Bell Telephone Company, and invented *photophone* and *graphophone*;

1877-91: Thomas Alva Edison, US inventor and physicist, invented the *gramophone, carbon granule microphone*, a *system for generating and distributing electricity*, designed the *first power plant*, and also invented *megaphone, storage battery, electric valve*, and *kinetoscope*;

1878: Sidney Gilchrist Thomas, English metallurgist, discovered how to remove phosphorus from steel by using dolomite for furnace lining, together with addition of lime to produce a basic slag allowing removal of both phosphorus and sulphur, this method being called *basic Bessemer process* in Great Britain, and *Thomas process* on Continent;

1879: Thomas Alva Edison, US inventor and physicist, and Joseph Wilson Swan, English inventor and industrialist, invented the *electric light bulb*, and founded the Edison & Swan Electric Light Company;

1880: Herman Hollerith, US inventor and computer scientist, realizing the need for automation in recording and processing data, devised the *punched-card machine* for storing and analysing data;

1880-1919: Elihu Thomson, US inventor, pioneered the electrical manufacturing industry in USA, co-operated in 700 patented electrical inventions, including the *three-phase alternating-current generator* and *arc lighting*; he also founded the Thomson-Houston Electric Company, and then the General Electric Company;

1884-97: Charles Algernon Parsons, Irish engineer, developed the *high-speed steam turbine*, and constructed the first *turbine-driven steamship* called 'Turbinia';

1884-1900: <u>George Eastman</u>, US inventor and philanthropist, produced the successful *roll-film*, *Kodak box camera*, and, together with TA Edison, experimented for moving-picture industry, then founded the *Eastman Kodak Company*, and developed the *Brownie camera*;

1885: <u>William Stanley</u>, US electrical engineer, invented the *electric transformer*, and a *long-range transmission system* for alternating current; and <u>Gottlieb Wilhelm Daimler</u>, German engineer and inventor, made the earliest roadworthy cars, using a *high-speed internal combustion engine*, and founded Daimler-Motoren-Gesellschaft in Cannstatt;

1885-88: <u>Nikola Tesla</u>, Croatian-born US physicist and electrical engineer, improved dynamos and electric motors, and invented the *high-frequency Tesla coil*, and *air-core transformer*;

1886: <u>George Westinghouse</u>, US engineer, used the *alternating current for distribution of electric power*, and founded Westinghouse Electrical Company; and <u>Paul Louis Toussaint Héroult</u>, French metallurgist, and independently <u>Charles Martin Hall</u>, US chemist, discovered a process for extracting aluminium by electrolysis of cryolite with a carbon-lined crucible cathode, as the first *economic method to obtain aluminium*;

1888: <u>Roland von Eötvös</u>, Hungarian physicist, researched terrestrial gravitation, demonstrated the equivalence of gravitational and inertial mass, and devised a sensitive *torsion balance* used in geophysical detection of oil reservoirs in the terrestrial crust;

1889-98: <u>James Dewar</u>, Scottish chemist and physicist, discovered *cordite*, accomplished the *liquefaction of hydrogen*, and invented the *vacuum flask*;

1891-96: <u>Edward Goodrich Acheson</u>, US chemist and inventor, developed the manufacture of *silicon carbide* 'carborundum', a useful abrasive, and devised the method of making *lubricants based on colloidal graphite*;

1897: <u>Rudolf Christian Karl Diesel</u>, German engineer, constructed the first *practical compression-ignition engine* 'diesel engine' with about twice efficiency that of comparable steam engines;

1897-1900: <u>Ferdinand von Zeppelin</u>, German army officer, constructed his *first airship*, a dirigible balloon of rigid type, called *zeppelin* after him;

1897-1929: <u>Richard Adolf Zsigmondy</u>, Austrian chemist, invented the

Jena milk glass, the *ultramicroscope*, leading to great advances in colloidal chemistry, and performed *ultrafiltration* in which substances to be separated are drawn through a membrane by a decrease in pressure;

1899-1908: <u>James Swinburne</u>, Scottish scientist and electrical engineer, pioneered the plastic industry, being known as 'the father of British plastics', experimenting with phenolic resins and resulting in a process for producing *synthetic resin* with its patent anticipated by one day by LH Baekeland;

1903-09: <u>Henry Ford</u>, US car engineer and manufacturer, founded the 'Ford Motor Company' pioneering modern *mass-production techniques* 'assembly line' for the famous 'Model T';

1904: <u>John Ambrose Fleming</u>, English physicist and electrical engineer, invented the thermionic rectifier or *diode valve* 'Fleming valve', which for half a century was a vital part of radio, television, and early computer circuitry; and <u>Piero Ginori Conti</u>, Italian prince, experimented with the transformation of heat into electricity, and constructed the first *geothermal power generator*;

1906: <u>Reginald Aubrey Fessenden</u>, US radio engineer and inventor, developed *amplitude modulation*, used for the first *radio broadcast*, and discovered the *heterodyne effect*, which soon was developed into *superheterodyne circuit* as an integral part in design of radio receivers;

1907: <u>Paul Louis Toussaint Héroult</u>, French metallurgist, obtained patent for the *arc furnace for melting iron and steel*, followed by its installation in USA and Great Britain;

1908: <u>Leo Hendrik Baekeland</u>, US chemist, invented the first synthetic phenolic resin, known as *Bakelite*, replacing hard rubber and amber as an insulator;

1908-10: <u>William Henry Hoover</u>, US industrialist, invented the *electric cleaning machine*, and established the Electric Suction Sweeper Company to manufacture and market this machine;

1911-18: <u>Elmer Ambrose Sperry</u>, US inventor and electrical engineer, invented the *gyroscopic compass* and *stabilizers for ships and aeroplanes*, devised a new type of *dynamo*, an *electrolytic process for obtaining pure caustic soda* from salt, and a *high-intensity arc searchlight*;

1914-34: <u>André Gustave Citroën</u>, French engineer and motor manufacturer, promoted the mass production of *armaments* during the

First World War, founded the *Citroën Company* and manufactured low-priced small cars;

1915-20: <u>Viktor Kaplan</u>, Austrian inventor, experimented a propeller turbine working at very low heads of water, discovered that its efficiency greatly improves when the angle of blades could be varied to suit operating conditions, and obtained patent for *Kaplan turbine*, used in most of world's low-head hydropower schemes;

1925-28: <u>Carl Bosch</u>, German industrial chemist, invented a process, bearing his name, in which *hydrogen is produced* on industrial scale *by passing steam and water gas over a catalyst* at high temperatures;

1925-55: <u>Ernst Frederick Werner Alexanderson</u>, Swedish-born US electrical engineer and inventor, developed the *Alexanderson alternator* for transoceanic communication, *antenna structures*, and *radio-receiving* and *transmitting systems*, and also a complete *television system* and a *colour television receiver*;

1934: <u>Ion Stefan Basgan</u>, Romanian engineer, obtained a patent for *sonic drilling*, which was applied in the USA for a spectacular increase of depths reached by drilling, especially in oil field exploration and development;

1934-50: <u>Ferdinand Porsche</u>, German car designer, provided plans for a revolutionary cheap car with rear engine, to which Nazis gave the name *Volkswagen* 'People's car', and then designed the *Beetle*, a record-breaking German export;

1934-73: <u>Soichiro Honda</u>, Japanese motor-cycle and car manufacturer, started with a piston-ring production factory, then produced *motor-cycles*, and became president of the Honda Corporation;

1935: <u>Wallace Hume Carothers</u>, US industrial chemist and inventor, provided a successful production of synthetic rubber, namely *neoprene*, and afterwards got patent for *nylon*;

1937-38: <u>Chester Floyd Carlson</u>, US inventor, made experiments with copying process using photoconductivity, and discovered the basic principles of *electrostatic xerography*, a non-chemical photographic process in which light discharges a charged dielectric surface;

1944-49: <u>Mikhail Timofeyevich Kalashnikov</u>, Russian military designer, experimented and developed a series of rifles, and designed the famous *AK-47* assault rifle, called *Kalashnikov*, extensively used by foreign armies and international terrorists;

1947: <u>Dennis Gabor</u>, Hungarian-born British physicist, conceived the technique of *holography*, as a method of photographically recording and reproducing *three-dimensional images*;

1960-70: <u>Jeorme Goldman</u>, Friede & Goldman, Inc.'s naval architect and marine engineer, pioneered the development of offshore technology, led the design of a substantial portion of offshore-drilling units, such as one of the first *jack-up drilling units*, first *catamaran drilling ship, semi-submersibles*, and introduced high standards of strength, stability and seaworthiness in *offshore engineering*;

1972-93: <u>Bertram Neville Brockhouse</u>, Canadian physicist, made experiments for development of *neutron spectroscopy*;

1985-95: <u>Carl Langner</u>, Shell E & P Technology Co.'s research engineer, innovated offshore technology in areas of *pipeline and subsea technology*, especially *articulated stinger for deep-water S-lay*, and used *J-lay pipelaying methods* and *steel catenary risers*;

2013-14: *Major countries by industrial output*, in billion US$, are: European Union countries 4265; China 4050; United States 3211; Japan 1317; Germany 1006; Russia 762; Brazil 576; United Kingdom 523; Canada 520; France 515; Italy 501; South Korea 477; Saudi Arabia 466; India 459; Mexico 454; Indonesia 408; Australia 406; Spain 358; Turkey 222; United Arab Emirates 218; and Norway 216.

Industry timeline is delimited, according to the development of human characteristics comprised in the table of chapter 7, by its beginning 10905 years ago, as the initial time, and its end 4350 years forward from present, as the final time

$t_\bullet$ = 10905 + 4350 = 15255 years from the origin of industry.

The industrial sequences can be reconstituted taking into consideration their times of transition from one to another, which are more trustily established for the later ones:

La Tène technology and steelmaking \ AD200 \
Steel weaponry \ AD800 \
Gunpowder and firearms \ AD1345 \
Initial industrial revolution \ AD1845 \ *Modern industry.*

These time limits correspond to 1814; 1214; 669; 169 years ago; or 10905 - 1814 ≈ 9091; 10905 - 1214 ≈ 9691; 10905 - 669 ≈ 10236; 10905 - 169 ≈ 10736 years from the origin of industry. As the successive ratios of transitional times from the origin of industry are 9091/9691 ≈ 0.938; 9691/10236 ≈ 0.947; 10236/10736 ≈ 0.953, and they can be identified in Background to time, Table *(l)* for *z/16 = 1* sequences, in the same order, as $t_{0.6875}/t_{0.75} = 0.9389507$; $t_{0.75}/t_{0.8125} = 0.9466172$; $t_{0.8125}/t_{0.875} = 0.9532060$; it follows that they can be calculated, in years from the origin of industry, as:

$$t_{0.6875} = t_{\bullet} \cdot tanh(0.6875) = (15255) \cdot (0.5963736) \approx 9098;$$
$$t_{0.75} = t_{\bullet} \cdot tanh(0.75) = (15255) \cdot (0.6351490) \approx 9689;$$
$$t_{0.8125} = t_{\bullet} \cdot tanh(0.8125) = (15255) \cdot (0.6709671) \approx 10236;$$
$$t_{0.875} = t_{\bullet} \cdot tanh(0.875) = (15255) \cdot (0.7039056) \approx 10738.$$

At full scale of the industrial development, the timeline, sequences, and intrinsic characteristics are presented in the tables below.

$z/16 = l$	Time (years))		Industrial sequences
	from origin $t_l = t_{\bullet} \cdot tanh(l)$	from present t_l - 10905	
	15255	+4350	
...	...	...	
0.9375	11198	+293 (AD2307)	–
0.875	10738	-167 (AD1847)	_Modern industry
0.8125	10236	-669 (AD1345)	_Initial industrial revolution
0.75	9689	-1216 (AD798)	_Gunpowder and firearms
0.6875	9098	-1807 (AD207)	_ Steel weaponry
0.625	8460	-2445 (431BC)	_La Tène technology, steelmaking
0.5625	7777	-3128 (1114BC)	_ Hallstatt technology, ironmaking
0.5	7050	-3855 (1841BC)	_ Bronze technology
0.4375	6279	-4626 (2612BC)	_Widespread bronze working
0.375	5467	-5438 (3424BC)	_Bronze metallurgy, chariots
0.3125	4618	-6287 (4273BC)	_Bronze Age and sail
0.25	3736	-7169(5155BC)	_ Bronze casting and plough
0.1875	2827	-8078 (6064BC)	_ Early bronze use and irrigation
0.125	1897	-9008 (6994BC)	_ Copper metallurgy
0.0625	952	-9953 (7939BC)	_ Experiments with copper ores
0	0	-10905	_ Discovery of copper ores

$z/16 = l$	Years from origin $t_l = t_{\bullet} \cdot tanh(l)$ (years)	Period $\tau_l = (t_{\bullet}^2 - t_l^2)/(2t_{\bullet})$ (years)	Frequency $f_l = (2t_{\bullet})/(t_{\bullet}^2 - t_l^2)$ (years)$^{-1}$	Angular speed $\omega_l = 2\pi/\tau_l = 2\pi \cdot f_l$ (years)$^{-1}$
	15255	0	∞	∞
...	...	...	...	...
0.9375	11198	3517.5	0.0002843	0.0017862
0.875	10738	3848.3	0.0002599	0.0016327
0.8125	10236	4193.4	0.0002385	0.0014984
0.75	9689	4550.6	0.0002198	0.0013807
0.6875	9098	4914.5	0.0002035	0.0012785
0.625	8460	5281.7	0.0001893	0.0011896
0.5625	7777	5645.1	0.0001771	0.0011130
0.5	7050	5998.4	0.0001667	0.0010475
0.4375	6279	6335.3	0.0001578	0.0009918

0.375	5467	6647.9	0,0001504	0,0009451
0.3125	4618	6928.5	0.0001443	0.0009069
0.25	3736	7170.0	0.0001395	0.0008763
0.1875	2827	7365.6	0.0001358	0.0008530
0.125	1897	7509.6	0.0001332	0.0008367
0.0625	952	7597.8	0.0001316	0.0008270
0	0	7627.5	0.0001311	0.0008238

22. Civilization

In general, the state of human society with high level of culture, science, and government is called *civilization* (Latin *civilis* 'civic, civil' from *civis* 'citizen', *civitas* 'citizenship, city'). The meaning of civilization was primarily referring to the material and instrumental side of human culture in hierarchical and urbanized societies; classically associated with civilized people to set them apart from barbarians, savages and primitive people or tribal society; and recently applied only to society with a certain set of characteristics, especially the founding of towns and cities.

Anthropologists, sociologists and other researchers make a distinction between earlier emerging *culture* (as: what we are; totality of traditions; and internal thoughts, feelings, ideas, values, etc. - like the 'soul' of an individual); and later emerging *civilization* (as: what we have; totality of great and little traditions; and expression and manifestation of grandness - like the 'body' of an individual).

Social scientists have named a number of traits that distinguish a civilization from other kinds of society, describing civilizations as being distinguished by their *means of subsistence*, types of *livehood*, *settlement* patterns, *government* forms, *social stratification, economic systems, literacy* and other cultural traits.

Civilization has been spread by colonization, invasion, religious conversion, the extension of bureaucratic control and trade, and by introducing agriculture and writing to non-literate peoples; while some non-civilized people may willingly adapt to civilized behaviour. Later, civilization also spread by the technical, material and social dominance that civilization engenders.

Because humans are travelling, trading, campaigning, working or migrating outside their native lands, they transmit and exchange social, cultural and technological heritage, so that *Civilizations* are interconnecting and spreading from local civilizations C_l through areal civilizations C_a and regional civilizations C_r to global civilization C_g. These civilizations can be integrated taking into consideration their successive *assimilations* of local into areal a^l_a, of the areal into regional a^a_r, and of regional into global a^r_g civilization, whereby the chain of integrations

$$C_l \rightarrow C_a = \Sigma_l \, a^l_a \cdot C_l \rightarrow C_r = \Sigma_a \, a^a_r \cdot C_a \rightarrow C_g = \Sigma_r \, a^r_g \cdot C_r,$$

which indicate the direction in development of civilization during

393

human history.

Civilization properly occurred between about 8000 and 7000 years ago in South-western, Southern and South-eastern Asia, in North central Africa, as well as in Central America. The development of civilization was marked by societies and civilizations, agriculture and irrigations, pottery and textiles, metal works and constructions, city-states and their cultural background, public and balneal places for meeting and changing ideas, artistic styles and cultural movements, and other achievements, in various parts of world; and the meaning and signification of civilization also evolved due to a series of personalities; as briefly displayed below.

9130BC: _Göbekli Tepe civilization_ - The first known civilization appeared in *Anatolia* (today Turkey), the Near-Middle East;

7500BC: _First Chinese civilization_ - *Pengtoushan culture* on the *Yangtze Valley*, and then on *Yellow River Valley*;

7000-5100BC: _First American civilization_ - *Norte Chico* civilization in Peru, and then *Mesoamerican* and *Balsas River* civilizations;

7000-3150BC: _First Egyptian civilization_ - originated with a stone circle at *Nabta Playa* in Egypt's Western Desert, acting as a calendar by the beginning of Saharan desiccation, on the route to the Nile Valley, where the civilization coalesced by the *political unification of Upper and Lower Egypt* under the first pharaoh;

7000-1900BC: _Indus Valley civilization_ - represented by sites such as the village and then city *Mehrgarh* (now in Dhadar, Balochistan, Pakistan), the fortified city *Harappa* (now in Punjab, East Pakistan), the city *Mohenjo-daro* 'Mound of the Dead' (now in Sindh, Pakistan), and town *Lothal* (in Saragwala, Gujarat, India);

5600-4900BC: _Agricultural civilizations_ in the Middle East and Near East, North-east Africa, South Asia, the Far East, and Europe; emergence of *irrigational-agricultural civilization* in Mesopotamia; and *pottery-producing civilizations* in Anatolia, the Near East, and Europe;

5000-2110BC: _Sumerian civilization_ - formed by the Udaid people who possibly originated in an area near the Black Sea and apparently came through a pastoral area near the Persian Gulf to settle in Sumer, on *the lower Tigris-Euphrates Valley*; founded cities such as Ur, Uruk, and Lagash; traded crops from their fertile soil for metal, stone and wood; invented cuneiform writing; made many discoveries, and achieved a relatively high cultural level; until it fell to Akkadians;

4900-4100BC: *Copper-working civilizations* in Anatolia, the Near East; Balkans, and Carpathian Basin; *city civilizations* in the Middle and Near East; occurrence of *Aeneolithic (Chalcolithic) civilizations*; and *bronze-casting civilizations* in the Near East;

4800-1750BC: *First Chinese civilizations* - Incised marks on pottery in Sian, Shensi, marking the *Yang-Shao culture*; and then *Xia civilization* descended from a wide-spread Yellow River region, known as the *Longshan culture*, famous for its black-lacquered pottery;

4100-3400BC: *City-state civilizations*, such as *Ur* and *Kish*, in South Mesopotamia; early *Sumerian, Egyptian* and *Chinese civilizations*;

3400-2700BC: *Dinastic-based civilizations* - Start of *Mayan civilization* in Mesoamerica; *Cycladic civilization* in the Aegean area; beginning of *dynastic-based civilization* in Egypt; urban civilization such as *Eridu* in South Mesopotamia; writing-based civilizations in Mesopotamia and Egypt; and start of *Old Kingdom civilization* in Egypt;

2700-2000BC: *Aegean type civilizations* - *Pyramid age civilization* of Old Kingdom in Egypt; *Harrapa* and *Mohenjo-Daro civilizations* on Indus Valley; *Agade (Sargon) Dynasty civilization* spread between Mediterranean and Persian Gulf; beginning of *Aegean civilization (Minoan)*, centred in Crete, with palaces and main public courts; and *metalworking civilization* in Peru;

2000-1300BC: *Mycenaean type civilizations* in Greece; *Shang civilization* in China; beginning of *New Kingdom civilization* in Egypt; and *Olmec civilization* in Mexico;

1300-700BC: *Greek type and Hallstatt civilizations* - Beginning of *Greek type civilization*, in Greece and Eastern Mediterranean, the Near and Middle East, Greece, Southern Europe, Western and central Asia, and Egypt; ascendancy of *Assyrian Empire civilization* in Mesopotamia; rise of *Chavín civilization* in Peru; and *Hallstatt civilization* in central and Western Europe;

700-80BC: *Ancient classic civilizations* - *East Mediterranean classic civilizations* with temples and agoras; *Celtic-La Tène civilization* in Western central Europe; emergence of *Ancient Roman civilization* with villas and forums; *Iron-working civilizations* on Nile Valley and in Nigeria; *Gallianazo* and *Salinar civilizations* in Peru; *Nok civilization* in Western Africa; and beginning of *Han dynasties civilizations* in China;

450-420BC: *Balneal practice* not only for use of *natural therapeutic factors*, mainly thermal or mineral water, but also for discussion and *change of ideas*, such as described by Herodotus in his history of Lydia, Persia, Babylon and Egypt;

200-100BC: *Hellenistic city Glanon* attracted people by its *miraculous fountain* and monumental temple dedicated to Apollo the Healer;

80BC-AD490: *Imperial civilizations* - *Roman civilization* in West, central and South Europe, in North Africa and in the Near East; *Indian civilization*; *Jin Dynasty civilization* in China; *Japanese civilization*; as well as *Teotihuacán civilization* in Mexico; and *South American civilization* in Peru;

c.AD100: Criton of Heraclea, Greek chief physician and procurator of Roman Emperor Trajan, wrote his book *Getica, a work on the history of the Getae*, collecting evidences during and after Roman-Dacian wars (AD101-106);

AD100-300: *Flourish of Roman city Aquae Sulis*, Britannia, on the site of the Celtic settlement 'Waters of Sul' developing elaborate *thermae* (baths), and becoming a place for gathering people;

106-271: *Thermae and balneal facilities* for public use at *Ad Mediam*, *Germisara*, and *Aquae* 'Hydata' (now Baile Herculane, Geoagiu, and Calan respectively, in Romania), which were mentioned by Claudius Ptolemy in his *Geography*;

490-1030: *Middle Ages civilizations* - *Byzantine civilization* in South-eastern Europe, Anatolia and East Mediterranean area; *Medieval European civilizations*; *Chinese-Gunpowder civilization*; and *Islamic civilization* extending from Persia in East, through Arabian peninsula and Northern Africa, to North-western Africa and South-western Europe, with mosques, medinas and public squares;

1030-1530: *Pre- and Renaissance civilizations* - *Moroccan-Andalusian civilization* with cities such as *Marrakesh* in High Atlas; *Agadir-building Berber civilization* in rural areas of Anti Atlas; beginning of *European-Renaissance civilizations*;

1348-1700: *German spa Carlsbad* (now Karlovy Vary, named after Czech *Var* 'Boiling'), was an internationally recognized place for meeting and *changing opinions* in Europe;

1377: Al-Ala'a Al-Hadrami, Tunisian Muslim philosopher and historian, wrote *Muqaddimah*, a monumental history of the Arabs, which influenced later theories of the analysis, growth and decline of *Islamic civilization*, suggesting that repeated invasions from nomadic peoples limited development and led to social collapse;

1400-1640: *Renaissance*, an intellectual movement traditionally seen as the end of mediaeval and beginning of modern times in Europe, with Spanish-like cathedrals and plazas;

1530-1992: *Pre- and Modern civilizations* - *Renaissance, Enlightenment, Neo-Classical, Romantic*, and *Modern civilizations*;

1693-1723: <u>Dimitrie Cantemir</u>, Moldavian prince and Enlightenment historian and writer, produced an ample study entitled *Historia incrementorum atque decrementorum Aulae Othomanicae* 'The History of Rise and Decline of the Ottoman Empire', explaining the change in Ottoman civilization;

1750-1820: *Enlightenment*, an intellectual movement, animated by social progress, liberation for rational and scientific knowledge, and principles of human natural rights in Europe and North America;

1750-1850: *Neo-Classicism*, a cultural movement aiming revival of classical style, which superseded *Rocco style* in Europe and the USA;

1756: <u>Victor de Riqueti, marquis de Mirabeau</u>, French economist, wrote and published *L'Ami des hommes ou Traité de la population*, approaching the meaning of *civilization*;

1759-67: <u>Adam Ferguson</u>, Scottish philosopher and historian, wrote *An Essay on the History of Civil Society*, published later by David Hume's advice, in which he mentioned the term *civilization* and noticed that 'Not only the individual advances from infancy to manhood, but the species itself from rudeness to civilization';

1760-90: *Classicism*, a style emphasizing qualities traditionally considered characteristic of ancient Greek and Roman forms of culture in Europe;

1762: <u>Jean Jacques Rousseau</u>, French political philosopher and educationist, in his novel *Émile, ou de l'éducation* 'A New System of Education' pointed out that *civilization*, being more rational and socially driven, is not fully in accordance with human nature, and made a distinction between culture and civilization;

1763-69: Development in *pottery* and *ceramics*, as in England where <u>Josiah Wedgwood</u> patented a cream-coloured *'Queen's' ware* and produced the *unglazed blue Jasper ware*;

1771: <u>John Millar</u>, Scottish jurist, used the term *civilization* in his writing *Origin of the Distinction and Ranks*, a pioneer work on sociology;

1776: <u>Adam Smith</u>, Scottish moral philosopher and pioneer of political economy, published *An Inquiry into the Nature and Causes of the Wealth of Nations*, where the term *civilization* was frequently used;

1800-1900: *Romantic century for bathing* and climax of *balneal resorts* seasonally attracting most of political, literary, and artistical personalities of time;

1820-50: *Romanticism*, an artistic style emphasizing imagination, emotions and creativity, and asserting importance of individual feels about world, and of natural and supernatural appearances, in Europe and the USA;

1845: <u>Domingo Faustino Sarmiento</u>, 7th President of Argentina, writer and publicist, was author of *Facundo: Civilización y Barbarie* 'Facundo: Civilization and Barbarism', describing the life of Juan Facundo Quiroga, a gaucho who had terrorized the country, and showing the contrast between civilization and barbarism as seen in Argentina of his time;

1854-56: <u>Christian Matthias Theodor Mommsen</u>, German classical scholar and historian, in his three-volume *History of Rome*, suggested that Rome collapsed with the collapse of the Western Roman Empire in AD476, and he also tended towards a biological analogy of 'genesis', 'growth', 'senescence', 'collapse', and 'decay';

1860-86: *Impressionism*, a painterly movement originating in France, aiming at realistic representation of light play in nature, purporting to render faithfully what artists actually saw, as well as dispensing with academic rules of composition and colouring in Europe and North America;

1871-90: <u>Heinrich Schliemann</u>, German archaeologist, excavated at the mound of Hisarlic in Asia Minor, and discovered nine superimposed city sites at the Homeric place of *Troy*; he also excavated at sites in Peloponnese's *Mycenae*, and in Ithaca, at *Orchomenos* and at *Tiryns*, unveiling the essential characteristics of *Troyan and Mycenaean civilizations*;

1890-1993: *Expressionism*, a style of painting, sculpture, and literature that expresses inner emotions, in Europe;

1895-1919: <u>William Henry Holmes</u>, US archaeologist and museum director, greatly contributed to unveil *Mesoamerican civilization*, magnificently illustrated in his work *Archaeological Studies among the Ancient Cities of Mexico*; and classified prehistoric ceramics and stone technology in North America, notably described in his *Aboriginal Pottery of the Eastern United States*, and *Handbook of Aboriginal American Antiquities*, informing about North American civilizations;

1899-1904: Arthur John Evans, English archaeologist, excavated the Bronze Age *Knossos* city, discovering and exhibiting the great remains of *Minoan civilization*;

1900-50: *Modernism*, a concern with form and exploration of technique, opposed to content and narrative, as conscious attempt to break with cultural traditions of the 19th century, in Europe and the USA;

1901-40: Nicolae Iorga, Romanian historian and politician, published *Geschichte des Osmanischen Reiches* 'History of the Ottoman Empire', and *La Révolution Française et le sud-est de l'Europe* 'The French Revolution and the South-east of Europe', analyzing the periods of Ottoman and French civilizations respectively, with their influences on East-European civilizations;

1912-34: Charles Leonard Woolley, English archaeologist, made excavations at *Carchemish*, *Al'Ubaid* and *Tell el-Amarna*, and then at *Ur* in Mesopotamia, which uncovered gold and lapis lazuli in royal tombs, and published *Digging up the Past*; he offered a further extension of information about *Eastern Mediterranean, Mesopotamian* and *Egyptian civilizations*;

1918-23: Oswald Arnold Gottfried Spengler, German historian and philosopher of history, published his two-volume *Der Untergang des Abendlandes* 'The Decline of the West', then revised and subtitled *Perspectives of World History*, rejecting the *Euro-centric view* of history, especially the division of history into the linear *'ancient-medieval-modern' rubric*, and recognizing *eight high cultures*; namely Babylonian, Egyptian, Chinese, Indian, Mexican (Mayan-Aztec), Classical (Greek-Roman), Arabian, and Western (European-American); the final stage of each culture being, in his word use, a *civilization*;

1920-50: *Ceramics and pottery* were further developed in England, where Susie Cooper created *ceramic objects of art*, and Roy Midwinter designed *modern pottery*;

1923: Albert Schweitzer, Alsatian medical missionary, theologian and philosopher, worked out in relation to the defects of European civilization as presented in his *Verfall und Wiederaufbau der Kultur* 'The Decay and Restoration of Civilization' and philosophical *Kultur*

399

und Ethik, in which he outlined the idea that there are dual opinions within society, one regarding civilization as purely *material* and another regarding civilization as both *ethical* and material; he defined civilization saying that it 'is the sum total of all progress made by man in every sphere of action and from every point of view in so far as the progress helps towards the spiritual perfecting of individuals as the progress of all progress';

1930: <u>Sigmund Freud</u>, Austrian neurologist and founder of psychoanalysis, published his *Das Unbehagen in der Kultur* 'The Uneasiness in Culture', later translated as *Civilization and Its Discontents*, which became a well-known book;

1934-61: <u>Arnold Joseph Toynbee</u>, British historian and philosopher of history, explored civilization process in his twelve-volume *A Study of History*, which traced the rise and, in most cases, the decline of 21 civilizations and 5 'arrested civilizations';

1935-75: <u>William James Durant</u> and his wife <u>Ariel Durant</u>, US historians, produced *The Story of Civilization*, an eleven-volume set of books covering Western history, and comprising *Our Oriental Heritage*; *The Life of Greece*; *Caesar and Christ*; *The Age of Faith*; *The Renaissance*, *The Reformation*; *The Age of Reason Begins*; *The Age of Louis XIV*; *The Age of Voltaire*; *Rousseau and Revolution*; and *The Age of Napoleon*;

1960-90: *Post-Modernism*, a cultural movement rejecting preoccupation of Modernism, and concerned with pure form and technique rather than content, in Europe and the USA;

1961: <u>Michel Foucault</u>, French philosopher, historian of ideas and social theorist, published a book entitled *Histoire de la folie à l'âge classique - Folie et déraison* 'Madness and Civilization: A History of Insanity in the Age of Reason', discussing how West European society had dealt with *madness*, and arguing that it was a social construct *distinct from mental illness*;

1975: <u>Paul MacKendrick</u>, US classicist, author and teacher, studying the Dacian civilization, he collected data assembled in the published work entitled *The Dacian Stones Speak*, Chapel Hill: University of North Carolina Press;

1989-95: *Falling Communism* and transition to free-market economies in Eastern European countries, starting of *new civilizations*; and tendency to a *globalized-like civilization*;

1992: *International type civilizations* emerged around the world;

1993-96: <u>Samuel Phillips Huntington</u>, US political scientist, wrote extensively on civilization, one of his published books being *Clash of Civilizations*, and defined civilization as 'the highest cultural grouping of people and the broadest level of cultural identity people have short of that which distinguishes humans from other species';

1997: <u>Phillip Lee Ralph</u>, <u>Robert Lerner</u>, <u>Standish Meacham</u>, <u>Allen Wood</u>, <u>Richard Hull</u> and <u>Edward McNall Burns</u>, US historians, were authors of *World Civilizations*, Ninth edition, incorporating new material that enriches the text's global coverage and comparative approach; <u>Marius Albu</u>, Romanian geologist and physicist, with <u>David Banks</u> and <u>Harriet Nash</u>, English hydrogeologists, completed a book entitled *Mineral and Thermal Groundwater Resources*, Chapman & Hall, London, including the properties, distribution and *use in earlier and later civilizations*;

1999-2011: <u>Philip Adler</u> and <u>Randall Pauwels</u>, US professors of history, published six editions of their work *World Civilizations*, Cengage Learning;

2001-12: <u>Peter Stearns</u>, US historian of civilizations, wrote books including *The Encyclopedia of World History: Ancient, Medieval and Modern - Chronologically Arranged*, and also *World History in Brief: Major Patterns of Change and Continuity*, Penguin Academic Edition;

2002: <u>Richard Velkley</u>, US philosopher and historian, published his work 'The Tension in the Beautiful: On Culture and Civilization in Rousseau and German Philosophy', included in *Being after Rousseau: Philosophy and Culture in Question*, University of Chicago Press;

2003-14: <u>Sarah Parcak</u>, US archaeologist and Egyptologist, using satellite imaging to identify potential archaeological sites, and then a combination of satellite imaging analysis and surface surveys, she detected a great number of archaeological sites, some dating back to 3000BC, and discovered many other pyramids, tombs and ancient settlements outside *Sa el-Hagar* in Egypt, resulting in a spectacular reduction of time and cost for identifying archaeological sites compared to surface detection, and opening a new way for searching *ancient civilizations*;

2004: <u>Jonathan Haas</u>, <u>Winifred Creamer</u> and <u>Alvaro Ruiz</u>, US dating researchers, published their work *Dating the Late Archaic occupation of the Norte Chico region in Peru*, Nature, 432; and <u>Arthur Demarest</u>, US anthropologist and archaeologist, was author of *Ancient Maya: The Rise and Fall of a Rainforest Civilization*, summarizing the results of his investigation of Mesoamerican civilization;

2005: <u>Bryan Ward-Perkins</u>, Italian-born British archaeologist and historian, in his book *The Fall of Rome and the End of Civilization*, Oxford University Press, unveiled the real horrors associated with the collapse of a civilization for the people who suffer its effects;

2005-11: <u>Charles Mann</u>, US journalist and science writer, produced two important books about the sophisticated culture of indigenous peoples of America, namely *1491: New Revelations of the Americas before Columbus*; and *1493: Uncovering the New World Columbus Created*; both edited by Knopf;

2006: <u>Derrick Jensen</u>, US author and environmental activist, published his *Endgame*: volume 1, *The Problem of Civilization*; and volume 2, *Resistance*; showing the inherent unsustainability of civilization, and arguing that modern civilization is intrinsically directed towards domination of the environment and humanity itself in a harmful and destructive fashion; and <u>Thomas Homer-Dixon</u>, Canadian political scientist for environment and business, in his work entitled *The Upside of Down: Catastrophe, Creativity, and the Renewal of Civilization*, Toronto: Knopf, considered the fall in the *energy return on investments*, the *energy expended to energy yield ratio* is central for limiting the survival of civilizations.

According to the table of chapter 7, civilization originated 7630 years ago = 5616BC, as its initial time, and will last until 4350 years from present into the future so that its final time will be

$t_{\bullet} = 7630 + 4350 = 11980$ years from the origin of civilization.

In order to establish the development of *civilization*, its sequences can be approximately delimited by the following transitional times:

Agricultural \ 4900BC \ *Copper-working* \ 4100BC \
City-state \ 3400BC \ *Dynastic-based* \ 2700BC \
Aegean type \ 2000BC \ *Mycenaean type* \ 1300BC \
Greek type and Hallstatt \ 700BC \ *Ancient classic* \ 80BC \
Imperial \ AD490 \ *Middle Ages* \ AD1030 \
Pre- and Renaissance \ AD1530 \ *Modern* \ AD1992 \
International type.

From these transitional times, the later and then more accurately established ones are referred, in years ago, as AD490 = 1524, AD1030 = 984, AD1530 = 484, AD1992 = 22; and in years from the origin of civilization as 7630 - 1524 ≈ 6106; 7630 - 984 ≈ 6646; 7630 - 484 ≈ 7146; 7630 - 22 ≈ 7608. The successive ratios of these times from the origin of civilization to the final time $t_\bullet$ = 11980 years from the same origin show the approximate values 6106/11980 ≈ 0.510; 6646/11980 ≈ 0.555; 7146/11980 ≈ 0.596; 7608/11980 ≈ 0.635; which can be recognized in Background to time, Table *(l)* for $z/16 = l$ sequences, in the same succession, as $t_{0.5625}/t_\bullet$ = $tanh(0.5625)$ = 0.5098300; $t_{0.625}/t_\bullet$ = $tanh(0.625)$ = 0.5545997; $t_{0.6875}/t_\bullet$ = $tanh(0.6875)$ = 0.5963736; $t_{0.75}/t_\bullet$ = $tanh(0.75)$ = 0.6351490, so that the accurate values of the selected times are calculated, in years from the origin of civilization, as follows:

$$t_{0.5625} = t_\bullet \cdot tanh(0.5625) = (11980) \cdot (0.5098300) \approx 6108;$$
$$t_{0.625} = t_\bullet \cdot tanh(0.625) = (11980) \cdot (0.5545997) \approx 6644;$$
$$t_{0.6875} = t_\bullet \cdot tanh(0.6875) = (11980) \cdot (0.5963736) \approx 7145;$$
$$t_{0.75} = t_\bullet \cdot tanh(0.75) = (11980) \cdot (0.6351490) \approx 7609.$$

Continuing the same procedure backward, the civilization timeline, sequences, and intrinsic characteristics are modelled as:

$z/16 = l$	Time (years)		Sequences of civilizations
	from origin $t_l = t_\bullet \cdot tanh(l)$	from present $t_l - 7630$	
	11980	+4350	
...	...	...	
0.8125	8038	+408 (AD2422)	–
0.75	7609	-21 (AD1993)	_International type
0.6875	7145	-485 (AD1529)	_Pre- and Modern
0.625	6644	-986 (AD1028)	_Pre- and Renaissance
0.5625	6108	-1522 (AD492)	_Middle Ages type
0.5	5536	-2094 (80BC)	_Imperial type
0.4375	4931	-2699 (685BC)	_Ancient classic
0.375	4293	-3337 (1323BC)	_Hellenistic type and Hallstatt
0.3125	3626	-4004 (1990BC)	_Mycenaean type
0.25	2934	-4696 (2682BC)	_Aegean type
0.1875	2220	-5410 (3396BC)	_Dynastic-based
0.125	1490	-6140 (4126BC)	_City-state
0.0625	748	-6882 (4868BC)	_Copper-working
0	0	-7630 (5616BC)	_Agricultural

$z/16 = l$	Time from origin $t_l = t_\bullet \cdot tanh(l)$ (years)	Period $\tau_l = (t_\bullet^2 - t_l^2)/(2t_\bullet)$ (years)	Frequency $f_l = (2t_\bullet)/(t_\bullet^2 - t_l^2)$ (years)$^{-1}$	Angular speed $\omega_l = 2\pi/\tau_l = 2\pi \cdot f_l$ (years)$^{-1}$
	11980	0	∞	∞
...	...	...	...	...
0.8125	8038	3293.4	0.0003036	0.0019078
0.75	7609	3573.6	0.0002798	0.0017582
0.6875	7145	3859.3	0.0002591	0.0016281
0.625	6644	4147.6	0.0002411	0.0015149
0.5625	6108	4432.9	0.0002256	0.0014174

0.5	5536	4710.9	0.0002123	0.0013338
0.4375	4931	4975.2	0.0002010	0.0012629
0.375	4293	5220.8	0.0001915	0.0012035
0.3125	3626	5441.3	0.0001838	0.0011547
0.25	2934	5630.7	0.0001776	0.0011159
0.1875	2220	5784.3	0.0001729	0.0010862
0.125	1490	5897.3	0.0001696	0.0010654
0.0625	748	5966.6	0.0001676	0.0010531
0	0	5990.0	0.0001669	0.0010489

23. Legislation

Law promulgated or enacted by a legislature or other governing body, or the process of making it is called *legislation* (Latin *legislatio* 'proposing of a law' from *lex* 'law', *legis* 'of law' and *latus* 'broad, wide, extensive, bringing'), which usually includes law-making by referendums, constitutional conventions, council orders or regulations. Legislation or *statutory law* can have many purposes, such as to regulate, authorize, proscribe, provide (funds), sanction; or to grant, declare, restrict. The concept of moral rightness, referred as *justice*, is based on ethics, rationality, law, natural law, religion, fairness, and equity, as well as punishment of ethical breaches.

At the base of justice, there are natural differences of moral features of character, which can be positive as good or right, and negative as bad or wrong. Any individual bears a set of positive characteristics $\{c^+_i\}$ and a set of negative characteristics $\{c^-_i\}$ which may be combined in a union $c^n_i = \{c^+_i\} \cup \{c^-_i\}$. Converted by factors a_+ and b_- in the manner $a_+ \cdot c^+_i$ and $b_- \cdot c^-_i$, these positive and negative characteristics participate together to form an individual moral character $C_i = a_+ \cdot c^+_i - b_- \cdot c^-_i$, as a component of social morality, and ultimately is subjected to the laws of justice in a given country, where its justice can be:

Utilitarianism, as a form of consequentialism in three ways – deterrence, rehabilitation, and security or incapacitation;

Retributive justice, with a military variant of the law of retaliation (lex talionis);

Restorative justice, concerned not only with retribution and punishment, but also with making the victim whole and reintegrating the offender into society;

Distributive justice, directed at the proper allocation of wealth, power, reward, and respect among different people;

Oppressive law, as an authoritarian approach to legislation; or

Egalitarianism, setting that justice can only exist within the co-ordinates of equality.

Theory and philosophy of law are applied as *jurisprudence*, comprising natural law, legal positivism, legal realism, and critical legal studies. The practical authority granted to a formally constituted legal body or to a political leader expresses *jurisdiction*, directed to deal with and make pronouncements on legal matters and to administer justice, which draws its substance from public international law, conflict of laws, constitutional law, and powers of the executive and legislative branches of government to allocate resources to best serve the needs of their own society.

At the dawn of the first states, after a period of necessary preparation until around 5000 years ago, the first social rules were established and implemented in Mesopotamia and Egypt; subsequently laws were diversified and enforced until the present time covering almost the entire continental and oceanic Earth's surface. The present level of justice and legislation was reached by consistent, complex, and evolutionary coverage with tables, settings, rules, laws, charters, codes, codices, constitutions, etc., along with works and achievements of thinkers, philosophers, statesmen, lawgivers, legislators and jurists, as summarily presented below.

3000-2700BC: *Orally transmitted social rules and norms* of first civilizations in the Middle East and North-western Africa;

2700-2400BC: *Social regulations in areas of civilization*;

2400-2100BC: First *Mesopotamian and Egyptian laws*;

2350BC: Urukagina, King of Lagash, is mainly known by his *Code of Laws*;

2047-2030BC: Ur-Nammu, ruler of Sumer, in Southern Mesopotamia, produced the code of laws, called *Code of Ur-Nammu*, consisting of casuistic statements, such as 'If a man commits a murder, that man must be killed';

1772-1760BC: Hammurabi, the 6th King of Babylon, left after him the famous *Code of Laws*, with partial copies on a human-sized stone stele and various clay tablets, consisting of 282 laws, comprising scaled punishments, and adjusting 'an eye for an eye, a tooth for a tooth', called *lex talionis*, as graded depending on social status, of slave versus free man, as inscribed on one of his tablets (now in Louvre, Paris);

1720BC: Niqmepuh, King of Iamhad (area of present-day Aleppo in Syria), ordered the inscription on a cuneiform tablet, still in its clay case, of the so-called *Legal case from Niqmepuh* (now in British Museum, London);

1260-1230BC: Moses (Môsheh), Old Testament Hebrew prophet and lawgiver, received the *Divine Law* and the *Ten Commandments* from God on his return from Egypt to North of Moab;

1050-1020BC: <u>David</u>, first king of Judean dynasty of Israel, is considered the traditional author of *several Psalms*, and builder of the *Ark of the Convenant* on *Zion* 'city of David';

932-922BC: <u>Solomon</u>, King of Israel, became well-known by his books including *Song of Solomon*, *Wisdom of Solomon*, *Proverbs*, and also *Psalms*;

715-673BC: <u>Pompilius Numa</u>, second ruler of Rome, according to *disciplina etrusca* 'the Etruscan discipline', he established the *religious code* in early Roman community;

670-650BC: <u>Lycurgus</u>, legendary lawgiver of Sparta, established the *military-oriented reformation of Spartan society*, respecting three virtues: equality among citizens, military fitness, and austerity;

621-594BC: <u>Draco</u>, Athenian legislator, promulgated the *harsh codification of law in Athens*, associated with the word 'draconian', which was largely abolished by Solon;

594-570BC: <u>Solon</u>, Athenian lawgiver, seeking for a *compromise between democracy and oligarchy*, he repealed the Draco's more stringent laws, except that of murder;

390-370BC: <u>Plato</u>, Greek philosopher, wrote the dialogue *Republic*, which opened with the question 'What is justice?', and developed precepts that the state is formed on rigid class structure;

340-323BC: <u>Aristotle</u>, Greek philosopher, founded *Natural law*, based on belief that human happiness is achievable by living in conformity with nature, sayings that 'It is clear that those constitutions which aim at the common good are right, as being in accord with absolute justice; while those which aim only at the good of the rulers are wrong', and 'Rule of one outstandingly good man, backed by just laws, is most desirable';

325-280BC: <u>Chanakya</u>, Indian political legislator, became known as the traditional founder of ancient Indian political legislation, presented in his treatise *Arthasastra*, as an authoritative legal guidance;

320-290BC: <u>Theophrastus</u>, Greek philosopher, leading the Peripatetic School (Lyceum) and being responsible for preserving many of Aristotle's works, he wrote *On the Laws*, containing a recapitulation of laws of various barbarian and Greek states, and intending to be a companion to Aristotle's outline of politics;

304BC: <u>Gnaeus Flavius</u>, Roman state functionary, presented to the Roman people his *Ius Flavianum*, an account of legal procedures in ancient Rome;

240-233BC: <u>Han Fei</u>, Chinese philosopher, developed the *School of Law or Legalism*, a doctrine concerning on control of state by three concepts - position of power, certain techniques, and laws;

AD140-170: <u>Gaius</u>, Roman jurist, completed the *Institutes* as the basis for those of Justinian I the Great, which remained the only substantial texts of *classical Roman law*;

180-215: <u>Yehudah HaNasi</u> (Judah the Prince), Jewish rabbi, redacted the *Mishnah*, the first major written work of Jewish oral tradition, called *Oral Law*;

529-540: <u>Justinian I the Great</u>, Emperor of Eastern Roman Empire, supervised the work *Corpus Juris Civilis*, including *Digesta* or *Pandectae*, based on the main concepts of Roman law, and *Novellae*, based on Gaius' Institutiones, which was immensely influential on the laws of nearly all European countries up to modern times;

662-1100: *Muslim religious law* was initiated by the Arab prophet <u>Muhammad</u> who received revelations from the angel Jibra'el (Gabriel) of the word of Allah, the one and only God, from which derived the *Qur'an* (Koran) 'Reading' as basis of Islam;

795-810: <u>Charlemagne</u> (<u>Charles the Great</u>), King of Franks and Christian Emperor of the West, created *stable administrations and good laws* to consolidate his vast empire extending from Ebro in North Spain to Elbe in East;

1214-15: <u>Stephen Langton</u>, English prelate, prepared the draft of *Magna Carta*, an English charter including most direct challenges to the monarch's authority;

1215: <u>King John</u> of England, under pressure from the rebel barons, he signed the *Magna Carta*, the basis of English political and personal liberty;

1340-57: <u>Bartolo da Sassoferrato</u>, Italian judge and jurist, was author of *Commentarius in Tria Digesta*; *Commentarius in libros IX Codicis priores*; *Commentarius super libris III posterioribus Codicis*; and *Lectura super Authenticis*;

1356: <u>Charles IV</u>, Holy Roman Emperor, supervised the constitutional framework for the Holy Roman Empire, called the *Golden Bull*, laying down the procedure for election of the monarch, excluding papal pretensions, and defining rights of the seven electors;

1530-60: <u>Süleyman the Magnificent</u>, Ottoman emperor, one of the greatest Ottoman sultans, promulgated *Kanuni* 'Lawgiver', by which he instituted a programme of internal reforms aimed to secure higher *standards of justice* and *administration*, and ensure *freedom of religion* throughout his empire;

1653-55: Oliver Cromwell, English soldier and statesman, based on his study of law in London, ruled as leader of the Puritan Convention, and then, supervised the implementation of a *new Constitution*, establishing Puritanism; his legacy re-directing the course of English and British parliament and law;

1690: John Locke, English empiricist philosopher, published *Two Treatises of Government*, outlining a theory of political or civil society based on natural rights and contract theory;

1710-37: Cornelis van Bynkershoek, Dutch judge and jurist, published *Observationum juris Romani* on Roman law, and *Quaestionum juris privati* on Roman-Dutch law; elaborated *De Domino Maris* on maritime law, and *De Foro Legatorum* on diplomatic rights; as well as *Quaestiones juris publici* on war and neutrality;

1726-34: John Ayliffe, English scholar in Roman law, wrote *Parergon Juris Canonici Anglicani*, and *A New Pandect of the Roman Civil Law*;

1730-50: Henri François d'Aguesseau, French jurist and Chancellor, reformed and codified *French law*, defending rights of people and Gallican Church, the author being pronounced by FMA Voltaire as the most learned magistrate that his country had ever possessed;

1734-48: Charles-Louis de Secondat Montesquieu, French philosopher and jurist, wrote *Causes de la grandeur des Romains et de leur decadence* 'Causes of the Greatness and Decadence of the Romans', and *Défense de l'esprit des lois* 'The Spirit of the Laws', widely influencing legislative activity in Europe and the USA;

1765-69: William Blackstone, English judge, jurist and Tory politician, published the celebrated *Commentaries on the Laws of England*, an influential treatise on common law, setting out the structure of English law and explaining its major principles;

1770-1800: John Adams, 2nd President of the USA, played a major role in the debate that resulted in *Declaration of Independence*, and published the three-volume *Defence of the Constitution of the United States*;

1784-87: James Madison, 4th President of USA, was an active delegate at the Constitutional Convention, winning the title 'Master builder of the constitution' because of his influence in the campaign for Ratification of *the USA Federal Constitution*;

1796-1825: Paul Johann Anselm Feuerbach, German jurist, published *Kritik des natürlichen Rechts*; *Anti-Hobbes*; and *Lehrbuch des gemeinen peinlichen Rechts*, and most importantly *Geschworenengericht*, maintaining that the verdict of a jury is not an adequate legal proof of crime;

1801-30: John Marshal, US jurist, established the power and independence of the Supreme Court, and the *fundamental principles of constitutional law*;

1803-07: Jean Jacques-Regis de Cambacérès, French jurisconsult, elaborated the *Projet de Code Civil*, which formed the base of Napoleon's Code Civil for France, and is still lasting as a foundation of French civil law;

1804-07: Napoleon I Bonaparte, as 1st Consul and then Emperor of France, initiated and supervised the *Code Civil* for France, which had a great influence in many European countries;

1811-36: Joseph Story, US jurist, was member of the Supreme Court of the USA and wrote *Commentaries on the Constitution of the US*; *The Conflicts of Laws*; and *Equity Jurisprudence*;

1831-35: Leopold I, King of Belgium, conducted with prudence and moderation and with constant regard to the legal principles contained in the *Belgian constitution*, serving as an example for other countries;

1859-69: John Stuart Mill, English philosopher and social reformer, wrote and published his works *On Liberty*, as the most popular, and *Subjection of Women*, provoking a great antagonism;

1870-71: Christopher Columbus Langdell, US legal scholar, exercised a powerful influence on legal education throughout the USA, initiating *case method* of teaching in which real cases are cited in text as examples, and compiled *Casebook on Contracts* which established a national trend;

1881-1902: Oliver Wendell Holmes, US jurist, originated *The Common Law*, being nicknamed 'the Great Dissenter' because of his dissension from opinions of his conservative colleagues;

1882-95: Frederich Pollock, English jurist, edited *Law Reports*, and *Law Quarterly Review*; and published *Principles of Contract*; *Digest of the Law of Partnership*; *Law of Torts*; and *History of English Law before Edward I*;

410

1905-10: <u>James Barr Ames</u>, US jurist, adopting the case method of instruction pioneered by CC Langdell, he founded the *Harvard Law Review*, and published *Lectures on Legal History*;

1916-61: <u>Quincy Wright</u>, US international lawyer, wrote *The Enforcement of International Law through Municipal Law in the US*; *The Causes of War and the Conditions of Peace*; *A Study of War*; *Problems of Stability and Progress in International Relations*; and *The Role of International Law in the Prevention of War*;

1920-40: <u>Hans Kelsen</u>, Austrian-US jurist and legal theorist, created the 'pure theory of law' that was exposed in his *Reine Rechtslehre*, and wrote *General Theory of Law and the State*, and *Pure Theory of Law*;

1933-50: <u>Hersch Lauterpacht</u>, British lawyer, worked on international laws, and published *The Function of Law in the International Community*; *Recognition in International Law*; and *International Law and Human Rights*;

1953-58: <u>Jean Monet</u>, French, <u>Robert Schuman</u>, Luxembourger-French, <u>Paul-Henri Spaak</u>, Belgian, and <u>Alcide de Gasperi</u>, Italian, juristic initiators, sustained and activated formation of the *European Community*, which later developed as the *European Union*;

1959-82: <u>Herbert Lionel Adolphus Hart</u>, English jurist, published *Causation in the Law*; *Law, Liberty and Morality*; and *The Concept of Law*, emphasizing that the fundamental rule of a legal system should be the identification of its other rules;

1962-93: <u>John Rawls</u>, US philosopher, wrote *A Theory of Justice*, described theoretical principles of theory of justice, their implications in detail for social institutions, and supports in moral psychology; as well as published *Justice as Fairness*, and *Political Liberalism*;

1976: <u>Leo Gross</u>, Austrian-US lawyer, edited two-volume *The Future of the International Court of Justice*, Dobbs Ferry, New York: Oceana Publications Inc.;

1978: <u>Min-Chuan Ku</u>, Chinese-born US bookseller, compiled and edited the two-volume *A Comprehensive Handbook of the United Nations*, New York: Monarch;

1986-96: Regarding the *International Court of Justice* (*ICJ*) - <u>Gerald Gray Fitzmaurice</u>, British barrister, produced *The Law and Procedure of the ICJ*, Cambridge: Grotius; <u>Lori Fisler Damrosch</u>, US jurist, wrote *The ICJ at a Crossroads*, Dobbs Ferry, New York: Transnational; <u>Thomas Bodie</u>, US Lab Supervisor at Kaser

Permanente Georgia, made confessions in his *Politics and the Emergence of an Activist ICJ*, Westport, Conn.: Praeger; and <u>Arthur Eyffinger</u>, Dutch classicist and legal historian in the Hague, published his extensive work *The ICJ, 1946-1996*, published by Boston: Kluwer Law International;

1989: <u>Terry Gill</u>, Dutch professor of Military Law and Public International Law, studied some cases of litigations at the level of *ICJ*, one of them being presented in his published work *Litigations Strategy at the International Court: A Case Study of the Nicaragua v. United States Dispute*, Boston: Nijhoff;

1998: <u>Peter Bekker</u>, Dutch-born US professor of law, compiled and edited *Commentaries on World Court Decisions (1987-1996)*, The Hague, Boston: Nijhoff;

2013: <u>Karen Alter</u>, US professor of political science and law, published his book *The New Terrain of International Law: Courts, Politics, Rights*, Princeton University Press; where he argued that international courts alter politics by providing legal, symbolic, and leverage resources that shift the political balance in favour of domestic and international actors who prefer policies more consistent with international law objectives.

Legislation timeline was delimited in the table of chapter 7, with its origin 5041 years ago, as the initial time, and its end 4350 years from present into the future, as the final time $t_\bullet = 5041 + 4350 = 9391$ years from the origin of legislation.

Data above show that in the course of legislation there were successive sequences, the later ones being better delimited by the approximate times as:

English Magna Carta \ AD1230...1400 ≈ 1315 \
Holy Roman Empire's Golden Bull \ AD1500...1580 ≈ 1540 \
Ottoman, English, Dutch and *French laws* \ AD1730...1790 ≈ 1760 \
US law, French Code Civil, Interstate laws \ AD1950...2000 ≈ 1975 \
European Union and *International laws*.

These approximate transitional times represent 699; 474; 254; 39 years ago; or $5041 - 699 \approx 4342$; $5041 - 474 \approx 4567$; $5041 - 254 \approx 4787$; $5041 - 39 \approx 5002$ years from the origin of legislation; and the last ones divided by the final time $t_\bullet = 9391$ years from the same origin result in the ratios $4342/9391 \approx 0.462$; $4567/9391 \approx 0.486$; $4787/9391 \approx 0.510$; $5002/9391 \approx 0.533$. Such successive ratios are given in Background to time, Table *(m)* for $z/32 = m$ sequences, as follows: $t_{0.5}/t_\bullet = tanh(0.5) = 0.4621172$; $t_{0.53125}/t_\bullet = tanh(0.53125) = 0.4863360$; $t_{0.5625}/t_\bullet = tanh(0.5625) = 0.5098300$; $t_{0.59375}/t_\bullet = than(0.59375) = 0.5325873$. According to these ratios, the above transitional times, in years from the origin of legislation, are:

$$t_{0.5} = t_\bullet \cdot tanh(0.5) \approx 4340; \quad t_{0.53125} = t_\bullet \cdot tanh(0.53125) \approx 4567;$$
$$t_{0.5625} = t_\bullet \cdot tanh(0.5625) \approx 4788; \quad t_{0.59375} = t_\bullet \cdot than(0.59375) \approx 5002.$$

Thus calibrated, the transitional times delimiting the earlier sequences can be calculated, in years from the origin of laws and legislation, as

$$t_{0.03125} = t_\bullet \cdot tanh(0.03125) \approx 293; \quad t_{0.0625} = t_\bullet \cdot tanh(0.0625) \approx 586;$$
$$t_{0.09375} = t_\bullet \cdot tanh(0.09375) \approx 878; \quad t_{0.125} = t_\bullet \cdot tanh(0.125) \approx 1168;$$
$$t_{0.15625} = t_\bullet \cdot tanh(0.15625) \approx 1456; \quad t_{0.1875} = t_\bullet \cdot tanh(0.1875) \approx 1740;$$
$$t_{0.21875} = t_\bullet \cdot tanh(0.21875) \approx 2022; \quad t_{0.25} = t_\bullet \cdot tanh(0.25) \approx 2300;$$
$$t_{0.28125} = t_\bullet \cdot tanh(0.28125) \approx 2574; \quad t_{0.3125} = t_\bullet \cdot tanh(0.3125) \approx 2843;$$
$$t_{0.34375} = t_\bullet \cdot tanh(0.34375) \approx 3107; \quad t_{0.375} = t_\bullet \cdot tanh(0.375) \approx 3365;$$
$$t_{0.40625} = t_\bullet \cdot tanh(0.40625) \approx 3618; \quad t_{0.4375} = t_\bullet \cdot tanh(0.4375) \approx 3865;$$
$$t_{0.46875} = t_\bullet \cdot tanh(0.46875) \approx 4106.$$

Finally the legislation timeline, sequences, and intrinsic characteristics are reconstituted in the next tables.

$z/32 = $ m	Time(years)		Sequences of legislation
	from origin $t_m = $ $t_\bullet \cdot tanh(m)$	from present $t_m - 5041$	
	9391	+4350	
...	...	...	
0.625	5208	+167 (AD2181)	_ European Union, International laws
0.59375	5002	-39 (AD1975)	_ US law, French Code, Interstate
0.5625	4788	-253 (AD1761)	_ Ottoman, English, Dutch, French
0.53125	4567	-474 (AD1540)	_ Holy Roman Empire's Golden Bull
0.5	4340	-701 (AD1313)	_ English Magna Carta
0.46875	4106	-935 (AD1079)	_ Muslim religious law
0.4375	3865	-1176 (AD838)	_ Charlemagne's laws
0.40625	3618	-1423 (AD591)	_ Justinian's Corpus Juris Civilis
0.375	3365	-1676 (AD338)	_ Jewish Oral Law
0.34375	3107	-1934 (AD80)	_ Classical Roman Law
0.3125	2843	-2198 (184BC)	_ Natural, Arthasastra, Chinese law
0.28125	2574	-2467 (453BC)	_ Greco-Roman laws
0.25	2300	-2741 (727BC)	_ Psalms of David and Solomon
0.21875	2022	-3019 (1005BC)	_ Moses' Divine Law
0.1875	1740	-3301 (1287BC)	_ Jewish religious regulations
0.15625	1456	-3585 (1571BC)	_ Hammurabi's Code of Laws
0.125	1168	-3873 (1859BC)	_ Sumerian Code of Ur-Nammu
0.09375	878	-4163 (2149BC)	_ Mesopotamian and Egyptian laws
0.0625	586	-4455 (2441BC)	_ Regulations in areas of civilization
0.03125	293	-4748 (2734BC)	_ Orally rules and norms
0	0	-5041 (3027BC)	_

$z/32 =$ m	Time from origin $t_m = t_\bullet \cdot tanh(m)$ (years)	Period $\tau_m =$ $(t_\bullet^2 - t_m^2)/(2t_\bullet)$ (years)	Frequency $f_m =$ $(2t_\bullet)/(t_\bullet^2 - t_m^2)$ (years)$^{-1}$	Angular speed $\omega_m =$ $2\pi/\tau_m = 2\pi \cdot f_m$ (years)$^{-1}$
	9391	0	∞	∞
...	...	...	...	...
0.625	5208	3251.4	0.0003076	0.0019325
0.59375	5002	3363.4	0.0002973	0.0018681
0.5625	4788	3474.9	0.0002878	0.0018082
0.53125	4567	3585.0	0.0002789	0.0017526
0.5	4340	3692.6	0.0002708	0.0017015
0.46875	4106	3797.9	0.0002633	0.0016544
0.4375	3865	3900.2	0.0002564	0.0016110
0.40625	3618	3998.6	0.0002501	0.0015714
0.375	3365	4092.6	0.0002443	0.0015352
0.34375	3107	4181.5	0.0002391	0.0015026
0.3125	2843	4265.2	0.0002345	0.0014731
0.28125	2574	4342.7	0.0002303	0.0014468
0.25	2300	4413.8	0.0002266	0.0014235
0.21875	2022	4477.8	0.0002233	0.0014032
0.1875	1740	4534.3	0.0002205	0.0013857
0.15625	1456	4582.6	0.0002182	0.0013711
0.125	1168	4622.9	0.0002163	0.0013592
0.09375	878	4654.5	0.0002148	0.0013499
0.0625	586	4677.2	0.0002138	0.0013434
0.03125	293	4690.9	0.0002132	0.0013394
0	0	4695.5	0.0002130	0.0013381

24. Philosophy

The study of general and fundamental problems, such as reality, existence, knowledge, values, reason, mind and language, is called *philosophy* (from Greek *φιλοσοφία* 'love of wisdom'). From early antiquity to the present-day, the main philosophical theories were *realism, nominalism, rationalism, empiricism, scepticism, idealism, pragmatism, phenomenology, existentialism, structuralism, post-structuralism*, and *analytic philosophical tradition*. Meanwhile, moral and political philosophy dealt with *human nature, political legitimacy, consequentialism, deontology*, and *Aretinian turn* (named after Guide Aretino, 1030-50).

The meaning of philosophy was extended to the most basic beliefs, concepts, and attitudes of an individual or group of individuals, including: *epistemology* (theory of knowledge); *logic* (science and art of correct reasoning); *metaphysics* (investigating the first principles of nature and thought) with central branches such as *ontology* (science of nature and essence of things and beings) and *cosmology* (science of the universe as a whole); *ethics* (moral philosophy, studying human character and conduct); *political philosophy* (study of topics such as politics, liberty, property, rights, etc.); and *aesthetics* (principles of taste and art). There are also in use terms such as *philosophy of reason, philosophy of mind, philosophy of language, philosophy of law* (jurisprudence), *philosophy of religion*, and so on.

By the mid-late 18th century, a part of metaphysics called *natural philosophy* (science of the physical properties of bodies) was differentiated as *science* (knowledge ascertained by observation and experiment, critically tested, systematized and brought under general principles), thus distinguishing from philosophy which afterwards resumed to a non-empirical approach of the nature of existence. After the Second World War, analytic philosophy was developed in two main directions: the use of language to avoid or re-describe traditional philosophical problems; and the resort of naturalism to cope with natural sciences. During the 20th century, philosophy became a professional discipline like other academic ones, and consequently less general and more specialized, being dominated by analytical philosophy and moving from a rejection of idealism to a new kind of empiricism connected with modern mathematical logic. Recent trends in experimental philosophy seem to be the re-ascension of its traditional problems by techniques practised in social science.

Therefore, ancestral philosophy ph_0 diversified in the course of history, generating a variety of descendent branches and sub-branches ph_1, ph_2,...ph_i,...ph_n which participate with corresponding coefficients c^1, c^2,...c^i,...c^n in making up present philosophy

$$ph = \Sigma^n_{i=1} \, c^i \cdot ph_i > ph_0.$$

As a subject of study, philosophy was established and developed by traditions, personalities and schools, such as those presented below with their thoughts and achievements.

1000-975BC: Dawn of *Ancient Egyptian and Babylonian philosophy*;

800-700BC: Beginning of *Early ancient Greek philosophy*;

580-560BC: <u>Thales of Miletus</u>, Ionian natural philosopher, proposed the first natural cosmology, identifying *water as original substance* and basis of the universe, and founded *Greek*, and therefore European, *philosophy*;

570-547BC: <u>Anaximander</u>, Ionian thinker, stated that first principle is *apeiron* 'boundless' conceiving of in both physical and theological terms;

550-500BC: Emergence of *Classic ancient philosophy*;

540-490BC: <u>Xenophanes</u>, Greek philosopher, poet and religious thinker, believed in a *single deity who energizes the world* 'without toil he shakes all things by thought of his mind', and speculated about *successive inundations* of Earth based on *observation of fossils*;

530-500BC: <u>Pythagoras</u>, Greek philosopher, mystic and mathematician, founded the *Pythagorean doctrines*, on the basis of his belief in the immortality and transmigration of the soul, which is imprisoned in the body; and emphasized moral asceticism and purification in association with ritual rules of abstinence; he also first introduced the term *philosophy* (from Greek *φιλοσοφία* 'love of wisdom');

490-470BC: <u>Heraclitus</u>, Greek philosopher, wrote the book *On Nature* from which only fragments remain, but seeming to have held that everything is in a state of flux and that world's apparent unity and stability conceals a dynamic tension between opposites measured and controlled by *Logos* 'Reason';

485-480BC: <u>Confucius</u>, Chinese philosopher, after a dozen years spent as a sage wandering from court to court seeking a sympathetic patron and attended by some of his disciples, he founded the school of *Confucianism*, an ethos forming the basis for Chinese education and social organization in the Far East, according to records within *Analects*;

465-450BC: Parmenides of Elea, Greek philosopher, produced the remarkable philosophical treatise *On Nature*, representing a radical departure from cosmologies of Ionian predecessors and setting an agenda of problems for subsequent pre-Socratic philosophers;

460-430BC: Anaxagoras, Ionian thinker, stated that *order is the rational subordination of the inferior to superior elements*, obtained by a difficult victory, when you must vanquish yourself or, even more, surpass yourself;

c.450BC: Empedocles, Greek philosopher and poet, wrote two long poems: *On Nature*, describing a cosmic cycle in which the basic elements Earth, Air, Fire and Water periodically combine and separate under the influence of dynamic forces akin to love and strife; and *Purifications*, describing the Fall of Man; and Antiphon, Greek philosopher and Sophist, was author of the works *On Truth*, and *On Concord*, dealing with Sophistic themes such as the relation of nature and convention, and the nature of language;

450-420BC: Protagoras, Greek philosopher, wrote *On the Gods*, with remaining maxim 'Man is the measure of all things'; and originated *Sophism* (Greek *σόφισμα* from *σοφός* 'wise man'), as a plausibly deceptive fallacy;

441-405BC: Sophocles, Athenian tragedian, initiated the *Dramatic style*, concerning the conflict of individual and state, the action of individuals to show their heroic stature, and the relation between an individual's character and behaviour;

427-390BC: Gorgias, Greek philosopher and rhetorician, developed a philosophy as an extreme form of *scepticism* or *nihilism*, considering that nothing exists if it is not knowable, and if it is knowable would be incommunicable to others, and we live in a *world of opinion*, manipulated by persuasion;

420-380BC: Democritus, Thracian-born Greek metaphysician, developed the *atomic theory of the universe* - all things originate from a vortex of atoms, differing only by shape and arrangement of their atoms;

406-399BC: Socrates, Greek philosopher, founded in Athens the *Socratic method*, consisting of asking for definitions of familiar concepts, such as justice, courage, or piety, eliciting contradictions in responses of interlocutors, and thus demonstrating their ignorance as

claimed to share;

388-348BC: <u>Plato</u>, Greek philosopher, founded *Platonism*, that was synthesized in his *Republic* as the theory of knowledge, defining *knowledge* as *justified true belief*, specifying that a statement must meet three criteria in order to be considered knowledge, namely it must be *justified*, *true*, and *believed*, and showing the dualism of the immortal soul and mortal body; he also wrote *Phaedo* as a basis of knowledge theory, and above all the theory of forms, or ideas, contrasting transient material things with Ideas that they reflect, which are true objects of knowledge;

348-314BC; <u>Xenocrates</u>, Greek philosopher and scientist, worked on natural science, astronomy and philosophy, resulting in statements that philosophy is subdivided into *logic*, *ethics* and *physics*, and that reality is divided into *objects of sensation, belief,* and *knowledge*;

c.330BC: <u>Aristotle</u>, Greek philosopher and scientist, defined *wisdom* as *understanding of causes*, i.e. knowing why things are a certain way, which is deeper than merely knowing that things are a certain way; and coined the term *metaphysics* (from Greek μετά 'beyond, upon, after' and φυσικά 'physics') as a traditional branch of philosophy concerned with explaining the fundamental principles of nature and thought;

330-290BC: <u>Mencius</u> (Mengzi), Chinese philosopher and sage, developed Confucian ideas, founded a school promoting their study, emphasized the cardinal virtues of magnanimity *ren*, sense of duty *yi*, politeness *li* and wisdom *zhi*, which later were compiled in *Book of Mengzi*, and proposed practical recommendations about taxes, road maintenance and poor law;

305-270BC: <u>Epicurus</u>, Greek philosopher, founded *Epicureanism*, a philosophical system consisting of belief that human happiness is only goal of morality, as freedom from pain and anxiety;

c.300BC: Appearance of *Late ancient philosophy*; <u>Zeno of Citium</u>, Greek Cypriot philosopher, founded the school of *Stoicism*, so named after a 'painted porch' (Greek *Stoa poikile*) used during his teaching in Athens, as a distinctive and coherent philosophy, especially in ethics;

100-50BC: Ascendancy of *Early imperial Roman philosophy*;

AD30-40: <u>Philo Judaeus</u>, Hellenistic Jewish philosopher, tried to bring together *Greek philosophy* and *Jewish scripture*, most work consisting of commentaries on Pentateuch;

30-50: <u>Lucius Annaeus Seneca</u>, Roman Stoic philosopher, statesman and tragedian, wrote *Epistolae morales ad Lucilium*, and *Apocolocyntosis divi Claudii* 'The Pumpkinification of the Divine Claudius' as a scathing satire;

40-90: <u>Apollonius of Tyana</u>, Greek philosopher and seer, studied miracles, then taught and opened the school of *Neo-Pythagoreanism* at Ephesus in Asia Minor;

70-90: <u>Epictetus</u>, Greek Stoic philosopher and moralist, taught *Stoicism* in Rome, until banishment by Emperor Domitian, his sayings being collected into a manual entitled *Enchiridion*, and into eight volumes of *Discourses*;

100-200: Rise of *Mid-imperial Roman philosophy*;

170-180: <u>Marcus Aurelius</u>, Roman emperor and philosopher, founded chairs of philosophy for each of chief schools: Platonic, Stoic, Peripatetic, and Epicurean, at Athens; and wrote the twelve-book *Meditationes*, resulting from the study of law and philosophy, especially from Stoicism;

310-330: <u>Iamblichus Chalcidensis</u> (Iamblichus of Apamea), Syrian Neo-Platonist philosopher, wrote *Protreptikos eis Philosophian* 'Summons to Philosophy', including excerpts from earlier philosophers, and also *De mysteriis* 'On Mysteries', defending relevance of magic to philosophy;

350-400: Appearance of *Late Roman-scholastic philosophy*;

523-524: <u>Anicius Manlius Severinus Boethius</u>, Roman philosopher and politician, wrote *De Consolatione Philosophie* 'The Consolation of Philosophy', most widely read after Bible for the next thousand years, the author being described as 'the last of the Roman philosophers, the first of the scholastic theologians';

c.600: Emergence of *Arabic-Islamic philosophy*;

622-632: <u>Muhammad</u>, Arab prophet and founder of Islam, promoted the *Qur'an* (*Koran*) stating 'He gives wisdom to whose He wills, and whoever has been given wisdom has certainly been given much good. And none will remember except those of understanding';

800-825: Affirmation of *Islamic philosophy*;

c.950: <u>Abu Nasr al-Farabi</u>, Islamic neo-Platonist philosopher, wrote a utopian political philosophy presented in his *The Perfect City*, showing Plato's influence;

1000-1050: Appearance of *Medieval philosophy*;

1030-50: <u>Giude Aretino</u> (Guide d'Arezzo), Italian monk and musical theorist, introduced the system of using syllables to name the notes of a scale, and his name was later given to the *Aretinian turn*, as a philosophical turn marking and steadily increasing frequency to direct appeals by American legal scholars to philosophical literatures, in search of explanation or support of positions they take with regard to matters of legal theory, legal institutions, and legal doctrine;

1050-70: <u>Avicebrón</u> (Solomon ben Yehuda ibn Gabriol), Jewish poet and philosopher, produced the famous ethical treatise *Fons vitae* 'The Fountain of Life' and founded a *pantheist system*;

1090-1110: <u>Muhammad al-Ghazali</u>, Islamic philosopher and theologian, wrote *The Intentions of the Philosophers*; *The Incoherence of the Philosophers*; *The Deliverance from Error*; and *The Revival of the Religious Sciences*, attempting to reconcile philosophy and Islamic dogma;

1125-32: <u>Peter Abelard</u>, French philosopher and scholar, was author of the works *Sic et non*; *Nosce te ipsum*; and *Historia Calamitatum Mearum* 'The Story of my Troubles'; as key texts in movement from faith to reason;

1130-40: <u>Adelard of Bath</u>, English philosopher, wrote *De Eodem et Diverso* 'On Identity and Difference', and *Perdifficiles Quaestiones Naturales*;

1185-95: <u>Ibn Rushd Averroës</u>, Islamic philosopher and physician, was author of *Commentaries* on Aristotle, produced in Marrakesh, and considered the most famous medieval Islamic work of philosophy;

1200-1280: Beginning of *Early European philosophy*;

1223-40: <u>William of Auvergne</u>, French philosopher and theologian, wrote *Magisterium divinale* 'The Divine Teaching', attempting to integrate classical Greek and Arabic philosophies with Christian theology;

1259-73: <u>St Thomas Aquinas</u>, Italian scholastic philosopher and theologian, produced *Summa contra Gentiles*, and *Summa Theologiae*, dealing with the principles of natural religion and 'five ways' or proofs of existence of God respectively;

1266-67: <u>Roger Bacon</u> (Doctor Mirabilis), English philosopher and scientist, compiled *Opus Majus* 'Greater Work', along with two other works, as a summary of all learning;

1270-80: <u>St Albertus Magnus</u>, German philosopher and cleric, was a faithfull follower of Aristotle as presented by Jewish, Arabians and Western commentators, and wrote the *Summa theologiae*, and *Summa de creaturiss*, comprising 13th century European knowledge of natural sciences, mathematics and philosophy;

1300-08: <u>John Duns Scotus</u>, Scottish Franciscan philosopher and theologian, was author of *Opus Parisiense* 'Parisian Lectures'; *Tractatus de Primo Principio* 'A Treatise on God as First Principle'; and *Quaestiones Quodlibetales*;

1328-47: <u>William of Ockham</u>, English philosopher, theologian and political writer, worked on logic, producing *Summa Logicae*; *Quodlibeta Septem*; *Dialogus de potestate Papae et Imperatoris*; and *Opus nonaginta dierum* 'Work of 90 Days'; as well as the principle of *Ockham's razor*, a rule of ontological economy to effect that 'entities are not to be multiplied beyond necessity', i.e. a theory should not propose existence of anything more than is needed for its explanation;

1400-1490: Ascendancy of *Renaissance philosophy*;

1500-23: <u>Desiderius Erasmus</u>, Dutch humanist and scholar, wrote *Adagia* 'Adages'; *Enchiridion Militis Christiani* 'Handbook of a Christian Soldier'; the famous *Encomium Moriae* 'In Praise of Folly'; the critical *Colloquia familiaria*; and *De Libero Arbitrio* attacking Martin Luther;

1513-32: <u>Niccolò Machiavelli</u>, Italian political philosopher, was author of *The Prince*, intending to be a handbook for rulers, who must be prepared to do evil if they judge that good will come of it; as well as of *The Art of War*, and *Mandragola*;

1531-38: <u>Juan Luis Vives</u>, Spanish philosopher and humanist, produced *De disciplinis libri XX* 'Twenty Books on Disciplines', and *De anima et vita libri III* 'Three Books on the Soul and on Life';

1591-1602: <u>Tommaso Campanella</u>, Italian philosopher, wrote the books *Philosophia sensibus demonstrata* 'Philosophy Demonstrated by the Senses', and *La Città del Sole* 'City of the Sun';

1600-1690: Emergence of *Post-Renaissance philosophy*;

1605-25: <u>Francis Bacon</u>, English philosopher, published *The Advancement of Learning*; *De Augmentis Scientiarum* 'On Scientific Advancement'; and *Novum Organum* 'New Organon'; showing the importance of experiment to interpret nature, and giving an impetus to future scientific investigation; he also wrote *Meditationes Sacrae*, including the famous aphorism *Nam et ipsa scientia potestas est* 'For also knowledge itself is power';

1615-16: <u>Lucilio Vanini</u>, Italian philosopher, wrote *Amphytheatrum aeternae providentiae*, and *De admirandis naturae reginae deaeque mortalium arcanis*, expressing a form of extreme, even materialistic, pantheism;

1637-49: <u>René Descartes</u>, French philosopher and mathematician, produced *Discours de la méthode* 'Discourse on Method' with the famous quotation *Dubito, ergo cogito; Cogito, ergo sum* 'I doubt, therefore I think; I think, therefore I am'; *Meditationes de prima Philosophia* 'Meditations on First Philosophy'; *Principia Philosophiae* 'Principles of Philosophy', by which he set out the fundamental Cartesian doctrines; and *Regulae ad directionem ingenii* 'Rules for the Direction of the Mind', composed but unfinished and published posthumously; as well as *Les Passions de l'âme* 'Passions of the Soul', thus becoming the founder of modern philosophy;

1640-58: <u>Thomas Hobbes</u>, English political philosopher, wrote and published *Elements of Law Natural and Politic*; *Leviathan*; *De Corpore*; and *De Homine*;

1662-77: <u>Benedict de Spinoza</u>, Dutch philosopher and theologian, was author of *Tractatus de Intellectus Emendatione* 'Treatise on the Correction of the Understanding'; *Tractatus Theologico-Politicus* 'Theological-Political Treatise'; and most importantly *Ethica Ordine Geometrico Demonstrata*, called simply 'Ethics', as a complete deductive metaphysical system, intended to be a proof of what is good for humans derived with mathematical certainty from axioms, theorems and definitions;

1670-90: <u>John Locke</u>, English philosopher, wrote *Two Treatises of Government*, as a social contract theory for constitutional law; and *Essay concerning Human Understanding*, an enquiry into nature and scope of human reason deriving from sensation;

1674-88: <u>Nicolas Malebranche</u>, French philosopher, was author of *De la recherche de la vérité* 'Search for the Truth', drawing on Descartes's dualism of mind and body; *Traité de la morale* 'A Treatise of Morality'; and *Entretiens sur la métaphysique et la religion* 'Dialogues on Metaphysics and on Religion';

1690-1710: <u>Gottfried Wilhelm Leibniz</u>, German philosopher and mathematician, elaborated the best-known doctrine that the world is composed of an infinity of simple, indivisible, immaterial, mutually isolated *monads* which form a hierarchy, the highest being God, and that monads do not interact causally but constitute a synchronized harmony with material phenomena; he also wrote *New Essays on*

Human Understanding; and *Discours de Métaphysique*;

1709-13: <u>George Berkeley</u>, Irish Anglican Bishop and philosopher, produced *Essay towards a New Theory of Vision*; *A Treatise concerning the Principles of Human Knowledge*; and *Three Dialogues between Hylas and Philonous*;

1709-25: <u>Giambattista Vico</u>, Italian historical philosopher, was author of *De nostri temporis studiorum ratione* 'On method in contemporary fields of study'; *De antiquissima Italorum sapientia* 'On the ancient wisdom of the Italians'; two-volume *On Universal Law*; and *Scienza Nuova* 'The New Science' explaining fundamental distinctions between scientific and historical settings and rejecting idea of single, fixed, human nature invariant over time;

1725-46: <u>Francis Hutcheson</u>, British philosopher, among his main works there are *An Inquiry into the Original of Our Ideas of Beauty and Virtue*; *An Essay on the Nature and Conduct of the Passions and Affections, with Illustrations on the Moral Sense*; and *System of Moral Philosophy*;

1729: <u>Christian von Wolff</u>, German philosopher, published his *Philosophia prima sive ontologia*, systematizing and popularizing GW Leibniz's conceptions;

1739-52: <u>David Hume</u>, Scottish philosopher and historian, was author of *A Treatise of Human Nature*; *Essays Moral and Political*; *Enquiry concerning Human Understanding*; *Dialogues concerning Natural Religion*; and *Political Discourses*;

1750-64: <u>Jean Jacques Rousseau</u>, French political philosopher and educationist, produced *Discours sur les arts et sciences* 'A Discourse on the Arts and Sciences'; *Discours sur l'origine et les fondements de l'inégalité parmi les hommes* 'Discourse on the Origin and Foundations of Inequality Among Men'; and the well-known *Du contrat social* 'A Treatise on the Social Contract', which contained the slogan 'Liberty, Equality, Fraternity' so much used during the French Revolution;

1754: <u>Jonathan Edwards</u>, American philosopher and theologian, wrote *Careful and Strict Enquiry into the Modern Prevailing Notions of that Freedom of the Will*;

1755-95: <u>Immanuel Kant</u>, German great figure in history of western thought, became well-known by his *Kritik der reinen Vernunft* 'Critique of Pure Reason'; *Kritik der praktischen Vernunft* 'Critique of Practical Reason'; *Kritik der Urteilskraft* 'Critique of Judgement';

and especially *Grundlagen zur Metaphysik der Sitten* 'Groundwork of the Metaphysic of Morals', pointing to 'Act only on that maxim which you can at the same time will to become a universal law', and to *transcendental (critical) idealism*, that exerted an enormous influence on European philosophy;

1759-76: <u>Adam Smith</u>, Scottish economist and philosopher, published *Theory of Moral Sentiments*, and *Inquiry into the Nature and Causes of the Wealth of Nations*, founding the theory of society in the tradition of Scottish moral philosophy;

1776-89: <u>Jeremy Bentham</u>, English philosopher, wrote *A Fragment on Government*, and *Introduction to the Principles of Morals and Legislation*, arguing that the aim of all actions and legislation should be 'the greatest happiness of the general number', and becoming a pioneer of *Utilitarianism*;

1785-1808: <u>Johann Gottlieb Fichte</u>, German philosopher, published *Wissenschaftslehre* 'The Science of Knowledge'; *Grundlage des Naturrechts* 'The Science of Rights'; *System der Sittenlehre* 'The Science of Ethics'; *Grundzüge des gegenwärtigen Zeitalters* 'The Characteristics of the Present Age'; *Anweisung zum seligen Leben und Religionslehre* 'The Way towards the Blessed Life'; and the patriotic lectures *Reden an die Deutsche Nation* 'Address to the German Nation' inciting to resistance to Napoleon I;

1800-1890: Dawn of *Modern philosophy*;

1807-21: <u>Georg Wilhelm Friedrich Hegel</u>, German philosopher, was author of *Phänomenologie des Geistes* 'The Phenomenology of Mind'; *Wissenschaft der Logik* 'Science of Logic'; *Encyclopädie der philosophischen Wissenschaften im Grundrisse* 'Encyclopaedia of the Philosophical Sciences, Comprising Logic, Philosophy of Nature and of Mind'; and *Grundlinien der Philosophie des Rechts* 'The Philosophy of Right', which had a great influence on Marxism, Positivism, and Existentialism;

1835-45: <u>Ludwig Andreas Feuerbach</u>, German philosopher, wrote *Das Wesen des Christentums* 'The Essence of Christianity', pointing that religion is 'the dream of the human mind', and influencing the works of K Marx and F Engels;

1841-49: <u>Søren Aabye Kierkegaard</u>, Danish philosopher and theologian, published *The Concept of Irony*; *Enten-Eller* 'Either-Or'; *Philisóphiske Smuler* 'Philosophical Fragments'; *Afsluttende uvidenskabelig Efterskrift* 'Concluding Unscientific Postscript'; *Frygt og Baeven* 'Fear and Trembling'; *Kjerlighedens Gjerninger* 'Works of

Love'; *Christelige Tales* 'Christian Discourses'; and *Sygdommen til Döden* 'The Sickness unto Death', contributing to the foundation of *Existentialism*;

1845-55: William Hamilton, Scottish philosopher and scholar, wrote *Lectures on Metaphysics and Logic*, presenting a view on perception and knowledge; and *Discussions on Philosophy and Literature, Education and University Reform*, reviving philosophy in Great Britain;

1848-85: Karl Marx and Friedrich Engels, German social theorists, founded *Marxism* 'dialectal materialism' with theses on class struggle, history and importance of economic factors in politics, as presented in *The Capital*, a magnum opus on surplus value, class conflict, and exploitation of working class;

1857-81: Hippolyte Adolphe Taine, French critic, historian and philosopher, strongly expressed Positivism in *Les Philosophes français du dix-neuvième siècle*; *Philosophie de l'art*; and *De l'intelligence*;

1869: Thomas Henry Huxley, English biologist, coined the term *agnostic*, evolving into *agnosticism* as the belief that existence of God cannot be proved, and the individual cannot know anything of what lies behind or beyond the world of natural phenomena;

1873-88: Friedrich Wilhelm Nietzsche, German philosopher, scholar and writer, published *Unzeitgemässe Betrachtungen* 'Thoughts Out of Season'; *Die Fröhliche Wissenschaft* 'The Joyful Wisdom'; *Also sprach Zarathustra* 'Thus Spake Zarathustra'; *Jenseits von Gut und Böse* 'Beyond Good and Evil'; *Zur Genealogie der Moral* 'A Genealogy of Morals'; and autobiographical *Ecce Homo*;

1885-1912: Bernard Bosanquet, English philosopher, wrote *Knowledge and Reality*; *Logic*; *History of Aesthetic*; *The Philosophical Theory of the State*; and *The Principle of Individuality and Value*;

1890-1909: William James, US philosopher and psychologist, was author of *Principles of Psychology*; *The Varieties of Religious Experience*; *The Will to Believe*; *Pragmatism*; and *The Meaning of Truth*;

1900-13: Edmund Gustav Albrecht Husserl, German philosopher and founder of Phenomenology (science of phenomena), published *Logische Untersuchungen* 'Logical Investigations', and *Ideen zu einer Phänomenologie und phänomenologischen Philosophie* 'Ideas:

General Introduction to Pure Phenomenology';

1903-57: <u>Bertrand Arthur William Russell</u>, English philosopher, mathematician and writer, was a productive writer, publishing works such as *The Principles of Mathematics*; *Tractatus Logico-philosophicus*; *The Problems of Philosophy*; *Introduction to Mathematical Philosophy*; *Theory and Practice of Bolshevism*; *On Education*; *Education and the Social Order*; *History of Western Philosophy*; and *Why I am not a Christian*; as well as *Human Knowledge: Its Scope and Limits*; developed the *theory of knowledge*; and also published *Logic and Knowledge: Essays*;

1917-22: <u>Vladimir Ilyich Lenin</u>, Russian revolutionary, inaugurated the 'dictatorship of proletariat', and modified Marxist doctrine as *Marxism-Leninism*, the basis of *communist ideology*, which was imposed in Russia and later in other countries;

1921-51: <u>Ludwig Josef Johann Wittgenstein</u>, Austrian-born British philosopher, investigated the tendencies in philosophy of his time, and published *Logisch-philosophische Abhandlung* 'Tractatus Logico-Philosophicus', and *Philosophical Investigations*;

1927-45: <u>Martin Heidegger</u>, German philosopher, wrote *Sein und Zeit* 'Being and Time', presenting an ontological classification of 'Being' and an examination of the distinctive human existence 'Dasein';

1928-50: <u>Rudolf Carnap</u>, German-born US philosopher and logician, published *Der logische Aufbau der Welt* 'The Logical Construction of the World'; *Logische Syntax der Sprache* 'Logical Syntax of Language'; *Meaning and Necessity*; and *The Logical Foundations of Probability*;

1931-36: <u>Feng Youlan</u>, Chinese philosopher, was author of the two-volume *History of Chinese Philosophy*, and leader of the movement to revive *neo-Confucianism*;

1931-67: <u>Charles Arthur Campbell</u>, Scottish philosopher, published *Scepticism and Construction*; *On Selfhood and Godhood*; and *In defence of Free Will*;

1934-73: <u>Emil Cioran</u>, Romanian philosopher and essayist, wrote books in Romanian - *On the Heights of Despair*; *The Book of Delusions*; *The Transfiguration of Romania*; and *Tears and Saints*; and in French - *Précis de décomposition* 'A Short History of Decay'; *Histoire et utopie* 'History and Utopia'; *La chute dans le temps* 'The Fall into Time'; *Le mauvais démiurge* 'The New Gods'; and *De l'inconvénient d'être né* 'The Trouble With Being Born';

1943-83: <u>Lucian Blaga</u>, Romanian philosopher, poet and playwright, wrote remarkable trilogies such as *The Philosophy of Knowledge* (The Dogmatic Aeon, Luciferian Knowledge, Transcendent Censorship); *The Philosophy of Culture* (Horizon and Style, The Mioritic Space, The Genesis of Metaphor and the Meaning of Culture); *The Trilogy of Values* (Science and Creation, Magical Thinking and Religion, Art and Value); and *The Cosmological Trilogy* (The Divine Differentials, Anthropological Aspects, Historical Existence);

1951-95: <u>William Van Orman Quine</u>, US philosopher and logician, published *Two Dogmas of Empiricism*; *From a Logical Point of View*; *Word and Object*; *The Roots of Reference*; *The Logic Sequences*; and *From Stimulus to Science*;

1965-85: <u>Hywel David Lewis</u>, Welsh philosopher of religion, was author of *Our Experience of God*; *The Self and Immortality*; *Persons and Life after Death*; and also of the trilogy *The Elusive Mind*, *The Elusive Self*, and *Freedom and Alienation*;

1969: <u>Michel Foucault</u>, French philosopher, wrote *L'Archéologie du savoir* 'The Archaeology of Knowledge', showing that prevailing social attitudes are manipulated by those in power;

1969-86: <u>Constantin Noica</u>, Romanian philosopher, essayist, and poet, after imprisonment by communists for nonconformist ideas, he published books including *Twenty-seven levels of the real*; *Platon: Lysis, Eminescu or Thoughts on the complete man of Romanian culture*; and *Letters on Hermes' logic*;

1979-91: <u>Richard McKay Rorty</u>, US philosopher, wrote *Philosophy and the Mirror of Nature*, constituting a forceful and dramatic attack on foundationalist, metaphysical aspirations of traditional philosophy, and showing that the demarcation of philosophy from science was made possible by considering philosophy's core as 'theory of knowledge'; and also published *Contingency, Irony and Solidarity*, and *Objectivity, Relativism and Truth*;

2012: <u>Helen Steward</u>, English professor of philosophy of mind and action, published the book *A Metaphysics for Freedom*, Oxford University Press, where she argued for a distinctive version of incompatibilism, based on the idea that there is a conflict not only between determinism and free human action, but also between determinism and the activities of a wide variety of animals;

2013: <u>Michael Huemer</u>, US philosopher, produced a work entitled *The Problem of Political Authority: An Examination of the Right to Coerce and the Duty to Obey*, Palgrave Macmillan.

According to the data above, philosophy developed by the following sequences, approximately delimited by transitional times as:

Ancient Egyptian and Babylonian \ 800...700BC ≈ 750BC \
Early-ancient Greek \ 550...500BC ≈ 525BC \
Classic ancient \ c.300BC \
Late ancient \ 100...50BC ≈ 75BC \
Early-imperial Roman \ AD100-200 ≈ AD150 \
Mid-imperial Roman \ AD350...400 ≈ AD375 \
Late Roman-scholastic \ c.AD600 \
Arabic-Islamic \ AD800...825 ≈ AD812 \
Islamic \ AD1000...1050 ≈ AD1025 \
Medieval \ AD1200...1280 ≈ AD1240 \
Early European \ AD1400...1490 ≈ AD1445 \
Renaissance \ AD1600...1690 ≈ AD1645 \
Post-Renaissance \ AD1800...1890 ≈ 1845 \ *Modern.*

The philosophy timeline was established in the table of chapter 7, with its origin 3001 years ago = 987BC, as the initial time, and its end 4350 years from present into the future, as the final time

$t_\bullet = 3001 + 4350 = 7351$ years from the origin of philosophy.

The sequences of philosophy have been drawn above, but only the later ones are better approximated by transitional times: AD1240 = 774 years ago; AD1445 = 569 years ago; AD1645 = 369 years ago; AD1845 = 169 years ago. Referring to the origin of philosophy, these time limits become 3001 - 774 ≈ 2227; 3001 - 569 ≈ 2432; 3001 - 369 ≈ 2632; 3001 - 169 ≈ 2832 years respectively, so that their ratios to the final time $t_\bullet = 7351$ years from the same origin are approximated as 2227/7351 ≈ 0.303; 2432/7351 ≈ 0.331; 2632/7351 ≈ 0.358; 2832/7351 ≈ 0.385. Such successive ratios can be found in Background to time, Table *(m)* for $z/32 = m$ sequences, where they are given as

$$t_{0.3125}/t_\bullet = tanh(0.3125) = 0.3027097;$$
$$t_{0.34375}/t_\bullet = tanh(0.34375) = 0.3308211;$$
$$t_{0.375}/t_\bullet = tanh(0.375) = 0.3583574;$$
$$t_{0.40625}/t_\bullet = tanh(0.40625) = 0.3852840.$$

Accordingly, the above transitional times are calculated, in years from the origin of philosophy, as follows:

$$t_{0.3125} = t_\bullet \cdot tanh(0.3125) = (7351) \cdot (0.3027097) \approx 2225;$$
$$t_{0.34375} = t_\bullet \cdot tanh(0.34375) = (7351) \cdot (0.3308211) \approx 2432;$$
$$t_{0.375} = t_\bullet \cdot tanh(0.375) = (7351) \cdot (0.3583574) \approx 2634;$$
$$t_{0.40625} = t_\bullet \cdot tanh(0.40625) = (7351) \cdot (0.3852840) \approx 2832.$$

Using the formula $t_m = t_\bullet \cdot tanh(m)$ to calculate the earlier transitional times; the timeline, sequences, and intrinsic characteristics in development of philosophy can be reconstituted as in the next tables.

z/32 = m	Time (years)		Sequences of philosophy
	from origin $t_m = t_\bullet \cdot tanh(m)$	from present t_m - 3001	
	7351	+4350	
...	...	...	
0.4375	3025	+24 (AD2038)	– Modern
0.40625	2832	-169 (AD1845)	– Post-Renaissance
0.375	2634	-367 (AD1647)	– Renaissance
0.34375	2432	-569 (AD1445)	– Early European
0.3125	2225	-776 (AD1238)	– Medieval
0.28125	2015	-986 (AD1028)	– Islamic
0.25	1800	-1201 (AD813)	– Arabic-Islamic
0.21875	1583	-1418 (AD596)	– Late Roman-scholastic
0.1875	1362	-1639 (AD375)	– Mid-imperial Roman
0.15625	1139	-1862 (AD152)	– Early-imperial Roman
0.125	914	-2087 (74BC)	– Late ancient
0.09375	687	-2314 (300BC)	– Classic ancient
0.0625	459	-2542 (528BC)	– Early-ancient Greek
0.03125	230	-2771 (757BC)	– Ancient Egyptian + Babylonian
0	0	-3001 (987BC)	

z/32 = m	Time from origin $t_m = t_\bullet \cdot tanh(m)$ (years)	Period $\tau_m = (t_\bullet^2 - t_m^2)/(2t_\bullet)$ (years)	Frequency $f_m = (2t_\bullet)/(t_\bullet^2 - t_m^2)$ (years)$^{-1}$	Angular speed $\omega_m = 2\pi/\tau_m = 2\pi \cdot f_m$ (years)$^{-1}$
	7351	0	∞	∞
...	...	...	...	...
0.4375	3025	3053.1	0.0003275	0.0020580
0.40625	2832	3130.0	0.0003195	0.0020074
0.375	2634	3203.6	0.0003121	0.0019613
0.34375	2432	3273.2	0.0003055	0.0019196
0.3125	2225	3338.8	0.0002995	0.0018819
0.28125	2015	3399.3	0.0002942	0.0018484
0.25	1800	3455.1	0.0002894	0.0018185
0.21875	1583	3505.1	0.0002853	0.0017926
0.1875	1362	3549.3	0.0002817	0.0017702
0.15625	1139	3587.3	0.0002788	0.0017515
0.125	914	3618.7	0.0002763	0.0017363
0.09375	687	3643.4	0.0002745	0.0017245
0.0625	459	3661.2	0.0002731	0.0017162
0.03125	230	3671.9	0.0002723	0.0017112
0	0	3675.5	0.0002721	0.0017095

25. Aero-astronautics

The science or art of aerial navigation related to the study, design, and manufacturing of airflight-capable machines, or the techniques of operating aircraft and rocketry within the atmosphere is called *aeronautics* (from Greek *ἀήρ* 'air' and *ναυτική* 'navigation'); and the theory and practice of travel in space concerned with navigation beyond the Earth's atmosphere is named *astronautics* or *cosmonautics* (from Greek *ἄστρον, ἀστήρ* 'star' or *κοσμος* 'order, world' and *ναυτική* 'navigation').

Ideas of flight appeared from attentive observation of flying birds, leaves, and other light objects carried by wind. Although these ideas were narrated in ancient mythologies, such as the Greek one telling about *Icarus* (Greek *Ικαρος*, Etruscan *Vikare*) who attempted to escape from Crete by means of wings made from feathers and wax, and drowned in the sea; as well as the Vedic Bana's Harsa-carita relating to *Yavana* who manufactured an aerial machine, to architect *Mandhata* who used an aerial car for travelling at distance, and to flying machines called *Vimanas* which were described in Sanskrit epics; the first man-lifting kites, which unveil the level reached in man's flying devices, were recorded in China only from AD550-700.

The emergence and development of aeronautics and astronautics are selectively presented below in chronological order.

AD550-680: *Kites and kite-like flying structures* were first experimented in China;

552: <u>Lu Ban</u>, Chinese philosopher, artisan and engineer, was named 'father of craftsmanship' because he created a *wooden bird*, essentially a kite, that remained flying in the air for three days;

559: <u>Yuan Huangtou</u>, Eastern Wei prince, used a *man-lifting test kite* specially designed to lift a person from the ground, succeeding to fly from the top of a tower near Ye as far as the Purple Way, China, where he came to earth; he was the only survivor of the prisoners condemned to death by the Chinese tyrant Gao Yang, Emperor Wenxuan of Northern Qi, who executed them by ordering them to fly with bird-shaped kite wings;

750-840: *Gliders and gliding wings* emerged in the Islamic world;

850-870: <u>Abbas Ibn Firnas</u>, Muslim Andalusian of Berber descent, polymath, inventor, engineer and aviator, who lived in the Emirate of Córdoba, was reputed to have flown from the hill Jabal al-'arus by using a rudimentary *glider*; the *crater Ibn Firnas* on the Moon being named in his honour;

950-1000: *Winged-man flying* began to be experimented in Europe;

1010: Eilmer of Malmesbury, English Benedictine monk, become known for his early attempt at a *gliding flight using wings* attached to his hands and feet, launching himself from the top of a tower at Malmesbury Abbey, flying more than a furlong (201 metres) for about 15 seconds, and then falling and breaking both his legs; he was an old lame man during the 1066 Norman invasion of England;

1020-37: Avicenna ('Abd Allah ibn Sina), Persian polymath, philosopher and physician, understood *inclination* as a permanent force whose effect is dissipated by external forces as *air resistance*;

1140-70: *Winged flapping* emerged in some places of the Old World;

1320-40: *Renaissance flying devices* became to be conceptualized in Italy;

1485: Leonardo da Vinci, Italian painter, sculptor, architect and engineer, designed many mechanical devices, including parachutes, studied the flight of birds, and most importantly drew a series of detailed plans for a *human-powered ornithopter*, a wing-flapping device for flying, and other such devices displayed in his twelve-volume *Codex Atlanticus* (now in Biblioteca Ambrosiana, Milan);

1500-20: *Man-lifting kites* began to be experimented in Japan;

1560-70: Ishikawa Goemon, semi-legendary Japanese outlaw, used a *man-lifting kite* to steal the golden scales from a pair of ornamental fish images which were mounted on the top of Nagoya Castle, manoeuvring a trapeze attached to the tail of a *giant kite*;

1660-1700: *Manned balloons* became attractive for French inventors;

1780-83: Jacques Alexandre César Charles, French inventor, scientist and mathematician, and Nicholas-Louis Robert, French engineer, constructed a *manned hydrogen balloon*, that was launched from the Jardin des Tuileries, Paris, ascended to a height of about 550 metres, flew 2 hour 5 minutes, and landed in Nesles-la-Vallée, covering a distance of 36 kilometres;

1782-83: Joseph Michel Montgolfier and his brother Jacques Étienne Montgolfier, French aeronautical inventors, constructed and launched a *manned balloon flight*, a tethered balloon with humans on board, at the Folie Titon, Paris, covering 12.1 kilometres in less than half an hour at a height of 915 metres, and carrying J-F Pilatre de Rozier and the Marquis d'Arlandes;

1783: French <u>Jean-François Pilâtre de Rozier</u>, scientist, <u>Jean-Baptiste Réveillon</u>, manufacture manager, and <u>André Giroud de Villette</u>, physicist and chemist, constructed a *balloon with a smoky fire* built on a grill attached to the bottom, performing another *flight carrying humans*, at the Folie Titon in Paris;

1796: <u>French Aerostatic Corps</u> used the reconnaissance balloon *L'Intrépide*, a roughly *spherical silk hull* with diameter of 9.8 metres, wooden gondola measuring 1.14 by 0.75 metres, and railing of 1.05 metres height, which participated at the Battle of Würzburg, where the French army was defeated by Austrian forces who captured the balloon and brought it to their capital (now in Heeresgeschichtliches Museum, Vienna);

1848: <u>John Stringfellow</u>, English inventor, built a *steam-powered model aeroplane*, which was tested indoors in a disused lace factory, in Chard, Somerset, England, and showed that it was capable of climbing flight under its own power;

1850-56: *Aircrafts and spacecrafts* started their ascendancy in Europe and the United States of America;

1852: <u>Henri Giffard</u>, French engineer, invented the *steam injector* and the powered airship *Giffard dirigible*, the first powered, controlled and sustained lighter-than-air craft, that flew 24 kilometres in France, with a steam engine driven mechanism;

1856: <u>Jean-Marie Le Bris</u>, French aviator, made the first flight higher than the point of departure, by his glider *L'Albatros artificiel* pulled by a horse on a beach, performing to fly 100 metre height over a distance of 200 metres;

1863-1900: <u>Ferdinand von Zeppelin</u>, German army officer, first flew as a balloon passenger with the Union Army of the Potomac during the American Civil War, and then constructed his own airship, a dirigible balloon of rigid type, named *zeppelin*, which first flew in 1900;

1871: <u>Alphonse Penaud</u>, French pioneer of aviation design and engineering, was originator of the use of twisted rubber to power the aircraft *Planophore*, the first truly successful automatically stable model flying in Paris;

1874: <u>Félix du Temple de la Croix</u>, French naval officer and inventor, built the *Monoplane*, a large plane made of aluminium in Brest,

France, with a wingspan of 13 metres and a weight of only 80 kilograms (without the driver), achieving lift off under its own power after a ski-jump run, glided for a short time and returned safely to the ground, thus making the *first successful manned powered flight* in history;

1882-83: <u>Charles Renard</u>, French military engineer, and <u>Arthur Constantin Krebs</u>, French officer and pioneer in automotive engineering, constructed and piloted respectively the French Army electric-powered airship *La France*, which made its maiden *first controlled free-flight* at Chalais Meudon, covering 8 kilometres in a 23 minute circular flight and returned to the place of take-off (exhibited at the Paris Exposition Universelle in 1889);

1882-97: <u>Ernst Mach</u>, Austrian physicist and philosopher, experimented with the supersonic projectiles and the flow of gases, obtaining remarkable photographs of shock waves and gas jets, his name being given to the ratio of speed of flow of a gas to speed of sound, called *Mach number*, and to the angle of a shock wave to direction of motion, called *Mach angle*; he wrote *Mechanik in ihrer Entwicklung* 'Thee Science of Mechanics', and *Beiträzur Analyse der Empfindung* 'Contributions to the Analysis of Sensation', laying the foundations of logical positivism;

1883-84: <u>Alexander Fedorovich Mozhaysky</u>, Russian naval officer, aviation pioneer and heavy-than-air craft designer, had achievements in flight control and propulsion, and constructed *Mozhaysky's monoplane*, that was tested for 20-30 metres near Krasnoye Selo, Russia, but failed to fly because of an impracticable angle of attack at its wings;

1890: <u>Clément Ader</u>, French inventor and engineer, constructed the *steam-powered Eole*, a 4-cylinder developing 20 horsepower and driving a four-blade propeller, which made a 50-metre flight near Paris, reached a height of 20 centimetres and flew uncontrolled for approximately 50 metres;

1895-1914: <u>Frederich William Lanchester</u>, English engineer, inventor and designer, built the first experimental motor car in Great Britain, founded the Lanchester Engine Company which produced the first *Lanchester car*, and laid the theoretical foundations of aircraft design as published in his two-volume *Aerial Flight*, and *Aircraft in Warfare*;

1903-09: <u>Orville Wright</u> and <u>Wilbur Wright</u>, US aviation pioneers, operating on a bicycle shop and being self-taught inventors, they became the first to fly in a *heavier-than-air machine* at Kitty Hawk,

North Carolina, and then formed an aircraft production company;

1903-29: <u>Konstantin Eduardovich Tsiolkovsky</u>, Russian astrophysicist and rocket pioneer, published his *Exploration of Cosmic Space by means of Reaction Devices*, which established author's reputation as 'the father of space flight theory', and founded *multi-stage rocket technology*;

1906: <u>Traian Vuia</u>, Romanian inventor, designed and constructed a fully self propelled, fixed wing aircraft *Vuia 1 monoplane* in the Bois de Boulogne, France, which was accelerated for about 50 metres, left the soil and flew at a height of about one metre on a distance of 12 metres in flight, and then landed; and <u>Alberto Santos-Dumont</u>, Brazilian aviation pioneer, designed and constructed the fixed-wing heavier-than-air craft *14-bis Oiseau de proie* 'Bird of pray', that established a world record by flying 220 metres in 21.5 seconds;

1907: <u>Paul Cornu</u>, Romanian-born French inventor, designed and built *Cornu helicopter*, the first truly manned flying machine to have risen from the ground using rotating wings instead of fixed wings, which lifted its inventor to one foot and remained aloft for 20 seconds;

1907-10: <u>Charles Stewart Rolls</u>, English car manufacturer and aviator, and <u>Henry Royce</u>, English engineer, based on their experience in making engines and cars, they joined Royce Ltd and CS Rolls & Co into *Rolls-Royce Ltd*, motor-car and aero-engine builders of Derby and London; then the younger CS Rolls became the fist man to make a non-stop double crossing of the English Channel by *plane*, taking 95 minutes, and was the first Briton to be killed in a flying accident;

1910: <u>Henri Coanda</u>, Romanian aeronautical engineer and inventor, designed and constructed the experimental aircraft *Coanda-1910*, as the world's *first jet*, using a 4-cylinder piston-engine to power a rotary compressor by combining suction at front and airflow out at rear, instead of a propeller;

1910-13: <u>Aurel Vlaicu</u>, Romanian engineer, airplane constructor and early pilot, built three original arrow-shaped airplanes, all being controlled in front, two coaxial propelled, ring around engined, and tricycle-landing geared with independent suspension and brakes: *Vlaicu I*, flying for the first time over the Cotroceni airfield near Bucharest; *Vlaicu II*, that won several prizes, was exhibited at 1912 Aspern Air Show near Vienna, and used to cross from Transylvania in flight over the Carpathian Mountains to the Kingdom of Romania, when the pilot Vlaicu died near Campina, becoming a national hero; and *Vlaicu III*, a two seated monoplane contracted by Marconi

Company for experiments with aerial wireless radio, that was partially built at his death, being completed in 1916 by his friends <u>Giovanni Marconi</u> and <u>Constantin Silisteanu</u>, and shipped to Germany during the German occupation of Bucharest in the First World War (then presented in 1942 Aviation Exhibition in Berlin);

1911: <u>Elmer Ambrose Sperry</u>, US inventor and electrical engineer, invented the *gyroscopic compass* and *stabilizers for ships and aeroplanes*;

1912-35: <u>Hugo Junkers</u>, German aircraft engineer, founded aircraft factories at Dessau, Magdeburg and Stassfurt, where famous *planes*, both civil and military, were produced, including *JU 87* 'Stuka' *dive bomber* used in the Second World War;

1913-43: <u>Igor Ivanovich Sikorsky</u>, Ukrainian-Russian US aeronautical engineer, designed and flew the world's first *multi-engine fixed-wing aircraft* 'Russky Vityaz', and the *first airliner* 'Ilya Muromets'; founded the Sikorsky Aircraft Corporation; developed the first of Pan American Airways' ocean-conquering *flying boats*, including the *American Clipper*; and also designed and flew Vought-Sikorsky *VS-300*, the first viable US helicopter, later modified into the *Sikorsky R-4*, the world's first *production helicopter*;

1916-27: <u>William Edward Boeing</u>, US aircraft manufacturer, formed the Pacific Aero Products Co, building *seaplanes*; then founded the *Boeing Airplane Company*, as the largest manufacturer of military and civilian *aircrafts*;

1916-36: <u>Reginald Joseph Mitchell</u>, English aircraft designer, invented *world-beating seaplanes* for the Schneider trophy races, and then devised the famous *Spitfire*, which was crucial in Battle for Britain during the Second World War;

1919-29: <u>Robert Hutchings Goddard</u>, US physicist, rocket engineer and inventor, published *A Method of Reaching Extreme Altitudes*, launched the first *liquid-fuel rocket*, and developed the first *instrument-carrying rocket* able to make observations in flight;

1920-23: <u>Juan de la Cierva y Codorniu</u>, Spanish aeronautical engineer and pilot, invented the *Autogiro* 'Autogyro', a single-rotor type of aircraft; and then developed the articulated rotor that was used for making the world's first successful flight of a *stable rotary-wing aircraft* with its *C.4 prototype*;

1924-29: <u>Umberto Nobile</u>, Italian aviator, constructed the airships *Norge* and *Italia*, flew across the North Pole in his 'Norge' with Roald

Amundsen and Lincoln Ellsworth, and then he was wrecked in his 'Italia' when returning from the North Pole;

1924-31: <u>Hugo Eckener</u>, German aeronautical engineer, flew on *ZR3*, latter called 'Los Angeles' from Friedrichshafen (Germany) to Lakehurst (New Jersey, SUA), so performing the first *flight by an airship* directly *from continental Europe across Atlantic to SUA*, as well as constructed the *Graf Zeppelin* which was piloted in a world's circumnavigation and in a polar expedition;

1926-39: <u>Lincoln Ellsworth</u>, US explorer, flew over both *North Pole* and *South Pole*, and in Antarctic expeditions, claiming thousands of square miles of territory for USA, now called *Ellsworth Land*;

1930: <u>Frank Whittle</u>, English aeronautical engineer and inventor, patented a *jet engine*, and searched into jet propulsion, followed by his first *successfully flown engine*;

1930-37: <u>Elie Carafoli</u>, Romanian engineer and aircraft designer, worked at Industria Aeronautica Romana (IAR) 'Romanian Aeronautical Industry', designed the *IAR CV-11*, a single-seat low-wing monoplane fighter aircraft, then *IAR 14* and *IAR 15* aircrafts, and finally the legendary *IAR 80* fighter aircraft; he published *High-speed aerodynamics, compressible flow*, New York: Pergamon Press;

1931-66: <u>Sergei Pavlovich Korolev</u>, Soviet aircraft engineer and rocket designer, led the *Moscow Group for Investigating Jet Propulsion*, launched USSR's *first liquid-propelled rocket*, worked on the *aircraft jet-assisted take-off systems*, designed a *Soviet spacecraft*, and directed the USSR's space programme with the historic first orbiting *Sputniks* and *first manned space flight* carrying Y Gagarin, then *Vostok* and *Voskhod* manned spacecraft, as well as *Cosmos* series of satellites;

1934: <u>Henri Coanda</u>, Romanian aeronautical engineer, patented his *Coanda effect* in fluid dynamic, as entrainment of a free jet alongside a curve surface, which then was largely applied for designing aircrafts;

1934-45: <u>Wilhelm Emil Messerschmitt</u>, German aircraft designer and manufacturer, established 'Messerschmitt aircraft' manufacturing works, designed the *ME-109* which set a world speed record, supplied Luftwaffe with foremost types of combat aircraft during the Second World War, and produced the *ME-262* fighter as the first *jet-plane flown in combat*, which set a world speed record;

1936-60: <u>Sergei Vladimirovich Ilyushin</u>, Soviet aircraft designer, successfully designed the *TSKB-30*, gaining several records and extensively used as bombardier in the Second World War, *IL-2 Shturmovik* dive-bomber, twin-engined passenger-carrying *IL-12*, *IL-28* jet bomber, *IL-18 Moskva* turboprop airliner, 182-passenger *IL-62* jet, and the wide-bodied successor *IL-86* airbus;

1936-70: <u>Wernher von Braun</u>, German and US rocket pioneer, constructed rockets at Peenemunde, where *V-1* and *V-2 rockets* were perfected and launched against Great Britain during the Second World War; and, after surrendering to the USA, constructed the first *US artificial earth satellite* 'Explorer I', and *Saturn rocket* that was used for 'Apollo 11' moon landing;

1942-50: <u>Roy Fedden</u>, English aero-engine designer, produced *Rolls-Royce aero-engines* during the First World War, and initiated the famous range of piston engines *Mercury*, *Pegasus*, *Perseus*, *Taurus*, *Hercules* and *Centaurus*, as well as uniquely developed the *sleeve-valve engine*;

1943-65: <u>Stanley George Hooker</u>, English aero-engine designer, working for Rolls-Royce, produced the jet engines *Welland*, *Nene*, *Derwent*, *Avon* and *Trent*; then for Bristol Aeroplane Company, where he designed the *Proteus*, *Olympus* for Concorde, and *Orpheus* and *Pegasus* jet engines;

1951-66: <u>James Arnot Hamilton</u>, Scottish aeronautical engineer, worked on aerodynamics of supersonic flight, directed the projects for *Jaguar* and *Tornado* aircraft, and finally the *Concorde project*, from early concepts to first flights of two prototypes;

1955-68: <u>Andrei Nikolayevich Tupolev</u>, Russian aeronautical engineer, constructed the first Soviet civil jet called *Tu-104*, and completed the first test flight of the supersonic passenger aircraft *Tu-144*;

1959-66: <u>USSR Luniks missions</u> were launched to contact and land on the Moon;

1961: <u>Yuri Alekseyevich Gagarin</u>, Soviet cosmonaut, was the first man to *travel in space* in USSR *Vostok spaceship satellite*, and completed a *full circuit of the Earth*;

1966-68: <u>Five US *Orbiters*</u> were launched to map the Moon surface, and then *Apollo 8*, *9* and *10* to test the lunar module;

1969: <u>Neil Alden Armstrong</u>, <u>Buzz Aldrin</u>, and <u>Michael Collins</u>, US astronomers, set out in the US *Apollo 11 mission*, making a successful *Moon-landing expedition*, and became famous for their *first men steps on the Moon*;

1970-75: USSR *Venera 7, 8* and *9 operations* for landing on *Venus* and taking pictures of its surface;

1973-74: US *Pioneer 10 and 11 missions* to explore *Jupiter*;

1973-75: Operation of US *Skylab space station*, as well as *Mariner 10 exploration* of *Mercury* and *Venus*;

1979-81: US *Voyager 1 mission* to explore *Jupiter*; and *Voyager 2 mission* to explore *Jupiter*, *Uranus*, and *Neptune*, and to bring information from preliminary *Pioneer 11* survey of *Saturn* and its satellite *Titan*;

1986: Operation of USSR *Mir space station*, as well as cruise of US *Voyager 2 mission* to *Neptune*, and beyond *Uranus*;

1990: Start of US *Hubble Space Telescope mission* launched from *Space Shuttle Discovery* and put into orbit around the Earth with 94-minute period at almost 600-kilometre distance for a preliminary estimated 15-years duration, by which it was already proved that objects once called spiral nebulae are actually independent star-systems;

1990-93: US *Magellan mission*, launched from *Shuttle Atlantis*, achieved radar mapping of *Venus*;

1994: US *Clementine mission* was launched for orbiting and surveying the *Moon*;

1996-97: US *Pathfinder mission* with *Sojourner* 'Rover' for landing on *Mars* and collecting rocks from there;

NASA current missions: *Voyager-The Interstellar Mission* (Study of the outermost edge of the Solar System and beyond into interstellar space), *Lunar Reconnaissance Orbiter* (Mission to find safe landing sites, locate potential resources, characterize the radiation environment, and demonstrate new technology), and *Aqua* (Earth science satellite mission to collect information about the Earth's water cycle).

Aero-astronautics started, according to the table in chapter 7, from AD616, or 2014 - 616 = 1398 years ago, as the initial time; and will have the same fate as other human abilities, about 4350 years from present into the future, as the final time $t_\bullet$ = 1398 + 4350 = 5748 years from the origin of aero-astronautics.

As usual, the most recent are the more accurately delimited sequences, so that the selected ones with their transitional times are presented as:

AD1320...1340 ≈ AD1330 \ *Renaissance flying devices*
\ AD1500...1520 ≈ AD1510 \ *Man-lifting kites*
\ AD1660...1700 ≈ AD1680 \ *Manned balloons*
\ AD1850...1856 ≈ AD1853 \ *Aircrafts and spacecrafts*;

i.e. 2014 - 1330 ≈ 684; 2014 - 1510 ≈ 504; 2014 - 1680 ≈ 334; 2014 - 1853 ≈ 161 years ago. Referring to the origin of aero-astronautics 1398 years ago, these transitional times become 1398 - 684 ≈ 714; 1398 - 504 ≈ 894; 1398 - 334 ≈ 1064; 1398 - 161 ≈ 1237 years; and then their successive ratios are 714/894 ≈ 0.799; 894/1064 ≈ 0.840; 1064/1237 ≈ 0.860, which can be found in Background to time, Table *(m)* for $z/32 = m$ sequences, in the same succession, with the corresponding values

$$t_{0.125}/t_{0.15625} = 0.8023256;$$
$$t_{0.15625}/t_{0.1875} = 0.8362814;$$
$$t_{0.1875}/t_{0.21875} = 0.8607086.$$

As the sequences are of the kind $z/32 = m$, and the final time is $t_• = 5748$ years from the origin of aero-astronautics, it follows that accurate transitional times for these sequences can be calculated, in years from the origin of aero-astronautics, as follows:

$$t_{0.125} = t_•·tanh(0.125) ≈ 715;$$
$$t_{0.15625} = t_•·tanh(0.15625) ≈ 891;$$
$$t_{0.1875} = t_•·tanh(0.1875) ≈ 1065;$$
$$t_{0.21875} = t_•·tanh(0.21875) ≈ 1238.$$

Proceeding in the same manner, the earlier transitional times of aero-astronautical development with its timeline, sequences and intrinsic characteristics, are calculated and displayed as:

$z/32 = $ m	Time (years)		Sequences
	from origin $t_m = $ $t_•·tanh(m)$	from present $t_m - 1398$	
	5748	+4350	
...	...	...	
0.25	1408	+10 (AD2024)	_ Aircrafts and spacecrafts
0.21875	1238	-160 (AD1854)	_ Manned balloons
0.1875	1065	-333 (AD1681)	_ Man-lifting kites
0.15625	891	-507 (AD1507)	_ Renaissance flying devices
0.125	715	-683 (AD1331)	_ Wing-flapping
0.09375	537	-861 (AD1153)	_ Winged-man flying
0.0625	359	-1039 (AD975)	_ Gliders and gliding wings
0.03125	180	-1218 (AD796)	_ Kites + kite-like flying structures
0	0	-1398 (AD616)	_

$z/32 =$ m	Time from origin $t_m = t_\bullet \cdot tanh(m)$ (years)	Period $\tau_m =$ $(t_\bullet^2-t_m^2)/(2t_\bullet)$ (years)	Frequency $f_m =$ $(2t_\bullet)/(t_\bullet^2-t_m^2)$ (years)$^{-1}$	Angular speed $\omega_m =$ $2\pi/t_m = 2\pi \cdot f_m$ (years)$^{-1}$
	5748	0	∞	∞
...	...	...	...	...
0.25	1408	2701.6	0.0003702	0.0023258
0.21875	1238	2740.7	0.0003649	0.0022926
0.1875	1065	2775.3	0.0003603	0.0022639
0.15625	891	2804.9	0.0003565	0.0022400
0.125	715	2829.5	0.0003534	0.0022206
0.09375	537	2848.9	0.0003510	0.0022055
0.0625	359	2862.8	0.0003493	0.0021948
0.03125	180	2871.2	0.0003483	0.0021884
0	0	2874.0	0.0003479	0.0021862

26. High technology

Technology (from Greek τεχνη 'skill, art' and λογος 'word, speech, story') generally refers to making, modification, usage, and knowledge of tools, machines, techniques, crafts, systems, and methods of organization, in order to solve problems, improve pre-existing solutions of problems, achieve goals, handle applied input/output relations, or perform specific functions; and *high technology* is the technology that uses highly sophisticated equipment and advanced engineering techniques, such as electrical engineering, automotive, nuclear physics, photonics, nanotechnology, telecommunications, information technology and systems, biotechnology, robotics, artificial intelligence, and so on.

The emergence of high technology was preceded by a series of scientific discoveries; especially those concerning the fundamental properties of waves, particles, materials; and the generation, transmission and reception of signals, as conveyors of energy and information, approachable, at a certain frequency band, according to the Fourier transform

$$\Phi(f) = \int_{-\infty}^{+\infty} \varphi(t) \cdot e^{-ift} \cdot dt.$$

For a rectangular pulse $\varphi(t) = 1/(2t')$ in a finite interval of time $-t' < t < +t'$, and $\varphi(t) = 0$ outside the interval, the above relationship becomes

$$\Phi(f) = [1/(2t')] \cdot \int_{-t'}^{+t'} e^{-ift} \cdot dt = [1/(2t')] \cdot sin(ft'),$$

where f is the signal frequency divided by 2π. This transform shows that $\Phi(0) \to 1$ when $f \cdot t' \to 0$; $\Phi(f) = 0$ when $f = \pm\pi/t', \pm2\pi/t',...$; whilst the amplitude of the carrier decreases when f increases and finally vanishes. In the time interval $2t'$, first annulment takes place when $f = \pi/t'$ and then most energy of the signal corresponds to frequencies less than π/t', so that the frequency band can be roughly estimated as $\Delta\varphi \sim 1/(2t')$. Taking into account that signal duration is $\Delta t = 2t'$, and meanwhile can be approximated that $\Delta\varphi \cdot \Delta t \sim 1$; it follows that a longer signal is associated with a shorter pulse spectrum, and in reverse. More rigorously, this relation can be expressed by the inequality $\Delta f \cdot \Delta t \geq 1$, which was established by Fourier (1810-22), and then used to show that the persistence of the information conveyed by a signal is mainly because of its reduced frequency band or spectrum. Subsequently, based on quantum theory formulated by Planck (1898-1900), wave-particle duality was developed by de Broglie (1923-28), showing that, in the propagation of electromagnetic waves with light speed c and, by virtue of their motion, associated with particles of relativistic momentum as pulse p [kilogram-metre per second], the wavelength $\lambda = c/f$ [metre] is given by the simple equation $\lambda = h/p$, where h is Planck's constant [$6.626 \cdot 10^{-34}$ joule-second]. This wave propagation is related not only to photons but also to electrons and other fundamental particles, including neutrinos and antineutrinos which are unrestrictedly crossing the entire universe. As asserted by wave-particle duality, the larger particles and bodies are also associated with waves, but their lengths are too short, i.e. frequencies too high, to be detected. Noticeably, the more harmonic, the more persistent the waves are. A more comprehensive understanding of wave propagation was possible due to the development of wave mechanics (Schrödinger, 1926-30) and quantum mechanics (Dirac, 1928-30). In the context of signal processing, the Dirac delta function, or unit impulse function, is regarded as a kind of a weak limit in a sequence of functions having a tall spike at the signal origin, and making it extremely useful in many natural processes involved in high technology.

Along with wave-particle duality, electromagnetism, radiation, quantum mechanics, and other scientific branches, high technology recorded a rapid evolution and diversification of its fields of study, including thermal radiation, cosmic microwave radiation, electromagnetic radiation, black-body radiation, and thermionics; radioactivity, radioelements, radioisotopy, radio-astronomy, radiochemistry, radiobiology, radiography, radiology, radioscopy, radiotherapy, radioimmunology; radiotelegraphy, radiotelephony, radio-communication, and radio-engineering; nuclear physics, nuclear collision, nuclear transmutation, nuclear fission, nuclear fusion, spectrology, spectroscopy, quasars, pulsars, string, and neutrinos; particle distribution, particle statistics, particle disintegration, particle acceleration, quarks, anti-particles, light shifting, photometry, superfluidity, superconductivity, cinematography, cybernetics, and television; as well as radar and sonar techniques, masers, lasers, computers, transistors, microprocessors, teleprocessors, telesoftware, and integrated circuits.

By its relatively short but spectacular ascendancy, the high technology covers today a wide range of branches, including:

Telecommunication technology - comprising technological means, particularly through electrical signals or electromagnetic waves (telephone, radio, television, internet, local and wide area networks);

Spatial technology - referring to any software or hardware that interacts with real world location, especially at the Earth's surface;

Electronics - dealing with electrical circuits (analogue circuits and digital circuits) that involve active electrical components, including vacuum tubes, transistors, diodes and integrated circuits, as well as associated passive interconnection technologies;

Biotechnology - using living systems and organisms to develop or make products, or more precisely any technological application that uses biological systems, living organisms or derivatives thereof, to make or modify products or processes for specific purposes;

Process technologies - as the machines, equipment and devices that contribute to an operation transforming materials and information, and customers to add value; these technologies including a wide range of building blocks (products, units and skids) such as heat transfer, membrane filtration; cheese and butter distillation and evaporation; blending and mixing technologies designated to operate at maximum efficiency in modern processing lines within dairy, food, beverage, brewery, etc.; as well as their linkage in processing lines;

Low-temperature and *high-temperature technologies* - introducing innovative solutions to provide and improve materials and equipments for working either at very low or very high temperatures, for example by using wolfram, quartz glass, ceramic materials and graphite to the processes which take place at temperatures more than 1000°C;

Robotics - dealing with the design, construction, operation and application of robots, as well as computer systems for their control, sensory feedback and information processing;

Nuclear technology - involving reactions of atomic nuclei, notably nuclear power, nuclear medicine and nuclear weaponry, which found applications from smoke detectors and gun sights to nuclear reactors and sophisticated nuclear weapons; and also industrial and commercial applications, food processing and agriculture;

Medical high-technology - including da Vinci-Si HD Surgical Robotic System (providing enhanced capabilities, including a magnified, high-definition, 3D), CyberKnife (as a method of delivering radiotherapy used for targeting treatment more accurately than standard radiotherapy), InnerCool RTx Endovascular System (for cooling and warming that provides advanced whole body temperature modulation therapy in a closed loop system from the inside out), Olympus VisiGlide (a single use guide wire for making ductal navigation easier and faster), PillCam (endoscopic camera/capsule for healthcare market), Siemens SOMATOM Definition Flash (for scanning at Flash speed, so the diagnostic images can be acquired from almost any patient), Antenna Pill (electronic device as a bar code that ingested emits a signal warning the patient to take his/her medicine);

Genetic engineering - the direct manipulation of an organism's genome using biotechnology, by methods of cloning genetically engineered molecules in foreign cells, creation of transgenetic organisms, etc.;

Computing - any goal-oriented activity requiring, benefiting from, or creating computers, including computer engineering, software engineering, computer science, information systems and information technology;

Direct-digital control - by which a single digital computer replaces a group of single-loop analogue controllers, increasing computational ability and permitting the application of more complex and advanced control techniques;

Information technology, or *Information and Communications Technology* (ITC) - the study, design, development, implementation, support, or management of computer-based information systems, and also unified communications and the integration of telecommunications (telephone lines and wireless signals), computers, as well as necessary enterprise software, middleware, storage and audio-visual systems, which enable users to access, store, transmit and manipulate information;

Satellite imagery, or *Satellite-Imaging technology* - consisting of images of the Earth or other planets collected by artificial satellites;

Artificial intelligence (AI) - the technology and a branch of computer science that studies and develops intelligent machines and software, with goals of deduction, reasoning, problem solving, knowledge representation, planning, learning, natural language processing, motion and manipulation, perception, social intelligence, creativity and general intelligence.

Based on scientific discoveries, experiments, technological innovations, etc., high technology emerged and ascended due to an increasing number of scientists, pioneers, promoters, inventors, engineers, organisations, teams and missions, some of them being mentioned below.

1871-73: James Clerk Maxwell, Scottish physicist, worked on the theory of electromagnetic radiation, formulated mathematically the *theory of electric and magnetic forces*, and published his *Treatise on Electricity and Magnetism*, providing the first conclusive evidence

444

that light consists of electromagnetic waves, suggesting that *electromagnetic waves can be generated in a laboratory*, and paving the way for the theory of relativity and quantum mechanics;

1877: <u>Ludwig Boltzmann</u>, Austrian physicist, extending JC Maxwell's theory of velocity distribution for colliding gas molecules, as a basis of *Maxwell-Boltzmann distribution* and *statistics*, he deduced the famous *Boltzmann equation*, showing how increasing entropy corresponds to increasing molecular randomness; and also theoretically stating the law for *black-body radiation*;

1879: <u>Josef Stefan</u>, Austrian physicist, experimentally established *Stefan's law*, asserting that energy radiated from a black body is proportional to absolute temperature, and applying it to estimate the Sun's surface temperature;

1886: <u>Eugene Goldstein</u>, German physicist, discovered the *Kanalstrahlen* 'canal rays', which emerge from channels or holes in anodes, eventually called *positive rays*, as positively charged particles of atomic mass;

1887-88: <u>Heinrich Rudolf Hertz</u>, German physicist, discovered *radio waves* 'Hertzian waves', confirming the existence of electromagnetic waves which behave like light waves;

1893: <u>Wilhelm Carl Werner Wien</u>, German physicist, worked on *black-body radiation*, showing that wavelength at which maximum energy is radiated is inversely proportional to absolute temperature of body;

1894-1905: <u>John William Strutt Rayleigh</u>, English physicist, and <u>James Hopwood Jeans</u>, English physicist and astronomer, developed a formula to describe spectral radiance of electromagnetic radiation and distribution of energy of enclosed radiation at long wavelengths from a black body, now known as *Rayleigh-Jeans law*, which accurately predicts long-wavelength radiation emitted by hot bodies;

1895: <u>Wilhelm Konrad von Röntgen</u>, German physicist, discovered *electromagnetic rays*, initially called 'X-rays' because of their unknown properties, and later named *Röntgen rays*; and <u>Guglielmo Marconi</u>, Italian physicist and inventor, successfully achieved *wireless telegraphy*, by converting electromagnetic waves into electricity, as the system of *radiotelegraphy*, and transmitted signals across the English Channel;

1895-1903: <u>Auguste Marie Louis Nicolas Lumière</u> and <u>Louis Jean Lumière</u>, French industrial physiological chemists and pioneers of motion photography, invented the *cinématographe* 'cinematograph', produced about 2,000 films, and also invented the *Autochrome screen plate* for colour photography;

1895-1930: <u>Ernest Rutherford</u>, New Zealand physicist, designed the first *wireless transmissions*, promoted *nuclear physics*, and searched on *submarine detection* for the Admiralty during the Second World War;

1896: <u>Antoine Henri Becquerel</u>, French physicist, discovered *radioactivity*, marking the beginning of the nuclear age, and noticing that 'Becquerel rays' were a property of atoms; and <u>Pieter Zeeman</u>, Dutch physicist, discovered the *Zeeman effect*, showing that the resultant broadening of spectral emission lines is due to the splitting of spectrum lines into two or three components;

1897: <u>Ferdinand Braun</u>, German physicist, constructed the first *cathode-ray oscilloscope* 'Braun tube' providing a basic component of television;

1898: <u>Marie Curie</u> and <u>Pierre Curie</u>, French physicists, worked on radioactivity, showing that the *rays emitted by radium* contain electrically positive, negative and neutral particles, and also discovered the elements *radium* and *polonium*;

1898-1900: <u>Max Karl Ernst Planck</u>, German theoretical physicist, deeply researching thermodynamics and black body radiation, he abandoned classical dynamical principles and formulated the *quantum theory*, assuming energy changes to take place in small discrete instalments or *quanta*, predicting phenomena inexplicable in classical Newtonian theory, and resulting in statement of *Planck's radiation law*;

1899-1900: <u>Ernest Rutherford</u>, New Zealand physicist, and <u>Paul Ulrich Villard</u>, French chemist and physicist, separated radiation into three types, eventually named *alpha*, *beta*, and *gamma rays*, based on penetration of objects and deflection by a magnetic field;

1899-1924: <u>John Joly</u>, Irish geologist and physicist, published books including *An Estimate of the Age of the Earth, Radium and the Geological Age of the Earth*, and *Radioactivity and Geology*;

1900: <u>Paul Ulrich Villard</u>, French chemist and physicist, discovered *gamma rays*, by investigating the radiation from a radium salt that escaped from a narrow aperture in a shielded container onto a photographic plate; and <u>Friedrich Ernst Dorn</u>, German chemist, noticing that the radium apparently becomes less radioactive if swept with a current of gas, discovered *radon*;

1900-12: <u>Max Theodor Felix von Laue</u>, German physicist, applied the concept of *entropy to optics*, relating the speed of light in flowing water to the theory of special relativity, and also discovered *how X-rays are diffracted by atoms in crystals*;

1902: <u>Oliver Heaviside</u>, English physicist and independently <u>Arthur Edwin Kennelly</u>, US engineer, predicted the existence of an ionized gaseous layer capable of reflecting radio waves, called 'Heaviside layer', now known as *ionosphere*;

1904: <u>Charles Glover Barkla</u>, English physicist, demonstrated that *X-rays could be polarized*, as *transverse waves*, and found that secondary X-rays consist of rays scattered from an incident beam and a fluorescent radiation characteristic of scattering substance; and <u>John Ambrose Fleming</u>, English physicist and electrical engineer, invented the thermionic rectifier or *diode valve* 'Fleming valve', which for half a century was a vital part of radio, television, and early computer circuitry;

1905-18: <u>Paul Langevin</u>, French physicist, established the *relation of paramagnetic molecular movement to absolute temperature*, predicted *paramagnetic saturation*, and applied *sonar techniques to detect submarines* during the First World War;

1905-19: <u>Albert Einstein</u>, German-Swiss-US mathematical physicist, discovered the *photoelectric effect*, a phenomenon resulting from absorption of photon energy by electrons, leading to their release from a surface, when photon energy exceeds the work function, or otherwise allowing conduction when incident energy exceeds an atomic binding energy; became famous for the *theories of relativity*: *Special Theory* postulated by relative motion and constant velocity of light or zero acceleration; and *General Theory* by which relativity was extended from constant to varying velocities or non-zero acceleration; and also renowned by his formula of *mass-energy equivalence* (energy is equal to mass multiplied by square speed of light);

1906: <u>Reginald Aubrey Fessenden</u>, US radio engineer and inventor, performed *amplitude modulation*, using it for the first *radio broadcast*, and also discovered the *heterodyne effect*, which soon was developed into the *superheterodyne circuit* as an integral part in the design of radio receivers;

1908-25: <u>Hans Wilhelm Geiger</u> and <u>Walther Müller</u>, German physicists, improved the particle counter, resulting in the modern form of *Geiger-Müller counter*, which also detects electrons and ionizing radiation;

1910-21: <u>Robert Andrews Millikan</u>, US physicist, refining the oil drop technique, showed that the charge on each droplet is always a multiple of the same basic unit as *charge of electron*; studied the photoelectric effect, confirming A Einstein's theoretical equation, and calculated an accurate value for *Planck's constant*;

1910-25: <u>Owen Willans Richardson</u>, English physicist, did famous work on *thermionics*, a term coined to describe the phenomenon of *emission of electricity from hot bodies*;

1911: <u>Ernest Rutherford</u>, New Zealand physicist, introduced the revolutionary concept of *nuclear atom*, in which mass is concentrated in nucleus surrounded by planetary electrons;

1912-15: <u>William Henry Bragg</u> and <u>William Lawrence Bragg</u>, British physicists, described the conditions for *X-ray diffraction by crystals*;

1912-33: <u>Peter Joseph Wilhelm Debye</u>, US physicist and physical chemist, worked on dielectric constants; on molecular dipole moments, known as *Debyes*; and on theory of *strong electrolytes*; as well as developed the *Debye-Scherrer X-ray diffraction powder method*, the theory of *X-ray scattering by gaseous molecules*, and theoretical treatments for electro-optical 'Kerr effect' and *adiabatic demagnetization*;

1913: <u>Frederick Soddy</u>, English radio chemist, demonstrated that uranium decays into radium, and discovered *isotopes*, which have fundamental importance to all physics and chemistry;

1913-16: <u>William David Coolidge</u>, US physical chemist and inventor, constructed the *hot cathode X-ray tube*, called the 'Coolidge tube', incorporating a tungsten cathode;

1913-22: <u>Niels Henrik David Bohr</u>, Danish physicist, extended the theory of atomic structure by explaining the spectrum of hydrogen, as *Bohr's model*; developed the collective model of the nucleus, which combined the quantum-mechanical shell model of the nucleus and classical liquid drop model; and introduced the *quantized energy level theory of the atom*;

1914-25: <u>James Franck</u>, US physicist, and <u>Gustav Ludwig Hertz</u>, German physicist, proved that atoms would only absorb a fixed amount of energy, and thus demonstrated the *quantized nature of the atom's electron energy levels*;

1917-19: <u>Károly Ereky</u>, Hungarian agricultural engineer, wrote and published the book *Biotechnologie der Fleisch-, Fett- und Milcherzeugung im landwirtschaftlichen Grossbetriebe* 'Biotechnology of Meat, Fat and Milk Production in an Agricultural Large-Scale Farm' in which he coined the term *biotechnology*;

1919-20: <u>Ernest Rutherford</u>, New Zealand physicist, discovered that *alpha-ray bombardments* induced *atomic transformation in atmospheric nitrogen*, liberating hydrogen nuclei; showed that *protons* are constituents of atomic nuclei, and proposed the existence of the 'neutron';

1919-36: <u>Johannes Stark</u>, German physicist, discovered the *Stark effect* concerning the splitting of spectrum lines by subjecting light source to a strong electrostatic field, and also *Doppler's effect in canal rays*;

1920-22: <u>Otto Stern</u> and <u>Walter Gerlach</u>, US physicists, projecting a beam of silver atoms through a non-uniform magnetic field, they produced *two distinct beams*, thus providing the quantum theory prediction that an atom's magnetic moment can only be oriented in two fixed directions relative to an external magnetic field;

1920-25: <u>Arthur Holly Compton</u>, US physicist, developed a theory to describe the *interaction of X-rays with matter*, measuring wavelength of X-rays scattered by a target;

1923-28: <u>Louis-Victor Broglie</u>, French physicist, formulated *wave-particle duality*, showing that particles can behave as waves, and thus opening the way to wave mechanics;

1923-37: <u>Karl Manne Georg Siegbahn</u>, Swedish physicist, produced X-rays of various wavelengths and penetrating power, by which reinforced the *shell model of the atom*;

1924: <u>Satyendra Nath Bose</u>, Indian physicist; and <u>Albert Einstein</u>, German-Swiss-US mathematical physicist, founded *Bose-Einstein statistics*, corresponding to the condensate phenomenon that some particles should appear at very low temperature;

1924-31: <u>Wolfgang Pauli</u>, Austrian-Swiss theoretical physicist, formulated *Pauli's exclusion principle*, according to which there are no two electrons in an atom with exactly same state, i.e. with the same quantum numbers, and predicted the existence of a low-mass neutral particle, later discovered as neutrino;

1925: <u>Werner Karl Heisenberg</u>, German theoretical physicist, reinterpreted classical mechanics introducing a *matrix-based quantum mechanics* where phenomena must be describable both in terms of wave theory and quanta;

1925-27: Charles Thomson Rees Wilson, Scottish pioneer of atomic and nuclear physics, devised the *cloud chamber* method of marking the track of alpha-particles and electrons, by which the movement and interaction of atoms could be directly observed and photographed;

1925-30: Chandrasekhara Venkata Raman, Indian physicist, demonstrated that the interaction of *vibrating molecules* with *photons passing through* alters the spectrum of scattered light, known as *Raman's effect*, which became an important spectroscopic technique;

1925-35: Victor Francis Hess, Austrian-born US physicist, discovered that the radiation intensity in the atmosphere increases with height, leading to the conclusion that *high-energy cosmic radiation* must originate *from outer space*;

1925-47: Patrick Maynard Stuart Blackett, English physicist, made the first photography of *nuclear collisions involving transmutation*, and developed CTR Wilson's cloud chamber, using it to confirm the existence of the *positron* (antiparticle of electron);

1925-55: Ernst Frederick Werner Alexanderson, Swedish-born US electrical engineer and inventor, constructed the *Alexanderson alternator* for transoceanic communication, *antenna structures* and *radio-receiving* and *transmitting systems*; and also a complete *television system* and a *colour television receiver*;

1926: Enrico Fermi, Italian-born US nuclear physicist, and independently Paul Adrien Maurice Dirac, English mathematical physicist, founded *Fermi-Dirac statistics*, describing energies of single particles in a system of many identical particles that obey W Pauli's exclusion principle, and applying to particles with half-integer spin, called *fermions*; and John Logie Baird, Scottish electrical engineer and television pioneer, had success in building a *television apparatus*, almost entirely from scrap materials, and made the first demonstration of a *television image*;

1926-30: Erwin Schrödinger, Austrian physicist, inspired by L-V Broglie's wave-particle duality, he originated the science of *wave mechanics* as a part of quantum theory, with celebrated *Schrödinger's wave equation*;

1927: Werner Karl Heisenberg, German theoretical physicist, formulated the revolutionary *principle of indeterminacy* or *uncertainty principle*, showing that there is a fundamental limit to the accuracy to which certain pairs of variables (such as position and momentum) can be determined;

1927-33: <u>Vladimir Aleksandrovich Fock</u>, Soviet theoretical physicist, working in quantum mechanics, generalized E Schrödinger's wave equation to a relativistic case, formulated the *Klein-Fock equation*; and developed quantum mechanics for *multi-particle systems* to solve the wave equation for multi-electron atoms, called *Hartree-Fock technique*;

1927-35: <u>Clinton Joseph Davisson</u>, and <u>Lester Halbert Germer</u>, US physicists, studied *electron scattering* from a block of nickel when their vacuum system broke down, and found familiar peaks of a diffraction pattern; as well as discovered *diffraction of electrons*, confirming L-V Broglie's theory of wave nature of particles;

1927-36: <u>George Paget Thomson</u>, English physicist, noticing that a beam of electrons could produce circular interference fringes; discovered *electron diffraction by crystals*;

1927-38: <u>Pascual Jordan</u>, German theoretical physicist, participated at formulation of the theory of *quantum mechanics in matrix representation*, showing how light could be interpreted as being composed of discrete quanta of energy, and contributed to foundation of *quantum electrodynamics*;

1927-60: <u>Eugene Paul Wigner</u>, US theoretical physicist, had the idea of *parity conservation in nuclear interactions*, demonstrated that *strong nuclear force has very short range*, participated at deduction of the *Breit-Wigner formula* which describes resonant nuclear reactions, and formulated the *Wigner theorem* concerning the conservation of angular momentum of electron spin;

1928-39: <u>Vladimir Kosma Zworykin</u>, Russian-born US physicist, patented the application of cathode-ray tube to television, and developed the *first practical television camera* and principles of the *electron microscope*;

1929: <u>Ernest Orlando Lawrence</u>, US physicist, invented the earliest operational *cyclotron particle accelerator* for production of *artificial radioactivity*, at the University of California, Berkley, which was fundamental to the development of the atomic bomb;

1930: <u>Paul Adrien Maurice Dirac</u>, English mathematician and physicist, completed his classic work *The Principles of Quantum Mechanics*, formulated the *relativistic wave equation*, and predicted the *existence of antimatter*; and <u>Robert Julius Trumpler</u>, US astronomer, studied the dimensions and brightnesses of open star clusters in the Milky Way, and explained the disproportionate faintness of more distant ones as the effect of *absorption of light in interstellar space*;

1930-35: <u>Max Born</u>, German physicist, using E Schrödinger's wave equation, he deduced that the state of a particle, e.g. its energy or position, could only be predicted in terms of probabilities, and from this the existence of *quantum jumps between discrete states*;

1931-33: <u>Ernst August Friedrich Ruska</u>, German electrical engineer, developed the world's first *electron microscope*, which continued to be improved in his subsequent work;

1932: <u>John Douglas Cockcroft</u>, English nuclear physicist, and <u>Ernest Thomas Sinton Walton</u>, Irish physicist, produced the first *artificial disintegration of a nucleus* by bombarding lithium with protons in the first successful use of a *particle accelerator*; <u>John Von Neumann</u>, US mathematician, carried out mathematical work on the theory of linear operators, a new axiomatization of set theory, and gave a precise description of *quantum theory*; <u>Carl David Anderson</u>, US physicist, discovered the *positron*, a positively charged electron-type particle, thus confirming the *existence of antimatter*; <u>Harold Clayton Urey</u>, US physical chemist, discovered *deuterium* 'heavy hydrogen', containing one proton and one neutron, by repeatedly distilling a sample of liquid hydrogen; and <u>James Chadwick</u>, English physicist, bombarding beryllium with alpha particles, discovered the *neutron*, a neutral particle whose mass is closed to that of proton;

1932-48: <u>Norbert Wiener</u>, US mathematician, worked on stochastic processes and harmonic analysis, introducing the concepts later called *Wiener integral* and *Wiener measure*, and formulated the *principles of cybernetics*, which were published in his work *Cybernetics, or control and communication in the animal and the machine*;

1932-58: <u>Douglas Rayner Hartree</u>, English mathematician and physicist, studied computational methods applied to a wide variety of problems ranging from atomic physics, resulting in invention of the *method of self-consistent field in quantum mechanics*, to the automated control of chemical plants; and developed the *differential analyser*, an analogue computer, used as an electronic digital computer;

1932-64: <u>Sin-Itiro Tomonaga</u>, Japanese physicist, wrote papers on positron creation and annihilation, and on high-energy neutrino-neutron scattering, and worked on relativistic quantum description of the interaction between a photon and an electron, producing the theory of *quantum electrodynamics*;

1933-34: <u>Irène Joliot-Curie</u> and <u>Frédéric Joliot-Curie</u>, French physicists, produced the first *artificial radioisotope* by alpha-particle bombardment, and also manufactured *a range of radioisotopes*, some of them proved to be indispensable in medicine, scientific research, and industry;

1934-42: <u>Leo Szilard</u>, US physicist, patented *nuclear fission as an energy source*, and, together with E Fermi, worked for the *first fission reactor* in Chicago;

1934-78: <u>Peter Kapitza</u>, Soviet physicist, constructed a helium liquefier to investigate extraordinary *superfluidity of helium-2*, and liquefied both hydrogen and helium;

1935: <u>Robert Alexander Watson-Watt</u>, Scottish physicist, perfected a *shortwave radio system*, called *radar* 'ra*dio *d*etection *a*nd *r*anging' able to locate flying aeroplanes;

1935-44: <u>Karl Manne Georg Siegbahn</u>, Swedish physicist, made accurate measurements of *X-ray wavelengths* produced by atoms of different elements, driving many developments in quantum theory and atomic physics; he also patented and developed the *Siegbahn pump* for higher vacuum levels;

1935-47: <u>Hideki Yukawa</u>, Japanese physicist, and <u>Cecil Frank Powel</u>, English physicist, suggested that a strong short-range attractive interaction between neutrons and protons would overcome the electrical repulsion between protons, and then discovered the intermediate particles propagating interaction as charged *p-mesons* or *pions*;

1935-50: <u>Frits Zernike</u>, Dutch physicist, developed the *phase-contrast technique* for microscopic examination of transparent objects, and invented the *coherent background technique* to detect phase variations in interference and diffraction patterns;

1935-60: <u>Robert Sanderson Mulliken</u>, US chemical physicist, worked on isotope effect in band spectra of diatomic molecules, created the *molecular orbital theory*, and studied donor-acceptor interactions and charge transfer spectra;

1937: <u>Pavel Cherenkov</u>, <u>Ilya Mikhailovich Frank</u> and <u>Igor Yevgenevich Tamm</u>, Soviet physicists, explained the *Cherenkov's effect*, which arises when a charged particle traverses a medium faster than the speed of light in that medium, relating the type of radiation observed to the particle mass and velocity;

1937-38: <u>Otto Hahn</u>, German radio chemist, and <u>Fritz Strassmann</u>, German chemist, researched the irradiation of uranium and thorium with neutrons, identified barium in the residue after bombarding uranium with neutrons, and discovered *nuclear fission*, later developed into atomic bombs; and <u>Chester Floyd Carlson</u>, US inventor, made an experiment with copying process using photoconductivity, and discovered the basic principles of *electrostatic xerography*, a non-chemical photographic process in which light discharges a charged dielectric surface;

1937-50: <u>Isidor Isaac Rabi</u>, US physicist, developed the resonance method for accurately determining the *magnetic moments of fundamental particles*, and also contributed to the development of radar, the nuclear bomb, laser and atomic clock;

1938: <u>Lise Meitner</u>, Austrian physicist, and <u>Otto Robert Frisch</u>, British physicist, studied the fission that occurred in experiments of neutron bombardment, proposed that the production of barium is a result of nuclear fission, and postulated that *mass has been converted into energy*;

1938-50: <u>Willis Eugene Lamb</u>, US physicist, discovered the *Lamb shift* related to two possible hydrogen energy states differing in energy by a very small amount, which led to a revision of the *theory of interaction of electron with electromagnetic radiation*, and ultimately to the theory of quantum electrodynamics;

1939-49: <u>Edward Victor Appleton</u>, English physicist, contributed to exploration of the ionosphere, revealing the existence of a layer of electrically charged particles, called the *Appleton layer*, in the upper atmosphere, which plays an essential part in making wireless communication between distant stations, and was fundamental in the *development of radar*;

1940: <u>Edwin Mattison McMillan</u>, US atomic scientist, and <u>Philip Hauge Abelson</u>, US physical chemist, synthesized *neptunium* by bombarding uranium with slow moving neutrons, at Berkeley Radiation Laboratory of University, California;

1940-41: <u>Glenn Theodore Seaborg</u> and <u>Edwin Mattison McMillan</u>, US atomic scientists, synthesized *plutonium*, as a secondary product of radioactive decay of neutron-bombarded uranium, which led to the development of the atomic bomb;

1941-46: <u>Robert Oppenheimer</u>, US nuclear physicist, studied electron-positron pairs, cosmic-ray theory, and deuteron reactions, as well as led the *Manhattan atomic bomb project*, set up at Los Alamos laboratory;

1941-50: Isaac Asimov, US novelist, critic and popular scientist, first used in print the word *robotics* in his science fiction short story *Liar!*, Astounding Science Fiction; and also wrote other short stories forming the collection *I, Robot*; by these works formulating his *Three Laws of Robotics*;

1942: John Vincent Atanasoff and Clifford Berry, US physicists and computer pioneers, constructed the *electronic calculating machine* 'ABC' (Atanasoff-Berry-Computer), an *early computer* using vacuum tubes;

1942-48: Felix Bloch, Swiss-born US physicist, working on radar, developed the technique of *nuclear magnetic resonance* 'NMR', an useful tool in chemistry and biology;

1942-61: James Rainwater, US physicist, contributed to the Manhattan atomic bomb project, directed the Nevis Cyclotron Laboratory, and developed a *collective model* combining the theory of nuclear particles arranged in concentric shells with the theory of the nucleus as analogous to a liquid drop;

1944-50: Glenn Theodore Seaborg, US atomic scientist, led a team who synthesized *americium* and *curium*, which were produced by bombarding uranium or plutonium with neutrons in a nuclear reactor;

1945-60: Lev Landau, Soviet physicist, worked on the theory of *superfluidity*, or *zero viscosity*, and of condensed matter, particularly helium, as a collective, rather than individual, behaviour of atoms in a liquid;

1945-65: Martin Ryle, English physicist radio astronomer, investigated the emission of *radio waves from the Sun*, and studied *radio waves from the universe*, pointing to an evolving universe that started with a 'Big Bang', and mapped radio sources by an ingenious method of the *aperture synthesis*;

1946: Theodor Ionescu and Vasile Mihu, Romanian physicists, built and tested the first *show boosted hydrogen device*, a precursor of the maser;

1947: Walter Houser Brattain, John Bardeen and William Bradford Shockley, US physicists, worked at the Bell Telephone Laboratories, trying to produce semiconductor devices for replacing the thermionic valves, and invented the *point-contact transistor*, using a thin germanium crystal;

1950-52: <u>Edward Teller</u>, US physicist, led the programme to build up and test the world's first *hydrogen bomb*, which works on the principle of nuclear fusion, resulting from thermonuclear release of energy when hydrogen nuclei are fused to form helium nuclei;

1950-54: <u>Hyman George Rickover</u>, US naval engineering officer, successfully adapted *nuclear reactors for ship propulsion*, resulting in the first vessel so equipped, called *USS Nautilus*, as the world's *first nuclear submarine*;

1950-60: <u>Robert Hofstadter</u>, US physicist, using a linear accelerator to probe nuclear structure, and investigating the nuclear charge distribution, he discovered that *protons and neutrons also contain inner structure*, now known to be due to quarks;

1950-77: <u>Rosalyn Yalow</u>, US biophysicist, researched diabetes, developing *radioimmunoassay* 'RIA', an ultrasensitive method of measuring concentrations of substances in the body, and suggesting that, in adult diabetics, antibodies which inactivate injected insulin are formed;

1951-53: <u>Maurice Hugh Frederick Wilkins</u>, British physicist, and <u>Rosalind Elsie Franklin</u>, English X-ray crystallographer, producing *X-ray diffraction picture of DNA*, they contributed to the double helix model of DNA by *X-ray data of DNA fibres*;

1952: <u>Nikolai Gennadevich Basov</u> and <u>Aleksandr Mikhailovich Prokhorov</u>, Russian physicists, stated the theoretical principles describing the operation of a *maser* 'microwave amplification by stimulated emission of radiation', and made pioneering research for developing the laser;

1952-65: <u>Alfred Kastler</u>, French scientist, worked to obtain precise information about *atomic structures* by probing energy levels within atoms, using visible light and radio waves to excite electrons in atoms, which then emitted radiation as they returned to lower energy states; he also used optical techniques to develop *optical pumping*, which laid the foundations for subsequent development of masers and lasers;

1953: <u>Charles Hard Townes</u> and <u>James Gordon</u>, US physicists, built the first *ammonia maser*, a device using stimulated emission in a stream of energized ammonia molecules to produce amplification of microwaves at a frequency of about 24 gigahertz;

1953-74: <u>Andrei Dimitriyevich Sakharov</u>, Soviet physicist, led the development of the *Soviet hydrogen bomb*, followed by his dissidence and internal exile;

1954-89: *European Organization for Nuclear Research* (CERN) was created and the Laboratory for Particle Physics was located just outside of Geneva, Switzerland, with 26.67 kilometres circumference circular underground tunnel that houses the *Large Electron-Positron* (LEP) *collider*, the world's largest and highest-energy synchrotron, which became operational and should produce proton-antiproton collisions in the energy range of 10-14 tera-electron-volts;

1955: Emilio Segrè and Owen Chamberlain, US physicists, discovered the *anti-proton*, an anti-particle of proton, with identical mass but negative electric charge;

1955-64: Richard Phillips Feynman, US physicist, initiated the *path integral approach* for describing quantum processes, by which he contributed to the development of quantum electrodynamics, and introduced *Feynman diagrams* providing a pictorial representation of particle interactions; and Julian Schwinger, US physicist, did fundamental work in *quantum electrodynamics*, with deep consequences for physics of elementary particles;

1955-80: Kai Mann Börje Siegbahn, Swedish physicist, studied the energies of electrons emitted from solids exposed to X-rays, which revealed sharp peaks at energies characteristic of materials, and led to the technique of *electron spectroscopy for chemical analyses*, applied for liquids and gases; and also contributed to development of *high-resolution electron spectroscopy*;

1956: Frederick Reines and Clyde Lorrain Cowan, US physicists, experimentally evidenced the *neutrino*, as nature's most elusive particle;

1957-58: Charles Hard Townes and Arthur Leonard Schawlaw, US physicists, studied *infrared radiation* which was then abandoned, and instead they concentrated upon *visible light* to build-up a laser;

1958-61: Rudolf Ludwig Mössbauer, German physicist, identified the *Mössbauer effect* as a narrow resonance in energy spectrum produced when whole of nuclear lattice, rather than just one nucleus, recoils from gamma radiation;

1958-70: Clive Marles Sinclair, English electronic engineer and inventor, developed a wide range of *calculators, miniature television sets* and *personal computers*, and then manufactured a small three-wheeled 'personal transport' vehicle powered by a washing-machine motor and rechargeable batteries;

1959: Gordon Gould, US physicist, worked on the energy levels of excited thallium, and coined the term *laser* in his published paper *The LASER, Light Amplification by Stimulated Emission of Radiation*;

1960: <u>Theodore Harold Maiman</u>, US physicist, created and operated the *first functioning laser*, useful in spectroscopy, surgical work, compact disc players, etc.; and <u>Allan Sandage</u> and <u>Thomas Matthews</u>, US astronomers, identified a faint optical object at the same location as the compact *radio source 3C 48* with an unusual spectrum, later known as 'quasar', and many other such objects, showing that most of them are not radio emitters;

1960-62: <u>Maria Goeppert-Mayer</u>, German-born US physicist, and independently <u>Johannes Hans Daniel Jensen</u>, German physicist, on the basis of stability of certain nuclei with 'magic numbers' of protons and neutrons, they developed the *nuclear shell model*;

1960-86: <u>Richard Edward Taylor</u>, Canadian physicist, investigated the structure of nucleons (protons and neutrons) by scattering high-energy electrons from nuclear targets, and established the *constituents of nucleons*, now known as quarks, as real entities, by determining some of their properties;

1962-70: <u>Leon Max Lederman</u> and <u>Melvin Schwartz</u>, US physicists, made experiments resulting in 20 muon events, and confirmed the existence of two distinct neutrino types, as the basis for the idea that fundamental particles come in generations, with electron, muon and tau-lepton all having associated neutrinos, thus demonstrating the doublet structure of leptons through the discovery of the *muon neutrino*;

1964: <u>Hong-Yee Chiu</u>, Chinese-born US astrophysicist, coined the term *quasar*, given to a quasi-stellar radio source, as high red shift origin of electromagnetic energy; and <u>Murray Gell-Man</u>, and independently <u>George Zweig</u>, US physicists, introduced the concept of *quarks* which have one-third integral charge and baryon number, the fundamental constituents of matter combining to form composite particles called 'hadrons' such as protons and neutrons;

1965: <u>Arno Allan Penzias</u>, US astrophysicist, and <u>Robert Woodrow Wilson</u>, US physicist, using a large radio telescope, they discovered *cosmic microwave background radiation*, a remnant of the Big Bang, coming from all directions with an energy distribution corresponding to that of a black body at a temperature of 2.725 kelvin;

1965-85: <u>Jerome Isaac Friedman</u>, US physicist, <u>Henry Way Kendal</u>, US physicist, and <u>Richard Edward Taylor</u>, Canadian physicist, working at the Stanford linear accelerator, they investigated electron-scattering from protons and neutrons, and studied the *properties of quarks*, establishing that quarks have spin and fractional charges of $+\frac{2}{3}$ and $\frac{1}{3}$ times electronic charge, thus providing the first incontrovertible *evidence for quarks as real entities* rather than abstract mathematical concepts;

1966: <u>*Stanford Linnear Accelerator*</u>, an *electron-positron collider*, became operational as the longest in the world, accelerating electrons to 30 giga-electron-volts in a 3 kilometre-long waveguide;

1968-89: <u>Douglas Carl Engelbart</u>, US computer scientist, invented the *computer mouse*, and contributed to the development of many features of modern computing, including *e-mail, groupware* and *hypermedia*;

1969-86: <u>Jack Steinberger</u>, German-born US physicist, proved the existence of a *neutral pion* by observations of its decay at the Berkeley synchrotron, measured the spin and parity of a charged pion, and established the existence of *two distinct neutrino types*;

1971: <u>Federico Faggin</u>, Italian-US, <u>Ted Hoff</u>, US, and <u>Stanley Mazor</u>, US, computer scientists, designed the world's first *microprocessor*, a computer central processing unit usually on one integrated-circuit chip, based on a very large-scale integration technology, called 'Intel 4004';

1971-83: <u>Carlo Rubbia</u>, Italian-born US physicist, and <u>Simon Van der Meer</u>, Dutch physicist, discovered the *W and Z bosons*, the field particles which transfer *weak nuclear interactions*;

1971-93: <u>Clifford Glenwood Shull</u>, US physicist, brought pioneering experimental contribution to the development of the *neutron diffraction technique*;

1972-93: <u>Bertram Neville Brockhouse</u>, Canadian physicist, made interesting experimental researches which led to the development of *neutron spectroscopy*;

1973: <u>Herbert Boyer</u>, US biochemist and genetic engineer, and <u>Stanley Cohen</u>, US professor of medicine, invented a method of *cloning genetically engineered molecules* in foreign cells, and created the first *transgenetic organism* by inserting *antibiotic resistance genes* into the plasmid of an Escherichia coli bacterium;

1974-76: <u>Stephen William Hawking</u>, English theoretical physicist, predicted that a black hole could evaporate through loss of thermal radiation; and that mass can escape from its gravitational pull, now known as *Hawking process*;

1975-77: <u>Victor Scheinman</u>, US pioneer in the field of robotics, invented the *Programmable universal manipulation arm* (PUMA), a design bought by *Unimation* and released at the Robot Type Work Handling Equipment factory;

1975-80: <u>Nicolaas Bloembergen</u>, Dutch-born US physicist, and <u>Arthur Leonard Schawlow</u>, US physicist, pioneered pumping methods to energize masers, introduced a modification to CH Townes's early design, *extending maser principle to light*, thereby establishing *feasibility of laser*, and developed *laser spectroscopy*; and <u>Martin Lewis Perl</u>, US physicist, working on experimental particle physics, established the existence of the elementary particle *tau lepton*;

1980: <u>James Watson Cronin</u>, and <u>Val Logsdon Fitch</u>, US physicists, discovered violations of fundamental symmetry principles in the *decay of neutral K-mesons*;

1980-92: <u>Norman Ramsey</u>, US physicist, developed the *separated field method*, in which an electromagnetic field, applied to a beam of atoms or molecules, induces transitions between specific energy states, which is useful in *hydrogen maser* and *atomic clocks*; and <u>Joseph Hooton Taylor</u>, US astronomer and physicist, and <u>Hulse Russel</u>, US physicist, did systematic research on *pulsars*, rapidly rotating dense stars which appear to emit regular pulses of radio waves, and first discovered a *binary pulsar*, a pulsar in orbit of another dense neutron star;

1981-96: <u>Steven Chu</u>, US physicist, <u>Claude Nessim Cohen-Tannoudji</u>, French physicist, and <u>William Daniel Phillips</u>, US physicist, developed *methods for cooling matter to very low temperatures* and *trapping atoms* by use of laser light;

1982-2000: <u>Bill Gates</u>, US computer scientist and businessman, had a licence for computer operating system to *International Business Machines* (IBM), fledgling *personal computer* (PC) industry; this system (MS-DOS) was phenomenally successful, and its updated versions, such as *Windows 2000*, allowed maintenance of Microsoft's PC hegemony, and showed how somebody could became a billionaire and world's most wealthy private individual;

1984-85: <u>Gerd Binnig</u>, German physicist, and <u>Heinrich Rohrer</u>, Swiss physicist, developed a high-resolution electron microscope, whose tip measured just one atom wide, and surface of a sample to profile sample's surface, thus designing the *scanning tunnelling microscope*;

1985-87: <u>Michael Boris Green</u>, <u>John Schwarz</u> and <u>Edward Witten</u>, English physicists, had the idea that the ultimate constituents of nature, when inspected at very small scales, do not exist as point-like particles but as *strings* in more than three dimensions, leading to foundation of *string theory*, later developed as superstring theory, and opening a new way for subatomic investigations and cosmic studies;

1985-97: <u>Robert Betts Laughlin</u>, US physicist, <u>Horst Ludwig Störmer</u>, German-born US physicist, and <u>Daniel Chee Tsui</u>, Chinese-born US physicist, discovered a *new form of quantum fluid* with functionally charged excitations;

1985-2003: <u>Roy Jay Glauber</u>, US theoretical physicist, contributed to the *quantum theory of optical coherence*;

1987-95: <u>George Smoot</u> and <u>John Mather</u>, US investigators of Cosmic Background Explorer (COBE), studied the cosmic microwave background radiation of the universe, and its *blackbody feature and anisotropy*, indicating 'ripples' which show that the early universe was not smooth and uniform, so that matter could concentrate to form galaxies and stars; <u>Zhores Ivanovich Alferov</u>, Russian physicist, and <u>Herbert Kroemer</u>, German physicist, developed *semiconductor heterostructures used in high-speed- and opto-electronics*, as basic work on information and communication technology; and <u>Jack St Clair Kilby</u>, US electrical engineer, contributed to the development of information and communication by participation in the invention of an *integrated circuit*;

1988-96: <u>Raymond Davis, Jr</u>, US chemist and astrophysicist, and <u>Masatoshi Koshiba</u>, Japanese physicist, contributed to astrophysics by pioneering the *detection of cosmic neutrinos*;

1988-2000: <u>Eric Allin Cornell</u>, US physicist, <u>Wolfgang Ketterle</u>, German physicist, and <u>Carl Edwin Wieman</u>, US physicist, achieved *Bose-Einstein condensation in dilute gases of alkali atoms* at very low temperatures, and made early fundamental studies of the *properties of condensates*;

1989-2004: <u>Riccardo Giacconi</u>, Italian-born US astrophysicist, discovered *cosmic X-ray sources*, pioneering the development of astrophysics; and <u>John Lewis Hall</u>, US physicist, and <u>Theodor Wolfgang Hänsch</u>, German physicist, contributed to the development of *laser-based precision spectroscopy*, including the *optical frequency comb technique*;

1990-2002: <u>Alexei Alexeyevich Abrikosov</u>, Russian theoretical physicist, <u>Vitaly Lazarevich Ginzburg</u>, Russian physicist, and <u>Anthony James Leggett</u>, English physicist, did pioneering works in the *theory of superfluids and superconductors*;

1993-2003: <u>David Jonathan Gross</u>, US physicist, <u>Hugh David Politzer</u>, US theoretical physicist, and <u>Frank Anthony Wilczek</u>, US theoretical physicist and mathematician, discovered the *asymptotic freedom in theory of strong interaction*;

1995-2005: <u>John Cromwell Mather</u> and <u>George Fitzgerald Smoot</u>, US astrophysicists and cosmologists, discovered the *blackbody form* and *anisotropy of cosmic microwave background radiation*;

1995-2007: <u>Charles Kuen Kao</u>, Chinese-born Hong Kong, US and British electrical engineer, had groundbreaking achievements concerning *transmission of light in fibres* for optical communication;

1996-2008: <u>Willard Sterling Boyle</u>, Canadian physicist, and <u>George Elwood Smith</u>, US scientist, invented an *imaging semiconductor circuit*, called *charge-coupled device* 'CCD' *sensor*;

2000-07: <u>Yoichiro Nambu</u>, Japanese-born US physicist, discovered the mechanism of *spontaneous broken symmetry in subatomic physics*; and <u>Makoto Kobayashi</u>, Japanese physicist, and <u>Toshihide Maskawa</u>, Japanese theoretical physicist, discovered the *origin of broken symmetry*, predicting the *existence of at least three families of quarks* in nature;

2005-11: <u>Serge Haroche</u>, French physicist, and <u>David Jeffrey Wineland</u>, US physicist, produced groundbreaking experimental *methods that enable to measure and manipulate individual quantum systems*;

2006-11: <u>Mihai Patrascu</u>, Romanian-born US computer scientist, working on the fundamental questions about basic data structures, became known by his papers published such as: with <u>Erik Demaine</u> *Logarithmic lower bounds in the cell-probe model*; with <u>Mikkel Thorup</u> *Higher lower bounds for near-neighbour and further rich problems*; and his own *Unifying the landscape of cell-probe lower bounds*; all in SIAM Journal on Computing, 35(4), 39(2), and 40(3) respectively;

2009: <u>Paul Baumann</u>, US geographer, published an important work titled *History of Remote Sensing, Satellite Imagery, Part II*;

<u>*NASA current missions*</u>: *Wide-field Infrared Survey Explorer* (Mission to study the Solar System, Milky Way and universe, including asteroids, the coolest and dimmest stars and the most luminous galaxies), *Fermi Gamma-ray Space Telescope* (Mission to answer questions about supermassive black hole systems, pulsars and the origin of cosmic rays), *Cosmic Hot Interstellar Plasma Spectrometer*

(Mission to use an extreme ultraviolet spectrograph for studying the 'Local Bubble' surrounding our Solar System), *Chandra X-ray Observatory* (Mission to provide X-ray images for unravelling the structure and evolution of the universe), and *Astro-E2/Suzaku* (Mission to discover more about the X-ray universe).

According to the data above, high technology emerged in AD1871-77 when the first theory of electromagnetic radiation and the first equation showing how entropy corresponds to increasing molecular randomness have been conceptualized, that is about 140 years ago, as the initial time; and, according to the table of human abilities (see chapter 7), the development of high technology will last until 4350 years from present into the future, as the final time

$$t_\bullet = 140 + 4350 = 4490 \text{ years from its origin.}$$

High technology had a spectacular ascension, first marked by syntheses of transuranic elements, the electronic calculating machine, nuclear tests, production of atomic bombs, biochemistry, wave-particle duality, radar, television, electron microscope, particles and antiparticles, and cybernetics. This fundamental stage lasted until 1943-45, being followed by an even more rapid ascension by discoveries of superfluidity, superconductivity, new particles and antiparticles, quarks an families of quarks, reality of neutrino and its types, bosons, tau leptons, cosmic microwave background radiation and its anisotropy, quasars, pulsars, cosmic X-ray sources, blackbody feature and anisotropy; invention and production of robots, transistors, nuclear submarines, calculators, personal computers, masers, lasers, hydrogen bomb, atomic clocks, microprocessors, integrated circuits, point-contact transistors and transgenetic organisms; affirmation and application of quantum electrodynamics, radioimmunology, high-resolution electron spectroscopy, neutron diffraction technique, neutron spectroscopy, particle accelerators, transmission of light in fibres, satellite imagery, information and communication systems; and so on.

Therefore, the development of high technology can be divided into two major sequences delimited by approximate times as

AD1871...1877 ≈ AD1874 \ *Fundamental stage*
\ AD1943...1945 ≈ AD1944 \ *Advanced stage* \ AD2014.

These transitional times represent 70, and 0 years ago; or 140 - 70 = 70; 140 - 0 = 140 years from the origin of high technology; and the last two of them divided by the final time $t_\bullet$ = 4490 years from the same origin result in the ratios 70/4490 ≈ 0.016; 140/4490 ≈ 0.031, which are close to the accurate values of times ratios given in Background to time, Table *(n)* for *z/64 = n* sequences, where they are given as: $t_{0.015625}/t_\bullet = tanh(0.015625) = 0.0156237$; $t_{0.03125}/t_\bullet = tanh(0.03125) = 0.0312398$;

whereby the transitional times are calculated, in years from the origin of high technology, as

$$t_{0.015625} = t_\bullet \cdot tanh(0.015625) = 70;$$
$$t_{0.03125} = t_\bullet \cdot tanh(0.03125) = 140.$$

By this calibrated model, the high technology timeline, sequences, and intrinsic characteristics can be reconstituted as shown in the next tables.

z/64 = n	Time (years)		Sequences of high technology
	from origin $t_n = t_\bullet \cdot tanh(n)$	from present $t_n - 140$	
	4490	+4350	
...	...	...	
0.046875	210	+70 (AD2084)	–
0.03125	140	0 (AD2014)	– Nearer future stage
0.015625	70	-70 (AD1944)	– Advanced stage
0	0	-140 (AD1874)	– Fundamental stage

z/64 = n	Time from origin $t_n = t_\bullet \cdot tanh(n)$ (years)	Period $\tau_n = (t_\bullet^2 - t_n^2)/(2t_\bullet)$ (years)	Frequency $f_n = (2t_\bullet)/(t_\bullet^2 - t_n^2)$ (years)$^{-1}$	Angular speed $\omega_n = 2\pi/\tau_n = 2\pi \cdot f_n$ (years)$^{-1}$
	4490	0	∞	∞
...	...	...	...	...
0.046875	210	2240.1	0.0004464	0.0028049
0.03125	140	2242.8	0.0004459	0.0028015
0.015625	70	2244.5	0.0004455	0.0027994
0	0	2245.0	0.0004454	0.0027987

Synopsis

Time function

$$t(\theta) = t_{\bullet} \cdot tanh(\theta/4\pi);$$

Discrete values delimiting *kinds of sequences*

$$t_z = t_{\bullet} \cdot tanh(z), z = 0, 1, 2, 3...,$$
$$t_i = t_{\bullet} \cdot tanh(i), z/2 = i = 0, 1/2, 1, 3/2...;$$
$$t_j = t_{\bullet} \cdot tanh(j), z/4 = j = 0, 1/4, 1/2, 3/4...,$$
$$t_k = t_{\bullet} \cdot tanh(k), z/8 = k = 0, 1/8, 1/4, 3/8...,$$
$$t_l = t_{\bullet} \cdot tanh(l), z/16 = l = 0, 1/16, 1/8, 3/16...,$$
$$t_m = t_{\bullet} \cdot tanh(m), z/32 = m = 0, 1/32, 1/16, 3/32...,$$
$$t_n = t_{\bullet} \cdot tanh(n), z/64 = n = 0, 1/64, 1/32, 3/64...$$

Table with *calibration* for 26 fields of interest or activity

Field of study or activity	Initial time (back from present)	Final time (from present in future)	Total duration	Kind of sequences
Creativity (years)	0	67.5	67.5	$z/2 = i$
Cosmology (10^6 years)	-13798	+15941	29739	$z/4 = j$
Geology (10^6 years)	-4568	+214	4782	$z/4 = j$
Climatology (10^6 years)	-4550	+210	4760	$z/4 = j$
Biology (10^6 years)	-4200	+200	4400	$z/4 = j$
Anthropology (10^6 years)	-5.29	+0.21	5.50	$z/8 = k$
Humanity (years)	-200000	+4350	204350	$z/8 = k$
Linguistics (years)	-200000	+4350	204350	$z/8 = k$
Art (years)	-174588	+4350	178938	$z/8 = k$
Theology (years)	-149951	+4350	154301	$z/8 = k$
Musicology (years)	-126770	+4350	131120	$z/8 = k$
Exploration (years)	-105566	+4350	109916	$z/8 = k$
Inhabitation (years)	-86668	+4350	91018	$z/8 = k$
Knowledge (years)	-70207	+4350	74557	$z/8 = k$
Navigation (years)	-56157	+4350	60507	$z/8 = k$
Meteorology (years)	-44368	+4350	48718	$z/16 = l$
Medicine (years)	-34619	+4350	38969	$z/16 = l$
Sociology (years)	-26653	+4350	31003	$z/16 = l$
Construction (years)	-20207	+4350	24557	$z/16 = l$
Science (years)	-15033	+4350	19383	$z/16 = l$
Industry (years)	-10905	+4350	15255	$z/16 = l$
Civilization (years)	-7630	+4350	11980	$z/16 = l$
Legislation (years)	-5041	+4350	9391	$z/32 = m$
Philosophy (years)	-3001	+4350	7351	$z/32 = m$
Aero-astronautics (years)	-1398	+4350	5748	$z/32 = m$
High technology (years)	-140	+4350	4490	$z/64 = n$

Extendibility within a super-cycle

$-2t_\bullet...-t_\bullet$: expanding quarter-super-cycle,

$-t_\bullet...0$: shrinking quater-super-cycle,

$0...+t_\bullet$: expanding quarter-super-cycle,

$+t_\bullet...+2t_\bullet$: shrinking quarter-super-cycle.